AF389345

Plant Essential Oil with Biological Activity

Plant Essential Oil with Biological Activity

Editors

Hazem Salaheldin Elshafie
Laura De Martino
Adriano Sofo

MDPI • Basel • Beijing • Wuhan • Barcelona • Belgrade • Manchester • Tokyo • Cluj • Tianjin

Editors
Hazem Salaheldin Elshafie
University of Basilicata
Italy

Laura De Martino
University of Salerno
Italy

Adriano Sofo
University of Basilicata
Italy

Editorial Office
MDPI
St. Alban-Anlage 66
4052 Basel, Switzerland

This is a reprint of articles from the Special Issue published online in the open access journal *Plants* (ISSN 2223-7747) (available at: https://www.mdpi.com/journal/plants/special_issues/Oil_Bioactivity).

For citation purposes, cite each article independently as indicated on the article page online and as indicated below:

LastName, A.A.; LastName, B.B.; LastName, C.C. Article Title. *Journal Name* **Year**, *Volume Number*, Page Range.

ISBN 978-3-0365-3845-7 (Hbk)
ISBN 978-3-0365-3846-4 (PDF)

Contents

Editorial

Plant Essential Oil with Biological Activity

Hazem S. Elshafie

School of Agricultural, Forestry, Food and Environmental Sciences, University of Basilicata, Via dell'Ateneo Lucano 10, 85100 Potenza, Italy; hazem.elshafie@unibas.it; Tel.: +39-0971-205522; Fax: +39-0971-205503

Abstract: Plant essential oils (PEOs), extracted from many aromatic and medicinal plants, are used in folk medicine and often represent an important part of the traditional pharmacopoeia: they have a long history of use in folk medicine as antimicrobial agents to control several human and phytopathogens. Many PEOs have been registered as effective alternatives to chemical and synthetic antimicrobials, and in the last few decades, they have also been effectively used in the food industry as antioxidants and anticarcinogens, thanks to the efforts of many research/medical institutions and pharmaceutical companies. This Special Issue discussed the chemical composition and biological-pharmaceutical activities of some important PEOs and their single constituents. Detailed information has been also covered in this Special Issue regarding the mechanisms, possible modes of action, and factors affecting these activities, such as geographical origins, environmental conditions, nutritional status, and the extraction methods used.

Keywords: natural products; antimicrobial activity; cytotoxicity; phytotoxicity; antioxidant; phytopathogens; food preservatives

Citation: Elshafie, H.S. Plant Essential Oil with Biological Activity. *Plants* **2022**, *11*, 980. https://doi.org/10.3390/plants11070980

Academic Editor: Filippo Maggi

Received: 31 March 2022
Accepted: 1 April 2022
Published: 4 April 2022

Publisher's Note: MDPI stays neutral with regard to jurisdictional claims in published maps and institutional affiliations.

1. Introduction

This Special Issue entitled 'Plant Essential Oil with Biological Activity' comprises 15 papers covering a wide range of different aspects, ranging from the biochemical characterization of some PEOs and their biopharmaceutical activities and possible agri-food and medical applications. In particular, some important published research papers have manipulated different aspects related to the biological investigation of crude essential oils (EOs), whereas other papers were related to the study of their specific single constituents. Some other research papers focused on the comparative chemical profiles of different species/ecospecies of the same plant. In addition, some studies have highlighted the seasonal changes in the production and utilization stages of some PEOs. Furthermore, some research studies are related to the use of EOs for the control of serious phytopathogens. On the other hand, our Special Issue has also covered some important review papers covering different aspects of bio-pharmaceutical properties and agri-food applications.

2. Antimicrobial, Phytotoxic Activity, and Agri-Food Applications

Recently, the use of some EOs as alternative antimicrobial and pharmaceutical agents has attracted considerable interest from scientists worldwide [1]. In particular, Abd-ElGawad et al. [2] reviewed the phytotoxic effect of some EOs and their chemical compositions as reported in bibliographic research from 1972 to 2020. The same authors used chemometric analysis to build a structure–activity relationship between phytotoxicity and EO chemical composition. In particular, the analysis of the collected data revealed that oxygenated terpenes and mono- and sesquiterpenes play principal roles in the phytotoxicity of EOs [2].

Another important study, conducted by Abd-ElGawad et al. [3], deals with the chemical composition of EO extracted from the aerial parts of *Persicaria lapathifolia* and its free radical scavenging activity and its herbicidal effect on the weed *Echinochloa colona*. The results obtained showed that the extracted EO of *P. lapathifolia* exhibited substantial allelopathic activity against the germination, seedling root, and shoot growth of the weed *E. colona* in a dose-dependent manner [3]. In addition, *P. lapathifolia* EO demonstrated promising antioxidant activity [3]. On the other hand, Abd-ElGawad et al. [4] evaluated the phytotoxic activity of EOs extracted from two ecospecies of *Pulicaria undulata* (Saudi and Egyptian) against the weeds *Dactyloctenium aegyptium* and *Bidens pilosa*. The results obtained showed that the EO of the Egyptian ecospecies showed more phytotoxic activity against *D. aegyptium* and *B. pilosa* than the Saudi ecospecies [4].

An important study conducted by Ibáñez and Blázquez [5] reviewed the agri-food applications of *Curcuma longa* L. rhizome EO. This review study focused on some interesting information regarding conventional and recent extraction methods of *C. longa* rhizome oil, their characteristics, and their suitability to be applied at the industrial scale [5].

An important study published in this Special Issue, conducted by Xylia et al. [6], evaluated the efficacy of an eco-friendly product based on rosemary and eucalyptus Eos and two different application methods (vapor and dipping) on the quality attributes of tomato fruits, and the obtained results indicated that eco-friendly products based on the studied EOs were able to maintain the quality of tomato fruits. In addition, Chrysargyris et al. [7] studied the vapor application of sage EO for maintaining tomato fruit, where the quality attributes were more affected in green fruits and were less affected red fruits. The results also showed that sage EO has a lowering effect of the total phenolics, acidity, total soluble solids, and fruit chroma, with no specific trend found in both breaker and red tomatoes.

Calvopiña et al. [8] studied the chemical analysis of a new sesquiterpene essential oil from the native Andean species *Jungia rugosa* and studied its cholinergic activity. The results showed that the volatile fraction of this EO was exclusively composed of sesquiterpenes, specially curcumene (more than 45%) and sesquiphellandrene (about 17%). This EO demonstrated weak inhibition activity against AChE.

Regarding the antimicrobial activity, Camele et al. [9] studied the potential microbicide activity of *Mentha piperita* cv. 'Kristinka' and its main constituents against some common phytopathogens (*Botrytis cinerea*, *Monilinia fructicola*, *Penicillium expansum*, and *Aspergillus niger*). The results obtained showed that the tested EO has promising antifungal activity against all tested fungi. In addition, Soliman et al. [10] studied the antifungal activity of *Mentha spicata* and *Mentha longifolia* EOs against *F. oxysporum*. The results obtained also showed that the single compounds (thymol, adapic acid, menthol, and menthyl acetate) possess antifungal effects through the malformation and degradation of the fungal cell wall [10].

3. Bio-Pharmaceutical Properties

Many plant EOs are being widely utilized in the pharmaceutical industry, aromatherapy, and other related medical applications. Regarding the cytotoxicity effect and antioxidant activity, several studies have been published in this Special Issue. Elgamal et al. [11] studied the chemical profiles of EOs of the above-ground parts of *Pluchea dioscoridis* (L.) DC. and *Erigeron bonariensis* (L.) in addition to their cytotoxic and anti-aging activities. The results obtained explicated that the terpenoids are the main constituents of both plants, with a relative concentration of 93.59% and 97.66%, respectively, mainly including sesquiterpenes (93.40% and 81.06%) [11]. Another study conducted by Shahin et al. [12] reported that the flowers' EO extracted from *Aerva javanica*, isolated during four seasons, is considered a good source of natural bioactive antioxidants. Khalil et al. [13] studied the chemical composition of EO isolated from *Anisosciadium lanatum* and evaluated its anti-cancer potential and mechanistic effect on HepG2 liver cancer cell lines. The obtained results showed that the studied EO was able to regulate the cell proliferation and cell viability in HepG2 liver cancer cells at a sub-lethal dose of 10 to 25 g/mL and displayed reductions in migration and invasion [13]. In addition, the treatment with *A. lanatum* EO indicated the mitigation of cancer activity by aborting the mRNA of pro-apoptotic markers such as BCL-2, CASPASE-3, CYP-1A1, and NF_kB [13].

4. Future Perspectives

Further studies are required to determine the consumer demand for agri-food product-based EOs; however, a balance must be struck between matching customer expectations and maximizing industrial production efficiency in accordance with Green Chemistry. On the other hand, reality simulation models are required in order to determine the composition of essential oils based on the environmental conditions surrounding plants; these models should aim to manage these variables and thus obtain high yield oils with the appropriate chemical composition for specific functions in the agri-food sector. In addition, the use of PEOs as food preservatives should be investigated further to determine the optimum application regarding the effective dose, the time of exposure, and the number of treatments. Regarding the herbicidal effect of PEOs, further future studies are recommended in order to characterize the allelochemical substances of many EOs against various weeds. In the same context, additional investigations are also needed to encapsulate EOs to maximize their biological activities and to examine the application of single active ingredients for several purposes.

5. Conclusions

As conclusion, the studies in this Special Issue prove that several studied PEOs have promising pharmaceutical, medicinal, and culinary benefits. In addition, other studies highlighted the potential applications of numerous EOs in the agri-food industry, where they have a strong antimicrobial activity against a broad spectrum of food spoilage microorganisms and extend food shelf-lives. Furthermore, the studied EOs and their two main constituents can be used successfully as possible natural alternatives to synthetic substances against several phytopathogens. Many researchers have also attributed the biological activity of many EOs to the characteristic chemical composition, usually sesquiterpenes and phenolic compounds. We hope that we were able to contribute to the enrichment of the scientific content of the reader in this important field of research.

Funding: This research received no external funding.

Institutional Review Board Statement: Not applicable.

Informed Consent Statement: Not applicable.

Acknowledgments: Personally, I would like to thank and express my deep gratitude to everyone who contributed and participated in this Special Issue. I am so grateful to the editorial team staff of this journal for their kind collaboration in managing all stages of this Special Issue.

Conflicts of Interest: The author declares no conflict of interest.

References

1. Elshafie, H.S.; Caputo, L.; De Martino, L.; Gruľová, D.; Zheljazkov, V.D.; De Feo, V.; Camele, I. Biological investigations of essential oils extracted from three *Juniperus* species and evaluation of their antimicrobial, antioxidant and cytotoxic activities. *J. Appl. Microbiol.* **2020**, *129*, 1261–1271. [CrossRef] [PubMed]
2. Abd-ElGawad, A.M.; El Gendy, A.E.-N.G.; Assaeed, A.M.; Al-Rowaily, S.L.; Alharthi, A.S.; Mohamed, T.A.; Nassar, M.I.; Dewir, Y.H.; Elshamy, A.I. Phytotoxic Effects of Plant Essential Oils: A Systematic Review and Structure-Activity Relationship Based on Chemometric Analyses. *Plants* **2021**, *10*, 36. [CrossRef] [PubMed]
3. Abd-ElGawad, A.M.; Bonanomi, G.; Al-Rashed, S.A.; Elshamy, A.I. *Persicaria lapathifolia* Essential Oil: Chemical Constituents, Antioxidant Activity, and Allelopathic Effect on the Weed Echinochloa colona. *Plants* **2021**, *10*, 1798. [CrossRef] [PubMed]
4. Abd-ELGawad, A.M.; Al-Rowaily, S.L.; Assaeed, A.M.; El-Amier, Y.A.; El Gendy, A.E.-N.G.; Omer, E.; Al-Dosari, D.H.; Bonanomi, G.; Kassem, H.S.; Elshamy, A.I. Comparative Chemical Profiles and Phytotoxic Activity of Essential Oils of Two Ecospecies of *Pulicaria undulata* (L.) CA Mey. *Plants* **2021**, *10*, 2366. [CrossRef] [PubMed]
5. Ibáñez, M.D.; Blázquez, M.A. Curcuma longa L. Rhizome Essential Oil from Extraction to Its Agri-Food Applications. A Review. *Plants* **2021**, *10*, 44. [CrossRef] [PubMed]
6. Xylia, P.; Ioannou, I.; Chrysargyris, A.; Stavrinides, M.C.; Tzortzakis, N. Quality Attributes and Storage of Tomato Fruits as Affected by an Eco-Friendly, Essential Oil-Based Product. *Plants* **2021**, *10*, 1125. [CrossRef] [PubMed]
7. Chrysargyris, A.; Rousos, C.; Xylia, P.; Tzortzakis, N. Vapour Application of Sage Essential Oil Maintain Tomato Fruit Quality in Breaker and Red Ripening Stages. *Plants* **2021**, *10*, 2645. [CrossRef] [PubMed]

8.   Calvopiña, K.; Malagón, O.; Capetti, F.; Sgorbini, B.; Verdugo, V.; Gilardoni, G. A New Sesquiterpene Essential Oil from the Native Andean Species *Jungia rugosa* Less (Asteraceae): Chemical Analysis, Enantiomeric Evaluation, and Cholinergic Activity. *Plants* **2021**, *10*, 2102. [CrossRef] [PubMed]

9.   Camele, I.; Grul'ová, D.; Elshafie, H.S. Chemical Composition and Antimicrobial Properties of *Mentha x piperita* cv. 'Kristinka' Essential Oil. *Plants* **2021**, *10*, 1567. [CrossRef] [PubMed]

10.   Soliman, S.A.; Hafez, E.E.; Al-Kolaibe, A.M.G.; Abdel Razik, E.-S.S.; Abd-Ellatif, S.; Ibrahim, A.A.; Kabeil, S.S.A.; Elshafie, H.S. Biochemical Characterization, Antifungal Activity, and Relative Gene Expression of Two Mentha Essential Oils Controlling *Fusarium oxysporum*, the Causal Agent of *Lycopersicon esculentum* Root Rot. *Plants* **2022**, *11*, 189. [CrossRef] [PubMed]

11.   Elgamal, A.M.; Ahmed, R.F.; Abd-ElGawad, A.M.; El Gendy, A.E.-N.G.; Elshamy, A.I.; Nassar, M.I. Chemical Profiles, Anticancer, and Anti-Aging Activities of Essential Oils of *Pluchea dioscoridis* (L.) DC. and *Erigeron bonariensis* L. *Plants* **2021**, *10*, 667. [CrossRef] [PubMed]

12.   Shahin, S.M.; Jaleel, A.; Alyafei, M.A.M. Yield and In Vitro Antioxidant Potential of Essential Oil from *Aerva javanica* (Burm. f.) Juss. Ex Schul. Flower with Special Emphasis on Seasonal Changes. *Plants* **2021**, *10*, 2618. [CrossRef] [PubMed]

13.   Khalil, H.E.; Ibrahim, H.-I.M.; Darrag, H.M.; Matsunami, K. Insight into Analysis of Essential Oil from *Anisosciadium lanatum* Boiss—Chemical Composition, Molecular Docking, and Mitigation of Hepg2 Cancer Cells through Apoptotic Markers. *Plants* **2022**, *11*, 66. [CrossRef] [PubMed]

Article

Chemical Composition and Cosmeceutical Potential of the Essential Oil of *Oncosiphon suffruticosum* (L.) Källersjö

Selena O. Adewinogo [1], Rajan Sharma [1], Charlene W. J. Africa [2], Jeanine L. Marnewick [3] and Ahmed A. Hussein [1,*]

[1] Chemistry Department, Bellville Campus, Cape Peninsula University of Technology, Symphony Road, Bellville 7535, South Africa; selenaorangoeunice@gmail.com (S.O.A.); sharmar@cput.ac.za (R.S.)

[2] Department of Medical Biosciences, University of the Western Cape, Bellville 7535, South Africa; cafrica@uwc.ac.za

[3] Applied Microbial and Health Biotechnology Institute, Cape Peninsula University of Technology, Symphony Rd., Bellville 7535, South Africa; marnewickj@cput.ac.za

* Correspondence: mohammedam@cput.ac.za; Tel.: +27-21-959-6193

Abstract: The South African medicinal plant *Oncosiphon suffruticosum* (L.) Källersjö is an important remedy used to treat chronic, respiratory, and skin ailments. From the essential oil (EO) extracted by the hydrodistillation, sixteen constituent components were identified with oxygenated monoterpenes: camphor (31.21%), filifolone (13.98%), chrysanthenone (8.72%), 1,8-cineole (7.85%), and terpinen-4-ol (7.39%) as predominant constituents. In the antibacterial activity study, the EO was found most susceptible against *Pseudomonas aeruginosa* with an MIC of 6.4 mg/mL; however, it showed the same activity against *Staphylococcus aureus* and *Escherichia coli* with an MIC value of 12.8 mg/mL. The sun protecting factor (SPF) of the EO was found to be 2.299 and thus establishing it as a potentially important cosmeceutical for sunscreen applications. This is the first report investigating the essential oil of *O. suffruticosum* for its chemical composition and skin-related in vitro biological activities viz antibacterial, antioxidant capacity, antityrosinase, and sun protection factor.

Keywords: essential oils; *Oncosiphon suffruticosum*; antioxidant; antibacterial; tyrosinase inhibition; sun protection factor

Citation: Adewinogo, S.O.; Sharma, R.; Africa, C.W.J.; Marnewick, J.L.; Hussein, A.A. Chemical Composition and Cosmeceutical Potential of the Essential Oil of *Oncosiphon suffruticosum* (L.) Källersjö. *Plants* **2021**, *10*, 1315. https://doi.org/10.3390/plants10071315

Academic Editors: Daniela Rigano, Hazem Salaheldin Elshafie, Laura De Martino and Adriano Sofo

Received: 15 May 2021
Accepted: 24 June 2021
Published: 28 June 2021

Publisher's Note: MDPI stays neutral with regard to jurisdictional claims in published maps and institutional affiliations.

1. Introduction

Essential oils (EOs) are aromatic oily liquids composed of a complex mixture of volatile compounds and are produced by aromatic plants as secondary metabolites. The volatile constituents of EOs have been important materials for preventing and treating human diseases since the early days [1]. Although mainly used for their agreeable scents, EOs present themselves as excellent candidates to meet the current beauty industry's demands for two principal reasons. Firstly, research backs up their efficacy as valuable cosmeceuticals. They have been shown to exhibit properties of antimicrobials [2,3], antioxidant agents [4,5], antityrosinase agents [6–8], sunscreens [9,10], natural preservatives [11], natural sources of fragrance [12], as well as inhibitors of skin's degradation enzymes (collagenase and elastase) [13]. Secondly, the small lipophilic molecules that make up their composition grant easy penetration through the skin layers [14].

South Africa (SA) is home to an important and rich botanical diversity. The country boasts over 30,000 flowering species with high endemism and is ranked third in biodiversity in the world [15,16]. A significant fraction of aromatic plant species contributes to this rich heritage. To date, oil-rich plant species recorded in South Africa belong to the Asteraceae family (2300 species), Rutaceae family (290 species), and Lamiaceae (235 species) family [16].

Oncosiphon Källersjö is an aromatic genus of the Asteraceae family and Anthemideae tribe that counts seven species. Some species of the genus were formerly classified in the *Pentzia* Thunb. and others in *Matricaria* L. genera. However, the *Oncosiphon* genus later arose due to the morphological differences recorded in the now-*Oncosiphon* species which were not present in the *Pentzia* genus. Most of the *Oncosiphon* species are native to the Greater Cape Floristic Region except for *O. piluliferum* (L.f.) Källersjö and *O. suffruticosus* (L.) Källersjö. These two species also grow in Australia and are respectively known as Globe Chamomile and Calomba Daisy. *Oncosiphon* species bear the Afrikaans name "stinkruid" which means stinkweed due to their pungent aroma. Among them, *O. piluliferum*, *O. suffruticosus*, and *O. africanum* are important materials of Cape Dutch ethnobotany and Khoi-San medicine [17].

The *O. suffruticosum* (L.) Källersjö herb features hairless and thin leaves (Figure 1). It bears a typical sharp and powerful scent like other *Oncosiphon* species. The herb grows up to 50 cm tall annually and is distributed in the southern part of Africa from the Western Cape to Namibia [18]. In traditional healing practices, oral administrations aim to treat asthma, gastric disorders, convulsions, diabetes, rheumatic fever, typhoid fever, colds, and influenza [19,20]. Additionally, the herb is used topically as a leaf poultice to treat scorpion stings and inflammation [20].

(a)   (b)

Figure 1. Photographs of *O. suffruticosum*. (**a**) Uprooted branch; (**b**) flower heads. These photographs were taken during the summer season (December, 2018) at the Cape Flats regions of Cape Town, South Africa.

Since time immemorial, plants have been renowned sources of bioactive materials used in traditional therapies and a reservoir for innovative cures in modern medicine. The use of plants ranges from culinary preparations, medicine, to perfume compositions [21]. However, only a few SA medicinal plants are explored commercially [22] and investigated scientifically [20]. According to the literature, the essential oil of *O. suffruticosum* has never been studied before. In the quest to explore the South African flora for novel cosmeceutical ingredients, the aim of the present research was to elucidate the chemical composition and study the biological studies, antimicrobial activity, antioxidant capacity, antityrosinase activity, and photoprotection of the essential oil of *O. suffruticosum*.

2. Results and Discussion

2.1. Chemical Composition of O. suffruticosum Essential Oil

The hydrodistillation of fresh aerial parts of *O. suffruticosum* gave an average essential oil yield of 0.23% (*v/w*). According to the present GC-MS analysis, sixteen components representing 85.09% of the EO in composition were identified (Table 1).

Table 1. GC-MS analysis of *O. suffruticosum* essential oil.

RT (Min)	Component Code	Mass Spectral Matching	Composition (%)	Experimental RI	Literature RI	Identification
9.214	1	α-Pinene	0.80	935	939 [A]	RI, MS
9.981	2	Camphene	2.17	950	950 [B]	RI, MS
11.374	3	Sabinene	0.54	974	973 [B]	RI, MS
13.928	4	α-Terpinene	0.71	1016	1017 [B]	RI, MS
14.508	5	*p*-Cymene	2.45	1026	1024 [B]	RI, MS
15.016	**6**	**1,8-Cineole**	**7.85**	**1035**	**1032** [B]	**RI, MS**
16.710	7	γ-Terpinene	1.48	1061	1060 [B]	RI, MS
20.058	**8**	**Filifolone**	**13.98**	**1109**	**1109** [Wb]	**RI**
20.372	9	Unknown	2.56	1114	-	-
20.560	10	Unknown	2.03	1117	-	-
21.426	**11**	**Chrysanthenone**	**8.72**	**1131**	**1125** [B]	**RI, MS**
23.039	**12**	**Camphor**	**31.21**	**1155**	**1156** [Wb]	**RI, MS**
23.683	13	Pinocarvone	0.29	1164	1164 [A]	RI, MS
25.032	**14**	**Terpinen-4-ol**	**7.39**	**1183**	**1177** [B]	**RI, MS**
26.745	15	Verbenone	0.56	1207	1206 [B]	RI, MS
29.015	16	Unknown	1.10	1243	-	-
35.372	17	Piperitenone	0.78	1339	1341 [B]	RI, MS
39.371	18	3,5-Heptadienal, 2-ethylidene-6-methyl-	5.71	1400	1395 [Wb]	RI
40.828	19	Unknown	3.75	1425	-	-
49.798	20	Caryophyllene oxide	0.45	1576	1580 [B]	RI, MS
Monoterpene hydrocarbons:			8.15			
Oxygenated monoterpenes:			76.49			
Total monoterpenoids:			84.64			
Sesquiterpene hydrocarbons:			0.00			
Oxygenated sesquiterpenes:			0.45			
Total sesquiterpenoids:			0.45			
Total identified:			85.09			
Unidentified:			9.44			
		Total	94.53			

[A] = Adams [23], [B] = Babushok et al. [24], [Wb] = NIST Chemistry WebBook [25], MS = In addition to RI, the MS of the analyzed compound matched with the MS of the compound in [23] and/or NIST Chemistry WebBook [25], Unknown = The MS of the compound could not be matched with the available literature data.

The major constituents of the EO were found to be the hydrocarbons and oxygenated monoterpenes amounting to 84.64%, of which the oxygenated monoterpenes were dominant by 76.49%. The only identified sesquiterpene was found to be caryophyllene oxide present as 0.45%. No hydrocarbon sesquiterpenes were detected. The major constituents were found to be oxygenated monoterpenes: camphor (31.21%), filifolone (13.98%), chrysanthenone (8.72%), 1,8-cineole (7.85%), and terpinen-4-ol (7.39%) (Figure 2).

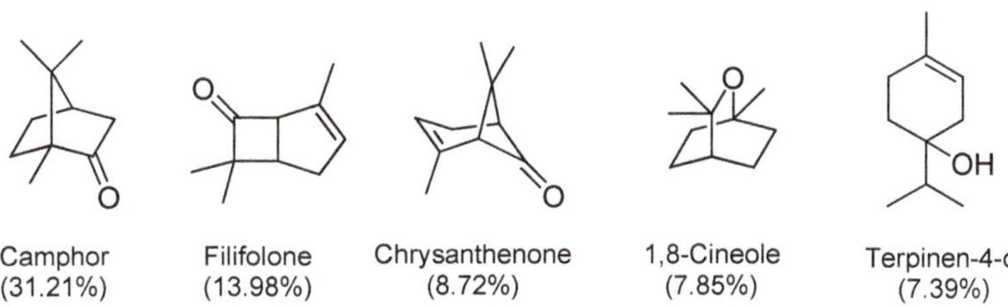

Figure 2. Major components detected in *O. suffruticosum* essential oil.

As per the literature, the *O. suffruticosum* essential oil had never been studied before as it is for other plants of the same genus. However, according to the results obtained, a chemical link to its historical classification in the *Pentzia* genus was observed. Like *O. suffru-*

ticosum EO, the chromatographed EOs of *Pentzia incana* [26] and *Pentzia punctata* [27] have shown to possess a significant content of camphor of up to 47.9% and 27.3%, respectively. Additionally, 1,8-cineole was also found as a major compound in *Pentzia incana* with up to 16.7% [26].

2.2. Antibacterial Activity: Minimum Inhibitory Concentration (MIC) Using the Broth Microdilution Method

The evaluation of the cutaneous antibacterial effect of *O. suffruticosum* essential oil was assessed against three bacterial strains, *Staphylococcus aureus*, *Pseudomonas aeruginosa*, and *Escherichia coli*, in the broth microdilution susceptibility assay. The results were taken as the lowest concentration inhibiting visible bacterial growth as detected by the *p*-iodonitrotetrazolium chloride (INT) reagent and expressed in mg/mL as presented in Table 2.

Table 2. MICs (mg/mL) of *O. suffruticosum* EO and control.

Sample	Micro-Organisms		
	S. aureus	*E. coli*	*P. aeruginosa*
O. suffruticosum	12.8	12.8	6.4
Ampicillin	<0.2	<0.2	R *

* R = resistant.

The MIC of *O. suffruticosum* EO was detected as 12.8 mg/mL for *S. aureus* and *E. coli*, whereas it was found twice as lower for *P. aeruginosa* and detected as 6.4 mg/mL. According to Van Vuuren [16], essential oils with an MIC $\leq$ 2 mg/mL can be taken as effective. Therefore, according to these results, *O. suffruticosum* EO may be classified to possess low to moderate antibacterial activity. These findings correlate well with the chemical composition of this EO. Indeed, it is known that the chemical structure of terpenoids parallels their activity [28], whereby the presence of an oxygen function in the framework enhances their antimicrobial properties [29]. The phenol and aldehydes are often characterized by the highest antibacterial activity [30] followed by the alcohols which are usually bactericidal rather than bacteriostatic, then the ketones and the terpene hydrocarbons which have weak activities [29]. In the EO of *O. suffruticosum*, phenols were not detected and only one aldehyde terpene was detected, 2-ethylidene-6-methyl-3,5-heptadienal, as 5.71%. The predominant functional moieties were ketones, alcohols, and terpene hydrocarbons by 78.93% which could explain the lower bacterial inhibitory activity.

2.3. Antioxidant Capacities

Free radicals chain reactions culminate in oxidative stress when the number of free radicals surpasses the number of systemic defenses, the antioxidants, in the target cell [31]. Oxidative stress in the skin is expressed by blotchy pigmentation, sagging skin, and wrinkles [32]. The strength of the antioxidative potential of *O. suffruticosum* essential oil was evaluated by four in vitro antioxidant capacity assays. The selection of the assays considered covering electron transfer (ET)- and hydrogen atom transfer (HAT)-based mechanisms. The ET-based methods selected were the 2,2-diphenyl-1-picrylhydrazyl (DPPH), 2,2′-Azino-bis(3-ethylbenzothiazoline-6-sulfonic acid) (ABTS), and Ferric reducing antioxidant power (FRAP) assays although DPPH and ABTS can involve both HAT and ET mechanisms [33,34]. The HAT-based method selected was Oxygen radical absorbance capacity (ORAC) assay [35,36]. The results are summarized in Table 3.

In the DPPH assay, the values of % radical scavenging activity (% RSA) of the essential oils were found to be extremely poor, 10.03 ± 1.02%, 8.38 ± 0.24%, and 7.06 ± 0.20% at 2, 1, and 0.5 mg/mL, respectively. Additionally, they were significantly lower than that of Trolox® positive control found as 94.94 ± 0.02%, 94.78 ± 0.06%, and 94.45 ± 0.04% at the respective concentrations tested. In the ABTS assay, the % RSA's were found to be higher and comparable to the gallic acid positive control. The values were found to be

$87.17 \pm 0.76\%$ to $71.46 \pm 0.04\%$ for the EO vs. $97.97 \pm 0.13\%$ to $98.05 \pm 0.03\%$ for the positive control in the 2 to 0.5 mg/mL concentration range. The higher performance of the EO in the ABTS assay was expected as ABTS•+ are more reactive than DPPH radicals [34]. However, the difference in antioxidant strength between the EO and gallic acid was evident in the discrepancy in Trolox® equivalent values which were 100-fold higher for gallic acid than those of the EO. Moreover, the EO was found to be -505.8 ± 80.8 µmol (AAE)/L at 2 mg/mL in the FRAP assay against being $635,500 \pm 4070.9$ µmol AAE/L for gallic acid positive control, and 6701.8 ± 57.2 µmol TE/L in the ORAC assay at the same concentration against $26,904 \pm 328.2$ µmol TE/L for EGCG positive control. The antioxidant capacity of a substance assesses the amount of antioxidant which reacts with the oxidant [37]. Overall, the EO was found to exhibit a much weaker performance than the reference controls. Therefore, the results indicated that the EO possesses poor to moderate antioxidant capacity.

Table 3. Antioxidant capacities of *O. suffruticosum* essential oil in the DPPH, ABTS, FRAP, and ORAC assays.

Sample	DPPH *		ABTS *			FRAP *	ORAC *
	mg/mL	% RSA6 min ± SD	% RSA6 min ± SD	TEAC (µmol TE/L ± SD)	mg/mL	FRAP (µmol AAE/L ± SD)	ORAC (µmol TE/L ± SD)
O. suffruticosum	2	10.03 ± 1.02	87.17 ± 0.76	9431.2 ± 81.5	2	−505.8 ± 80.8	6701.8 ± 57.2
	1	8.38 ± 0.24	81.13 ± 0.51	8784.6 ± 54.5			
	0.5	7.06 ± 0.20	71.46 ± 0.04	7750.1 ± 4.5			
Trolox®	2	94.94 ± 0.02	−	−	−	−	−
	1	94.78 ± 0.06					
	0.5	94.45 ± 0.04					
Gallic acid	2	−	97.97 ± 0.13	605,840 ± 27,811.3	2	635,500 ± 4070.9	−
	1		97.96 ± 0.16	355,740 ± 7127.6			
	0.5		98.05 ± 0.03	195,220 ± 6241.5			
EGCG **	−	−	−	−	2	−	26,904 ± 328.2

* Average values of triplicate measurements (n = 3); RSA: radical scavenging activity; SD = standard deviation; RSD = relative standard deviation; TE: Trolox® equivalent; AAE: ascorbic acid equivalent. ** EGCG: (-)-epigallocatechin gallate.

2.4. Tyrosinase Inhibition

Tyrosinase (EC 1.14.18.1), also known as polyphenol oxidase, is a copper-containing enzyme that has a central role in the production of melanin, the pigment responsible for the color of the skin. It catalyzes the first two steps of the multiphase process of melanogenesis, the biosynthesis of melanin. Today, tyrosinase inhibitors are increasingly prevalent cosmeceuticals' ingredients aiming to treat hyperpigmentation problems caused by the overproduction of melanin in the skin [38]. In the present work, the *O. suffruticosum* essential oil was tested in the tyrosinase inhibition assay exploring the monophenolase activity of the enzyme by monitoring the absorbance of L-DOPA (λ_{490} nm) using L-tyrosine as a substrate. The essential oils were tested at 200 µg/mL and 50 µg/mL and compared to kojic acid, a standard tyrosinase inhibitor used in cosmetics, at the same concentrations. The results were obtained as presented in Table 4.

Table 4. Summary of the tyrosinase inhibition assay results of *O. suffruticosum* EO at 200 µg/mL and 50.

	Tyrosinase Inhibition (%)	
Samples	at 200 µg/mL	at 50 µg/mL
O. suffruticosum	61.46 ± 11.0	26.14 ± 3.74
Kojic acid	96.24 ± 3.62	98.34 ± 0.80

O. suffruticosum EO exhibited significantly lower tyrosinase inhibition values than kojic acid at both concentrations tested. At 200 µg/mL, the EO was found to be $61.46 \pm 11.00\%$ against $96.24 \pm 3.62\%$ for kojic acid and at 50 µg/mL, the EO was found to be $26.14 \pm 3.74\%$

against 98.34 ± 0.80% for kojic acid. These values indicate that the enzyme inhibition is concentration dependent, and *O. suffruticosum* EO is a relatively weak tyrosinase inhibitor.

2.5. Sun Protection Factor (SPF)

Solar UV rays are recognized as the main contributor to extrinsic cutaneous aging in humans [39–41]. Chronic exposure to ultraviolet radiation (UVR) induces various dermatological problems including skin cancer [42]. Herein, the SPF of *O. suffruticosum* essential oil was determined by measuring the absorbance of a dilute hydroalcoholic solution of EO (0.1% *v/v*) at 290–320 nm at 5 nm interval then calculated using the equation given by Mansur et al. [43]. The results are presented in Table 5.

Table 5. Spectrophotometric absorbances of hydroalcoholic aliquots of *O. suffruticosum* essential oil and its calculated SPF.

Wavelength (nm)	EE(λ) x I(λ) ** Employed	Absorbance *
290	0.0150	0.2844 ± 0.0075
295	0.0817	0.2759 ± 0.0023
300	0.2874	0.2647 ± 0.0065
305	0.3278	0.2340 ± 0.0053
310	0.1864	0.1919 ± 0.0049
315	0.0837	0.1501 ± 0.0038
320	0.0180	0.1115 ± 0.0030
Calculated SPF		2.299

* Values represent mean absorbance values ± standard deviation of triplicate measurements, n = 3; ** constant values erythemogenic effect (EE) of radiation with wavelength λ x solar intensity (I) at wavelength λ determined by Sayre et al. [44].

According to the study, the essential oil of *O. suffruticosum* was found to possess an SPF value of 2.299. It has been reported that an SPF value above 2 is noteworthy [45,46]. Such a substance may block UV radiation by approximately 57% [45–47]. Therefore, the results establish *O. suffruticosum* EO as an important cosmeceutical for sunscreen formulation.

In an attempt to compare the biological activities of the plants which are rich in the major components found in the *O. suffruticosum* EO, the essential oils of *Cinnamomum camphora*, *Artemisia herba-alba*, *Eucalyptus globulus*, and *Melaleuca alternifolia* were selected as representative examples with camphor, chrysanthenone, 1,8-cineole, and terpinen-4-ol as respective major components. The results from the literature search indicated that mainly these essential oils have been studied for their antibacterial and antioxidant properties and they showed variable degree of activities.

The *C. camphora* essential oil contains camphor as the main constituent. The sample collected from Pantnagar, India, was effective against *Pasturella multocida* but not against *Salmonella enterica enterica* and *Escherichia coli* [48]. *C. camphora* oil from Nepal also showed marginal activity against *B. cereus* and *S. aureus*, with a MIC = 313 µg/mL [49]. During an antioxidant study by the DPPH assay, the IC_{50} value of the *C. camphora* essential oil was found to be 31.85 µL/mL, whereas that for the reference butylated hydroxytoluene (BHT) was reported to be 7.6 µg/mL [50].

Essential oil of *A. herba-alba* from Makther Seliana, Tunisia having camphor and chrysanthenone as major components displayed MIC (µg/mL) values of 100, 50, and >100 against *S. aureus*, *E. coli*, and *P. aeruginosa*, respectively [51]. The antioxidant activity of *A. herba-alba* EO by DPPH assay showed an IC_{50} of 2.66 µg/mL whereas that for the synthetic antioxidant butylated hydroxyanisole (BHA) was 1.66 µg/mL [52]. At a concentration of 1 mg/mL, the *A. herba-alba* EO exhibited a tyrosinase inhibition of 31.35%, which was much lower than that of the standard inhibitor kojic acid (87.54% at 0.05 mg/mL) [53].

EOs of *E. globulus* collected from Skoura, Morocco, presented excellent activity on *E. coli* in the agar disc diffusion assay with inhibition zone diameter (izd) = 48.15 mm compared to *S. aureus* (izd = 13.5 mm) and *S. intermedius* (izd = 10.26). The MIC for *E. coli* corresponded to 0.15 mg/mL while for *S. aureus* and *S. intermedius* the values corresponded to 0.75 mg/mL and 1.08 mg/mL, respectively [54]. The main component of *E. globulus* EOs is 1,8-cineole and it has been demonstrated that this compound has antimicrobial activity against several microorganisms including *S. aureus* and *E. coli* [55]. In an antioxidant study of this plant with the DPPH method, its methanolic extract exhibited the strongest free radical-scavenging activity with an IC_{50} value of 23 µg/mL, followed by the ethyl acetate extract (IC_{50} = 29 µg/mL) and hexane extract (IC_{50} = 65 µg/mL). However, the essential oil did not show any noticeable activity with the DPPH method [56]. This activity may be attributed to the high content of phenolic compounds (542.42 mg GAE/g) in methanol extract from *E. globulus*.

Essential oil from *M. alternifolia* is referred to as tea tree oil, the major component of which is terpinen-4-ol present at least 30% of total oil [57]. *M. alternifolia* essential oil obtained from a commercial source in Germany inhibited the growth of *S. aureus, E. coli,* and *P. aeruginosa* at a concentration of 5% *w/v* [58]. An antioxidant activity study by the DPPH method indicated that *M. alternifolia* EO at a concentration of 10 µL/mL produced 80% free radical scavenging activity which was equivalent to that of 30 mM BHT [59].

No tyrosinase inhibition studies are reported for essential oil of these plants except *A. herba-alba*. There was also no report in the literature regarding the SPF studies of the essential oils of these plants. As per the above discussed results from the literature, no direct correlation could be ascertained among the biological activity and the major component of the essential oil, suggesting that the biological activity of the essential oils is because of the synergism among the components of the essential oil rather than any one of the major constituents.

3. Materials and Methods

3.1. Plant Material

The plant material (3.0 kg) was wildly harvested from the Cape Flats Nature Reserve of the University of the Western Cape in December 2018. A voucher specimen was authenticated by Hlokane Mabela and deposited at the Horticultural Sciences Department of the Cape Peninsula University of Technology.

3.2. Extraction of Essential Oil

The fresh aerial parts (leaves, stems, and flowers) of *O. suffruticosum* were subjected to hydrodistillation using the Clevenger-type apparatus for 3 h as per the European Pharmacopeia's guidelines [60]. The essential oil was recovered by decantation in glass vials and stored in the dark at 4 °C until further use. The oil yield was expressed as the average percentage of volume in mL per weight in g (% *v/w*) of triplicate analyses.

3.3. Gas Chromatography-Mass Spectrometry (GC-MS) Analysis

The GC-MS analyses were carried out according to the in-house method and the procedure previously reported by Kuiate et al. [61] with some adjustments. The instrument consisted of an Agilent GC-7820A fitted with an HP-5MS fused silica column (30 m × 0.25 mm i.d. × 0.25 µm film thickness) and coupled with an Agilent 5977E mass selective compartment (Agilent Technologies, Inc., Santa Clara, CA, USA). The oven temperature was programmed at 50 °C for 5 min, 50–220 °C at a rate of 2 °C.min^{-1} then 220 °C temperature hold for 5 min for the first ramp. For the second ramp, the temperature was set to 300 °C at a rate of 25 °C.min^{-1}. Helium was used as a carrier gas at 1 mL.min^{-1} flowrate and pressure of 7.6522 psi. Sample injection of 1 µL of 1% (*v/v*) solution diluted in n-hexane was splitless and operated at 250 °C. A reference standard of homologous n-paraffin series of C8-C20 (Sigma-Aldrich®, St. Louis, MO, USA, Cat no. 04070) was prepared and co-injected under identical experimental conditions as the samples for the determination of retention indices (RIs). The MS spectra were obtained on electron impact at 70 eV scanning from 30.0 to 650 *m/z*.

The identification of the constituents was achieved by computerized matching (MassHunter software, Agilent Technologies, Inc., Santa Clara, CA, USA) of each mass spectrum generated with authentic samples (if available) and with those stored in the instrument's built-in mass spectral libraries (National Institute of Standards and Technologies, NIST), comparing of the experimental RIs [62] with those of the NIST online data collection [25] and literature [23,24]. The relative amounts of individual constituents were calculated automatically based on the total ion count detected by the GC-MS and expressed as percentage composition.

3.4. Antibacterial Assay

3.4.1. Micro-Organisms

The essential oil was tested against three skin pathogenic bacterial strains obtained from the Medical Bioscience Department at the University of the Western Cape. These were one gram-positive strain, wild-type (WT) *S. aureus*, and two gram-negative strains, wild-type (WT) *E. coli* and wild-type (WT) *P. aeruginosa*.

3.4.2. Preparation of the Media

The bacterial species were resuscitated by inoculation into Brain heart infusion (BHI) broth (Oxoid UK, Cat. no. CM1135) and incubated at 37 °C for 24 h after which, each strain was streaked aseptically onto Tryptone soya agar for single colony formation and incubated at 37 °C for 24 h. The cell suspensions were performed in sterile saline, standardized at 0.5 McFarland standard (Remel™, Kansas, Cat. no. R20410) at 1.5×10^8 colony forming units (CFU)/mL. Then, the working suspensions were obtained by a second 1:100 dilution onto BHI to approximately 10^6 CFU/mL.

3.4.3. Broth Microdilution Susceptibility Assay

The broth microdilution test was performed as previously described by Lourens et al. [63] and Sartoratto et al. [64] with slight adjustments. An EO stock solution of 51.2 mg/mL was prepared with a BHI:dimethyl sulphoxide (DMSO) (1:1) solution. In a 96-well plate, 100 µL of BHI was added to the experimental wells in triplicate except in well 1. Then, 200 µL of EO stock solution was added to well 1, from which a serial dilution was performed to the last experimental well. Subsequently, 100 µL of cell suspension was added to establish the two-fold 25.6–0.2 mg/mL sample concentration range and a bacterial cell suspension of approximately 5×10^5 CFU/mL. The plate was incubated at 37 °C for 20 h. After incubation, the antimicrobial activity was detected by adding 40 µL of 0.2 mg/mL INT (Sigma-Aldrich®, Cat no. I10406) aqueous solution. The plates were incubated at 37 °C for 1 h. The MICs were defined as the lowest concentration of essential oil that inhibited visible growth, as indicated by the color change of INT. Ampicillin (Sigma-Aldrich®, Cat no. A9393) was used as a positive control.

3.5. Antioxidant Capacity Assays

The antioxidant capacity of the *O. suffruticosum* EO was studied by the following four antioxidant assays to cover both HAT and ET mechanisms.

3.5.1. 2,2-Diphenyl-1-Picrylhydrazyl (DPPH) Assay

The DPPH assay was performed according to the method previously described by Bondet et al. [65] with slight modifications. In a clear 96-well plate, 275 µL of DPPH reagent (Sigma-Aldrich®, Cat no. D9132) (absorbance of 2.0 ± 0.1 at 517 nm) was added to 25 µL of EO sample and Trolox® (Sigma-Aldrich®, Cat no. 238831) positive control (2.0, 1.0, and 0.5 mg/mL). For the blank, ethanol was added instead of the sample. The total volume of the assay was 300 µL. The absorbance was read at 517 nm and 37 °C at the 6 min time point. The EO/Trolox® sample was read in triplicate (n = 3). The percentage radical scavenging activity (% RSA) of the samples was calculated using Equation (1).

$$\% \, RSA_{6 \, min} = 1 - \frac{Abs_{sample}}{Abs_{blank}}, \tag{1}$$

where Abs_{sample} is the absorbance signal of the EO sample and Abs_{blank} is the absorbance signal of the DPPH solution (ethanol in place of the sample) at 517 nm after 6 min. The results were expressed as the mean percentage of triplicate measurements ($\pm$standard deviation, SD).

3.5.2. 2,2′-Azino-bis(3-Ethylbenzothiazoline-6-Sulfonic Acid) (ABTS) Assay

The ABTS assay was performed according to Re et al. [66] with slight modifications. The ABTS radical cation (ABTS•+) (Sigma-Aldrich®, Cat no. A1888) stock reagent was produced by reacting 5 mL of freshly prepared 7 mM ABTS solution with 88 µL of a freshly prepared 140 µM $K_2S_2O_8$ (Merck, Cat no. 105091) then allowing the mixture to sit overnight for 16 h in the dark at room temperature. In a clear 96-well plate, 275 µL of ABTS•+ reagent (absorbance of 2.0 $\pm$ 0.1 at 734 nm) was added to 25 µL of each ethanolic Trolox® working standard (50 µM, 100 µM, 150 µM, 250 µM, and 500 µM) and EO sample (2.0, 1.0, and 0.5 mg/mL). Gallic acid (Sigma-Aldrich®, Cat no. G7384) was used as a positive control. For the blank, ethanol was added instead of the sample. The total volume of the assay was 300 µL. The absorbance was read at 734 nm and 37 °C at the 6 min time point. The EO sample, working standard, and gallic acid sample were read in triplicate (n = 3). The percentage of radical scavenging activity (% RSA) of each EO or positive control working solution was calculated using Equation (1), where Abs_{sample} is the absorbance signal of the EO sample/positive control and Abs_{blank} is the absorbance signal of the ABTS•+ solution (ethanol in place of the sample) at 734 nm. The results were expressed as the mean percentage of triplicate measurements ($\pm$standard deviation, SD). The Trolox® equivalent capacity assay (TEAC) values were reduced from the linear regression ($R^2 = 0.9980$) of Trolox® concentrations (µM) and the absorbance readings at 734 nm at 6 min and expressed as mean ($\pm$SD) of triplicate measurements in µmol Trolox® equivalents per liter of the sample tested (µmol TE/L).

3.5.3. Oxygen Radical Absorbance Capacity (ORAC) Assay

The ORAC assay was performed according to the method described by Prior et al. [67] with slight modifications. In a black 96-well plate, 12 µL of the Trolox® working solutions (83 µM, 167 µM, 250 µM, 333 µM, and 417 µM prepared with phosphate buffer at pH 7.4) and EO sample (2.0 mg/mL) were added in triplicate (n = 3). Subsequently, 138 µL of fluorescein solution was added followed by 50 µL of freshly prepared by dissolving 2,2′-Azobis (2-methylpropionamidine) dihydrochloride (AAPH) (Sigma-Aldrich®, Cat no. 440914) in phosphate buffer (150 mg of AAPH in 6 mL buffer). (-)-Epigallocatechin gallate (EGCG) (Sigma-Aldrich®, Cat no. E4143) was used as a positive control. For the blank, the phosphate buffer was added in place of the sample. The total volume of the assay was 200 µL and the temperature was set at 37 °C. Readings of the EO/EGCG samples (2.0 mg/mL) and Trolox® working standard solutions were taken using the excitation wavelength set at 485 nm and the emission wavelength at 530 nm for 2 h at 1 min reading interval. After analysis, the data points of the blank, EO sample, EGCG sample, and Trolox® working standards were summed up over time to obtain the area under the fluorescence decay curve (AUC). The ORAC values were calculated using the linear regression ($R^2 = 0.9861$) equation (Y = aX + c) between Trolox® concentration (Y) (µM) and the net area (blank-corrected) under the fluorescence decay curve (X). The results were expressed as the mean ($\pm$SD) of triplicate measurements in µmol of Trolox® equivalents per liter of the sample tested (µmol TE/L).

3.5.4. Ferric Reducing Antioxidant Power (FRAP) Assay

The FRAP assay was conducted as recommended by Benzie and Strain [68] with slight adjustments. Firstly, the fresh blue FRAP reagent was achieved by mixing 30 mL of acetate buffer, 3 mL of 2,4,6-tris[2-pyridyl]-s-triazine (TPTZ) (Merck, Cat no. T1253) with 3 mL of $FeCl_3$ solution and 6.6 mL of distilled water. Then, an L-ascorbic acid (Sigma-Aldrich®, Cat no. A5960) standard series of 50 µM, 100 µM, 200 µM, 500 µM, and 1000 µM was prepared from a 1 mM of L-ascorbic acid stock in distilled water. Lastly, in a clear 96-well plate, 300 μL of the FRAP reagent was added to 10 μL of L-ascorbic acid working standard solutions and EO sample (2.0 mg/mL) in triplicate (n = 3). Gallic acid was used as a positive control. For the blank, the phosphate buffer (pH 3.6) was added instead of the sample. The total volume of the assay was 310 μL. The absorbance of TPTZ-Fe (II) in the samples was read at 593 nm at 37 °C for 30 min. The results were calculated using the linear regression (R^2 = 0.9965) of the L-ascorbic acid (AA) standard series concentrations (μM) and absorbance signals expressed as mean (±SD) of triplicate measurements in μmol L-ascorbic acid equivalents per liter of the sample tested (μmol AAE/L).

3.6. Antityrosinase Assay

3.6.1. Essential Oils Samples and Positive Control Preparation

A total of 10 mg/mL of EO working solution was prepared with a DMSO: Tween®20 (1:1) solution to facilitate dispersion of the oils then further diluted to 1 mg/mL working solutions with methanol. A 10 mg/mL kojic acid working solution was made up with 100% DMSO and then diluted to 1 mg/mL with methanol.

3.6.2. Tyrosinase Inhibition Assay

The tyrosinase inhibition assay was performed as described previously by Popoola et al. [69] and Cui et al. [70] with slight modifications. The concentrations of EO sample and kojic acid chosen, 200 μg/mL and 50 μg/mL, were achieved by setting up the 96-well plate in the following order: 70 μL of the sample (1 mg/mL) then 30 μL of tyrosinase enzyme (500 U/mL). Each concentration of the sample and positive control was set up in two different wells, whereby one of the wells received enzyme and the other well had no enzyme volume added. All volume deficits were compensated by adding excess buffer. The negative controls, 10% *v/v* of 1:1 DMSO: Tween®20 in methanol for the EO and 10% *v/v* DMSO in methanol for kojic acid were treated the same way. Then, the plate was incubated at 37 °C (±2.0 °C) for 5 min. Thereafter, the reaction was initiated by adding 110 μL of L-tyrosine (2 mM) and subsequently incubated at 37 °C (±2.0 °C) for 30 min. The absorbance of L-DOPA was read at 490 nm on a Multiskan™ spectrum plate reader (Thermo Fisher Scientific, Waltham, MA, USA). Two independent experiments were carried out in triplicate and the percentage tyrosinase inhibition was calculated using Equation (2).

$$Tyrosinase\ inhibition\ (\%) = \frac{(A - B) - (C - D)}{(A - B)} \times 100, \tag{2}$$

where A is the negative control with an enzyme, B is the negative control without enzyme, C is the EO sample or kojic acid with enzyme and D is the EO sample or kojic acid without enzyme. The inhibition percentages were expressed as the mean (±standard deviation) of duplicate measurements. One-way ANOVA was used to compare the absorbance values of the two groups ($p < 0.05$).

3.7. Sun Protection Factor (SPF)

The protocol used for this assay was conducted as per Kaur and Saraf [71]. The solubility of the EO in different ratios of ethanol and water was tested by taking 10% to 50% of ethanol in distilled water. The maximum solubility was detected at ethanol: water in a 40:60 ratio above which turbidity developed. Thereafter, an initial stock solution of 1% *v/v* was prepared by making up 10 μL of the EO to 1 mL of ethanol: water (40:60). Then out of this stock, 0.1% *v/v* was prepared. Subsequently, 100 μL of the EO aliquot and the blank (ethanol: water, 40:60) were injected in the 96-well plate and read in triplicate (n = 3) over the 290–320 nm range at 5 nm interval. The SPF value of the essential oil was calculated following the method by Mansur et al. [43]. The mean of the observed absorbance values was multiplied by their respective erythemogenic effect times solar intensity at wavelength λ values, $EE\ (\lambda) \times I\ (\lambda)$, then their summation was obtained and multiplied with the correction factor (=10). The calculation is described as Equation (3).

$$\text{SPF}_{\text{spectrophotometric}} = CF \times \sum\nolimits_{290}^{320} EE\ (\lambda) \times I\ (\lambda) \times Abs\ (\lambda), \qquad (3)$$

where CF is the correction factor (=10), $EE\ (\lambda)$ is the erythemogenic effect of radiation at wavelength λ, $I\ (\lambda)$ is the solar intensity at wavelength λ, and $Abs\ (\lambda)$ represents the spectrometric absorbance value at wavelength λ. The values of $EE\ (\lambda) \times I\ (\lambda)$ are constant values that were determined by Sayre et al. [44] as shown in Table 6.

Table 6. Relationship between erythemogenic effect and radiation intensity.

Wavelength (nm)	EE X I (Normalized)
290	0.0150
295	0.0817
300	0.2874
305	0.3278
310	0.1864
315	0.0837
320	0.0180
Total	1

4. Conclusions

The present work is the first report to investigate the chemical composition of *O. suffruticosum* essential oil and its biological activities to explore its cosmeceutical potential in selected biological activities of dermatological relevance. The GC-MS analysis served to identify sixteen constituents (**1–8, 11–15, 17, 18, 20**) totaling 85.09% of the composition. The monoterpenoids predominated the chemical composition of the essential oil by 84.64%. The major compounds were found to be ketone and alcohol monoterpenes, camphor (**12**) 31.21%, filifolone (**8**) 13.98%, chrysanthenone (**11**) 8.72%, 1,8-cineole (**6**) 7.85%, and terpinen-4-ol (**14**) 7.39%. According to the in vitro biological evaluations conducted, *O. suffruticosum* essential oil possessed low tyrosinase inhibitory activity, low to moderate antibacterial and antioxidant activity, but a promising sun protection ability as per the calculated SPF value. It is further proposed that the therapeutic properties of this essential oil can be improved by the application of nanotechnologies such as nanoencapsulation and nanostructured lipid carriers. This study establishes that the *O. suffruticosum* essential oil can be used as a complementary ingredient to boost the performance of cosmeceuticals with a prominent potential to be used in sunscreen formulations.

Author Contributions: All the authors have participated and contributed substantially to this manuscript. S.O.A.: methodology, investigation, data curation, writing the original draft. R.S.: supervision, validation, writing—review & editing. C.W.J.A.: methodology, validation, resources-biology-antioxidants. J.L.M.: methodology, validation, resources—biology-antioxidants. A.A.H.: conceptualization, methodology, supervision, resources, writing—review & editing. All authors have read and agreed to the published version of the manuscript.

Funding: This study was supported by National Research Foundation, South Africa (grant number 106055), and "The APC was funded by the Cape Peninsula University of Technology and University of the Western Cape".

Acknowledgments: Thanks to the Cape Flat Natural Reserve for the collection of the plant materials.

Conflicts of Interest: The authors declare no conflict of interest.

References

1. Properzi, A.; Angelini, P.; Bertuzzi, G.; Venanzoni, R. Some biological activities of essential oils. *Med. Aromat. Plants* **2013**, *2*, 136.
2. Dreger, M.; Wielgus, K. Application of essential oils as natural cosmetic preservatives. *Herba Pol.* **2013**, *59*, 142–156. [CrossRef]
3. Orchard, A.; van Vuuren, S. Commercial essential oils as potential antimicrobials to treat skin diseases. *Evid. Based Complement. Altern. Med.* **2017**, *2017*, 4517971. [CrossRef]
4. Shaaban, H.A.E.; El-Ghorab, A.H.; Shibamoto, T. Bioactivity of essential oils and their volatile aroma components: Review. *J. Essent. Oil Res.* **2012**, *24*, 203–212. [CrossRef]
5. Tu, P.T.B.; Tawata, S. Anti-oxidant, anti-aging, and anti-melanogenic properties of the essential oils from two varieties of *Alpinia zerumbet*. *Molecules* **2015**, *20*, 16723–16740. [CrossRef]
6. Manosroi, A.; Manosroi, J. Free radical scavenging and tyrosinase inhibition activity of aromatic volatile oil from Thai medicinal plants for cosmetic uses. *Acta Hortic.* **2005**, *680*, 97–100. [CrossRef]
7. Salleh, W.W.; Ahmad, F.; Yen, K.H.; Zulkifli, R.M. Chemical compositions and biological activities of essential oils of *Beilschmiedia glabra*. *Nat. Prod. Commun.* **2015**, *10*, 1297–1300. [CrossRef]
8. El Khoury, R.; Michael Jubeli, R.; El Beyrouthy, M.; Baillet Guffroy, A.; Rizk, T.; Tfayli, A.; Lteif, R. Phytochemical screening and antityrosinase activity of carvacrol, thymoquinone, and four essential oils of Lebanese plants. *J. Cosmet. Dermatol.* **2018**, *18*, 944–952. [CrossRef]
9. IRC Conference on Science, Engineering and Technology. Available online: http://ircset.org/anand/2017papers/IRC-SET_2017 _paper_S3-3.pdf (accessed on 4 June 2021).
10. Mali, S.S.; Killedar, S.G. Formulation and in vitro evaluation of gel for SPF determination and free radical scavenging activity of turpentine and lavender oil. *Pharma Innov. J.* **2018**, *7*, 85–90.
11. Androutsopoulou, C.; Christopoulou, S.D.; Hahalis, P.; Kotsalou, C.; Lamari, F.N.; Vantarakis, A. Evaluation of essential oils and extracts of rose geranium and rose petals as natural preservatives in terms of toxicity, antimicrobial, and antiviral activity. *Pathogens* **2021**, *10*, 494. [CrossRef]
12. Sharmeen, J.B.; Mahomoodally, F.M.; Zengin, G.; Maggi, F. Essential oils as natural sources of fragrance compounds for cosmetics and cosmeceuticals. *Molecules* **2021**, *26*, 666. [CrossRef]
13. Aumeeruddy-Elalfi, Z.; Lall, N.; Fibrich, B.; van Staden, A.B.; Hosenally, M.; Mahomoodally, M. Selected essential oils inhibit key physiological enzymes and possess intracellular and extracellular antimelanogenic properties in vitro. *J. Food Drug Anal.* **2018**, *26*, 232–243. [CrossRef]
14. Sarkic, A.; Stappen, I. Essential oils and their single compounds in cosmetics-a critical review. *Cosmetics* **2018**, *5*, 11. [CrossRef]
15. Western Cape Government-Biodiversity. Available online: https://www.westerncape.gov.za/text/2005/12/04_biodiversity_ optimised.pdf. (accessed on 13 April 2018).
16. Van Vuuren, S.F. Antimicrobial activity of South African medicinal plants. *J. Ethnopharmacol.* **2008**, *119*, 462–472. [CrossRef]
17. Kolokoto, R.; Magee, A.R. Cape stinkweeds: Taxonomy of Oncosiphon (Anthemideae, Asteraceae). *S. Afr. J. Bot.* **2018**, *117*, 57–70. [CrossRef]
18. Magee, A.R. *Oncosiphon suffructicosum*. 2011. Available online: http://pza.sanbi.org/oncosiphon-suffructicosum (accessed on 13 April 2018).
19. Scott, G.; Springfield, E.P.; Coldrey, N. A pharmacognostical study of 26 South African plant species used as traditional medicines. *Pharm. Biol.* **2004**, *42*, 186–213. [CrossRef]
20. Lall, N.; Kishore, N. Are plants used for skin care in South Africa fully explored? *J. Ethnopharmacol.* **2014**, *153*, 61–84. [CrossRef]
21. Goudjil, M.B.; Zighmi, S.; Hamada, D.; Mahcene, Z.; Bencheikh, S.E.; Ladjel, S. Biological activities of essential oils extracted from *Thymus capitatus* (Lamiaceae). *S. Afr. J. Bot.* **2020**, *128*, 274–282. [CrossRef]
22. Van Wyk, B.E. The potential of South African plants in the development of new medicinal products. *S. Afr. J. Bot.* **2011**, *77*, 812–829. [CrossRef]
23. Adams, R.P. *Identification of Essential Oil Components by Gas Chromatography/Mass Spectrometry*, 4th ed.; Allured Publishing Corp.: Carol Stream, IL, USA, 2007.
24. Babushok, V.I.; Linstrom, P.J.; Zenkevich, I.G. Retention indices for frequently reported compounds of plant essential oils. *J. Phys. Chem. Ref. Data* **2011**, *40*, 043101. [CrossRef]
25. National Institute of Standards and Technologies. Available online: https://webbook.nist.gov/chemistry/name-ser/ (accessed on 16 October 2019).
26. Hulley, I.M.; Sadgrove, N.J.; Tilney, P.M.; Özek, G.; Yur, S.; Özek, T.; Başer, K.H.C.; van Wyk, B.E. Essential oil composition of *Pentzia incana* (Asteraceae), an important natural pasture plant in the Karoo region of South Africa. *Afr. J. Range Forage Sci.* **2018**, *35*, 137–145. [CrossRef]
27. Hulley, I.M.; Özek, O.G.; Sadgrove, N.J.; Tilney, P.M.; Özek, T.; Başer, K.H.C.; van Wyk, B.E. Essential oil composition of a medicinally important Cape species: *Pentzia punctata* (Asteraceae). *S. Afr. J. Bot.* **2019**, *127*, 208–212. [CrossRef]
28. Ludwiczuk, A.; Skalicka-Woźniak, K.; Georgiev, M.I. Terpenoids. In *Pharmacognosy: Fundamentals, Applications and Strategies*; Badal, S., Delgoda, R., Eds.; Academic Press: San Diego, CA, USA, 2017; p. 251.
29. Dorman, H.J.D.; Deans, S.G. Antimicrobial agents from plants: Antibacterial activity of plant volatile oils. *J. Appl. Microbiol.* **2000**, *88*, 308–316. [CrossRef] [PubMed]

30. Dhifi, W.; Bellili, S.; Jazi, S.; Bahloul, N.; Mnif, W. Essential oils' chemical characterization and investigation of some biological activities: A critical review. *Medicines* **2016**, *3*, 25. [CrossRef] [PubMed]
31. Poljšak, B.; Dahmane, R. Free radicals and extrinsic skin aging. *Dermatol. Res. Pract.* **2012**, *2012*, 135206. [CrossRef]
32. Garg, C.; Khurana, P.; Garg, M. Molecular mechanisms of skin photoaging and plant inhibitors. *Int. J. Green Pharm.* **2017**, *11*, 17–33.
33. Santos-Sánchez, N.F.; Salas-Coronado, R.; Villanueva-Cañongo, C.; Hernández-Carlos, B. Antioxidant Compounds and Their Antioxidant Mechanism. In *Antioxidants*; Shalaby, E., Ed.; Intech: Rijeka, Croatia, 2019; pp. 1–29.
34. Gulcin, İ. Antioxidants and antioxidant methods: An updated overview. *Arch. Toxicol.* **2020**, *94*, 651–715. [CrossRef] [PubMed]
35. Huang, D.; Ou, B.; Prior, R.L. The chemistry behind antioxidant capacity assays. *J. Agric. Food Chem.* **2005**, *3*, 1841–1856. [CrossRef] [PubMed]
36. Gupta, D. Methods for determination of antioxidant capacity: A review. *Int. J. Pharm. Sci. Res.* **2015**, *6*, 546–566.
37. Schaich, K.M.; Tian, X.; Xie, J. Reprint of "Hurdles and pitfalls in measuring antioxidant efficacy: A critical evaluation of ABTS, DPPH, and ORAC assays". *J. Funct. Foods* **2015**, *18*, 782–796. [CrossRef]
38. Saeio, K.; Chaiyana, W.; Okonogi, S. Antityrosinase and antioxidant activities of essential oils of edible Thai plants. *Drug Discov. Ther.* **2011**, *5*, 144–149. [CrossRef]
39. Baumann, L. Skin ageing and its treatment. *J. Pathol.* **2007**, *211*, 241–251. [CrossRef] [PubMed]
40. Mbanga, L.; Mulenga, M.; Mpiana, P.T.; Bokolo, K.; Mumbwa, M.; Mvingu, K. Determination of sun protection factor (SPF) of some body creams and lotions marketed in Kinshasa by ultraviolet spectrophotometry. *Int. J. Adv. Res. Chem. Sci.* **2014**, *1*, 7–13.
41. Tobin, D. Introduction to skin aging. *J. Tissue Viability* **2017**, *26*, 37–46. [CrossRef] [PubMed]
42. Lohani, A.; Mishra, A.K.; Verma, A. Cosmeceutical potential of geranium and calendula essential oil: Determination of antioxidant activity and in vitro sun protection factor. *J. Cosmet. Dermatol.* **2019**, *18*, 550–557. [CrossRef] [PubMed]
43. Mansur, J.S.; Breder, M.N.R.; Mansur, M.C.A.; Azulay, R.D. Determinação do fato de proteção solar por espectrofotometrica. *An. Bras. Dermatol.* **1986**, *61*, 121–124.
44. Sayre, R.M.; Agin, P.P.; LeVee, G.J.; Marlowe, E. Comparison of in vivo and in vitro testing of sun screening formulas. *Photochem. Photobiol.* **1979**, *29*, 559–566. [CrossRef] [PubMed]
45. Kale, S.; Gaikwad, M.; Bhandare, S. Determination and comparison of in vitro SPF of topical formulation containing Lutein ester from *Tagetes erecta* L. flowers, *Moringa oleifera* Lam seed oil and *Moringa oleifera* Lam seed oil containing Lutein ester. *Int. J. Res. Pharm. Biomed. Sci.* **2011**, *2*, 1220–1224.
46. Imam, S.; Azhar, I.; Mahmood, Z.A. In-vitro evaluation of sun protection factor of a cream formulation prepared from extracts of *Musa accuminata* (L.), *Psidium gujava* (L.) and *Pyrus communis* (L.). *Asian J. Pharm. Clin. Res.* **2015**, *8*, 234–237.
47. Khor, P.Y.; Na'im Mohamed, F.S.; Ramli, I.; Nor, N.F.A.M.; Razali, S.K.C.M.; Zainuddin, J.A.; Jaafar, N.S.M. Phytochemical, antioxidant and photo-protective activity study of Bunga Kantan (*Etlingera elatior*) essential oil. *J. Appl. Pharm. Sci.* **2017**, *7*, 209–213.
48. Agarwal, R.; Pant, A.K.; Prakash, O. Chemical composition and biological activities of essential oils of *Cinnamomum tamala*, *Cinnamomum zeylenicum* and *Cinnamomum camphora* growing in Uttarakhand. In *Chemistry of Phytopotentials: Health, Energy and Environmental Perspectives*; Khemani, L.D., Srivastava, M.M., Srivastava, S., Eds.; Springer: Berlin/Heidelberg, Germany, 2012; pp. 87–92.
49. Satyal, P.; Paudel, P.; Poudel, A.; Dosoky, N.S.; Pokharel, K.K.; Setzer, W.N. Bioactivities and compositional analyses of Cinnamomum essential oils from Nepal: *C. camphora*, *C. tamala*, and *C. glaucescens*. *Nat. Prod. Commun.* **2013**, *8*, 1777–1784. [CrossRef] [PubMed]
50. Prakash, B.; Kedia, A.; Singh, A.; Yadav, S.; Singh, A.; Yadav, A.; Dubey, N.K. Antifungal, antiaflatoxin and antioxidant activity of plant essential oils and their in vivo efficacy in protection of chickpea seeds. *J. Food Qual.* **2016**, *39*, 36–44. [CrossRef]
51. Amri, I.; De Martino, L.; Marandino, A.; Lamia, H.; Mohsen, H.; Scandolera, E.; De Feo, V.; Mancini, E. Chemical composition and biological activities of the essential oil from *Artemisia herba-alba* growing wild in Tunisia. *Nat. Prod. Commun.* **2013**, *8*, 407–410. [CrossRef] [PubMed]
52. Bouzidi, N.; Mederbal, K.; Raho, B. Antioxidant activity of essential oil of *Artemisia herba alba*. *J. Appl. Environ. Biol. Sci.* **2016**, *6*, 59–65.
53. Cheraif, K.; Bakchiche, B.; Gherib, A.; Bardaweel, S.K.; Çol Ayvaz, M.; Flamini, G.; Ascrizzi, R.; Ghareeb, M.A. Chemical composition, antioxidant, anti-tyrosinase, anti-cholinesterase and cytotoxic activities of essential oils of six Algerian plants. *Molecules* **2020**, *25*, 1710. [CrossRef]
54. Derwich, E.; Benziane, Z.; Boukir, A. GC/MS analysis of volatile constituents and antibacterial activity of the essential oil of the leaves of *Eucalyptus globulus* in Atlas Median from Morocco. *Adv. Nat. Appl. Sci.* **2009**, *3*, 305–314.
55. Barbosa, L.C.; Filomeno, C.A.; Teixeira, R.R. Chemical variability and biological activities of *Eucalyptus* spp. essential oils. *Molecules* **2016**, *21*, 1671. [CrossRef] [PubMed]
56. Jerbi, A.; Derbali, A.; Elfeki, A.; Kammoun, M. Essential oil composition and biological activities of *Eucalyptus globulus* leaves extracts from Tunisia. *J. Essent. Oil Bear. Plants* **2017**, *20*, 438–448. [CrossRef]
57. Sharifi-Rad, J.; Salehi, B.; Varoni, E.M.; Sharopov, F.; Yousaf, Z.; Ayatollahi, S.A.; Kobarfard, F.; Sharifi-Rad, M.; Afdjei, M.H.; Sharifi-Rad, M.; et al. Plants of the Melaleuca genus as antimicrobial agents: From farm to pharmacy. *Phytother. Res.* **2017**, *31*, 1475–1494. [CrossRef] [PubMed]

58. Mickiene, R.; Bakutis, B.; Baliukoniene, V. Antimicrobial activity of two essential oils. *Ann. Agric. Environ. Med.* **2011**, *18*, 139–144. [PubMed]

59. Kim, H.J.; Chen, F.; Wu, C.; Wang, X.; Chung, H.Y.; Jin, Z. Evaluation of antioxidant activity of Australian tea tree (*Melaleuca alternifolia*) oil and its components. *J. Agric. Food Chem.* **2004**, *52*, 2849–2854. [CrossRef]

60. Europarådet; European Pharmacopoeia Commission. *European Pharmacopoeia*; Maisonneuve: Sainte-Ruffine, France, 1975; Volume 3, pp. 68–80.

61. Kuiate, J.R.; Amvam Zollo, P.H.; Nguefa, E.H.; Bessière, J.M.; Lamaty, G.; Menut, C. Composition of the essential oils from the leaves of *Microglossa pyrifolia* (Lam.) O. Kuntze and *Helichrysum odoratissimum* (L.) Less. growing in Cameroon. *Flavour Fragr. J.* **1999**, *14*, 82–84. [CrossRef]

62. Lucero, M.; Estell, R.; Tellez, M.; Fredrickson, E. A retention index calculator simplifies identification of plant volatile organic compounds. *Phytochem. Anal.* **2009**, *20*, 378–384. [CrossRef]

63. Lourens, A.C.U.; Reddy, D.; Başer, K.H.C.; Viljoen, A.M.; Van Vuuren, S.F. In vitro biological activity and essential oil composition of four indigenous South African *Helichrysum* species. *J. Ethnopharmacol.* **2004**, *95*, 253–258. [CrossRef] [PubMed]

64. Sartoratto, A.; Machado, A.L.M.; Delarmelina, C.; Figueira, G.M.; Duarte, M.C.T.; Rehder, V.L.G. Composition and antimicrobial activity of essential oils from aromatic plants used in Brazil. *Braz. J. Microbiol.* **2004**, *35*, 275–280. [CrossRef]

65. Bondet, V.; Brand-Williams, W.; Berset, C. Kinetics and mechanisms of antioxidant activity using the DPPH$^\bullet$ free radical method. *Lebensm.-Wiss. Technol.* **1997**, *30*, 609–615. [CrossRef]

66. Re, R.; Pellegrini, N.; Proteggente, A.; Pannala, A.; Yang, M.; Rice-Evans, C. Antioxidant activity applying an improved ABTS radical cation decolorization assay. *Free. Radic. Biol. Med.* **1999**, *26*, 1231–1237. [CrossRef]

67. Prior, R.L.; Hoang, H.A.; Gu, L.; Wu, X.; Bacchiocca, M.; Howard, L.; Hampsch-Woodill, M.; Huang, D.; Ou, B.; Jacob, R. Assays for hydrophilic and lipophilic antioxidant capacity (oxygen radical absorbance capacity (ORACFL)) of plasma and other biological and food samples. *J. Agric. Food Chem.* **2003**, *51*, 3273–3279. [CrossRef] [PubMed]

68. Benzie, I.F.F.; Strain, J.J. The ferric reducing ability of plasma (frap) as a measure of "antioxidant power": The FRAP assay. *Anal. Biochem.* **1996**, *238*, 70–76. [CrossRef]

69. Popoola, O.K.; Marnewick, J.L.; Rautenbach, F.; Ameer, F.; Iwuoha, E.I.; Hussein, A.A. Inhibition of oxidative stress and skin aging-related enzymes by prenylated chalcones and other flavonoids from *Helichrysum teretifolium*. *Molecules* **2015**, *20*, 7143–7155. [CrossRef]

70. Cui, H.; Duan, F.; Jia, S.; Cheng, F.; Yuan, K. Antioxidant and tyrosinase inhibitory activities of seed oils from *Torreya grandis* Fort. ex Lindl. *BioMed Res. Int.* **2018**, *2018*, 1–10.

71. Kaur, C.D.; Saraf, S. In vitro sun protection factor determination of herbal oils used in cosmetics. *Pharmacogn. Res.* **2010**, *2*, 22–25.

Article

Persicaria lapathifolia Essential Oil: Chemical Constituents, Antioxidant Activity, and Allelopathic Effect on the Weed *Echinochloa colona*

Ahmed M. Abd-ElGawad [1,2,*], **Giuliano Bonanomi** [3,4], **Sarah A. Al-Rashed** [5] **and Abdelsamed I. Elshamy** [6]

[1]   Plant Production Department, College of Food & Agriculture Sciences, King Saud University, P.O. Box 2460, Riyadh 11451, Saudi Arabia

[2]   Department of Botany, Faculty of Sciences, Mansoura University, Mansoura 35516, Egypt

[3]   Department of Agricultural Sciences, University of Naples Federico II, Via Università 100, 80055 Portici, Italy; giuliano.bonanomi@unina.it

[4]   Task Force on Microbiome Studies, University of Naples Federico II, 80131 Naples, Italy

[5]   Department of Botany and Microbiology, College of Science, King Saud University, P.O. 2455, Riyadh 11451, Saudi Arabia; salrashed@ksu.edu.sa

[6]   Chemistry of Natural Compounds Department, National Research Centre, 33 El Bohouth St., Dokki, Giza 12622, Egypt; elshamynrc@yahoo.com

*   Correspondence: aibrahim2@ksu.edu.sa; Tel.: +966-562680864

Citation: Abd-ElGawad, A.M.; Bonanomi, G.; Al-Rashed, S.A.; Elshamy, A.I. *Persicaria lapathifolia* Essential Oil: Chemical Constituents, Antioxidant Activity, and Allelopathic Effect on the Weed *Echinochloa colona*. *Plants* **2021**, *10*, 1798. https://doi.org/10.3390/plants10091798

Academic Editors: Hazem Salaheldin Elshafie, Laura De Martino and Adriano Sofo

Received: 9 August 2021
Accepted: 24 August 2021
Published: 29 August 2021

Publisher's Note: MDPI stays neutral with regard to jurisdictional claims in published maps and institutional affiliations.

Abstract: The exploration of new green, ecofriendly bioactive compounds has attracted the attention of researchers and scientists worldwide to avoid the harmful effects of chemically synthesized compounds. *Persicaria lapathifolia* has been reported to have various bioactive compounds, while its essential oil (EO) has not been determined yet. The current work dealt with the first description of the chemical composition of the EO from the aerial parts of *P. lapathifolia*, along with studying its free radical scavenging activity and herbicidal effect on the weed *Echinochloa colona*. Twenty-one volatile compounds were identified via GC–MS analysis. Nonterpenoids were the main components, with a relative concentration of 58.69%, in addition to terpenoids (37.86%) and carotenoid-derived compounds (1.75%). *n*-dodecanal (22.61%), α-humulene (11.29%), 2,4-dimethylicosane (8.97%), 2*E*-hexenoic acid (8.04%), γ-nonalactone (3.51%), and limonene (3.09%) were characterized as main compounds. The extracted EO exhibited substantial allelopathic activity against the germination, seedling root, and shoot growth of the weed *E. colona* in a dose-dependent manner, showing IC_{50} values of 77.27, 60.84, and 33.80 mg L^{-1}, respectively. In addition, the *P. lapathifolia* EO showed substantial antioxidant activity compared to ascorbic acid as a standard antioxidant. The EO attained IC_{50} values of 159.69 and 230.43 mg L^{-1}, for DPPH and ABTS, respectively, while ascorbic acid exhibited IC_{50} values 47.49 and 56.68 mg L^{-1}, respectively. The present results showed that the emergent leafy stems of aquatic plants such as *P. lapathifolia* have considerably low content of the EO, which exhibited substantial activities such as antioxidant and allelopathic activities. Further study is recommended to evaluate the effects of various environmental and climatic conditions on the production and composition of the EOs of *P. lapathifolia*.

Keywords: pale smartweed; essential oil; phytotoxicity; green chemistry; herbicides

1. Introduction

Essential oils (EOs) are complex mixtures of small compounds, and they have been documented to have persuasive medicinal potentialities including antioxidant [1,2], antibacterial [3,4], anti-inflammatory [5], anticancer [6], antiaging [2] antipyretic [5], and other activities. Because of these activities, EOs have been integrated into the food and cosmetics industries [7]. In the agriculture sector, EOs can be integrated as green, ecofriendly bioherbicides, where they can substitute the synthetic and chemical herbicides that are responsible for environmental pollution and affect human health [8].

Allelopathy is defined as the chemical interference between plants [9]. Understanding this biological phenomenon aids the development of applications for natural and agricultural systems [10]. Allelopathic interactions are considered to be crucial for the success of many invasive plants and are a key element in determining species distribution and abundance within plant ecosystems. Temperature, water content, salinity, nutrition availability, competitive stress, and microbiota are among the most important variables influencing allelopathy [11–14]. EOs from various wild plants have been reported as allelochemicals that interfere with the germination and growth of weeds [6,12,14,15]. These EOs have promising characteristics against weeds, as they are ecofriendly, have low resistance from weeds, and possess a wide allelopathic spectrum [8]. Over 3000 plant species have been studied for their EO composition, and hundreds of these EOs have been produced commercially [16]. In consequence, the exploration of new EOs from various wild plants with biological activities has attracted the attention of researchers and scientists worldwide.

Polygonaceae is an important edible families of flowering plants and has a worldwide distribution with 46 genera and 1100 species [17,18]. In the flora of Egypt, there are 28 species of Polygonaceae, belonging to eight genera [17]. *Persicaria* is commonly known as smartweeds or knotweeds, and this genus comprises about 150 herbaceous cosmopolitan species. Most species are found in temperate regions, with a few others are found in tropical and subtropical regions from sea level to a range of different altitudes [19]. Seven *Persicaria* species have been recorded in the flora of Egypt. *Persicaria lapathifolia* (L.) Delarbre (synonym *Polygonum lapathifolium*) is commonly known as pale persicaria, pale smartweed, or curlytop knotweed (Figure 1).

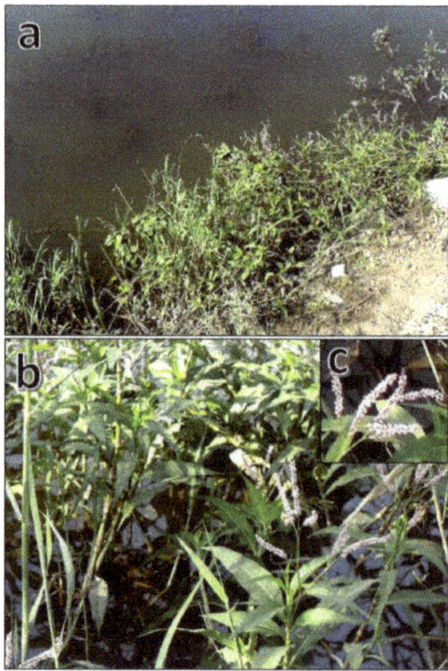

Figure 1. *Persicaria lapathifolia* (L.) Delarbre: (**a**) overview of the aerial parts growing on the canal bank habitat, (**b**) close view of the aerial parts, and (**c**) close view of the inflorescence.

Persicaria lapathifolia is an annual herb that grows up to 80 cm and has reddish stems and alternate leaves [17]. This plant has grown worldwide (cosmopolitan) and inhabits various moist habitats and in agricultural fields. This species is considered a troublesome weed. *Persicaria lapathifolia* has many uses in traditional medicine such as antibacterial, antiviral, anti-inflammatory, astringent, antiseptic, anti-stomach complaint, hepatoprotective, and antifungal uses in addition to its use for the treatment of dysentery, burns, and fevers [20]. Several studies have revealed that different extracts and isolated metabolites of this plant have antimicrobial [21], anthelmintic and antiemetic [20], anticancer [22], antioxidant, α/β-glucosidase inhibitory, and anticholinesterase activity [23]. These activities were ascribed to the numerous phytoconstituents in this plant such as flavonoids, chalcones, acylated flavonoids, ferulate esters, and phenolic acids [21–25]. However, to our knowledge, the EO of *P. lapathifolia* has not been described yet, and in consequence, its EO's allelopathic effect against weeds has not been studied. *Echinochloa colona* (L.) Link is commonly recognized as jungle rice because of its excellent competition with rice [26]. It is a noxious worldwide distributed weed that infests many crops (rice, maize, sugarcane, and cotton) as well as other habitats such as roadsides, gardens, disturbed sites, fallow lands, and pastures [27]. This weed is hardly controlled in the agroecosystem because of its diverse ecotypes, vast production of seeds, short seed dormancy, fast growth rate, high competitive perspective, allelopathic interference, and herbicide resistance [26].

Thereby, the present study aimed to analyze for the first time the chemical composition of the volatile oil extracted from the aerial parts of *P. lapathifolia*, evaluate its allelopathic activity of the EO on the weed *E. colona*, and determine its antioxidant activity.

2. Results and Discussion

2.1. Chemical Composition of P. lapathifolia EO

The hydrodistillation of the aerial parts of *P. lapathifolia* via the Clevenger apparatus produced 0.18% (v/w) of a dark yellow oil accompanied by a slight scent. This yield was lower than those reported from other *Persicaria* species such as Vietnamese *P. odorata* (0.41%) [28], Malaysian *P. odorata* (0.64%) [29], and *P. hydropiper* (0.70%) [15]. Overall, twenty-one volatile components were identified via GC–MS analysis, representing 98.3% of the total mass of oil (Table 1).

Eight classes of the compounds were characterized for the EO, including monoterpenes (hydrocarbons and oxygenated), sesquiterpenes (hydrocarbons and oxygenated), diterpenes (hydrocarbons and oxygenated), carotenoid-derived compounds, and other nonterpenoid compounds (Table 1). Among these components, nonterpenoid constituents represented the predominant type of compounds, with a relative concentration of 58.69%. This result is consistent with the previously described essential oils from Vietnamese [28] and Australian [30] *P. odorata*. In contrast, Dũng, et al. [31] described that sesquiterpene hydrocarbons were the main compounds of the Vietnamese *P. odorata* along with the oxygenated constituents. Herein, 11 nonterpenoids were identified, comprising *n*-dodecanal (22.61%), 2,4-dimethylicosane (8.97%), 2*E*-hexenoic acid (8.04%), and γ-nonalactone (3.51%) as the abundant compounds, while *n*-tricosane (1.18%) was determined as a minor compound. The preponderance of the alkyl aldehyde, *n*-dodecanal, in the EO of *P. lapathifolia* is in harmony with the published data on the EOs of Vietnamese and Australian *P. odorata* [28,30].

Monoterpenes were determined with a relative concentration of 17.23% of the EO mass. They can be divided into monoterpene hydrocarbons (6.80%) and oxygenated monoterpenes (10.43%). Three monoterpene hydrocarbon compounds were assigned, of which limonene (3.09%) was the major and α-pinene was a minor compound. Limonene is not a widely distributed compound in the EOs of *Persicaria* plants, although it has been reported as a major constituent of the EOs of several species such as *Schinus terebinthifolius* [32], *Callistemon viminalis* [33], *Artemisia scoparia* [34], *Heterothalamus psiadioides* [35], and *Carum carvi* [36].

On the other side, 2-methyl butyl isovalerate was the only identified oxygenated monoterpene, with a relative concentration of 10.43%. Isovalerate derivatives have been described in the EOs of several plants such as *Eucalyptus brockwayii* [37], Algerian *Daucus gracilis* [38], and *Chamaemelum fuscatum* [39].

Table 1. Essential oil chemical composition from aerial parts of *Persicaria lapathifolia*.

No	Rt [a]	Relative Conc. (%) [b]	Compound Name	Type	Identification [c]	
					KI$_{Observed}$ [d]	KI$_{Published}$ [e]
1	3.23	1.74 ± 0.02	α-Pinene	MH	932	935
2	4.14	1.97 ± 0.05	Sabinene	MH	969	966
3	5.43	8.04 ± 0.06	2E-Hexenoic acid	Others	1007	1009
4	13.84	3.09 ± 0.05	Limonene	MH	1024	1028
5	14.54	10.43 ± 0.14	2-Methyl butyl isovalerate	OM	1104	1104
6	18.26	1.98 ± 0.03	Hydrocinnamyl alcohol	Others	1227	1232
7	20.12	3.51 ± 0.06	γ-Nonalactone	Others	1361	1367
8	21.67	1.29 ± 0.02	*trans*-Caryophyllene	SH	1408	1403
9	22.64	22.61 ± 0.16	*n*-Dodecanal	Others	1408	1415
10	23.07	11.29 ± 0.13	α-Humulene	SH	1452	1459
11	32.46	2.69 ± 0.04	2-Ethyl chromone	Others	1614	1618
12	32.84	3.78 ± 0.04	Phytane	DH	1810	1818
13	36.88	1.75 ± 0.02	Hexahydrofarnesyl acetone	Car	1845	1847
14	38.16	3.19 ± 0.03	Carissone	OS	1927	1934
15	39.42	8.97 ± 0.08	2,4-Dimethylicosane	Others	2080	2087
16	40.02	1.08 ± 0.04	Abienol	OD	2150	2156
17	42.56	2.27 ± 0.06	*n*-Docosane	Others	2200	2202
18	46.02	1.18 ± 0.03	*n*-Tricosane	Others	2300	2300
19	46.2	2.72 ± 0.05	*n*-Nonacosane	Others	2900	2905
20	49.22	3.01 ± 0.04	*n*-Hentriacontane	Others	3100	3101
21	51.14	1.71 ± 0.03	*n*-Dotriacontane	Others	3200	3203
		6.80 ± 0.07	Monoterpene Hydrocarbons (MH)			
		10.43 ± 0.14	Oxygenated Monoterpenes (OM)			
		12.58 ± 0.12	Sesquiterpene Hydrocarbons (SH)			
		3.19 ± 0.03	Oxygenated Sesquiterpenes (OS)			
		3.78 ± 0.04	Diterpene Hydrocarbons (DH)			
		1.08 ± 0.04	Oxygenated Diterpenes (OD)			
		1.75 ± 0.02	Carotenoid derived compounds (Car)			
		58.69 ± 0.13	Other compounds (Others)			
		98.3	Total			

[a] Rt: retention time, [b] average value ± standard deviation, [c] the identification of EO constituents was based on the comparison of mass spectral data and Kovats indices (KI) with those of the NIST Mass Spectral Library (2011) and Wiley Registry of Mass Spectral Data 8th edition and literature, [d] KI$_{published}$: reported Kovats retention indices; [e] KI$_{Observed}$: experimentally calculated Kovats index relative to C$_8$–C$_{28}$ *n*-alkanes.

Sesquiterpenes made up a relative concentration of 15.77% of the total oil mass. They consisted of sesquiterpene hydrocarbons (12.58%) and oxygenated sesquiterpene (3.19%). α-Humulene (11.29%), and *trans*-caryophyllene (1.29%) were the only identified sesquiterpene hydrocarbons. The two compounds are popular in the EOs of the plants belonging to *Persicaria* such as Vietnamese *P. odorata* [28,31] and Australian *P. odorata* [30]. In the EO of *P. lapathifolia*, diterpenes represented 4.86% of the total oil mass, including one diterpene hydrocarbon, phytane (3.78%), and one oxygenated compound, abienol (1.08%). Diterpenes have been recorded in the EOs of *Persicaria* species as traces [28,30,31]. The scarcity of diterpenoids in EOs derived from plants is a common phenomenon with a few exceptions such as *Araucaria heterophylla* [5] and *Calotropis procera* [14].

Carotenoid-derived components were represented by only the common compound, hexahydrofarnesyl acetone (1.75%), which has been characterized in the EOs of many plants such as *Launaea mucronata*, *Launaea nudicaulis* [40], and *Heliotropium curassavicum* [13].

2.2. Allelopathic Activity of P. lapathifolia EO

The extracted EO of *P. lapathifolia* exhibited substantial allelopathic activity against the germination, seedling root, and shoot growth of the weed *E. colona* in a dose-dependent manner (Figure 2). At the highest concentration (100 mg L^{-1}), the germination, seedling root, and shoot growth were inhibited by 64.25, 82.48, and 95.25%, respectively. Additionally, the EO showed IC_{50} values of 77.27, 60.84, and 33.80 mg L^{-1}, respectively (Figure 2). It is clear that the seedling growth was more inhibited than germination. However, the root growth of *E. colona* was more sensitive to the EO than the shoots, and this observation was in harmony with other studies [13,41]. This could be attributed to the permeability of the root cells and the direct contact with the medium that contained the EO [14,41,42].

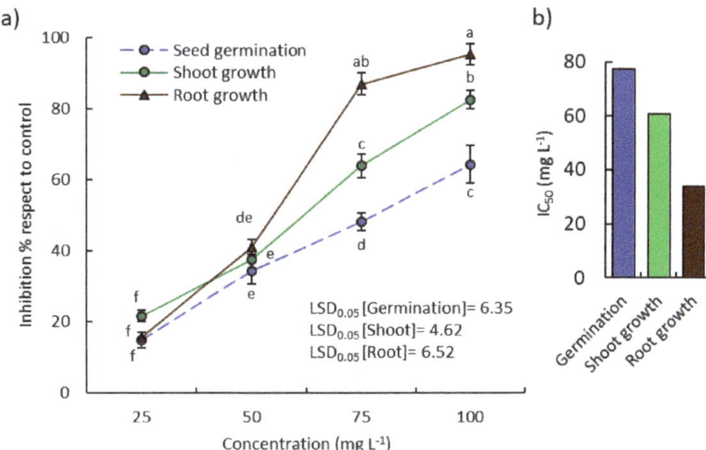

Figure 2. Allelopathic activity of the *Persicaria lapathifolia* essential oil. (**a**) Inhibitory effect on the germination, seedling root, and shoot growth of the weed *Echinochloa colona*; (**b**) IC_{50} values. Different letters mean a significant difference in values after Tukey's HSD test ($p < 0.05$).

To our knowledge, the allelopathic activity of *P. lapathifolia* EO has not been described yet. Herein, the EO of *P. lapathifolia* was found to exhibit allelopathic effects on *E. colona*. Many published data have revealed the principal and direct relationship between the allelopathic properties of EOs derived from plants and their chemical compositions [4,14,40–42]. Therefore, the observed allelopathic activity of *P. lapathifolia* EO might be ascribed to its chemical constituents, especially the main compounds *n*-dodecanal, α-humulene, 2,4-dimethylicosane, and 2*E*-hexenoic acid, γ-nonalactone, and limonene. These compounds could act either singularly or in a synergistic manner as allelochemicals that inhibit germination and seedling growth [41]. The allelopathic effect of allelochemicals may occur by inhibition of cell division, reduction of respiration, affecting photosynthesis, inhibition of enzymatic systems, affecting nucleic acid, or induction of reactive oxygen species in plant cells [43–45]. Humulene has been reported as a major compound in the EO of *Symphyotrichum squamatum*, which exhibited significant allelopathic activity on the weed *Bidens pilosa* [12]. Also, limonene has been reported as a major compound in the EOs of various plants that exhibited considerable allelopathic activity such as *Heterothalamus psiadioides* [35], *Agastache rugosa* [46], *Schinus terebinthifolius* [32], *Callistemon viminalis* [33], *Artemisia scoparia* [34], *Pinus pinea* [47], and *Carum carvi* [36].

It is worth mentioning here that *E. colona* has been reported to have allelopathic activity against various crops and weeds such as rice, soybean [48], and *Avena fatua* [49]. Also, the allelopathic activity of various plant extracts, such as those of *Sorghum bicolor*, *Helianthus annuus*, *Brassica campestris* [50], *Mikania micrantha*, *Clidemia hirta*, *Dicranopteris linearis*, and *Ageratum conyzoides* [51], has been studied against *E. colona*. However, no study has revealed the activity of the essential oil of *P. lapathifolia* against this weed. In this line, the present study showed that the EO of the aerial parts of *P. lapathifolia* could be used in the management of weeds as a green, ecofriendly herbicide, particularly against species of *Echinochloa*, including *E. colona*, which has been reported to be resistant against herbicides [52].

2.3. Antioxidant Activity of P. lapathifolia EO

The EO of *P. lapathifolia* was tested for antioxidant activity via the reduction of the free radicals DPPH and ABTS, and it showed a substantial antioxidant activity compared to ascorbic acid as a standard antioxidant (Figure 3). By increasing the concentration of the EO, the reduction of radicals was increased. At the highest concentration of the EO (400 mg L^{-1}), the DPPH and ABTS were reduced by 70.68 and 67.23%, respectively. The EO showed IC_{50} values of 159.69 and 230.43 mg L^{-1}, respectively (Figure 3a), while ascorbic acid exhibited IC_{50} values 47.49 and 56.68 mg L^{-1}, respectively (Figure 3b).

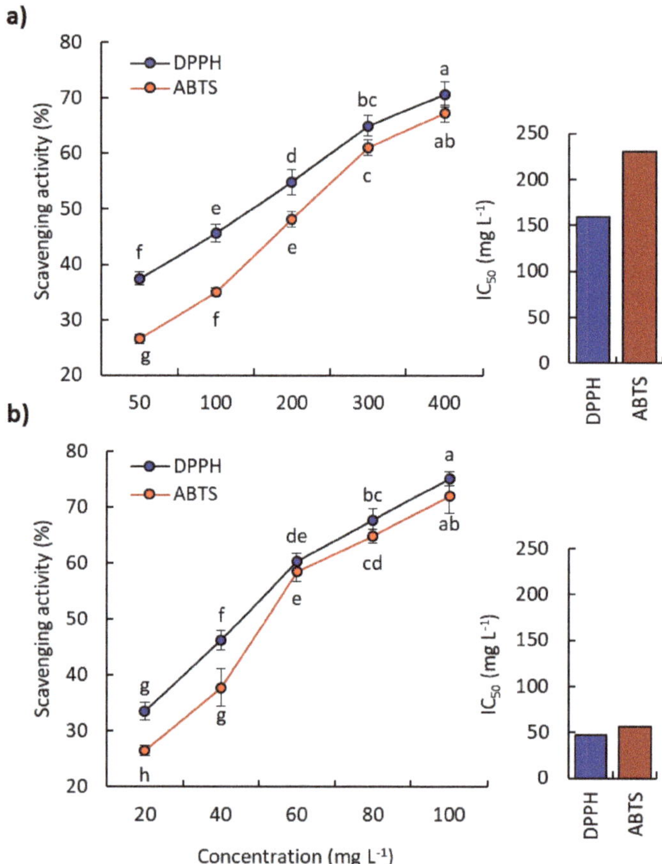

Figure 3. Antioxidant activity of *Persicaria lapathifolia* essential oil (**a**) and ascorbic acid as a standard antioxidant (**b**). Different letters mean a significant difference in values after Tukey's HSD test ($p < 0.05$).

The antioxidant activity of the EO is usually correlated to the oxygenated compounds in the EO profile [8,12,41]. Thereby, the substantial antioxidant activity of EO of *P. lapathifolia* could be attributed to its high content of oxygenated components (> 70%), especially oxygenated hydrocarbons. Additionally, terpenoid compounds have been stated to have important functions as free radical scavengers, especially the oxygenated derivatives [12,40]. In the present study, about 38% of the EO mass consisted of terpenoids, including 15% oxygenated derivatives that could be responsible for the scavenging of free radicals. Many EOs of plants have been proven to exhibit significant antioxidant effects via different methods because of their high percentages of terpenoids, such as those of *Euphorbia mauritanica* [1], *Deverra tortuosa* [3], and *Launaea* species [40].

3. Materials and Methods

3.1. Plant Materials Collected and Preparation

The aerial parts of *Persicaria lapathifolia* were collected in May from a population growing on the canal bank habitat of an irrigation canal near the city of Mansoura, Egypt (31.0708553N 31.4394701E). The collected aerial parts were air dried at room temperature (25 ± 3 °C), crushed gently by hand, and packed in a paper bag till further analyses. A plant voucher specimen was deposited in the herbarium of the Department of Botany, Faculty of Science, Mansoura University, Egypt.

3.2. Extraction of EO, GC-MS Analysis, and Identification of Chemical Constituents

About 250 g of the air-dried powder aerial parts of *P. lapathifolia* was subjected to hydrodistillation over Clevenger-type apparatus for three hours, and the dark yellow, oily layer was separated by n-hexane via a separating funnel. The EO chemical composition was analyzed and identified based on gas chromatography–mass spectrometry (GC–MS) using the instrument at the Medicinal and Aromatic Plants Research Dept., National Research Center, Egypt. The instrument comprises a TRACE GC Ultra Gas Chromatographs (THERMO Scientific Corp., Miami, CA, USA) and a thermo mass spectrometer detector (ISQ Single Quadrupole Mass Spectrometer; Model ISQ spectrometer, THERMO Scientific Corp., Miami, CA, USA). The system was equipped with a TR-5 MS column with specifications of 30 m $\times$ 0.32 mm i.d. and 0.25 μm film thickness. Helium was used as a carrier gas at a flow rate of 1.0 mL min^{-1} and a split ratio of 1:10. The temperature program started at 60 °C for one minute and was then raised 4.0 °C every minute to 240 °C, then held for one minute. Electron ionization (EI) at 70 eV with a spectral range of m/z 40–450 was used for performing mass spectra. The chemical constituents of the EO were tentatively identified by their retention indices (relative to n-alkanes C_8-C_{22}) and mass spectrum matching to the Wiley spectral library collection and the NSIT library database [41].

3.3. Allelopathic Bioassay

The extracted EO from the aerial parts of *P. lapathifolia* was tested in vitro for its allelopathic activity against the germination and seedling growth of the weed *E. colona*. The ripened seeds of *E. colona* were collected from a cultivated field near the city of Manzala, Al-Dakahlya Governorate, Egypt (31.1691466N 31.896379E). The homogenized seeds in size and color were sterilized with 0.3% NaClO and dried in a sterilized condition. The viability of seeds was preliminarily tested and found to be 94.56% $\pm$ 3.25. For bioassay, different concentrations (0, 25, 50, 75, and 100 mg L^{-1}) of the extracted EO were prepared in 1% Tween® 80 (Sigma-Aldrich, Darmstadt, Germany) as an emulsifier. Twenty sterilized seeds were arranged over a filter paper (Whatman No. 1) in Petri dishes. About four mL of each concentration and Tween® 80 (as control) were poured over the filter paper, and the dishes were sealed with Parafilm® tape (Sigma, St. Louis, MO, USA) to avoid the loss of EO [41]. For each concentration, five dishes were tested, and the experiment was repeated three times. A total of 75 dishes (5 treatments (4 concentrations + control) $\times$ 5 dishes (replications) $\times$ 3 times) were prepared and incubated at 27 °C in a growth chamber with adjusted light conditions of 16 h light and 8 h dark. After ten days of incubation, the number of germinated *E. colona* seeds was counted, and the lengths of the seedling root and shoot of the weed were measured. The inhibition of germination and growth was calculated with respect to control according to the following equation:

$$\text{Inhibition (\%)} = 100 \times \frac{\left(\dfrac{N_{control}}{L_{control}} - \dfrac{N_{treatment}}{L_{treatment}} \right)}{\dfrac{N_{treatment}}{L_{control}}}$$

where N is the number of germinated seeds and L is the length of the seedling root or shoot.

3.4. Antioxidant Activity Estimation

The antioxidant activity of the extracted EO was assessed by its ability to reduce the free radicals 2,2-diphenyl-1-picrylhydrazyl (DPPH, Sigma-Aldrich, Germany) and 2,2′-azinobis(3-ethylbenzothiazoline-6-sulfonic acid) (ABTS, Sigma-Aldrich, Germany). According to Miguel [53], a range of concentrations of the EO (50–400 mg L^{-1}) were prepared in MeOH. This range was selected based on the observed scavenging percentage, i.e., to be suitable to determine the IC_{50} (the concentration of EO necessary to scavenge the radical by 50%) [3]. In glass tubes, 2 mL of each concentration was mixed vigorously with 2 mL of freshly prepared 0.3 mM DPPH. Negative control was performed using MeOH treated with DPPH-like treatment. The reaction mixtures were kept in dark conditions at room temperature for 30 min, and the absorbance was measured immediately at 517 nm by spectrophotometer (Spectronic 21D model).

For confirmation, the scavenging of ABTS was performed following the method of Re et al. [54]. Briefly, the ABTS radical was prepared by mixing about 7 mM ABTS (1/1, v/v) with 2.45 mM potassium persulfate, and this mixture was stored in dark conditions at 25 ± 2 °C for 16 h. In glass tubes, 2 mL of the prepared ABTS radical and 0.2 mL of each EO concentration (50–400 mg L^{-1}) were mixed well and kept for 6 min at room temperature. The absorbance was measured by a spectrophotometer at 734 nm. Moreover, the antioxidant activity of ascorbic acid as a standard antioxidant was determined following the same procedures for DPPH and ABTS using a range of 20–100 mg L^{-1}. The scavenging activity was calculated according to the following equation:

$$Radical\ scavenging\ activity\ (\%) = 100 \times \left[1 - \left(Absorbance_{sample} / Absorbance_{control} \right) \right]$$

3.5. Statistical Analysis

The experiment of allelopathic activity was performed three times with five replicas per treatment. The mean values and standard error were calculated, while the raw data were subjected to one-way ANOVA followed by Tukey's HSD test. Moreover, the data of antioxidant activity of the EO and ascorbic acid with three replicates were subjected to one-way ANOVA followed by Tukey's HSD test as well. The analysis was accomplished in the CoStat program (version 6.311, CoHort Software, Monterey, CA, USA). The IC_{50} for allelopathic and antioxidant assays were calculated graphically using MS Excel.

4. Conclusions

The current study showed for the first time the chemical composition of the EO from the aerial parts of *P. lapathifolia*. The EO had 21 compounds, with 58.69% as nonterpenoids and 37.86% as terpenoids. The main compounds were *n*-dodecanal, α-humulene, 2,4-dimethylicosane, 2E-hexenoic acid, γ-nonalactone, and limonene. The EO of *P. lapathifolia* showed substantial herbicidal activity against the weed *E. colona*; 100 mg L^{-1} of this EO inhibited the germination, seedling root, and shoot growth by 64.25, 82.48, and 95.25%, respectively. Also, the EO exhibited considerable antioxidant activity compared to ascorbic acid as a standard. The present results showed that the emergent leafy stems of aquatic plants such as *P. lapathifolia* have considerably low content of the EO, which exhibited substantial activities such as antioxidant and allelopathic activities. Further study is recommended to evaluate the effects of various environmental and climatic conditions on the production and composition of the EOs of *P. lapathifolia*.

Author Contributions: Conceptualization, A.M.A.-E. and A.I.E.; validation, A.M.A.-E., G.B., and A.I.E.; formal analysis, A.M.A.-E. and A.I.E.; investigation, A.M.A.-E., G.B., and A.I.E.; writing—original draft preparation, A.M.A.-E., G.B., and A.I.E.; writing—review and editing, A.M.A.-E., G.B., S.A.A.-R., and A.I.E. All authors have read and agreed to the published version of the manuscript.

Funding: The Researchers Supporting Project number (RSP-2021/200) King Saud University, Riyadh, Saudi Arabia.

Institutional Review Board Statement: Not applicable.

Informed Consent Statement: Not applicable.

Data Availability Statement: Not applicable.

Acknowledgments: The authors extend their appreciation to The Researchers Supporting Project number (RSP-2021/200) King Saud University, Riyadh, Saudi Arabia.

Conflicts of Interest: The authors declare no conflict of interest.

Sample Availability: Samples of the compounds are not available from the authors.

References

1. Essa, A.F.; El-Hawary, S.S.; Abd-ElGawad, A.M.; Kubacy, T.M.; El-Khrisy, E.E.-D.A.; Elshamy, A.; Younis, I.Y. Prevalence of diterpenes in essential oil of *Euphorbia mauritanica* L.: Detailed chemical profile, antioxidant, cytotoxic and phytotoxic activities. *Chem. Biodivers.* **2021**, *18*, e2100238. [CrossRef] [PubMed]
2. Elgamal, A.M.; Ahmed, R.F.; Abd-ElGawad, A.M.; El Gendy, A.E.-N.G.; Elshamy, A.I.; Nassar, M.I. Chemical profiles, anticancer, and anti-aging activities of essential oils of *Pluchea dioscoridis* (L.) DC. and *Erigeron bonariensis* L. *Plants* **2021**, *10*, 667. [CrossRef] [PubMed]
3. Fayed, E.M.; Abd-ElGawad, A.M.; Elshamy, A.I.; El-Halawany, E.S.F.; El-Amier, Y.A. Essential oil of *Deverra tortuosa* aerial parts: Detailed chemical profile, allelopathic, antimicrobial, and antioxidant activities. *Chem. Biodivers.* **2021**, *18*, e2000914. [CrossRef] [PubMed]
4. Saleh, I.; Abd-ElGawad, A.; El Gendy, A.E.-N.; Abd El Aty, A.; Mohamed, T.; Kassem, H.; Aldosri, F.; Elshamy, A.; Hegazy, M.-E.F. Phytotoxic and antimicrobial activities of *Teucrium polium* and *Thymus decussatus* essential oils extracted using hydrodistillation and microwave-assisted techniques. *Plants* **2020**, *9*, 716. [CrossRef] [PubMed]
5. Elshamy, A.I.; Ammar, N.M.; Hassan, H.A.; Al-Rowaily, S.L.; Ragab, T.I.; El Gendy, A.E.-N.G.; Abd-ElGawad, A.M. Essential oil and its nanoemulsion of *Araucaria heterophylla* resin: Chemical characterization, anti-inflammatory, and antipyretic activities. *Ind. Crops Prod.* **2020**, *148*, 112272. [CrossRef]
6. Ali, H.; Al-Khalifa, A.R.; Aouf, A.; Boukhebti, H.; Farouk, A. Effect of nanoencapsulation on volatile constituents, and antioxidant and anticancer activities of Algerian *Origanum glandulosum* Desf. essential oil. *Sci. Rep.* **2020**, *10*, 1–9. [CrossRef]
7. Silvestre, W.; Livinalli, N.; Baldasso, C.; Tessaro, I. Pervaporation in the separation of essential oil components: A review. *Trends Food Sci. Technol.* **2019**, *93*, 42–52. [CrossRef]
8. Abd-ElGawad, A.M.; El Gendy, A.E.-N.G.; Assaeed, A.M.; Al-Rowaily, S.L.; Alharthi, A.S.; Mohamed, T.A.; Nassar, M.I.; Dewir, Y.H.; Elshamy, A.I. Phytotoxic effects of plant essential oils: A systematic review and structure-activity relationship based on chemometric analyses. *Plants* **2021**, *10*, 36. [CrossRef]
9. Blum, U. Plant–plant allelopathic interactions. In *Plant-Plant Allelopathic Interactions*; Springer: Dordrecht, The Netherlands, 2011; pp. 1–7.
10. Serra, S.N.; Shanmuganathan, R.; Becker, C. Allelopathy in rice: A story of momilactones, kin recognition, and weed management. *J. Exp. Bot.* **2021**, *72*, 4022–4037. [CrossRef]
11. Bonanomi, G.; Zotti, M.; Idbella, M.; Mazzoleni, S.; Abd-ElGawad, A.M. Microbiota modulation of allelopathy depends on litter chemistry: Mitigation or exacerbation? *Sci. Total Environ.* **2021**, *776*, 145942. [CrossRef] [PubMed]
12. Abd-ElGawad, A.M.; Elshamy, A.I.; El-Amier, Y.A.; El Gendy, A.E.-N.G.; Al-Barati, S.A.; Dar, B.A.; Al-Rowaily, S.L.; Assaeed, A.M. Chemical composition variations, allelopathic, and antioxidant activities of *Symphyotrichum squamatum* (Spreng.) Nesom essential oils growing in heterogeneous habitats. *Arab. J. Chem.* **2020**, *13*, 4237–4245. [CrossRef]
13. Abd-ElGawad, A.M.; Elshamy, A.I.; Al-Rowaily, S.L.; El-Amier, Y.A. Habitat affects the chemical profile, allelopathy, and antioxidant properties of essential oils and phenolic enriched extracts of the invasive plant *Heliotropium curassavicum*. *Plants* **2019**, *8*, 482. [CrossRef] [PubMed]
14. Al-Rowaily, S.L.; Abd-ElGawad, A.M.; Assaeed, A.M.; Elgamal, A.M.; Gendy, A.E.-N.G.E.; Mohamed, T.A.; Dar, B.A.; Mohamed, T.K.; Elshamy, A.I. Essential oil of *Calotropis procera*: Comparative chemical profiles, antimicrobial activity, and allelopathic potential on weeds. *Molecules* **2020**, *25*, 5203. [CrossRef] [PubMed]
15. Aziman, N.; Abdullah, N.; Bujang, A.; Mohd Noor, Z.; Abdul Aziz, A.; Ahmad, R. Phytochemicals of ethanolic extract and essential oil of *Persicaria hydropiper* and their potential as antibacterial agents for food packaging polylactic acid film. *J. Food Saf.* **2021**, *41*, e12864. [CrossRef]
16. Sankarikutty, B.; Narayanan, C.S. Essential oils isolation and production. In *Encyclopedia of Food Sciences and Nutrition*, 2nd ed.; Caballero, B., Ed.; Academic Press: Oxford, UK, 2003; pp. 2185–2189.
17. Boulos, L. *Flora of Egypt*; All Hadara Publishing: Cairo, Egypt, 1999; Volume 1.
18. Tackholm, V. *Students' Flora of Egypt*, 2nd ed.; Cairo University Press: Cairo, Egypt, 1974.
19. Heywood, V.H.; Brummitt, R.; Culham, A.; Seberg, O. *Flowering Plant Families of the World*; Kew Publishing: London, UK, 2007; Volume 88.

20. Bulbul, L.; Sushanta, S.M.; Uddin, M.J.; Tanni, S. Phytochemical and pharmacological evaluations of *Polygonum lapathifolium* stem extract for anthelmintic and antiemetic activity. *Int. Curr. Pharm. J.* **2013**, *2*, 57–62. [CrossRef]

21. Hailemariam, A.; Feyera, M.; Deyou, T.; Abdissa, N. Antimicrobial Chalcones from the Seeds of *Persicaria lapathifolia*. *Biochem. Pharmacol.* **2018**, *7*, 2167-0501. [CrossRef]

22. Takasaki, M.; Konoshima, T.; Kuroki, S.; Tokuda, H.; Nishino, H. Cancer chemopreventive activity of phenylpropanoid esters of sucrose, vanicoside B and lapathoside A, from *Polygonum lapathifolium*. *Cancer Lett.* **2001**, *173*, 133–138. [CrossRef]

23. Kubínová, R.; Pořízková, R.; Bartl, T.; Navrátilová, A.; Čížek, A.; Valentová, M. Biological activities of polyphenols from *Polygonum lapathifolium*. *B. Latinoam. Caribe Plantas Med. Aromát.* **2014**, *13*, 506–516.

24. Park, S.-H.; Oh, S.R.; Jung, K.Y.; Lee, I.S.; Ahn, K.S.; Kim, J.H.; Kim, Y.S.; Lee, J.J.; Lee, H.-K. Acylated flavonol glycosides with anti-complement activity from *Persicaria lapathifolia*. *Chem. Pharm. Bull.* **1999**, *47*, 1484–1486. [CrossRef]

25. Smolarz, H.D. Flavonoids from *Polygonum lapathifolium* ssp. *tomentosum*. *Pharm. Biol.* **2002**, *40*, 390–394. [CrossRef]

26. Peerzada, A.M.; Bajwa, A.A.; Ali, H.H.; Chauhan, B.S. Biology, impact, and management of *Echinochloa colona* (L.) Link. *Crop Prot.* **2016**, *83*, 56–66. [CrossRef]

27. Holm, L.G.; Plucknett, D.L.; Pancho, P.V.; Herberger, J.P. *The World's Worst Weeds: Distribution and Biology*; University Press of Hawaii: Honolulu, HI, USA, 1991.

28. Řebíčková, K.; Bajer, T.; Šilha, D.; Houdková, M.; Ventura, K.; Bajerová, P. Chemical composition and determination of the antibacterial activity of essential oils in liquid and vapor phases extracted from two different southeast Asian herbs—*Houttuynia cordata* (saururaceae) and *Persicaria odorata* (polygonaceae). *Molecules* **2020**, *25*, 2432. [CrossRef]

29. Almarie, A.; Mamat, A.; Wahab, Z.; Rukunudin, I. Chemical composition and phytotoxicity of essential oils isolated from Malaysian plants. *Allelopathy J.* **2016**, *37*, 55–70.

30. Hunter, M.V.; Brophy, J.J.; Ralph, B.J.; Bienvenu, F.E. Composition of *Polygonum odoratum* Lour. from southern Australia. *J. Essent. Oil Res.* **1997**, *9*, 603–604. [CrossRef]

31. Dũng, N.X.; Van Hac, L.; Leclercq, P.A. Volatile constituents of the aerial parts of Vietnamese *Polygonum odoratum* L. *J. Essent. Oil Res.* **1995**, *7*, 339–340. [CrossRef]

32. Pinheiro, P.F.; Costa, A.V.; Tomaz, M.A.; Rodrigues, W.N.; Fialho Silva, W.P.; Moreira Valente, V.M. Characterization of the Essential Oil of Mastic Tree from Different Biomes and its Phytotoxic Potential on Cobbler's Pegs. *J. Essent. Oil-Bear. Plants* **2016**, *19*, 972–979. [CrossRef]

33. Bali, A.S.; Batish, D.R.; Singh, H.P.; Kaur, S.; Kohli, R.K. Chemical characterization and phytotoxicity of foliar volatiles and essential oil of *Callistemon viminalis*. *J. Essent. Oil-Bear. Plants* **2017**, *20*, 535–545. [CrossRef]

34. Kaur, S.; Singh, H.P.; Mittal, S.; Batish, D.R.; Kohli, R.K. Phytotoxic effects of volatile oil from *Artemisia scoparia* against weeds and its possible use as a bioherbicide. *Ind. Crops Prod.* **2010**, *32*, 54–61. [CrossRef]

35. Silva, E.R.; Overbeck, G.E.; Soares, G.L.G. Phytotoxicity of volatiles from fresh and dry leaves of two Asteraceae shrubs: Evaluation of seasonal effects. *S. Afr. J. Bot.* **2014**, *93*, 14–18. [CrossRef]

36. De Almeida, L.F.R.; Frei, F.; Mancini, E.; De Martino, L.; De Feo, V. Phytotoxic activities of Mediterranean essential oils. *Molecules* **2010**, *15*, 4309–4323. [CrossRef] [PubMed]

37. Zhang, J.; An, M.; Wu, H.; Li Liu, D.; Stanton, R. Chemical composition of essential oils of four *Eucalyptus* species and their phytotoxicity on silverleaf nightshade (*Solanum elaeagnifolium* Cav.) in Australia. *Plant Growth Regul.* **2012**, *68*, 231–237. [CrossRef]

38. Benyelles, B.; Allali, H.; Dib, M.E.A.; Djabou, N.; Paolini, J.; Costa, J. Chemical composition variability of essential oils of *Daucus gracilis* Steinh. from Algeria. *Chem. Biodivers.* **2017**, *14*, e1600490. [CrossRef] [PubMed]

39. Fernández-Cervantes, M.; Pérez-Alonso, M.J.; Blanco-Salas, J.; Soria, A.C.; Ruiz-Téllez, T. Analysis of the essential oils of *Chamaemelum fuscatum* (Brot.) Vasc. from Spain as a contribution to reinforce its ethnobotanical use. *Forests* **2019**, *10*, 539. [CrossRef]

40. Elshamy, A.; Abd-ElGawad, A.M.; El-Amier, Y.A.; El Gendy, A.; Al-Rowaily, S. Interspecific variation, antioxidant and allelopathic activity of the essential oil from three *Launaea* species growing naturally in heterogeneous habitats in Egypt. *Flavour Fragr. J.* **2019**, *34*, 316–328. [CrossRef]

41. Abd El-Gawad, A.M. Chemical constituents, antioxidant and potential allelopathic effect of the essential oil from the aerial parts of *Cullen plicata*. *Ind. Crops Prod.* **2016**, *80*, 36–41. [CrossRef]

42. Assaeed, A.; Elshamy, A.; El Gendy, A.E.-N.; Dar, B.; Al-Rowaily, S.; Abd-ElGawad, A. Sesquiterpenes-rich essential oil from above ground parts of *Pulicaria somalensis* exhibited antioxidant activity and allelopathic effect on weeds. *Agronomy* **2020**, *10*, 399. [CrossRef]

43. El-Shora, H.M.; Abd El-Gawad, A.M. Physiological and biochemical responses of *Cucurbita pepo* L. mediated by *Portulaca oleracea* L. allelopathy. *Fresenius Environ. Bull.* **2015**, *24*, 386–393.

44. El-Shora, H.M.; El-Gawad, A.M.A. Response of *Cicer arietinum* to allelopathic effect of *Portulaca oleracea* root extract. *Phyton-Ann. Rei Bot.* **2015**, *55*, 215–232.

45. Reigosa, M.J.; Sánchez-Moreiras, A.; González, L. Ecophysiological approach in allelopathy. *Crit. Rev. Plant Sci.* **1999**, *18*, 577–608. [CrossRef]

46. Kim, J. Phytotoxic and antimicrobial activities and chemical analysis of leaf essential oil from *Agastache rugosa*. *J. Plant Biol.* **2008**, *51*, 276–283. [CrossRef]

47. Amri, I.; Gargouri, S.; Hamrouni, L.; Hanana, M.; Fezzani, T.; Jamoussi, B. Chemical composition, phytotoxic and antifungal activities of *Pinus pinea* essential oil. *J. Pest Sci.* **2012**, *85*, 199–207. [CrossRef]
48. Chopra, N.; Tewari, G.; Tewari, L.M.; Upreti, B.; Pandey, N. Allelopathic effect of *Echinochloa colona* L. and *Cyperus iria* L. Weed extracts on the seed germination and seedling growth of rice and soyabean. *Adv. Agric.* **2017**, *2017*, 5748524. [CrossRef]
49. Hegab, M.; Abdelgawad, H.; Abdelhamed, M.; Hammouda, O.; Pandey, R.; Kumar, V.; Zinta, G. Effects of tricin isolated from jungle rice (*Echinochloa colona* L.) on amylase activity and oxidative stress in wild oat (*Avena fatua* L.). *Allelopathy J.* **2013**, *31*, 345–354.
50. Khaliq, A.; Matloob, A.; Cheema, Z.A.; Farooq, M. Allelopathic activity of crop residue incorporation alone or mixed against rice and its associated grass weed jungle rice (*Echinochloa colona* [L.] Link. *Chil. J. Agric. Res.* **2011**, *71*, 418–423. [CrossRef]
51. Lim, C.J.; Basri, M.; Ee, G.C.L.; Omar, D. Phytoinhibitory activities and extraction optimization of potent invasive plants as eco-friendly weed suppressant against *Echinochloa colona* (L.) Link. *Ind. Crops Prod.* **2017**, *100*, 19–34. [CrossRef]
52. Cook, T.; Storrie, A.; Moylan, P.; Adams, B. Field testing of glyphosate-resistant awnless barnyard grass (*Echinochloa colona*) in northern NSW. In Proceedings of the Proceedings of the 16th Australian Weeds Conference, Brisbane, QL, Australia, 2008; pp. 20–23.
53. Miguel, M.G. Antioxidant activity of medicinal and aromatic plants. *Flavour Fragr. J.* **2010**, *25*, 219–312. [CrossRef]
54. Re, R.; Pellegrini, N.; Proteggente, A.; Pannala, A.; Yang, M.; Rice-Evans, C. Antioxidant activity applying an improved ABTS radical cation decolorization assay. *Free Radic. Biol. Med.* **1999**, *26*, 1231–1237. [CrossRef]

Article

Comparative Chemical Profiles and Phytotoxic Activity of Essential Oils of Two Ecospecies of *Pulicaria undulata* (L.) C.A.Mey

Ahmed M. Abd-ELGawad [1,2,*], Saud L. Al-Rowaily [1], Abdulaziz M. Assaeed [1], Yasser A. EI-Amier [2], Abd El-Nasser G. El Gendy [3], Elsayed Omer [3], Dakhil H. Al-Dosari [1], Giuliano Bonanomi [4], Hazem S. Kassem [5] and Abdelsamed I. Elshamy [6]

[1] Plant Production Department, College of Food and Agriculture Sciences, King Saud University, P.O. Box 2460, Riyadh 11451, Saudi Arabia; srowaily@ksu.edu.sa (S.L.A.-R.); assaeed@ksu.edu.sa (A.M.A.); dadosari@gmail.com (D.H.A.-D.)

[2] Department of Botany, Faculty of Science, Mansoura University, Mansoura 35516, Egypt; yasran@mans.edu.eg

[3] Medicinal and Aromatic Plants Research Department, National Research Centre, 33 El Bohouth St., Dokki, Giza 12622, Egypt; aggundy_5@yahoo.com (A.E.-N.G.E.G.); sayedomer2001@yahoo.com (E.O.)

[4] Department of Agriculture, University of Naples Federico II, Portici, 80055 Naples, Italy; giuliano.bonanomi@unina.it

[5] Department of Agricultural Extension and Rural Society, College of Food and Agriculture Sciences, King Saud University, Riyadh 11451, Saudi Arabia; hskassem@ksu.edu.sa

[6] Department of Natural Compounds Chemistry, National Research Centre, 33 El Bohouth St., Dokki, Giza 12622, Egypt; elshamynrc@yahoo.com

[*] Correspondence: aibrahim2@ksu.edu.sa; Tel.: +966-562-680-864

Citation: Abd-ELGawad, A.M.; Al-Rowaily, S.L.; Assaeed, A.M.; EI-Amier, Y.A.; El Gendy, A.E.-N.G.; Omer, E.; Al-Dosari, D.H.; Bonanomi, G.; Kassem, H.S.; Elshamy, A.I. Comparative Chemical Profiles and Phytotoxic Activity of Essential Oils of Two Ecospecies of *Pulicaria undulata* (L.) C.A.Mey. *Plants* **2021**, *10*, 2366. https://doi.org/10.3390/plants10112366

Academic Editors: Hazem Salaheldin Elshafie, Laura De Martino and Adriano Sofo

Received: 11 October 2021
Accepted: 28 October 2021
Published: 3 November 2021

Publisher's Note: MDPI stays neutral with regard to jurisdictional claims in published maps and institutional affiliations.

Abstract: The Asteraceae (Compositae) family is one of the largest angiosperm families that has a large number of aromatic species. *Pulicaria undulata* is a well-known medicinal plant that is used in the treatment of various diseases due to its essential oil (EO). The EO of both Saudi and Egyptian ecospecies were extracted via hydrodistillation, and the chemical compounds were identified by GC–MS analysis. The composition of the EOs of Saudi and Egyptian ecospecies, as well as other reported ecospecies, were chemometrically analyzed. Additionally, the phytotoxic activity of the extracted EOs was tested against the weeds *Dactyloctenium aegyptium* and *Bidens pilosa*. In total, 80 compounds were identified from both ecospecies, of which 61 were Saudi ecospecies, with a preponderance of β-pinene, isoshyobunone, 6-epi-shyobunol, α-pinene, and α-terpinolene. However, the Egyptian ecospecies attained a lower number (34 compounds), with spathulenol, hexahydrofarnesyl acetone, α-bisabolol, and τ–cadinol as the main compounds. The chemometric analysis revealed that the studied ecospecies and other reported species were different in their composition. This variation could be attributed to the difference in the environmental and climatic conditions. The EO of the Egyptian ecospecies showed more phytotoxic activity against *D. aegyptium* and *B. pilosa* than the Saudi ecospecies. This variation might be ascribed to the difference in their major constituents. Therefore, further study is recommended for the characterization of authentic materials of these compounds as allelochemicals against various weeds, either singular or in combination.

Keywords: allelopathy; *Pulicaria crispa*; chemometric analysis; chemotype; Asteraceae

1. Introduction

Taxa belonging to *Pulicaria* (Asteraceae Family) are widely distributed in Asia, Africa, and Europe. These plants are considered very important medicinal plants due to their traditional applications around the world, in addition to the presence of interesting metabolites comprising mono-, sesqui-, and diterpenoids, as well as phenolic and flavonoids [1–4].

The Egyptian widespread desert plant, *Pulicaria undulata* (L.) (syn. *Pulicaria crispa* (Forssk.) Benth et Hook), was documented as a very important traditional plant for the treatment of diabetes, abscesses, cardiac and skin diseases, and chills [5]. In Egypt, this plant was used as a herbal tea for inflammation treatment, in addition to insect repellent [4].

Numerous pharmaceutical activities were described for different extracts and ingredients of this plant such as antioxidant [6–8], neuroprotective [9], antiulcer [10], anti-acetylcholinesterase [8], anticancer [11], and α-glucosidase inhibitory activity [12]. These biological activities of *P. undulata* were ascribed to different classes of identified chemical compounds such as terpenes [4,12–15], flavonoids [7,16], and sterols [11]. The essential oil (EO) of *P. undulata* exhibited various potent biological activities such as antiproliferative, antioxidant [6,15], anticancer [8], antibacterial [13], and cytotoxic [12].

Many documents have been published concerning the chemical characterization EOs of different ecospecies of *P. undulata* from different countries such as Sudan [6,17], Iran [15,18,19], Algeria [14], Yemen [13], and Egypt [8,16,20]. However, by comparing all these ecospecies, there was evidence that their EOs were different either in quality or quantity. This deduced that the biosynthesis of the natural metabolites including EOs in the plant kingdom is correlated with environmental and climatic conditions, in addition to genetic variability [21–23]. The present study aimed to analyze and compare the EO profiles of two ecospecies of *P. undulata* growing in Saudi Arabia and Egypt, assess phytotoxicity against the noxious weeds *Dactyloctenium aegyptium* and *Bidens pilosa,* as well as holistically categorize their EOs with other reported ecospecies using chemometric tools.

2. Results and Discussion

2.1. Yields and Chemical Constituents of P. undulata EOs

The aerial parts of Saudi and Egyptian *P. undulata* (150 g each) were subjected separately to the hydrodistillation for 3 h in Clevenger-type apparatus, provided pale yellow oil with an average yield of 0.43% and 0.36% (*v*/*w*), respectively. The yields of EOs in our study were comparable to those reported from other Egyptian ecospecies (0.23–0.60%) [9,16,20]. However, the yield of the present *P. undulata* ecospecies was lower than that reported in previous studies for other ecospecies such as Yemeni (2.10%) [13], Iranian (0.50–1.34%) [15,19,24], Sudanian (1.40–2.50%) [6,17], and Algerian ecospecies (1.20%) [14]. These variations in the yield of the EOs might be attributed to the difference in the geographical region, in addition to the environmental conditions such as soil, climate, as well as genetic pool [22,25–27].

In total, 80 compounds were characterized depending upon GC–MS analysis of the two EOs of *P. undulata* including 61 and 34 compounds from Saudi and Egyptian ecospecies, respectively. The identified constituents were classified into eight classes—namely, (i) monoterpene hydrocarbons, (ii) oxygenated monoterpenes, (iii) sesquiterpene hydrocarbons, (iv) oxygenated sesquiterpenes, (v) carotenoid-derived compounds, (vi) apocarotenoid-derived compounds, (vii) nonoxygenated hydrocarbons, and (viii) oxygenated hydrocarbons, (Figure 1a). Oxygenated sesquiterpenes were the most represented class—they represented 55.03% and 40.34% of the total oil of the Egyptian and Saudi ecospecies, respectively. Monoterpenes were determined with a high content (39.46%) of the EO of Saudi ecospecies, while it represented a minor class in Egyptian ecospecies (6.70%). Additionally, hydrocarbons represented 13.32% of the total EO content of the Egyptian eco-sample, while completely absent in the Saudi plant sample. Overall, the Egyptian ecospecies had oxygenated compounds as the main elements, while non-oxygenated compounds were represented as the main constituents of Saudi ecospecies (Figure 1b). The identified compounds, accounting for 97.22% and 97.61%, respectively, of overall EO mass, in addition to their retention times, and literature and experimental Kovats indices are presented in Table 1.

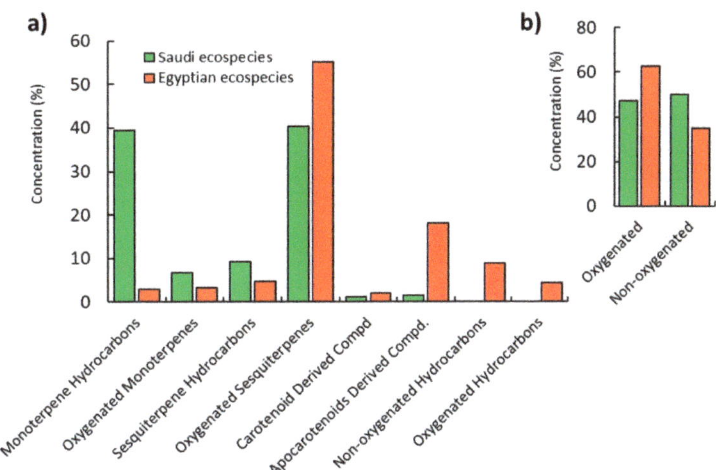

Figure 1. Classification of the chemical compounds of *Pulicaria undulata* EOs of Saudi and Egyptian ecospecies. (**a**) various classes and (**b**) oxygenated and non-oxygenated compounds.

Table 1. Chemical constituents of the EOs of the aerial parts of Saudi and Egyptian ecospecies of *Pulicaria undulata*.

No.	Rt [a]	Relative Conc. (%)		KI		Compound Name	Identification
		SA [b]	EG [c]	Lit. [d]	Exp. [e]		
				Monoterpene Hydrocarbons			
1	4.05	0.14 ± 0.01	—	931	931	α-Thujene	KI and MS
2	4.20	5.08 ± 0.06	0.91 ± 0.01	939	940	α-Pinene	KI and MS
3	4.58	0.15 ± 0.02	—	953	951	Camphene	KI and MS
4	5.07	0.76 ± 0.04	—	976	975	Sabinene	KI and MS
5	5.21	21.14 ± 0.12	1.32 ± 0.05	980	980	β-Pinene	KI and MS
6	6.19	0.65 ± 0.01	—	1031	1030	Limonene	KI and MS
7	6.50	7.70 ± 0.08	—	1064	1063	γ-Terpinene	KI and MS
8	8.09	3.84 ± 0.05	0.81 ± 0.03	1088	1086	α-Terpinolene	KI and MS
				Oxygenated Monoterpenes			
9	5.50	0.46 ± 0.03	—	991	990	Dehydro-1,8-cineole	KI and MS
10	5.92	0.22 ± 0.01	—	1005	1005	α-Phellandrene	KI and MS
11	6.01	0.13 ± 0.01	—	1129	1129	p-2-Menthen-1-ol	KI and MS
12	9.45	0.27 ± 0.02	—	1131	1132	*trans-p*-Mentha-2,8-dienol	KI and MS
13	9.96	0.15 ± 0.01	—	1137	1138	β-Nopinone	KI and MS
14	10.05	0.27 ± 0.02	—	1139	1139	Pinocarveol	KI and MS
15	10.23	0.21 ± 0.01	—	1140	1140	*cis*-Verbenol	KI and MS
16	10.72	0.27 ± 0.03	—	1143	1145	Camphor	KI and MS
17	10.99	0.28 ± 0.02	—	1162	1161	Pinocarvone	KI and MS
18	11.28	0.28 ± 0.01	—	1165	1167	*endo*-Borneol	KI and MS
19	11.82	1.20 ± 0.04	—	1177	1179	Terpinen-4-ol	KI and MS
20	13.04	0.20 ± 0.02	—	1194	1193	Myrtenal	KI and MS
21	13.51	0.11 ± 0.01	—	1228	1229	α-Citronellol	KI and MS
22	16.79	0.14 ± 0.01	—	1321	1319	Isopulegol acetate	KI and MS
23	17.33	0.48 ± 0.01	—	1354	1356	Citronellyl acetate	KI and MS
24	17.84	0.26 ± 0.01	3.36 ± 0.07	1258	1259	Carvotanacetone	KI and MS
25	20.21	2.50 ± 0.05	—	1326	1326	Myrtenyl acetate	KI and MS

Table 1. *Cont.*

No.	Rt [a]	Relative Conc. (%)		KI		Compound Name	Identification
		SA [b]	EG [c]	Lit. [d]	Exp. [e]		
				Sesquiterpene Hydrocarbons			
26	16.34	0.58 ± 0.01	—	1377	1375	Berkheyaradulen	KI and MS
27	17.48	3.63 ± 0.04	1.17 ± 0.06	1409	1410	α-Gurjunene	KI and MS
28	17.70	0.62 ± 0.03	—	1418	1418	*trans*-Caryophyllene	KI and MS
29	18.47	0.78 ± 0.02	—	1439	1437	α-Guaiene	KI and MS
30	18.69	0.88 ± 0.04	0.81 ± 0.01	1455	1456	α-Humulene	KI and MS
31	19.63	0.13 ± 0.01	2.76 ± 0.05	1473	1472	γ-Gurjunene	KI and MS
32	19.85	0.18 ± 0.01	—	1480	1480	Germacrene-D	KI and MS
33	20.77	0.37 ± 0.01	—	1483	1484	α-Curcumene	KI and MS
34	21.21	0.51 ± 0.01	—	1499	1500	α-Muurolene	KI and MS
35	21.78	1.49 ± 0.05	—	1524	1525	δ-Cadinene	KI and MS
				Oxygenated Sesquiterpenes			
36	20.95	1.54 ± 0.03	—	1515	1514	Shyobunone	KI and MS
37	21.67	6.51 ± 0.07	2.31 ± 0.02	1517	1517	6-epi-Shyobunol	KI and MS
38	22.37	0.12 ± 0.01	—	1518	1518	6-epi-Shyobunone	KI and MS
39	23.19	0.41 ± 0.01	—	1563	1562	Citronellyl iso-valerate	KI and MS
40	23.47	3.33 ± 0.08	0.87 ± 0.01	1564	1564	*trans*-Nerolidol	KI and MS
41	23.66	7.67 ± 0.05	1.63 ± 0.04	1571	1572	Isoshyobunone	KI and MS
42	23.78	3.43 ± 0.04	30.86 ± 0.12	1575	1575	Spathulenol	KI and MS
43	24.13	0.17 ± 0.01	—	1581	1582	Caryophyllene oxide	KI and MS
44	24.52	4.82 ± 0.09	0.95 ± 0.03	1584	1586	7-Hydroxyfarnesen	KI and MS
45	24.62	0.51 ± 0.01	1.25 ± 0.02	1595	1595	Salvial-4(14)-en-1-one	KI and MS
46	24.78	0.84 ± 0.02	—	1596	1598	Veridiflorol	KI and MS
47	24.85	2.41 ± 0.05	—	1608	1610	Humuladienone	KI and MS
48	24.97	0.55 ± 0.01	—	1613	1613	Longifolenaldehyde	KI and MS
49	25.12	0.16 ± 0.01	—	1625	1627	Isospathulenol	KI and MS
50	25.29	0.88 ± 0.04	—	1621	1620	Fonenol	KI and MS
51	25.43	0.74 ± 0.03	—	1641	1640	Cubenol	KI and MS
52	25.58	0.76 ± 0.02	3.65 ± 0.07	1642	1642	τ-Cadinol	KI and MS
53	25.65	0.38 ± 0.02	—	1643	1644	τ-Muurolol	KI and MS
54	25.98	1.39 ± 0.06	—	1649	1650	β-Eudesmol	KI and MS
55	26.88	0.29 ± 0.01	—	1653	1654	α-Cadinol	KI and MS
56	27.04	1.74 ± 0.08	—	1668	1668	Cedr-8-en-13-ol	KI and MS
57	27.5	0.40 ± 0.02	—	1671	1670	Calarene epoxide	KI and MS
58	28.61	0.18 ± 0.01	—	1682	1680	Ledene oxide-(I)	KI and MS
59	28.89	1.11 ± 0.03	6.34 ± 0.05	1683	1683	α-Bisabolol	KI and MS
60	30.66	—	1.53 ± 0.03	1690	1693	6-Isopropenyl-4,8a-dimethyl-1,2,3,5,6,7,8,8a-octahydro-naphthalen-2-ol	KI and MS
61	31.41	—	4.68 ± 0.07	2257	2259	4,4-Dimethyl-tetracyclo[6.3.2.0(2,5).0(1,8)]tridecan-9-ol	KI and MS
62	33.35	—	0.96 ± 0.01	2462	2463	Isocalamendiol	KI and MS
				Carotenoid Derived Compounds			
63	16.10	—	0.97 ± 0.04	1279	1280	Vitispirane	KI and MS
64	16.45	—	1.06 ± 0.04	1288	1287	Dihydroedulan II	KI and MS
65	23.29	0.64 ± 0.03	—	1444	1445	Citronellyl propionate	KI and MS
				Apocarotenoid Derived Compounds			
66	38.29	1.51	18.12 ± 0.11	1845	1845	Hexahydrofarnesyl acetone	KI and MS

Table 1. *Cont.*

No.	Rt [a]	Relative Conc. (%)		KI		Compound Name	Identification
		SA [b]	EG [c]	Lit. [d]	Exp. [e]		
				Non-oxygenated Hydrocarbons			
67	32.32	——	1.08 ± 0.06	1533	1535	2,6,10-Trimethyl-tetradecane	KI and MS
68	33.97	——	0.97 ± 0.01	1885	1883	2,6,10,15-Tetramethyl-heptadecane	KI and MS
69	39.53	——	1.57 ± 0.05	1900	1900	*n*-Nonadecane	KI and MS
70	44.39	——	0.66 ± 0.04	2200	2200	*n*-Docosane	KI and MS
1	46.11	——	1.05 ± 0.03	2300	2300	*n*-Tricosane	KI and MS
72	46.35	——	1.28 ± 0.07	2500	2500	*n*-Pentacosane	KI and MS
73	52.20	——	1.49 ± 0.05	2900	2900	*n*-Nonacosane	KI and MS
74	57.64	——	0.42 ± 0.04	3000	3000	*n*-Triacontane	KI and MS
75	57.71	——	0.46 ± 0.03	3200	3200	*n*-Dotriacontane	KI and MS
				Oxygenated Hydrocarbons			
76	37.85	——	3.73 ± 0.07	1942	1945	*cis*-9-Hexadecenoic acid	KI and MS
77	47.39	——	0.29 ± 0.01	2135	2132	9,12-Octadecadienoic acid	KI and MS
78	47.42	——	0.32 ± 0.01	2243	2246	9-hexyl-Heptadecane	KI and MS
	Total	98.55	99.64				

[a] Rt: retention time; [b] values are mean (n = 2) ± SD of Saudi ecospecies; [c] Egyptian ecospecies; [d] literature Kovats retention index; [e] experimental Kovats retention index; MS: mass spectral data of compounds; KI: Kovats indices with those of Wiley Spectral Library collection and National 104 Institute of Standards and Technology (NIST) Library database.

The analysis of the data revealed that the EOs of the two plant samples were very rich with terpenoids, with respective concentrations of 95.78% and 66.17% in addition to carotenoids (2.77% and 20.15%, respectively). The variations in the quantitative and qualitative analysis of EOs of the two plant samples were attributed directly to the environmental and climate variations between the Saudi and Egyptian environments [28,29].

More in-depth data indicated that the EO of the Saudi *P. undulata* contained mainly terpenoids, including almost equal concentrations of mono (46.27%) and sesquiterpenes (49.51%) with traces of carotenoids and a complete absence of diterpenoids and hydrocarbons. In comparison, the chemical characterization of the EO of the Egyptian plant showed that terpenoids were the major compounds, including minor elements of monoterpenes (6.40%) and abundance of sesquiterpenes (59.77%), as well as a high concentration of carotenoids. Similarly, the EO of the Egyptian plant was characterized by the complete absence of diterpenes and the presence of a remarkable concentration of hydrocarbons. The sesquiterpenes were found as major constituents of the EOs of both ecospecies (Saudi and Egyptian); this result was different than those reported for Yemeni leaves (2.1%) [13], Iranian aerial parts (0.5%) [15], and Egyptian aerial parts (0.6%) [20] of *P. undulata*. The sesquiterpenes in the EOs of Saudi and Egyptian ecospecies were categorized as sesquiterpene hydrocarbons (9.17% and 4.74%), and oxygenated sesquiterpenes (40.34% and 55.03%). Isoshyobunone (7.67%), 6-epi-Shyobunol (6.51%), spathulenol (3.43%), and *trans*-nerolidol (3.33%) represented the main oxygenated sesquiterpene of EO of the Saudi plant. In comparison, spathulenol (30.86%), α-bisabolol (6.34%), 4,4-dimethyl-tetracyclo[6.3.2.0(2,5).0(1,8)]tridecan-9-ol (4.68%), and τ–cadinol (3.65%) were found to be the abundant oxygenated sesquiterpenes of EO of Egyptian ecospecies. Most of the studied ecospecies of *P. undulata* have been described as non-rich of sesquiterpene [13,15,20]. However, EOs of other *Pulicaria* species such as *P. somalensis* [1], *P. dysenterica* [30], and *P. gnaphalodes* [31] were reported as rich in sesquiterpene.

Numerous *Pulicaria* plants were described to have spathulenol as minor and/or major compounds of their EOs such as *P. somalensis* [1] and *P. stephanocarpa* [32]. α-Bisabolol was detected as the main sesquiterpene in EOs derived from some *Pulicaria* species such as *P. somalensis* [1], *P. dysenterica* [30], and *P. gnaphalodes* [31]. Moreover, the major sesquiterpene, cadinol, in this study has been described as a major component in EO derived from aerial parts of the *P. undulata* collected from the Algerian Sahara [14], while it was reported as minor or trace element in other ecospecies.

Monoterpenes were reported as the main constituents of several *Pulicaria* ecospecies [13,33]. Saudi *P. undulata* was found to be in harmony with the reported documents where the monoterpenes represented around half of the total oil (46.27%) including hydrocarbons (39.46%) and oxygenated (6.81%) forms of monoterpene. However, the monoterpenes were identified as trace elements (6.40%) in the EO of Egyptian plants, including traces of non-oxygenated and oxygenated forms, with respective relative concentrations of 3.04% and 3.36%. In the EO from the Saudi sample, β-pinene (21.14%), α-pinene (5.08%), and α-terpinolene (3.84%) were assigned as the main monoterpene hydrocarbons, while myrtenyl acetate (2.50%) and terpinen-4-ol (1.20%) were characterized as main oxygenated monoterpenes. Only four monoterpenes were identified from overall compounds of EO of Egyptian ecospecies. β-Pinene (1.32%) was assigned as the main monoterpene hydrocarbons, and carvotanacetone (3.36%) was the only identified oxygenated one. Carvotanacetone was stated as the main monoterpene of *P. undulata* collected from Yemen [13] and from the Egyptian Western Desert region [8,20]. The present data revealed that the variations in the components in EO of Egyptian and Saudi samples might be attributed to the variations in collection areas, in addition to the environmental conditions such as soil, climate, as well as their genetic pool [23]. The abundance of pinene and myrtenyl derivatives, α-terpinolene, terpinen-4-ol was in complete harmony with the data reported from the Iranian *P. undulata* [15,18].

Carotenoid-derived compounds were represented as trace constituents in the EO of the Saudi ecospecies, with a concentration of 2.77%, comprising carotenoids (1.26%) and apocarotenoid-derived compounds (1.51%). Hexahydrofarnesyl acetone was found as the main component in all characterized carotenoid-derived compounds. By contrast, carotenoid-derived compounds derived from the Egyptian EO sample were characterized by high concentration (20.12%), representing carotenoid-derived compounds (2.03%) and apocarotenoid-derived compounds (18.12%). Additionally, hexahydrofarnesyl acetone represented the predominated compound in all overall carotenoid-derived compounds. Hexahydrofarnesyl acetone is a common apocarotenoid-derived compound in EOs derived from the plant kingdom such as *Hildegardia barteri* [34], *Stachys tmolea* [35], and *Bassia muricata* [36].

The hydrocarbons represented 13.32% of the total identified oil of the Egyptian plant involved non-oxygenated (8.98%) and oxygenated (4.34%) forms. *n*-nonadecane (1.57%) and *n*-nonacosane (1.49%) were identified as the majors of non-oxygenated hydrocarbons, while *cis*-9-hexadecenoic acid (3.73%) represented the main oxygenated hydrocarbons. Hydrocarbons were completely absent from the EO of the Saudi plant, and this result was found in agreement with Iranian *P. undulata* [15,18].

2.2. Chemometric Analysis of the EOs of Pulicaria Ecospecies

The application of the EOs profiles of the 2 ecotypes of *P. undulata* and the other 11 ecotypes were subjected to principal component multivariate data analysis (PCA) and agglomerative hierarchical clustering (AHC). The cluster analysis revealed that the EOs could be categorized into four clusters. Cluster-I consisted of the Iran–Baluchestan ecotype, while the EOs of the presently studied ecospecies (Saudi and Egyptian) were grouped as cluster-II. Further, the Egypt–Elba Mountain-2 and Egypt–Sinai ecospecies showed a close correlation, and therefore, they were grouped as cluster-III. Finally, cluster-IV contained Iranian (Iran–Baluchestan, Iran–Fars, and Iran–Hormozgan samples), Algerian, Sudanian, Yemeni, Egyptian (Elba Mountain-2, and Sadat) ecospecies (Figure 2a).

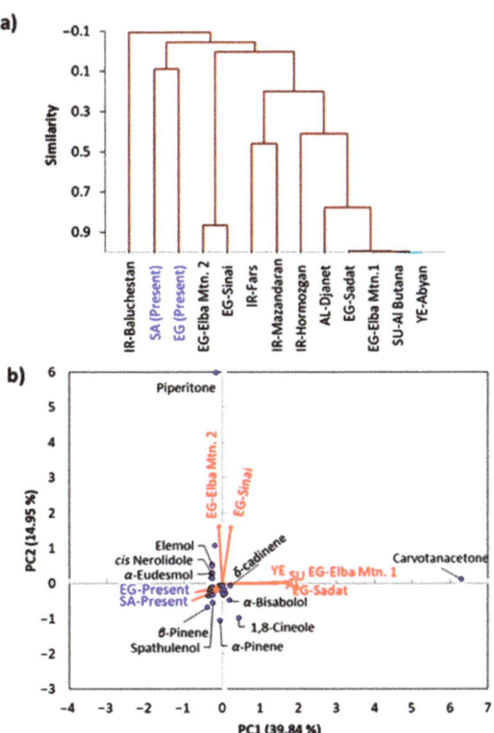

Figure 2. Chemometric analysis of the essential oil of different *Pulicaria undulata* ecospecies: (**a**) agglomerative hierarchical clustering (AHC) and (**b**) principal components analysis (PCA). SA: Suadi, EG: Egyptian, IR: Iranian, AL: Algerian, SU: Sudanian, and YE: Yemeni. The blue color represents the present samples.

The PCA score plot showed the distant separation of Egypt–Elba Mountain-2 and Egypt–Sinai ecospecies in the PC2, while Egypt–Elba Mountain-2, Egypt Sadat, Yemeni, Algerian, and Sudanian ecospecies were distantly distributed along the right side of the PC1 (Figure 2b). Conversely, the present samples (Saudi and Egyptian) as well as Iranian and Algerian were clustered together in the center of the PCA and had positive score values. In addition, the examination of the loading plot showed that piperitone was the most correlated/abundant compound in Egypt–Elba Mountain-2 and Egypt–Sinai ecospecies. However, carvotanacetone showed an abundance in Egypt–Elba Mountain-2, Egypt Sadat, Yemeni, Algerian, and Sudanian ecospecies. The detected variation among different ecospecies could be ascribed to the effect of climatic and environmental conditions, as well as the genetic characteristics [22,25,26,37].

2.3. Phytotoxic Activity of P. undulata EOs

The EOs of both Saudi and Egyptian ecotypes of *P. undulata* showed significant phytotoxic activity against seed germination and seedling growth of the noxious weed *B. pilosa* (Figure 3). At the highest concentration (100 µL L^{-1}), EOs of Saudi ecospecies showed inhibition of germination, shoot growth, and root growth of *B. pilosa* by 66.67%, 74.59%, and 83.47%, respectively, while the Egyptian species showed inhibition values of 86.67%, 79.23%, and 94.17%, respectively (Figure 3). Based on the IC$_{50}$, the Saudi ecospecies showed IC$_{50}$ values of 72.83, 72.84, and 44.55 µL L^{-1} regrading germination, shoot growth, and root growth of *B. pilosa*, respectively. However, the Egyptian ecospecies showed IC$_{50}$ values of 42.42, 65.71, and 40.70 µL L^{-1}, respectively (Figure 3).

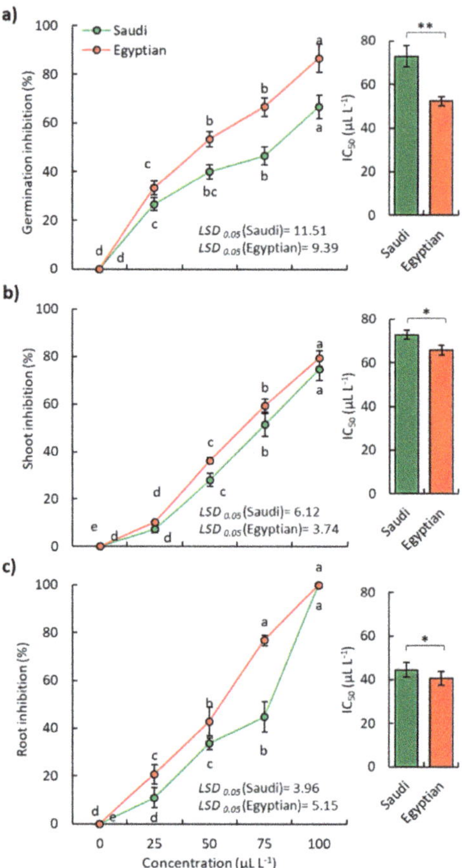

Figure 3. Phytotoxic effect of the EOs extracted from the aerial parts of both Saudi and Egyptian ecotypes of *P. undulata* on the (**a**) germination of seeds, (**b**) shoot growth, and (**c**) root growth of the weed *Bidens pilosa*. Different letters on each line mean significant differences (one-way randomized blocks ANOVA). Data are mean value ($n = 3$), and the bars represent the standard error. * $p < 0.05$, ** $p < 0.01$.

It was evident that the Egyptian ecospecies were more effective against *B. pilosa* than Saudi ecospecies, which could be ascribed to the variation in the quality and quantity of the chemical composition of the EO. In this study, the Egyptian ecospecies were richer in oxygenated compounds than the Saudi ones. EOs rich in oxygenated compounds have been reported to possess more activity [38–41]. The phytotoxic activity of the EO from Egyptian ecospecies might be attributed to its major compounds such as spathulenol, hexahydrofarnesyl acetone, α-bisabolol, and τ–cadinol. Additionally, the Saudi ecospecies had β-pinene, isoshyobunone, 6-epi-shyobunol, α-pinene, and α-terpinolene as major compounds. Moreover, τ–cadinol was identified as a major compound in the EO of *Cullen plicata*, where it showed strong phytotoxic activity against *B. pilosa* and *Urospermum picroides* [38]. Additionally, τ–cadinol was reported in a high concentration of the EO of *Rhynchosia minima*, which showed significant allelopathic activity against *Dactyloctenium aegyptium* and *Rumex dentatus* [42]. However, α-bisabolol, as a major compound of the Egyptian ecospecies in the present study, has not been reported to possess phytotoxicity; therefore, further study is recommended for its characterization as an allelochemical compound.

In the Egyptian ecospecies, the major compound, spathulenol (30.86%), has also been reported as major compounds of EOs with substantial phytotoxic activity such as *Launaea mucronata* [26], *Xanthium strumarium* [37], *Eucalyptus camaldulensis* [43], *Teucrium arduini* [44], and *Symphyotrichum squamatum* [25]. Moreover, hexahydrofarnesyl acetone (18.12%), was determined in a high concentration of the EO, which exhibited strong phytotoxicity such as *Heliotropium curassavicum* [23], *Launaea nudicaulis*, *Launaea mucronata* [26], and *Bassia muricata* [36].

Otherwise, the main compound in the EO of Saudi ecospecies, β-pinene (21.14%), has been reported as the main compound of EOs of various plants that have exhibited phytotoxic activity such as *Schinus terebinthifolius* [45], *Symphyotrichum squamatum* [25], *Pinus brutia*, *Pinus pinea* [46], *Lavandula angustifolia* [44], and *Heterothalamus psiadioides* [47]. The other major compounds of the Saudi ecospecies have also been reported in EOs with significant phytotoxicity [1,46,48]. Additionally, the present data showed that the roots were more sensitive to the EO than shoots since roots were directly exposed to the EO. Moreover, root cells have more permeability than the cells of the shoot [22,38].

Results also indicated that the EOs of Saudi and Egyptian ecospecies showed more inhibitory activity against the weed *D. aegyptium* than *B. pilosa* (Figure 4).

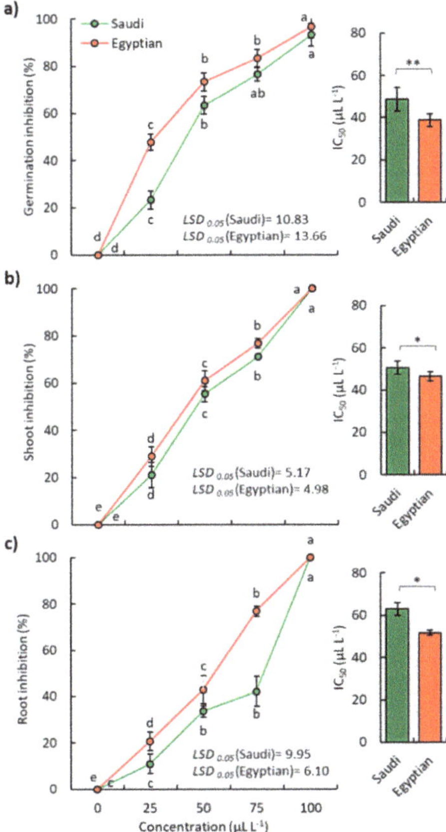

Figure 4. Phytotoxic effect of the EOs extracted from the aerial parts of both Saudi and Egyptian ecotypes of *P. undulata* on the (**a**) germination of seeds, (**b**) shoot growth, and (c) root growth of the weed *Dactyloctenium aegyptium*. Different letters on each line mean significant differences (one-way randomized blocks ANOVA). Data are mean value (*n = 3*) and the bars represent the standard error. * $p < 0.05$, ** $p < 0.01$.

At the highest concentration of the Saudi EOs (100 μL L^{-1}), the *D. aegyptium* seedling growth was completely inhibited. However, the germination was reduced by 93.33%, while the Egyptian ecospecies showed 96.67%. Based on the IC$_{50}$ values, the EO of the Saudi ecospecies showed IC$_{50}$ values of 48.61, 50.49, and 62.92 μL L^{-1} for germination, shoot growth, and root growth of *D. aegyptium*, respectively, while the Egyptian ecospecies attained IC$_{50}$ values of 38.84, 46.59, and 51.87 μL L^{-1}, respectively.

3. Materials and Methods

3.1. Plant Samples Collection and Preparation

The aerial parts of Saudi *P. undulata* were collected from the Wadi Alsahbaa, Alkharj, Riyadh region (24°16′34.1″ N 47°56′11.3″ E), while the Egyptian sample was collected from Wadi Hagoul, the Eastern Desert, Egypt (30°00′38.2″ N 32°05′35.5″ E), during spring of 2019. The specimens were authenticated according to Tackholm [49] and Boulos [50]. Voucher specimens were prepared and deposited in the herbarium of the Department of Botany, Faculty of Science, Mansoura University with No. Mans.001162117 and Mans.001162118.

The samples were collected from two populations of *P. undulata* in separate plastic bags and immediately transferred to the lab. The samples were dried in a shaded place at room temperature (25 $\pm$ 3 °C) for 7 days, crushed into powder using a grinder (IKA® MF 10 Basic Microfine Grinder Drive, Breisgau, Germany) at a dimension of 3.0 mm, and packed in paper bags.

3.2. EOs Extraction, GC–MS Analysis, and Chemical Compounds Identification

About 150 g of the prepared samples of *P. undulata* were extracted with hydrodistillation via a Clevenger-type apparatus for 3 h. The oils were collected, water was removed using 0.5 g of anhydrous sodium sulfate, and stored in glass vials in the fridge (−4 °C) till further analysis [29]. Two samples of the plant were extracted by the same protocol afforded two samples of EOs. The two extracted EOs were analyzed via gas chromatography–mass spectrometry (GC-MS) at the National Research Center, Giza, Egypt, as described in our previously documented work [25,26,48,51]. Briefly, the apparatus has TRACE GC Ultra Gas Chromatographs (THERMO Scientific™ Corporate, Waltham, MA, USA), together with Thermo Scientific ISQ™ EC single quadrupole mass spectrometer. The GC–MS system is equipped with a TR-5 MS column (0.25 μm film thickness, 30 m × 0.32 mm internal diameter). Helium was used as a carrier gas at a flow rate of 1.0 mL min^{-1}, with a divided ratio of 1:10. The temperature program was 60 °C for 1 min, rising by 4.0 °C min^{-1} to 240 °C, and held for 1 min. An aliquot of 1 μL of the EO sample in hexane was injected at a ratio of 1:10 (*v*/*v*), and the detector and injector were adjusted at 210 °C. Mass spectra were recorded by electron ionization (EI) at 70 eV, using a spectral range of *m*/*z* 40–450. The chemical compounds identification was accomplished by Automated Mass spectral Deconvolution and Identification (AMDIS) software, as well as Wiley Spectral Library collection, NIST Library database (Gaithersburg, MD, USA; Wiley, Hoboken, NJ, USA), which were used for retention indices relative to n-alkanes (C$_8$–C$_{22}$), or appraisal of the mass spectrum with authentic standards.

3.3. Phytotoxic Activity Estimation of the EOs

The extracted EOs were tested for their phytotoxicity against two noxious weeds *Dactyloctenium aegyptium* and *Bidens pilosa*. The seeds of *D. aegyptium* were collected from cultivated fields near the Mediterranean coast, at Gamasa City, northern Egypt (31°27′03.9″ N 31°27′44.8″ E), while the seeds of *B. pilosa* were collected from a garden in Mansoura University campus, Mansoura, Egypt (31°02′40.2″ N 31°21′18.4″ E). The homogenous and ripe seeds were selected, sterilized with 0.3% sodium hypochlorite, rinsed with distilled and sterilized water, dried, and stored in sterilized vials.

The phytotoxicity experiments were conducted in vitro following the methodology described by Abd El-Gawad et al. [38]. In brief, 20 seeds of the weed were transferred to a

Petri plate lined with Whatman No. 1 filter paper wetted with 4 mL of each concentration of the EOs (25, 50, 75, and 100 µL L^{-1}). Different concentrations of the EOs were prepared using 1% Tween$^{®}$ 80 (Sigma-Aldrich, Darmstadt, Germany) as an emulsifier. The plates were sealed with Parafilm$^{®}$ tape and incubated in a growth chamber adjusted with a temperature of 25 °C and light/dark cycle of 12/12 h. Besides, Tween$^{®}$ 80 was used as a control treatment. After seven days of incubation, the germinated seeds were counted and the length of shoots and roots of the seedlings were measured. The inhibition of germination and seedling growth were calculated based on the following equation:

$$\text{Inhibition \%} = 100 \times \frac{(\text{Length}/\text{Number}_{\text{Control}} - \text{Length}/\text{Number}_{\text{Treatment}})}{(\text{Length}/\text{Number}_{\text{Control}})}$$

The IC$_{50}$ (the concentration of the EO required to reduce the germination or growth by 50%) was calculated using MS-Excel.

3.4. Data Analysis

The experiment of phytotoxicity was repeated three times with three replications. The data of the inhibition were subjected to one-way ANOVA, followed by Duncan's test using CoStat program (version 6.311, CoHort Software, Monterey, CA, USA), while the IC$_{50}$ values were subjected to a two-tailed t-test using MS-EXCEL. To make a holistic categorization of the EOs of the two studied ecospecies (Saudi and Egyptian) and other reported ecospecies (Algerian, Egyptian, Iranian, Sudanian, and Yemeni), we constructed a data matrix of the 30 major chemical compounds, with concentration > 3%, from 11 ecospecies. The matrix was subjected to Principal component multivariate data analysis (PCA) and agglomerative hierarchical clustering (AHC) using the XLSTAT Statistical Software package (version 2018, Addinsoft Inc., New York, NY, USA).

4. Conclusions

The EO composition of the Saudi and Egyptian ecospecies of *P. undulata* showed substantial variation in both quantity and quality. The Saudi ecospecies had 61 compounds, with β-pinene, isoshyobunone, 6-epi-shyobunol, α-pinene, and α-terpinolene as major compounds, while the EO of the Egyptian ecospecies attained a lower number (34 compounds), with spathulenol, hexahydrofarnesyl acetone, α-bisabolol, and τ–cadinol as main compounds. This variation could be attributed to the difference in the environmental and climatic conditions. The EO of the Egyptian ecospecies showed more phytotoxic activity against *D. aegyptium* than *B. pilosa*, as well as more phytotoxic, compared with the Saudi ecospecies. This variation might be ascribed to the difference in their major constituents. Therefore, further study is recommended for the characterization of authentic materials of these compounds as allelochemicals against various weeds, either singular or in combination.

Author Contributions: Conceptualization, A.M.A.-E., A.M.A., Y.A.E.-A. and A.I.E.; formal analysis, A.M.A.-E., Y.A.E.-A., A.E.-N.G.E.G., D.H.A.-D. and A.I.E.; investigation, A.M.A.-E., S.L.A.-R., A.M.A., A.E.-N.G.E.G., G.B., Y.A.E.-A. and A.I.E.; writing—original draft preparation, A.M.A.-E., Y.A.E.-A., and A.I.E.; writing—review and editing, A.M.A.-E., S.L.A.-R., A.M.A., Y.A.E.-A., A.E.-N.G.E.G., E.O., D.H.A.-D., G.B., H.S.K. and A.I.E. All authors have read and agreed to the published version of the manuscript.

Funding: This study was funded by the Researchers Supporting Project, Number (RSP-2021/403), King Saud University, Riyadh, Saudi Arabia.

Institutional Review Board Statement: Not applicable.

Informed Consent Statement: Not applicable.

Data Availability Statement: Not applicable.

Acknowledgments: The authors extend their appreciation to the Researchers Supporting Project, Number (RSP-2021/403), King Saud University, Riyadh, Saudi Arabia.

Conflicts of Interest: The authors declare no conflict of interest.

Sample Availability: Samples of the compounds are not available from the authors.

References

1.   Assaeed, A.; Elshamy, A.; El Gendy, A.E.-N.; Dar, B.; Al-Rowaily, S.; Abd-ElGawad, A. Sesquiterpenes-rich essential oil from above ground parts of *Pulicaria somalensis* exhibited antioxidant activity and allelopathic effect on weeds. *Agronomy* **2020**, *10*, 399. [CrossRef]
2.   Al-Maqtari, Q.A.; Mahdi, A.A.; Al-Ansi, W.; Mohammed, J.K.; Wei, M.; Yao, W. Evaluation of bioactive compounds and antibacterial activity of *Pulicaria jaubertii* extract obtained by supercritical and conventional methods. *J. Food Meas. Charact.* **2020**, *15*, 449–456. [CrossRef]
3.   Elshamy, A.I.; Mohamed, T.A.; Marzouk, M.M.; Hussien, T.A.; Umeyama, A.; Hegazy, M.E.F.; Efferth, T. Phytochemical constituents and chemosystematic significance of *Pulicaria jaubertii* E. Gamal-Eldin (Asteraceae). *Phytochem. Lett.* **2018**, *24*, 105–109. [CrossRef]
4.   Hegazy, M.-E.F.; Matsuda, H.; Nakamura, S.; Yabe, M.; Matsumoto, T.; Yoshikawa, M. Sesquiterpenes from an Egyptian herbal medicine, *Pulicaria undulate*, with inhibitory effects on nitric oxide production in RAW264. 7 macrophage cells. *Chem. Pharm. Bull.* **2012**, *60*, 363–370. [CrossRef]
5.   Hammiche, V.; Maiza, K. Traditional medicine in Central Sahara: Pharmacopoeia of Tassili N'ajjer. *J. Ethnopharmacol.* **2006**, *105*, 358–367. [CrossRef]
6.   Mohammed, A.B.; Yagi, S.; Tzanova, T.; Schohn, H.; Abdelgadir, H.; Stefanucci, A.; Mollica, A.; Mahomoodally, M.F.; Adlan, T.A.; Zengin, G. Chemical profile, antiproliferative, antioxidant and enzyme inhibition activities of *Ocimum basilicum* L. and *Pulicaria undulata* (L.) CA Mey. grown in Sudan. *S. Afr. J. Botany* **2020**, *132*, 403–409. [CrossRef]
7.   Hussein, S.R.; Marzouk, M.M.; Soltan, M.M.; Ahmed, E.K.; Said, M.M.; Hamed, A.R. Phenolic constituents of *Pulicaria undulata* (L.) CA Mey. sub sp. *undulata* (Asteraceae): Antioxidant protective effects and chemosystematic significances. *J. Food Drug Anal.* **2017**, *25*, 333–339. [CrossRef] [PubMed]
8.   Mustafa, A.M.; Eldahmy, S.I.; Caprioli, G.; Bramucci, M.; Quassinti, L.; Lupidi, G.; Beghelli, D.; Vittori, S.; Maggi, F. Chemical composition and biological activities of the essential oil from *Pulicaria undulata* (L.) CA Mey. Growing wild in Egypt. *Nat. Prod. Res.* **2020**, *34*, 2358–2362. [CrossRef]
9.   Issa, M.Y.; Ezzat, M.I.; Sayed, R.H.; Elbaz, E.M.; Omar, F.A.; Mohsen, E. Neuroprotective effects of *Pulicaria undulata* essential oil in rotenone model of parkinson's disease in rats: Insights into its anti-inflammatory and anti-oxidant effects. *S. Afr. J. Bot.* **2020**, *132*, 289–298. [CrossRef]
10. Fahmi, A.A.; Abdur-Rahman, M.; Naser, A.F.A.; Hamed, M.A.; Abd-Alla, H.I.; Shalaby, N.M.; Nasr, M.I. Chemical composition and protective role of *Pulicaria undulata* (L.) CA Mey. subsp. *undulata* against gastric ulcer induced by ethanol in rats. *Heliyon* **2019**, *5*, e01359.
11. Abdallah, H.M.; Mohamed, G.A.; Ibrahim, S.R.M.; Asfour, H.Z.; Khayat, M.T. Undulaterpene A: A new triterpene fatty acid ester from *Pulicaria undulata*. *Pharmacogn. Mag.* **2019**, *15*, 671–674. [CrossRef]
12. Rasool, N.; Rashid, M.A.; Khan, S.S.; Ali, Z.; Zubair, M.; Ahmad, V.U.; Khan, S.N.; Choudhary, M.I.; Tareen, R.B. Novel α-glucosidase activator from *Pulicaria undulata*. *Nat. Prod. Commun.* **2013**, *8*, 757–759. [CrossRef]
13. Ali, N.A.A.; Sharopov, F.S.; Alhaj, M.; Hill, G.M.; Porzel, A.; Arnold, N.; Setzer, W.N.; Schmidt, J.; Wessjohann, L. Chemical composition and biological activity of essential oil from *Pulicaria undulata* from Yemen. *Nat. Prod. Commun.* **2012**, *7*, 257–260. [CrossRef] [PubMed]
14. Boumaraf, M.; Mekkiou, R.; Benyahia, S.; Chalchat, J.-C.; Chalard, P.; Benayache, F.; Benayache, S. Essential oil composition of *Pulicaria undulata* (L.) DC.(Asteraceae) growing in Algeria. *Int. J. Pharmacogn. Phytochem. Res.* **2016**, *8*, 746–749.
15. Ravandeh, M.; Valizadeh, J.; Noroozifar, M.; Khorasani-Motlagh, M. Screening of chemical composition of essential oil, mineral elements and antioxidant activity in *Pulicaria undulata* (L.) CA Mey from Iran. *J. Med. Plants Res.* **2011**, *5*, 2035–2040.
16. Ahmed, S.; Ibrahim, M. Chemical investigation and antimicrobial activity of *Francoeuria crispa* (Forssk) Grown Wild in Egypt. *J. Mater. Environ. Sci.* **2017**, *9*, 1–6. [CrossRef]
17. EL-Kamali, H.H.; Yousif, M.O.; Ahmed, O.I.; Sabir, S.S. Phytochemical analysis of the essential oil from aerial parts of *Pulicaria undulata* (L.) Kostel from Sudan. *Ethnobot. Leafl.* **2009**, *2009*, 467–471.
18. Nematollahi, F.; Rustaiyan, A.; Larijani, K.; Nadimi, M.; Masoudi, S. Essential oil composition of *Artemisia biennisz* Willd. and *Pulicaria undulata* (L.) CA Mey., two compositae herbs growing wild in Iran. *J. Essent. Oil Res.* **2006**, *18*, 339–341. [CrossRef]
19. Javadinamin, A.; Asgarpanah, J. Essential oil composition of *Francoeuria undulata* (L.) Lack. growing wild in Iran. *J. Essent. Oil Bear. Plants* **2014**, *17*, 875–879. [CrossRef]
20. Dekinash, M.F.; Abou-Hashem, M.M.; Beltagy, A.M.; El-Fiky, F.K. GC/MS profiling, in-vitro cytotoxic and antioxidant potential of the essential oil of *Pulicaria crispa* (Forssk) growing in Egypt. *Int. J. Pharmacogn. Chin. Med.* **2019**, *3*, 000175.
21. Hassan, E.M.; El Gendy, A.E.-N.G.; Abd-ElGawad, A.M.; Elshamy, A.I.; Farag, M.A.; Alamery, S.F.; Omer, E.A. Comparative chemical profiles of the essential oils from different varieties of *Psidium guajava* L. *Molecules* **2021**, *26*, 119. [CrossRef] [PubMed]

22. Abd-ElGawad, A.M.; El-Amier, Y.A.; Assaeed, A.M.; Al-Rowaily, S.L. Interspecific variations in the habitats of *Reichardia tingitana* (L.) Roth leading to changes in its bioactive constituents and allelopathic activity. *Saudi J. Biol. Sci.* **2020**, *27*, 489–499. [CrossRef]

23. Abd-ElGawad, A.M.; Elshamy, A.I.; Al-Rowaily, S.L.; El-Amier, Y.A. Habitat affects the chemical profile, allelopathy, and antioxidant properties of essential oils and phenolic enriched extracts of the invasive plant *Heliotropium curassavicum*. *Plants* **2019**, *8*, 482. [CrossRef] [PubMed]

24. Seidi Damyeh, M.; Niakousari, M. Ohmic hydrodistillation, an accelerated energy-saver green process in the extraction of *Pulicaria undulata* essential oil. *Ind. Crop. Prod.* **2017**, *98*, 100–107. [CrossRef]

25. Abd-ElGawad, A.M.; Elshamy, A.; El-Amier, Y.A.; El Gendy, A.; Al-Barati, S.; Dar, B.; Al-Rowaily, S.; Assaeed, A. Chemical composition variations, allelopathic, and antioxidant activities of *Symphyotrichum squamatum* (Spreng.) Nesom essential oils growing in heterogeneous habitats. *Arab. J. Chem.* **2020**, *13*, 237–4245. [CrossRef]

26. Elshamy, A.I.; Abd-ElGawad, A.M.; El-Amier, Y.A.; El Gendy, A.E.N.G.; Al-Rowaily, S.L. Interspecific variation, antioxidant and allelopathic activity of the essential oil from three *Launaea* species growing naturally in heterogeneous habitats in Egypt. *Flavour Fragr. J.* **2019**, *34*, 316–328. [CrossRef]

27. Fayed, E.M.; Abd-ElGawad, A.M.; Elshamy, A.I.; El-Halawany, E.S.F.; El-Amier, Y.A. Essential oil of *Deverra tortuosa* aerial parts: Detailed chemical profile, allelopathic, antimicrobial, and antioxidant activities. *Chem. Biodivers.* **2021**, *18*, e2000914. [CrossRef] [PubMed]

28. Abd-ElGawad, A.M.; El Gendy, A.E.-N.G.; Assaeed, A.M.; Al-Rowaily, S.L.; Alharthi, A.S.; Mohamed, T.A.; Nassar, M.I.; Dewir, Y.H.; Elshamy, A.I. Phytotoxic effects of plant essential oils: A systematic review and structure-activity relationship based on chemometric analyses. *Plants* **2021**, *10*, 36. [CrossRef]

29. Al-Rowaily, S.L.; Abd-ElGawad, A.M.; Assaeed, A.M.; Elgamal, A.M.; Gendy, A.E.-N.G.E.; Mohamed, T.A.; Dar, B.A.; Mohamed, T.K.; Elshamy, A.I. Essential oil of *Calotropis procera*: Comparative chemical profiles, antimicrobial activity, and allelopathic potential on weeds. *Molecules* **2020**, *25*, 5203. [CrossRef]

30. Mumivand, H.; Rustaii, A.-R.; Jahanbin, K.; Dastan, D. Essential oil composition of *Pulicaria dysenterica* (L.) Bernh from Iran. *J. Essent. Oil Bear. Plants* **2010**, *13*, 717–720. [CrossRef]

31. Bashi, D.S.; Ghani, A.; Asili, J. Essential oil composition of *Pulicaria gnaphalodes* (Vent.) Boiss. growing in Iran. *J. Essent. Oil Bear. Plants* **2013**, *16*, 252–256. [CrossRef]

32. Ali, N.A.A.; Crouch, R.A.; Al-Fatimi, M.A.; Arnold, N.; Teichert, A.; Setzer, W.N.; Wessjohann, L. Chemical composition, antimicrobial, antiradical and anticholinesterase activity of the essential oil of *Pulicaria stephanocarpa* from Soqotra. *Nat. Prod. Commun.* **2012**, *7*, 113–116. [CrossRef]

33. Fawzy, G.A.; Al Ati, H.Y.; El Gamal, A.A. Chemical composition and biological evaluation of essential oils of *Pulicaria jaubertii*. *Pharmacogn. Mag.* **2013**, *9*, 28–33. [CrossRef] [PubMed]

34. Balogun, O.S.; Ajayi, O.S.; Adeleke, A.J. Hexahydrofarnesyl acetone-rich extractives from *Hildegardia barteri*. *J. Herbs Spices Med. Plants* **2017**, *23*, 393–400. [CrossRef]

35. Demiray, H.; Tabanca, N.; Estep, A.S.; Becnel, J.J.; Demirci, B. Chemical composition of the essential oil and n-hexane extract of *Stachys tmolea* subsp. *tmolea* Boiss., an endemic species of Turkey, and their mosquitocidal activity against dengue vector *Aesdes aegypti*. *Saudi Pharm. J.* **2019**, *27*, 877–881. [PubMed]

36. Abd-ElGawad, A.; El Gendy, A.E.-N.; El-Amier, Y.; Gaara, A.; Omer, E.; Al-Rowaily, S.; Assaeed, A.; Al-Rashed, S.; Elshamy, A. Essential oil of *Bassia muricata*: Chemical characterization, antioxidant activity, and allelopathic effect on the weed *Chenopodium murale*. *Saudi J. Biol. Sci.* **2020**, *27*, 1900–1906. [CrossRef] [PubMed]

37. Abd El-Gawad, A.M.; Elshamy, A.I.; El Gendy, A.E.-N.; Gaara, A.; Assaeed, A.M. Volatiles profiling, allelopathic activity, and antioxidant potentiality of *Xanthium strumarium* leaves essential oil from Egypt: Evidence from chemometrics analysis. *Molecules* **2019**, *24*, 584. [CrossRef]

38. Abd El-Gawad, A.; El Gendy, A.; Elshamy, A.; Omer, E. Chemical composition of the essential oil of *Trianthema portulacastrum* L. Aerial parts and potential antimicrobial and phytotoxic activities of its extract. *J. Essent. Oil Bear. Plants* **2016**, *19*, 1684–1692. [CrossRef]

39. Elshamy, A.I.; Ammar, N.M.; Hassan, H.A.; Al-Rowaily, S.L.; Ragab, T.I.; El Gendy, A.E.-N.G.; Abd-ElGawad, A.M. Essential oil and its nanoemulsion of *Araucaria heterophylla* resin: Chemical characterization, anti-inflammatory, and antipyretic activities. *Ind. Crop. Prod.* **2020**, *148*, 112272. [CrossRef]

40. Elshamy, A.I.; Abd-ElGawad, A.M.; El Gendy, A.E.N.G.; Assaeed, A.M. Chemical characterization of *Euphorbia heterophylla* L. essential oils and their antioxidant activity and allelopathic potential on *Cenchrus echinatus* L. *Chem. Biodivers.* **2019**, *16*, e1900051. [CrossRef]

41. Abd El-Gawad, A.M.; El-Amier, Y.A.; Bonanomi, G. Essential oil composition, antioxidant and allelopathic activities of *Cleome droserifolia* (Forssk.) Delile. *Chem. Biodivers.* **2018**, *15*, e1800392. [CrossRef] [PubMed]

42. Abd El-Gawad, A.M.; El-Amier, Y.A.; Bonanomi, G. Allelopathic activity and chemical composition of *Rhynchosia minima* (L.) DC. essential oil from Egypt. *Chem. Biodivers.* **2018**, *15*, e1700438. [CrossRef]

43. Verdeguer, M.; Blázquez, M.A.; Boira, H. Phytotoxic effects of *Lantana camara*, *Eucalyptus camaldulensis* and *Eriocephalus africanus* essential oils in weeds of Mediterranean summer crops. *Biochem. Syst. Ecol.* **2009**, *37*, 362–369. [CrossRef]

44. De Almeida, L.F.R.; Frei, F.; Mancini, E.; De Martino, L.; De Feo, V. Phytotoxic activities of Mediterranean essential oils. *Molecules* **2010**, *15*, 4309–4323. [CrossRef]

45. Pinheiro, P.F.; Costa, A.V.; Tomaz, M.A.; Rodrigues, W.N.; Fialho Silva, W.P.; Moreira Valente, V.M. Characterization of the essential oil of mastic tree from different biomes and its phytotoxic potential on Cobbler's pegs. *J. Essent. Oil Bear. Plants* **2016**, *19*, 972–979. [CrossRef]
46. Ulukanli, Z.; Karabörklü, S.; Bozok, F.; Burhan, A.; Erdogan, S.; Cenet, M.; Karaaslan, M.G. Chemical composition, antimicrobial, insecticidal, phytotoxic and antioxidant activities of Mediterranean *Pinus brutia* and *Pinus pinea* resin essential oils. *Chin. J. Nat. Med.* **2014**, *12*, 901–910. [CrossRef]
47. Silva, E.; Overbeck, G.; Soares, G. Phytotoxicity of volatiles from fresh and dry leaves of two Asteraceae shrubs: Evaluation of seasonal effects. *S. Afr. J. Botany* **2014**, *93*, 14–18. [CrossRef]
48. Abd-ElGawad, A.M.; Elshamy, A.I.; El-Nasser El Gendy, A.; Al-Rowaily, S.L.; Assaeed, A.M. Preponderance of oxygenated sesquiterpenes and diterpenes in the volatile oil constituents of *Lactuca serriola* L. revealed antioxidant and allelopathic activity. *Chem. Biodivers.* **2019**, *16*, e1900278. [CrossRef]
49. Tackholm, V. *Students' Flora of Egypt*, 2nd ed.; Cairo University Press: Cairo, Egypt, 1974.
50. Boulos, L. *Flora of Egypt*; Al Hadara Publishing: Cairo, Egypt, 2002; Volume 3.
51. Saleh, I.; Abd-ElGawad, A.; El Gendy, A.E.-N.; Abd El Aty, A.; Mohamed, T.; Kassem, H.; Aldosri, F.; Elshamy, A.; Hegazy, M.-E.F. Phytotoxic and antimicrobial activities of *Teucrium polium* and *Thymus decussatus* essential oils extracted using hydrodistillation and microwave-assisted techniques. *Plants* **2020**, *9*, 716. [CrossRef] [PubMed]

Article

Chemical Composition and Antimicrobial Properties of *Mentha × piperita* cv. 'Kristinka' Essential Oil

Ippolito Camele [1], Daniela Gruľová [2] and Hazem S. Elshafie [1,*

[1]   School of Agricultural, Forestry, Food and Environmental Sciences (SAFE), University of Basilicata, Viale dell'Ateneo Lucano 10, 85100 Potenza, Italy; ippolito.camele@unibas.it

[2]   Department of Ecology, Faculty of Humanities and Natural Sciences, University of Prešov, 17. Novembra 1, 08001 Prešov, Slovakia; daniela.grulova@unipo.sk

*   Correspondence: hazem.elshafie@unibas.it; Tel.: +39-0971-205522

Abstract: Several economically important crops, fruits and vegetables are susceptible to infection by pathogenic fungi and/or bacteria postharvest or in field. Recently, plant essential oils (EOs) extracted from different medicinal and officinal plants have had promising antimicrobial effects against phytopathogens. In the present study, the potential microbicide activity of *Mentha × piperita* cv. 'Kristinka' (peppermint) EO and its main constituents have been evaluated against some common phytopathogens. In addition, the cell membrane permeability of the tested fungi and the minimum fungicidal concentrations were measured. The antifungal activity was tested against the following postharvest fungi: *Botrytis cinerea*, *Monilinia fructicola*, *Penicillium expansum* and *Aspergillus niger*, whereas antibacterial activity was evaluated against *Clavibacter michiganensis*, *Xanthomonas campestris*, *Pseudomonas savastanoi* and *P. syringae* pv. *phaseolicola*. The chemical analysis has been carried out using GC-MS and the main components were identified as menthol (70.08%) and menthone (14.49%) followed by limonene (4.32%), menthyl acetate (3.76%) and β-caryophyllene (2.96%). The results show that the tested EO has promising antifungal activity against all tested fungi, whereas they demonstrated only a moderate antibacterial effect against some of the tested bacteria.

Keywords: medicinal plants; GC-MS; postharvest diseases; biological control; cell membrane permeability

1. Introduction

Many microorganisms cause different plant diseases in field and/or postharvest. Without proper treatments, they can cause losses or decrease the shelf life of fruits and vegetables [1,2]. Although synthetic pesticides efficiently control diseases, their application is restricted, particularly postharvest, because of consumer concern for human health conditions, the harmful effects on the environment and the development of new resistant strains [3–5].

There are strict regulations worldwide regarding the minimum pesticide residue levels in the edible portion of the fresh vegetable and fruits for protecting human health and the environment [1,5]. On the other hand, in Europe, synthetic fungicides are prohibited in postharvest applications. For that reason, the discovery of new natural substances, such as plant essential oils (EOs), for controlling phytopathogens, especially in postharvest conditions, has attracted great interest recently. Several research projects reported the antifungal efficacy of plant EOs against postharvest fruit pathogens, being considered natural, safe and biodegradable alternatives [6–9].

Mentha × piperita L. (peppermint) is a perennial plant that is widespread throughout the Mediterranean region [10]. Peppermint, a plant in the Family *Lamiaceae*, has long been considered an economically important [11]. It was already known in Egyptian, Greek and Roman medicine for its wide benefits for human health, especially for digestive and diuretic problems and as a remedy for coughs and colds [12]. Peppermint has several medicinal uses such as treating stomach-aches, chest pains and for treating irritable bowel

Citation: Camele, I.; Gruľová, D.; Elshafie, H.S. Chemical Composition and Antimicrobial Properties of *Mentha × piperita* cv. 'Kristinka' Essential Oil. *Plants* **2021**, *10*, 1567. https://doi.org/10.3390/plants10081567

Academic Editor: Ki Hyun Kim

Received: 8 July 2021
Accepted: 27 July 2021
Published: 30 July 2021

Publisher's Note: MDPI stays neutral with regard to jurisdictional claims in published maps and institutional affiliations.

syndrome [13]. Peppermint EO can be extracted from the aerial parts of the flowering plant, from dried leaves or from fresh flowers [11,14]. Many studies reported the chemical composition of peppermint EO, which is composed mainly of menthol, menthone, menthofuran, 1,8-cineole, and menthyl acetate [15]. Previous studies revealed that most peppermint EO is rich in pulegone, menthon, menthol, carvone, 1,8-cineole, limonene and β-caryophyllene [16]. Regarding the chemical composition of peppermint EO, some previous studies have analysed its chemical composition and principal single constituents and found that the respective percentage of different peppermint species varied depending upon the origins of the plant, species as well as the possible variation within the same species [16]. In addition, there are some factors, such as physiological and environmental conditions, genetic and evolution that can also determine the chemical variability of peppermint EO [17].

Bibliographic research revealed that the plant EOs from different species of peppermint possess potential antimicrobial activity against different plant pathogens [18] as well as insecticidal activity against stored product [19]. Several researchers have also reported the promising biological activities of peppermint EO against different phyto- and food pathogens, especially against Gram-positive bacteria such as *Staphylococcus aureus* and *Enterococcus faecalis*, as reported by Jirovetz et al. [11]. Researchers at the University of Prešov bred a variety of peppermint with a very high content of the main constituents— menthol and menthone [20].

The current study aims to evaluate the in vitro antimicrobial activity of peppermint EO against some common postharvest fungal pathogens and some pathogenic bacteria. This research aims also to study the mode of antimicrobial actions and the minimum fungicidal concentrations, both for the EO and for its main components.

In particular, the main objectives of the current study were to: (i) identify the main components of the *Mentha × piperita* cv. 'Kristinka' in the harvest season 2020 cultivated in Prešov, Slovakia; (ii) screen the antifungal effect of the extracted EO against *Monilinia fructicola*, *Aspergillus niger*, *Penicillium expansum* and *Botrytis cinerea*; (iii) evaluate the antibacterial affect against *Clavibacter michiganensis*, *Xanthomonas campestris*, *Pseudomonas savastanoi* and *P. syringae* pv. *phaseolicola*; (iv) study the effect of EO and its two main constituents (menthol an menthone) on the fungal cell membrane permeability (CMP); (v) determine the minimum fungicidal concentration (MFC) of the studied EO and its two main constituents.

2. Results

2.1. Identification of M. piperita EO Components

Essential oil of *M. × piperita* cv. 'Kristinka' was hydrodistilled and qualitatively analysed using GC-MS for determining the main components as mentioned below. Average amount of EO was 0.4 ± 0.02% from plant materials. Qualitative parameters are summarised in Table 1. The most principal component identified was menthol (70.08 ± 0.05%), followed by menthone (14.49 ± 0.01%). This is the typical characteristic of the new cultivar Kristinka, where menthol is the dominant component with higher quantity. Other dominant components were limonene (4.32 ± 0.03%), menthyl acetate (3.76 ± 0.01%) and β-caryophyllene (2.96 ± 0.04%). Oxygenated monoterpenes presented 89.13% of the identified chemical group. Sesquiterpenes hydrocarbons (5.46%) and monoterpenes hydrocarbons (5.26%) followed with the almost the same quantity. Among the different chemical groups, oxygenated sesquiterpenes, such as spathulenol compound, were also present in very low quantities (0.03%).

Table 1. Identification of *Mentha* × *piperita* cv. 'Kristinka' EO components.

Components	*Mentha* × *piperita* cv. 'Kristinka'				
	Ki Exp	Ki Lit.	%	Formula	Chem. Group
α-Pinene	935	936	0.47 ± 0.01	$C_{10}H_{16}$	MH
β-Pinene	976	978	0.45 ± 0.01	$C_{10}H_{16}$	MH
Limonene	1025	1025	4.32 ± 0.03	$C_{10}H_{16}$	MH
cis-β-Ocimene	1028	1029	0.02 ± 0.00	$C_{10}H_{16}$	MH
Menthone	1146	1136	14.49 ± 0.01	$C_{10}H_{18}O$	OM
Menthol	1174	1172	70.08 ± 0.05	$C_{10}H_{20}O$	OM
α-Terpineol	1177	1176	0.18 ± 0.00	$C_{10}H_{18}O$	OM
Carvone	1210	1214	0.01 ± 0.00	$C_{10}H_{14}O$	OM
Piperitone	1223	1226	0.60 ± 0.02	$C_{10}H_{16}O$	OM
Isopulegol acetate	1271	1263	0.01 ± 0.00	$C_{12}H_{20}O_2$	OM
Menthyl acetate	1278	1280	3.76 ± 0.01	$C_{12}H_{22}O_2$	OM
α-Cubebene	1354	1355	0.01 ± 0.00	$C_{15}H_{24}$	SH
Clovene	1365	1365	0.03 ± 0.00	$C_{15}H_{24}$	SH
Isoledene	1370	1382	0.01 ± 0.00	$C_{15}H_{24}$	SH
β-Bourbonene	1386	1378	0.07 ± 0.01	$C_{15}H_{24}$	SH
β-Elemene	1389	1389	0.03 ± 0.00	$C_{15}H_{24}$	SH
β-Cubebene	1390	1390	0.17 ± 0.02	$C_{15}H_{24}$	SH
Longifolene	1404	1411	0.04 ± 0.00	$C_{15}H_{24}$	SH
α-Gurjunene	1410	1413	0.20 ± 0.01	$C_{15}H_{24}$	SH
(Z)-β-Farnesene	1420	1420	0.50 ± 0.00	$C_{15}H_{24}$	SH
β-Caryophyllene	1421	1421	2.96 ± 0.04	$C_{15}H_{24}$	SH
Aristolene	1422	1423	0.02 ± 0.00	$C_{15}H_{24}$	SH
Aromadendrene	1435	1443	0.01 ± 0.00	$C_{15}H_{24}$	SH
α-Humulene	1448	1455	0.07 ± 0.00	$C_{15}H_{24}$	SH
Allo-Aromadendrene	1460	1462	0.20 ± 0.01	$C_{15}H_{24}$	SH
γ-Gurjunene	1472	1472	0.14 ± 0.01	$C_{15}H_{24}$	SH
β-Chamigrene	1474	1474	0.03 ± 0.00	$C_{15}H_{24}$	SH
γ-Muurolene	1475	1474	0.01 ± 0.00	$C_{15}H_{24}$	SH
α-Amorphene	1477	1477	0.03 ± 0.00	$C_{15}H_{24}$	SH
Germacrene D	1479	1479	0.03 ± 0.00	$C_{15}H_{24}$	SH
Ledene	1489	1491	0.53 ± 0.02	$C_{15}H_{24}$	SH
Valencene	1493	1494	0.06 ± 0.00	$C_{15}H_{24}$	SH
α-Muurolene	1496	1496	0.06 ± 0.01	$C_{15}H_{24}$	SH
γ-Cadinene	1507	1507	0.09 ± 0.01	$C_{15}H_{24}$	SH
δ-Cadinene	1520	1520	0.16 ± 0.02	$C_{15}H_{24}$	SH
Spathulenol	1565	1572	0.03 ± 0.00	$C_{15}H_{24}O$	OS
Total			99.88		

Percentage was calculated as an average from three replication ± SD; Ki–Kovats index calculated by researchers; Ki lit. Kovats index from literature (MS Finder) for comparison.

2.2. Antibacterial Activity

The results of the antibacterial activity assay showed that the positive control (Tetracycline 1.6 mg/mL) demonstrated the highest inhibition against all tested phytopathogenic bacteria (Table 2). However, peppermint EO showed the highest significant antibacterial activity against *P. syringae* pv. *phaseolicola* (the diameter of inhibition zone was 39.5 mm) similar to tetracycline (1.6 mg/mL) where the diameter of its inhibition zone was 40 mm. In addition, there was a moderate activity against *C. michiganensis* at 10 mg/mL and low activity against *P. savastanoi* only at 10 mg/mL. On the other hand, there was no activity against *X. campestris*.

Table 2. Antibacterial activity of *Mentha × piperita* cv. 'Kristinka' EO.

Tested EOs	Diameter of Inhibition Zones (mm)				
	Conc.	*X. campestris*	*C. michiganensis*	*P. syr.* **pv.** *phaseolicola*	*P. savastanoi*
Peppermint EO	10 mg/mL	0.0 ± 0.0 b	27.5 ± 2.8 b	39.5 ± 0.5 a	09.0 ± 1.2 b
	1 mg/mL	0.0 ± 0.0 b	17.0 ± 2.3 c	26.5 ± 1.6 b	0.0 ± 0.0 c
	0.1 mg/mL	0.0 ± 0.0 b	09.0 ± 1.1 d	19.5 ± 0.6 c	0.0 ± 0.0 c
Tetracycline (1.6 mg/mL)		23.5 ± 1.70 a	39.5 ± 0.6 a	40.0 ± 1.6 a	37.0 ± 2.2 a

Values followed by different letters for each tested bacterium in each column are significantly different at $p < 0.05$ according to two-way ANOVA combined with Duncan post hoc test. Data are expressed as the mean of three replicates ± SD.

2.3. Antifungal Activity

The results of the fungicidal activity of peppermint EO are presented in Figure 1, where it showed the highest significant inhibition the mycelium growth of *B. cinerea* and *P. expansum* in plates at the two tested concentrations (1 and 5 mg/mL), whereas peppermint EO at 5 mg/mL demonstrated the highest inhibition against *M. fructicola* and *A. niger*. The lowest inhibition effect was observed in the case of the tested concentration 0.1 mg/mL against all tested pathogenic fungi, which was insignificantly different from the positive control (Azoxystrobin, 0.8 μL/mL). In fact, the inhibition of fungal mycelium growth in plates is considered as a general indication of the efficacy of the tested treatments, whereas the MFC assay is considered a more accurate test for determining the lowest concentration required to inhibit the visible growth of the microorganism.

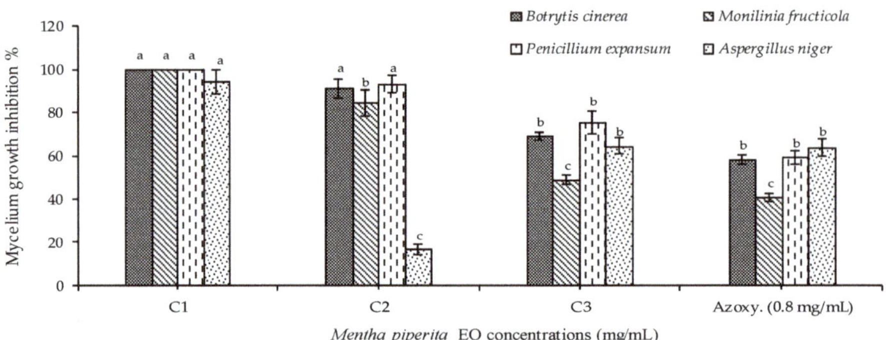

Figure 1. Antifungal activity of *M. piperita* EO. Where: C1: 5.0 mg/mL; C2: 1.0 mg/mL; C3: 0.1 mg/mL. Bars with different letters for each tested fungus indicate mean values significantly different at $p < 0.05$ according to two-way ANOVA combined with Duncan post hoc multiple comparison test.

2.4. Fungal Cell Membrane Permeability Assay

This assay was carried out to explain the possible mechanism of the antifungal activity of the tested EO. In general, the fungicidal effect of EO depends on the destruction of the fungal cell membrane that increases the cell permeability. For that reason, the current assay was performed to investigate the effect of mint EO and its main single constituents on the CMP of the tested phytopathogenic fungi by measuring their electric conductivity (EC) [21,22].

Figure 2 showed the effect of peppermint EO at different doses on the mycelium electrical conductivity (MEC) as indication of the cell membrane permeability (CMP) of the four tested fungi. Generally, the effect of the studied EO on the CMP of all tested fungi was dose-dependent. In particular, the highest tested concentration (7.0 mg/mL) showed the EC values 87.2, 85.3, 92.3 and 85.1 S/cm corresponding to the CMP of *M. fructicola*, *B. cinerea*, *A. niger* and *P. expansum*, respectively. On the other hand, the concentration 7.0 mg/mL showed a significant increase in the CMP in the case of *M. fructicola* and *B. cinerea*, whereas there was no significant difference between the two doses 5.0 and 7.0 mg/mL regarding *A. niger* and *P. expansum*. In addition, there was a dramatical increase in the CMP in the case of *P. expansum* after treatment with 5.0 mg/mL.

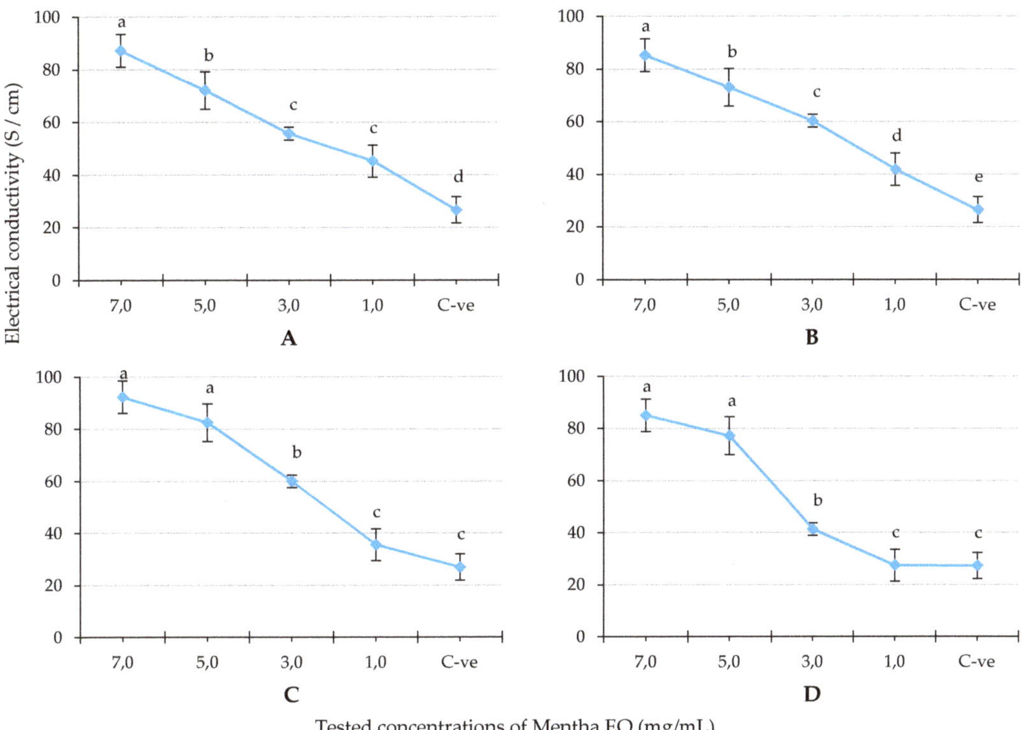

Figure 2. The effect of peppermint EO on mycelium electrical conductivity of the tested fungi. Where, (**A**): *Monilinia fructicola*; (**B**): *Botrytis cinerea*; (**C**): *Aspergillus niger*; (**D**): *Penicillium expansum*. C-ve: negative control (potato dextrose broth). Differences between the tested concentrations for each tested fungus indicate mean values significantly different at $p < 0.05$ according to one-way ANOVA for each fungus combined with Duncan post hoc multiple comparison test.

Figure 3 showed the effects of two single constituents of peppermint EO (menthol and menthone) at different doses on the MEC of the four tested fungi. In the case of menthol, it showed a dose-dependent effect on CMP of *M. fructicola*, *B. cinerea* and *A. niger*, whereas the CMP was highly decreased after treatment with 0.8 mg/mL in the case of *P. expansum*. The EC values were 35.9, 37.2, 39.0 and 45.5 S/cm for *M. fructicola*, *B. cinerea*, *A. niger* and *P. expansum*, respectively. Regarding the menthone, it showed a dose-dependent effect on the CMP against *M. fructicola*, *B. cinerea* and *P. expansum*. The EC values were 38.9, 44.5, 37.8 and 34.0 S/cm for *M. fructicola*, *B. cinerea*, *A. niger* and *P. expansum*, respectively.

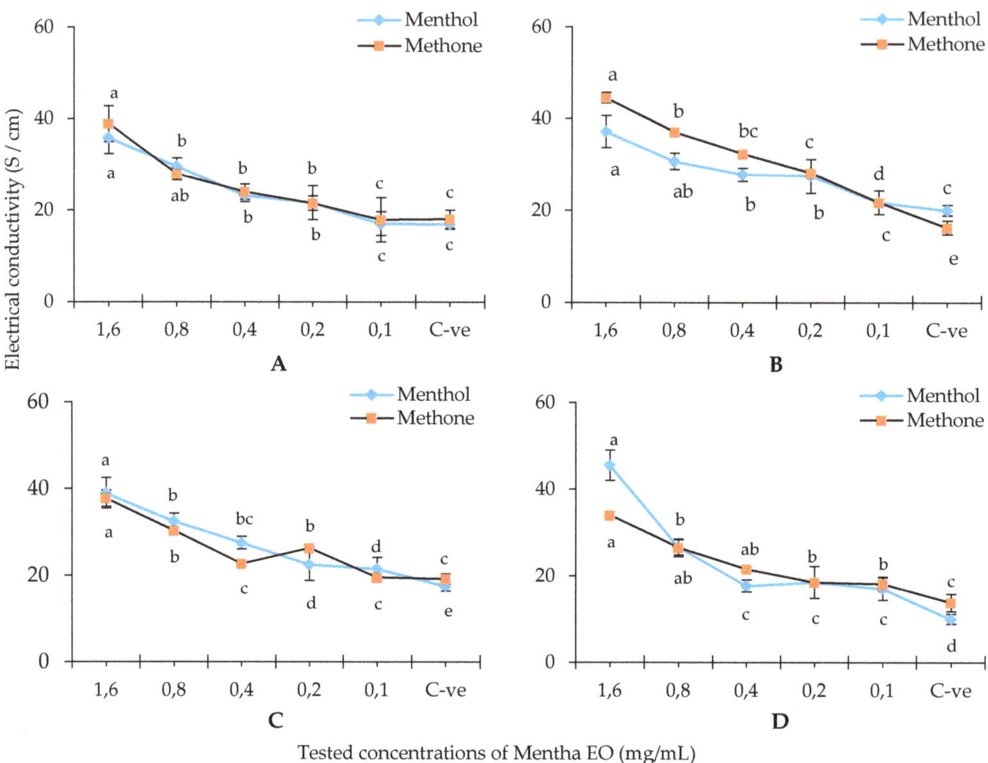

Figure 3. The effect of single constituents of peppermint EO on mycelium electrical conductivity of the tested fungi. Where (**A**): *Monilinia fructicola*; (**B**): *Botrytis cinerea*; (**C**): *Aspergillus niger*; (**D**): *Penicillium expansum*. C-ve: negative control (potato dextrose broth). Differences between the tested concentrations for each tested fungus indicate mean values significantly different at $p < 0.05$ according to one-way ANOVA for each fungus combined with Duncan post hoc multiple comparison test.

2.5. Fungicidal Microdilution Broth Assay (96-Microplate)

This assay was carried out to determine the minimum fungicidal concentration (MFC) which is defined as the lowest concentration of the tested antimicrobial agent that can inhibit the growth of fungi significantly differently to the growth of the negative control, as reported by Arikan [23]. The results of the fungicidal effect of mint EO and its main single constituents on mycelium growth percentage are reported in Table 3, whereas the MFC values of peppermint EO and its two main constituents are reported in Table 4 using the tendency-line formula of the chart in Microsoft Excel. The studied EO showed 4.78, 2.91, 5.40 and 4.98 mg/mL, corresponding to the inhibition of 50% visible growth of fungal mycelium of *M. fructicola*, *B. cinerea*, *A. niger* and *P. expansum*, respectively.

Regarding menthol, the MFC values were 0.85, 1.40, 1.45 and 1.21 mg/mL against *M. fructicola*, *B. cinerea*, *A. niger* and *P. expansum*, respectively. In the case of menthone, the MFC values were 1.31, 1.37, 1.90 and 1.69 mg/mL, against *M. fructicola*, *B. cinerea*, *A. niger* and *P. expansum*, respectively.

Table 3. Fungicidal effect of EO and single constituents on mycelium growth (%) in broth culture.

Tested Concentrations (mg/mL)	Mycelium Growth Percentage (MGP)			
	M. fructicola	*B. cinerea*	*A. niger*	*P. expansum*
M. piperita 7.0	33.3 ± 4.0 d	31.6 ± 6.0 d	35.1 ± 4.1 d	18.6 ± 3.0 c
M. piperita 5.0	47.8 ± 4.0 c	42.8 ± 2.4 bc	71.3 ± 3.4 c *	48.6 ± 4.0 b
M. piperita 3.0	60.5 ± 6.0 c *	48.9 ± 0.6 bc	74.3 ± 0.5 ab	50.1 ± 3.0 b
M. piperita 1.0	77.8 ± 4.1 ab	66.1 ± 5.9 b *	78.6 ± 1.9 ab	52.4 ± 0.5 b *
C-ve: PDB + F	100.0 ± 0.0 a	100.0 ± 0.0 a	100.0 ± 0.0 a	100.0 ± 0.0 a
Principal Single Components				
Menthol 1.6	40.0 ± 5.6 c	41.7 ± 4.3 d	58.1 ± 6.2 c	51.9 ± 7.4 cd
Menthol 0.8	50.5 ± 6.6 c	66.1 ± 5.7 c *	60.0 ± 5.8 c	52.8 ± 7.3 cd
Menthol 0.4	60.1 ± 4.3 b	78.2 ± 1.3 ab	62.1 ± 4.4 c	66.4 ± 5.0 c
Menthol 0.2	65.0 ± 5.0 b *	78.7 ± 3.3 ab	71.2 ± 3.8 b *	76.7 ± 5.3 b *
Menthol 0.1	69.1 ± 2.9 ab	81.4 ± 6.3 ab	93.8 ± 3.5 a	85.2 ± 4.2 ab
Menthone 1.6	29.2 ± 4.4 d	42.9 ± 2.0 d	45.1 ± 3.0 d	33.6 ± 3.3 d
Menthone 0.8	55.8 ± 4.0 c	66.0 ± 3.7 c	63.5 ± 4.5 c	60.6 ± 6.7 c
Menthone 0.4	64.2 ± 6.2 b *	70.4 ± 3.4 b *	71.3 ± 3.3 b	68.9 ± 5.8 b *
Menthone 0.2	69.6 ± 2.2 ab	89.2 ± 4.7 a	76.6 ± 4.2 b	79.2 ± 4.3 ab
Menthone 0.1	75.7 ± 1.5 ab	97.1 ± 0.6 a	77.0 ± 4.2 b *	83.2 ± 2.3 ab
C-ve: PDB + F	100.0 ± 0.0 a	100.0 ± 0.0 a	100.0 ± 0.0 a	100.0 ± 0.0 a

Values followed by different letters in each column for each tested fungi are significantly different at $p < 0.05$ according to one-way ANOVA combined with *Tukey* B post hoc test. (*) are the mycelium growth percentages corresponding to the MFC values. Data are expressed as the mean of three replicates ± SD and presented for peppermint EO and the two single substances.

Table 4. MFC values of fungicidal effect of studied EO and single constituents.

	MFC (mg/mL)			
	M. fructicola	*B. cinerea*	*A. niger*	*P. expansum*
M. piperita EO	4.78	2.91	5.40	4.98
Menthol	0.85	1.40	1.45	1.21
Menthone	1.31	1.37	1.90	1.69

3. Discussion

The studied EO in the current study was hydrodistilled from the new cultivar "Kristinka" of *M. piperita*. The parameters of EO differ from the standard ones for *M. piperita*. The cultivar Kristinka was bred and certificated to obtain a higher amount of the main component, menthol [24]. The newly bred cultivar of *M.* × *piperita* is characterised by a higher amount of menthol than found in other commercial cultivars [20,25]. The amount of EO depends on external factors influencing the vegetation season (environmental and climatic conditions) and may vary [25,26]. Plant biodiversity is also represented by different amounts of the main chemical components. This variation present great opportunity for the research to study the different effects of antibacterial activity.

Generally, the explanation of components of EO analysed by the GCMS is by peak area and it is explained in %. In particular, the major components of the EO of *M.* × *piperita* in different publications are menthol, menthone, menthofuran and menthyl acetate in the amounts of about 40, 30, 7 and 10%, respectively, of the whole amount of EO content [2,7,27–29]. Another study conducted by Kamatou et al. [18] on *Mentha canadensi* EO reported that the main components were identified as isomenthone (27.4%), menthol (24.3%), menthone (9.2%), limonene (5.8%), 1,8-cineole (5.6%), menthofuran (4.4%) and isomenthol (3.2%).

Many studies have highlighted the promising antibacterial and antifungal activity of peppermint EO against some human- and phytopathogens such as *Botrytis cinerea*, *Cladosporium cladosporioides*, *Penicillium aurantiogriseum*, *Staphylococcus aureus*, *Streptococcus pyogenes*,

Escherchia coli and *Klebsiella pneumonia* [2,7,27]. On the other hand, the antimicrobial activity of peppermint EO might be correlated to its chemical composition due to the hydrophobic nature of those above-mentioned compounds, which allows them to interact with microbial membranes causing cell lysis, interrupting the proton's motor force, electron flow and transport activity, and inhibiting protein synthesis [30,31]. Particularly, the obtained results of the current study of the bioactivity of menthol and menthone have confirmed their role in antimicrobial activity, as previously hypothesized.

Regarding the antifungal activity of peppermint EO, Tsao and Zhou [32] concluded that menthol was able to inhibit the postharvest fungi *Botrytis cinerea* and *Monilinia fructicola* [32]. Furthermore, different stereoisomers of menthol were active against *Fusarium verticillioides*, commonly reported as fungal species infecting maize (*Zea mays*) [33]. Tyagi et al. [34] have reported the efficacy of peppermint oil and its vapours against yeasts causing food spoilage in fruit juice such as *Saccharomyces cerevisiae*, *Zycosaccharomyces bailii*, *Aureobasidium pullulans*, *Candida diversa*, *Pichia fermentans*, *Pichia kluyveri*, *Pichia anomala* and *Hansenula polymorpha*.

A recent study conducted by Hsouna et al. [5] underlined the antibacterial activity of peppermint EO against *Agrobacterium tumefaciens*, the causal agent of crown gall disease in over 140 species of eudicots, where the tested concentration 200 mg/mL was able to completely inhibit the formation of tumours on tomato plants when inoculated with *A. tumefaciens* ATCC 23308^T [5].

In the current research, peppermint EO showed promising antifungal activity against the postharvest tested pathogenic fungi by measuring the growth of mycelium in plates. In addition, the studied EO explicated moderate to acceptable antibacterial activity, especially against *P. syringae* pv. *phaseolicola* and *C. michiganensis*, by measuring the diameter of the inhibition zones compared to the respective positive controls.

The results of the inhibitory effect against some phytopathogens are in agreement with some other important studies, especially those conducted by Afridi et al. [35]. The latter authors have attributed the biological activity of peppermint EO to their ability to penetrate the plasma membranes and cell walls of fungal cells, increasing their permeability, causing a significant decomposition of the walls, and later leading to the death of the fungal cells [32]. The last interpretation is what we tried to clarify through this research by conducting the CMP assays of cell membranes and their rate of electrical conductivity in the broth culture media of the tested fungi [36,37]. This, in turn, gave a clear indication of a change in the normal rate of permeability of the cell wall compared to the control cells due to the influence of biological oil as well as mono active compounds.

Consistently with this interpretive context, Ultee et al. [38] have also attributed the promising biological activity of peppermint EO to its rich content of menthol and its related compounds. These compounds can destabilize the cytoplasmic membrane and act as a proton exchanger to reduce pH gradient. Therefore, this action can destroy the proton motive force cause the depletion of ATP and hence increasing the possibility of cell death.

4. Materials and Methods

4.1. Plant Materials and Extraction of Essential Oil

Peppermint plant was grown in the experimental field belonging to University of Prešov in 2020. It was harvested in the flowering developmental stage, then dried in the shade for several days. When the plant materials were able to be crushed in the hands, they were placed into the Clevenger apparatus for hydrodistillation of EO. Fifty grams of plant materials were placed into glass flask, covered with water and connected to the Clevenger apparatus. After 3 h of hydrodistillation, the EO was reached. The procedures were repeated a few times to obtain amount of EO necessary for the successive chemical and biological analysis. Quantitative parameter was calculated as an average amount of all hydrodistillation, recalculated as a percentage of amount of utilized plant materials.

4.2. Gas Chromatography-Mass Spectrometry Analysis

Three samples of peppermint EO were analysed by a gas chromatography/mass spectrometry (GC/MS) for qualitative properties in laboratory at University of Prešov. GC/MS analyses were carried out on devices Varian 450-GC and 220-MS. Separation was conducted on a capillary column BR 5ms (30 m × 0.25 mm ID, 0.25 μm film thickness). Injector type 1177 was heated to a temperature of 220 °C. Injection mode was splitless (1 μL of a 1:1000 n-hexane solution). Helium was used as a carrier gas at a constant column flow rate of 1.2 mL min^{-1}. Column temperature was programmed as follows: initial temperature was 50 °C for 10 min, then increased to 100 °C for 3 min, maintained isothermally for 5 min and then increased to 150 °C for last 10 min. The total time for analysis was 46.67 min. The MS trap was heated to 200 °C, manifold 50 °C and transfer line 270 °C. Mass spectra were scanned every 1 s in the range 40–650 m/z. The retention indices were determined in relation to the Rt values of a homologous series of n-alkanes (C10–C35) under the same operation conditions. Constituents were identified by comparison of their retention indices (RI) with published data in different literature. Further identification was made by comparison of the mass spectra with those stored either in NIST 02 library or with those from the literature [39]. Components relative concentrations were obtained by percentage of peak area normalization.

4.3. Preliminary Screening of Antimicrobial Activity

4.3.1. Tested Bacterial and Fungal Isolates

The tested bacterial strains are: *Clavibacter michiganensis* Smith, and *Xanthomonas campestris* Pammel, *Pseudomonas savastanoi* Janse (Gardan) and *P. syringae* pv. *phaseolicola* Van Hall were used in this assay, whereas four postharvest phytopathogenic fungi: *Monilinia fructicola* (G. Winter) Honey, *Aspergillus niger* van Tieghem, *Penicillium expansum* Link, and *Botrytis cinerea* Pers were used for the antifungal activity assay. All tested isolates were identified by classical and molecular methods and conserved as pure culture in the collection of the School of Agricultural, Forestry, Food and Environmental Sciences (SAFE), University of Basilicata, Potenza, Italy.

4.3.2. Antibacterial Activity

Disc diffusion assay. The antibacterial test was carried out following the disc diffusion method [40,41] using the King B nutrient media (KB) [42]. A bacterial suspension of each tested bacteria was prepared in sterile distilled water adjusted at 10^6 CFU/mL (OD ≈ 0.2 nm) using a turbidimetry instrument (Biolog, Hayward, CA, USA). Four millilitres of bacterial suspension mixed with soft agar (0.7%) at ratio 9:1 (*v/v*) were poured over each plate (90 mm diameter). Blank discs of 6 mm (OXOID, Milan, Italy) were then placed over the KB-plate surfaces and about 20 μL from each tested EO concentration at 0.1, 1 and 10 mg/mL was carefully applied over discs. Tween 20 was added to each tested EO concentration at 0.2% for accelerating the oil solubility. Tetracycline (1.6 mg/mL) was used as a positive control. The antibacterial activity was estimated by measuring the diameter of inhibition zone in mm ± SDs around each treated disc compared to the positive control ones.

4.3.3. Antifungal Activity

Incorporation assay. The possible fungicidal activity of the studied EO was evaluated at three different doses (0.1, 1 and 5 mg/mL) following the incorporation method [43–45] as explained below. The EO was incorporated into Potato Dextrose Agar (PDA) medium at 45 ± 2 °C. Fungal disks (0.5 cm) from each of the phytopathogenic fungi (96 h fresh culture) were deposited in the centre of each Petri dish. All plates were incubated at 22 ± 2 °C for 96 h in darkness. As negative control, PDA plates without any treatments were inoculated only with each fungus. The diameter of fungal mycelium growth was measured in mm ± Standard Deviations (SDs) between the three replicates [46] and the percentage of growth inhibition (PGI%) was calculated using Equation (1) [47] compared to synthetic fungicides Azoxystrobin (0.8 µL/mL), a large spectrum fungicide, as control according to the international limit of microbicide standards.

$$\text{PGI (\%)} = \frac{\text{GC} - \text{GT}}{\text{GC}} \times 100 \tag{1}$$

where PGI is the percentage of growth inhibition, GC is the average diameter of fungal mycelium in PDA negative control and GT is the average diameter of fungal mycelium on the oil-treated PDA dish.

4.4. Cell Membrane Permeability

The CMP effect of the mint EO and its two main principals was determined by measuring the potential of electrical current transport through water as molar conductivity (MC) or electrolytic conductivity (EC) as reported by Elshafie et al. [21]. This assay was performed by transferring five mycelial discs (0.5 cm diameter) from fresh culture of each tested fungus into Potato Dextrose Broth (PDB) medium and incubated under shaking condition (180 rpm/min), at 28 °C for 96 h. A gram and half of dried mycelia from each fungal species was re-suspended into 20 mL of each tested EO concentration at 1, 3, 5 and 7 mg/mL or single components at 0.1, 0.2, 0.4, 0.8 and 1.6 mg/mL and incubated at 22 ± 2 °C. EC values have been measured after 72 h of incubation. The IP% of EC value was calculated following Equation (2) [21].

$$\text{IP (\%)} = \frac{\text{EC t}}{\text{EC ctrl}} \times 100 \tag{2}$$

where EC t. is the electrical conductivity of the treated sample and EC ctrl. is the electrical conductivity of the PDB broth culture.

4.5. Fungicidal Microdilution Broth Assay

The MFC was carried out against the four tested pathogenic fungi using a 96-well microplate (Nunc MaxiSorp®, Vedbaek, Denmark) by a micro-dilution method [22,48]. Four millilitres of liquid suspension from fresh fungal cultures (96 h) was prepared at 10^8 spore/mL. The tested EO was dissolved in PDB at 1.0, 3.0, 5.0 and 7.0 mg/mL, whereas the tested concentrations for menthol and menthone were 0.1, 0.2, 0.4, 0.8 and 1.6 mg/mL. The proposed concentrations of this assay were selected according to the obtained results from the preliminary in vitro antifungal assay. One hundred microlitres/well from each prepared concentration of EO and 100 µL/well of PDB media were added into the 96-well microplates then 30 µL/well of fungal suspension from each tested fungus was uploaded per all wells. All plates were incubated at 24 ± 2 °C. The absorbance was measured at $\lambda = 450$ nm using microplate reader instrument (DAS s.r.l., Rome, Italy) after 24 h. The MGP percentage was calculated using Equation (3). The whole experiment was repeated in triplicate $\pm$ SDs.

$$\text{MGI (\%)} = \frac{\text{Abs. t}}{\text{Abs. c}} \times 100 \tag{3}$$

where Abs. t: is the value of the absorbance at 450 nm for each treatment; Abs. c: is the value of the absorbance at 450 nm for the PDA + fungi as control.

4.6. Statistical Analysis

The obtained results of the biological assays were subjected to one-way ANOVA for the statistical analysis. Then, the significance level of the findings was checked by applying *Tukey* B Post Hoc multiple comparison test with a probability of $p < 0.05$ using statistical Package for the Social Sciences (SPSS) version 13.0 (Prentice Hall: Chicago, IL, USA, 2004).

5. Conclusions

The EO from *Mentha* × *piperita* cv. 'Kristinka' demonstrated promising antifungal activity against some serious phytopathogenic fungi even at low concentrations and this result is very interesting, especially for controlling postharvest fungi. In addition, the studied EO showed acceptable antibacterial activity against the following pathogenic bacteria: *P. syringae* pv. *phaseolicola* and *C. michiganensis*; however, it showed little effect against *P. savastanoi*, and no activity against *X. campestris*. On the other hand, the biological activity of the studied EO can be highly attributed to its rich content of menthol (70.08%) and menthone (14.49%). The MFC value of the fungicidal effect achieved by the peppermint EO was 2.9 mg/mL against *B. cinerea* and the MFC value in the case of menthol was 0.8 mg/mL against *M. fructicola*, whereas menthone achieved MFC values equal to 1.3 mg/mL against *M. fructicola* and *B. cinerea*.

The obtained promising results of the antimicrobial activity of peppermint EO proved its potential to control several fungal and bacterial pathogens. Furthermore, the studied EO and its two main constituents can be used successfully in the chemical industries as natural alternatives to synthetic substances against several phytopathogens.

Author Contributions: Conceptualization, I.C., D.G. and H.S.E.; methodology, H.S.E., D.G. and I.C.; investigation, I.C. and D.G.; resources, D.G. and H.S.E. data curation, I.C. writing—original draft preparation, H.S.E. and D.G.; writing—review and editing, H.S.E. and I.C. and supervision, I.C. All authors have read and agreed to the published version of the manuscript.

Funding: This research received no external funding.

Institutional Review Board Statement: Not applicable.

Informed Consent Statement: Not applicable.

Data Availability Statement: Not applicable.

Conflicts of Interest: The authors declare no conflict of interest.

References

1. Selvam, S.P.; Dharini, S.; Puffy, S. Antifungal activity and chemical composition of thyme, peppermint and citronella oils in vapour phase against avocado and peach postharvest pathogens. *J. Food Saf.* **2013**, *33*, 86–93.
2. Plavšić, D.V.; Škrinjar, M.M.; Psodorov, Đ.B.; Pezo, L.L.; Milovanović, I.L.J.; Psodorov, D.Đ.; Kojić, P.S.; Kocić, T.D. Chemical structure and antifungal activity of mint essential oil components. *J. Serb. Chem. Soc.* **2020**, *85*, 1149–1161. [CrossRef]
3. Ippolito, A.; Schena, L.; Pentimone, I.; Nigro, F. Control of postharvest rots of sweet cherries by pre and postharvest applications of *Aureobasidium pullulans* in combination with calcium chloride or sodium bicarbonate. *Postharvest Biol. Technol.* **2005**, *36*, 245–252. [CrossRef]
4. Bautista-Baños, S.; Hernandez-Lauzardo, A.N.; Velázquez-Del Valle, M.G.; Hernandez-Lopez, M.; Ait-Barka, E.; Bosquez-Molina, E.; Wilson, C.L. Chitosan as a potential natural compound to control pre and postharvest diseases of horticultural commodities. *Crop Prot.* **2006**, *25*, 108–118. [CrossRef]
5. Hsouna, A.B.; Touj, N.; Hammami, I.; Dridi, K.; Al-Ayed, A.S.; Hamdi, N. Chemical composition and in vivo efficacy of the essential oil of *Mentha piperita* L. in the suppression of crown gall disease on tomato plants. *J. Oleo Sci.* **2019**, *68*, 419–426. [CrossRef]
6. Pedrotti, C.; da Silva Ribeiro, R.T.; Schwambach, J. Control of postharvest fungal rots in grapes through the use of *Baccharis trimera* and *Baccharis dracunculifolia* essential oils. *Crop Prot.* **2019**, *125*, 1–7. [CrossRef]
7. Filho, J.G.; Silva, G.; de Aguiar, A.C.; Cipriano, L.; de Azeredo, H.M.C.; Junior, S.B.; Ferreira, M.D. Chemical composition and antifungal activity of essential oils and their combinations against *Botrytis cinerea* in strawberries. *J. Food Meas. Charact.* **2021**, *15*, 1815–1825. [CrossRef]
8. Tomazoni, E.Z.; Pauletti, G.F.; da Silvaibeiro, R.T.; Moura, S.; Schwambach, J. In vitro and in vivo activity of essential oils extracted from *Eucalyptus staigeriana*, *Eucalyptus globulus* and *Cinnamomum camphora* against *Alternaria solani* Sorauer causing early blight in tomato. *Sci. Hortic.* **2017**, *223*, 72–77. [CrossRef]
9. Gruľová, D.; Caputo, L.; Elshafie, H.S.; Baranová, B.; De Martino, L.; Sedlák, V.; Camele, I.; De Feo, V. Thymol chemotype *Origanum vulgare* L. essential oil as a potential selective bio-based herbicide on monocot plant species. *Molecules* **2020**, *25*, 595. [CrossRef] [PubMed]
10. Raymond, H.; Atkins, S.; Budantsev, A.; Cantino, P.; Conn, B.; Grayer, R. The Families and Genera of Vascular Plants, Lamiales. In *The Families and Genera of Vascular Plants*; Kadereit, J.W., Ed.; Springer: Berlin/Heidelberg, Germany, 2004; Volume VII, pp. 9–38.
11. Jirovetz, L.; Buchbauer, G.; Bail, S.; Denkova, Z.; Slavchev, A.; Stoyanova, A.; Schmidt, E.; Geissler, M. Antimicrobial activities of essential oils of mint and peppermint as well as some of their main compounds. *J. Essen. Oil Res.* **2009**, *21*, 363–366. [CrossRef]
12. *The Wealth of India: A Dictionary of Indian Raw Materials and Industrial Products*; Raw Material Series; CSIR: New Delhi, India, 1962; Volume VI, pp. 342–343.

13. Alammar, N.; Wang, L.; Saberi, B.; Nanavati, J.; Holtmann, G.; Shinohara, R.T.; Mullin, G.E. The impact of peppermint oil on the irritable bowel syndrome: A meta-analysis of the pooled clinical data. *BMC Compl. Altern. Med.* **2019**, *19*, 21. [CrossRef]

14. Briggs, C. Peppermint: Medicinal herb and flavouring agent. *Can. Pharm. J.* **1993**, *126*, 89–92.

15. Beigi, M.; Torki-Harchegani, M.; Pirbalouti, A.G. Quantity and chemical composition of essential oil of peppermint (*Mentha × piperita* L.) leaves under different drying methods. *Int. J. Food Prop.* **2018**, *21*, 267–276. [CrossRef]

16. Singh, P.; Pandey, A.K. Prospective of essential oils of the genus mentha as biopesticides: A review. *Front. Plant Sci.* **2018**, *9*, 1295. [CrossRef]

17. Figueiredo, A.C.; Barroso, J.G.; Pedro, L.G.; Scheffer, J.C. Factors affecting secondary metabolite production in plants: Volatile components and essential oils. *Flavour Fragr. J.* **2008**, *23*, 213–226. [CrossRef]

18. Kamatou, G.P.P.; Vermaak, I.; Viljoen, A.M.; Lawrence, B.M. Menthol: A simple monoterpene with remarkable biological properties. *Phytochemistry* **2013**, *96*, 15–25. [CrossRef]

19. Mishra, R.C.; Krumar, J. Evaluation of Mentha piperita L. oil as a fumigant against red flour beetle, *Tribolium castaneum* (Herbst). *Indian Perfum.* **1983**, *27*, 73–76.

20. Fejér, J.; Gruľová, D.; De Feo, V.; Ürgeová, E.; Obert, B.; Preťová, A. *Mentha × piperita* L. nodal segments cultures and their essential oil production. *Indust. Crops Prod.* **2018**, *112*, 550–555. [CrossRef]

21. Elshafie, H.S.; Caputo, L.; De Martino, L.; Gruľová, D.; Zheljazkov, V.D.; De Feo, V.; Camele, I. Biological investigations of essential oils extracted from three *Juniperus* species and evaluation of their antimicrobial, antioxidant and cytotoxic activities. *J. Appl. Microbiol.* **2020**, *129*, 1261–1271. [CrossRef] [PubMed]

22. Elshafie, H.S.; Gruľová, D.; Baranová, B.; Caputo, L.; De Martino, L.; Sedlák, V.; Camele, I.; De Feo, V. Antimicrobial activity and chemical composition of essential oil extracted from *Solidago canadensis* L. growing wild in Slovakia. *Molecules* **2019**, *24*, 1206. [CrossRef]

23. Arikan, S. Current status of antifungal susceptibility testing methods. *Med. Mycol.* **2007**, *45*, 569–587. [CrossRef]

24. Jozef, F.; Gruľová, D.; De Feo, V. Biomass production and essential oil in a new bred cultivar of peppermint (*Mentha × piperita* L.). *Indus. Crop. Prod.* **2017**, *109*, 812–817.

25. Grulová, D.; De Martino, L.; Mancini, E.; Salamon, I.; De Feo, V. Seasonal variability of the main components in essential oil of *Mentha × piperita* L. *J. Sci. Food Agric.* **2015**, *95*, 621–627. [CrossRef]

26. Fejér, J.; Gruľová, D.; Eliašová, A.; Kron, I.; De Feo, V. Influence of environmental factors on content and composition of essential oil from common juniper ripe berry cones (*Juniperus communis* L.). *Plant Biosys.* **2018**, *152*, 1227–1235. [CrossRef]

27. Singh, R.; Shushni, M.A.; Belkheir, A. Antibacterial and antioxidant activities of *Mentha piperita* L. *Arab. J. Chem.* **2015**, *8*, 322–328. [CrossRef]

28. Clark, R.K.; Menory, R.C. Environmental effects or peppermint (*Mentha piperita*). *Aust. J. Plant Physiol.* **1980**, *7*, 685–692.

29. Nagarjuna, R.D.; Jabbar, A.A.; Mukul, S.; Mary, M.; Ramachandra, R.G.; Mohammed, A. Chemical constituents, in vitro antibacterial and antifungal activity of *Mentha Piperita* L. (peppermint) essential oils. *J. King Saud Uni. Sci.* **2019**, *31*, 528–533.

30. Burt, S. Essential oils: Their antibacterial properties and potential applications in foods—A review. *Int. J. Food Microbiol.* **2004**, *94*, 223–253. [CrossRef]

31. Nerio, L.S.; Olivero-Verbel, J.; Stashenko, E. Repellent activity of essential oils: A review. *Bioresour. Technol.* **2010**, *101*, 372–378. [CrossRef] [PubMed]

32. Tsao, R.; Zhou, T. Antifungal activity of monoterpenoids against postharvest pathogens *Botrytis cinerea* and *Monilinia fructicola*. *J. Essen. Oil Res.* **2000**, *12*, 113–121. [CrossRef]

33. Dambolena, J.S.; Lopez, A.G.; Rubinstein, H.R.; Zygadlo, J.A. Effects of menthol stereoisomers on the growth, sporulation and fumonisin B1 production of *Fusarium verticillioides*. *Food Chem.* **2010**, *123*, 165–170. [CrossRef]

34. Tyagi, A.K.; Gottardi, D.; Malik, A.; Guerzoni, M.E. Anti-yeast activity of mentha oil and vapours through in vitro and in vivo (real fruit juices) assays. *Food Chem.* **2013**, *137*, 108–114. [CrossRef]

35. Afridi, M.S.; Ali, J.; Abbas, S.; Rehman, S.U.; Khan, F.A.; Khan, M.A.; Shahid, M. Essential oil composition of *Mentha piperita* L. and its antimicrobial effects against common human pathogenic bacterial and fungal strains. *Pharmacol. Online* **2016**, *3*, 90–97.

36. Elshafie, H.S.; Ghanney, N.; Mang, S.M.; Ferchichi, A.; Camele, I. An in vitro attempt for controlling severe phyto and human pathogens using essential oils from Mediterranean plants of genus Schinus. *J. Med. Food* **2016**, *19*, 266–273. [CrossRef] [PubMed]

37. Elshafie, H.S.; Camele, I. An overview of the biological effects of some Mediterranean essential oils on human health. *BioMed Res. Int.* **2017**, *2017*, 1–14. [CrossRef]

38. Ultee, A.; Bennik, M.; Moezelaar, R. The phenolic hydroxyl group of carvacrol is essential for action against the food-borne pathogen *Bacillus cereus*. *Appl. Environ. Microbiol.* **2002**, *68*, 1561–1568. [CrossRef] [PubMed]

39. Adams, R.P. *Identification of Essential Oil Components by Gas Chromatography/Mass Spectroscopy*, 4th ed.; Allured Publishing Corporation: Carol Stream, IL, USA, 2007.

40. Bhunia, A.; Johnson, M.C.; Ray, B. Purification, characterization and antimicrobial spectrum of a bacteriocin produced by *Pediococcus acidilactici*. *J. Appl. Bacteriol.* **1988**, *65*, 261–268. [CrossRef]

41. Elshafie, H.S.; Sakr, S.H.; Sadeek, S.A.; Camele, I. Biological investigations and spectroscopic studies of new Moxifloxacin/Glycine-Metal complexes. *Chem. Biodiver.* **2019**, *16*, e1800633. [CrossRef]

42. King, E.O.; Ward, M.K.; Raney, D.E. Two simple media for the demonstration of pyocyanin and fluorescin. *J. Lab. Clin. Med.* **1954**, *44*, 301–307.

43. Sofo, A.; Elshafie, H.S.; Scopa, A.; Mang, S.M.; Camele, I. Impact of airborne zinc pollution on the antimicrobial activity of olive oil and the microbial metabolic profiles of Zn-contaminated soils in an Italian olive orchard. *J. Trace Elem. Med. Biol.* **2018**, *49*, 276–284. [CrossRef]
44. Elshafie, H.S.; Camele, I.; Sofo, A.; Mazzone, G.; Caivano, M.; Masi, S.; Caniani, D. Mycoremediation effect of *Trichoderma harzianum* strain T22 combined with ozonation in diesel-contaminated sand. *Chemosphere* **2020**, *252*, 126597. [CrossRef]
45. Elshafie, H.S.; Mancini, E.; Sakr, S.; De Martino, L.; Mattia, C.A.; De Feo, V.; Camele, I. In vivo antifungal activity of two essential oils from Mediterranean plants against postharvest brown rot disease of peach fruit. *J. Med. Food* **2015**, *18*, 929–934. [CrossRef] [PubMed]
46. Elshafie, H.S.; Sakr, S.; Mang, S.M.; De Feo, V.; Camele, I. Antimicrobial activity and chemical composition of three essential oils extracted from Mediterranean aromatic plants. *J. Med. Food* **2016**, *19*, 1096–1103. [CrossRef] [PubMed]
47. Zygadlo, J.A.; Guzmán, C.A.; Grosso, N.R. Antifungal properties of the leaf oils of *Tagetes minuta* L. and *T. filifolia* Lag. *J. Essent. Oil Res.* **1994**, *6*, 617–621. [CrossRef]
48. Sakr, S.H.; Elshafie, H.S.; Camele, I.; Sadeek, S.A. Synthesis, spectroscopic, and biological studies of mixed ligand complexes of gemifloxacin and glycine with Zn(II), Sn(II), and Ce(III). *Molecules* **2018**, *23*, 1182. [CrossRef]

Article

Chemical Profiles, Anticancer, and Anti-Aging Activities of Essential Oils of *Pluchea dioscoridis* (L.) DC. and *Erigeron bonariensis* L.

Abdelbaset M. Elgamal [1,*], **Rania F. Ahmed** [2], **Ahmed M. Abd-ElGawad** [3], **Abd El-Nasser G. El Gendy** [4], **Abdelsamed I. Elshamy** [2,*] **and Mahmoud I. Nassar** [2]

[1] Department of Chemistry of Microbial and Natural Products, National Research Centre, 33 El-Bohouth St., Dokki, Giza 12622, Egypt
[2] Chemistry of Natural Compounds Department, National Research Center, 33 El Bohouth St., Dokki, Giza 12622, Egypt; rf.ali@nrc.sci.eg (R.F.A.); mnassar_eg@yahoo.com (M.I.N.)
[3] Department of Botany, Faculty of Science, Mansoura University, Mansoura 35516, Egypt; aibrahim2@ksu.edu.sa
[4] Medicinal and Aromatic Plants Research Department, National Research Centre, 33 El Bohouth St., Dokki, Giza 12622, Egypt; ag.el-gendy@nrc.sci.eg
* Correspondence: algamalgene@yahoo.com (A.M.E.); elshamy@yahoo.com (A.I.E.); Tel.: +20-100-155-8689 (A.M.E.); +20-100-552-5108 (A.I.E.)

Citation: Elgamal, A.M.; Ahmed, R.F.; Abd-ElGawad, A.M.; El Gendy, A.E.-N.G.; Elshamy, A.I.; Nassar, M.I. Chemical Profiles, Anticancer, and Anti-Aging Activities of Essential Oils of *Pluchea dioscoridis* (L.) DC. and *Erigeron bonariensis* L. *Plants* **2021**, *10*, 667. https://doi.org/10.3390/plants10040667

Academic Editors: Daniela Rigano and Hazem Salaheldin Elshafie

Received: 3 March 2021
Accepted: 24 March 2021
Published: 31 March 2021

Publisher's Note: MDPI stays neutral with regard to jurisdictional claims in published maps and institutional affiliations.

Abstract: Plants belonging to the Asteraceae family are widely used as traditional medicinal herbs around the world for the treatment of numerous diseases. In this work, the chemical profiles of essential oils (EOs) of the above-ground parts of *Pluchea dioscoridis* (L.) DC. and *Erigeron bonariensis* (L.) were studied in addition to their cytotoxic and anti-aging activities. The extracted EOs from the two plants via hydrodistillation were analyzed by gas chromatography-mass spectroscopy (GC-MS). GC-MS of EO of *P. dioscoridis* revealed the identification of 29 compounds representing 96.91% of the total oil. While 35 compounds were characterized from EO of *E. bonariensis* representing 98.21%. The terpenoids were found the main constituents of both plants with a relative concentration of 93.59% and 97.66%, respectively, including mainly sesquiterpenes (93.40% and 81.06%). α-Maaliene (18.84%), berkheyaradulen (13.99%), dehydro-cyclolongifolene oxide (10.35%), aromadendrene oxide-2 (8.81%), β-muurolene (8.09%), and α-eudesmol (6.79%), represented the preponderance compounds of EO of *P. dioscoridis*. While, trans-α-farnesene (25.03%), O-ocimene (12.58%), isolongifolene-5-ol (5.53%), α-maaliene (6.64%), berkheyaradulen (4.82%), and α-muurolene (3.99%), represented the major compounds EO of *E. bonariensis*. A comparative study of our results with the previously described data was constructed based upon principal component analysis (PCA) and agglomerative hierarchical clustering (AHC), where the results revealed a substantial variation of the present studied species than other reported ecospecies. EO of *P. dioscoridis* exhibited significant cytotoxicity against the two cancer cells, MCF-7 and A-549 with IC_{50} of 37.3 and 22.3 μM, respectively. While the EO of the *E. bonariensis* showed strong cytotoxicity against HepG2 with IC_{50} of 25.6 μM. The EOs of *P. dioscoridis*, *E. bonariensis*, and their mixture (1:1) exhibited significant inhibitory activity of the collagenase, elastase, hyaluronidase, and tyrosinase comparing with epigallocatechin gallate (EGCG) as a reference. The results of anti-aging showed that the activity of mixture (1:1) > *P. dioscoridis* > *E. bonariensis* against the four enzymes.

Keywords: horseweed; wavy-leaf fleabane; sesquiterpenes; cytotoxicity; anti-senility

1. Introduction

Natural products derived from the plant kingdom represented potent resources for foods, cosmetics, and traditional medicines [1,2]. Many scientists focused on the study of the chemical characterization of essential oils (EOs) along with their pharmaceutical effects for many decades [3–6]. Due to the complicated composition from different iso-prenoids based compounds [7], EOs exhibited several significant biological effects including anti-inflammatory, antipyretic [8], antioxidant [9,10], allelopathy [8,11–17], antiulcer [18], antimicrobial [8,19], and hepatoprotective [20]. EOs have been reported as potent agents against degenerative diseases via inhibition of oxidative stress due to the strong free radical scavenging activity [21].

Plants belonging to *Conyza* genus (Family Asteraceae), including around 150 plant species [22], were described as important traditional medicinal plants in the treatment of toothache, skin diseases, rheumatism, haemorrhoidal, diarrhoeal, and injuries bleeding [23,24]. Some members of Asteraceae were named *Conyza* formerly; however, some taxa names have been changed later based on taxonomic criteria. From these taxa, *Conyza dioscoridis* (L.) Desf. that its accepted name now is *Pluchea dioscoridis* (L.) DC., and *Conyza linifolia* (Willd.) Täckh. that now have been accepted as *Erigeron bonariensis* L. [25,26]. Studying various criteria of the plant species such as morphological, anatomical, molecular, and chemical properties, providing valuable information for taxonomists, thereby, some taxa names have been changed [27,28]. EOs analysis has been reported to provide profitable information for chemotaxonomy [27,29].

Pluchea dioscoridis (L.) DC. (syn. *Conyza dioscoridis* (L.) Desf.) is a widely distributed wild plant in the Nile delta, Mediterranean coast, Sinai Peninsula, Western Desert, and Eastern Desert [25]. This plant was described in folk medicines for the treatment of some diseases as ulcer, colic, carminative, epilepsy in children, rheumatic pains, and cold [30]. Many documents described that the different extracts of this plant have several potent biological activities comprising anti-inflammation, antiulcer, antidiabetic, antinociceptive, antipyretic, ant-diarrheal, antibacterial, antifungal, and free radical scavenging activities, along with diuretic effect [30–33]. Many metabolites were isolated and characterized from *P. dioscoridis* including steroids, triterpenes [30], flavonoids, and phenolic acids [30,32].

The chemical constituents of EO of *E. bonariensis* collected from Alexandria, Egypt has been reported in addition to its antimicrobial and insecticidal activities [34]. In this study, 25 compounds were identified from EO of *E. bonariensis* including sesqui- and monoter-penes. From the total of this oil, α-bergamotene, and D-limonene represented the mains with concentrations of 27.4 and 22.5%. In the same report, EO of *C. linifolia* was documented to exhibit antibacterial potentiality against *B. subtilis* with MIC of 125 mg/mL [34,35]. Little reports concerning the chemical profiles as well as biological activities of *Erigeron bonariensis* L. (syn. *Conyza linifolia* (Willd.) Täckh.) have been recorded.

We hypothesized that these two plant species were formerly named *Conyza*, and their names were changed. Therefore, the chemical characterization of their EOs could be valu-able in their chemotaxonomy. Herein, this work aimed to (i) identify the chemical profiles of EOs from *P. dioscoridis* and *E. bonariensis*, collected from Egypt, (ii) establish comparative profiles of the two plants based upon chemometric analysis with other reported ecospecies, (iii) study the cytotoxic activity of the EOs of the two plants against several human cancer cell lines, and (iv) assess in vitro anti-aging potentialities of the EOs of the two plants.

2. Results and Discussion

2.1. Chemical Compositions of EO of P. dioscoridis

The chemical characterization of *P. dioscoridis* EO was extracted via hydro-distillation afforded golden yellow (0.037%). The chemical profiles of the extracted EO were assigned depending upon the GC-MS analysis. The GC-MS chromatogram of the EO is presented in Figure 1 exhibiting the main peaks from all identified components. Twenty-nine com-pounds were identified from the EO of *P. dioscoridis* represented 96.91% of the total oil. All the identified compounds along with their chemical and physical properties were summarized in Table 1.

Table 1. Components of essential oils of *Pluchea dioscoridis* and *Erigeron bonariensis*.

No	Rt [a]	Compound Name	MF	KI_Lit [b]	KI_Exp [c]	Relative Concentration %	
						P. discodirdis	*E. bonariensis*
		Monoterpenes				**0.19%**	**14.16%**
1	4.13	α-Pinene	$C_{10}H_{16}$	933	934	0.19 ± 0.01	0.18 ± 0.01
2	5.43	α-Myrcene	$C_{10}H_{16}$	991	990	——	0.20 ± 0.01
3	6.41	O-Ocimene	$C_{10}H_{16}$	1012	1012	——	12.58 ± 0.09
4	6.91	D-Limonene	$C_{10}H_{16}$	1035	1036	——	1.20 ± 0.04
		Sesquiterpenes				**93.4%**	**81.06%**
5	16.22	β-Caryophyllene	$C_{15}H_{24}$	1418	1420	4.95 ± 0.07	2.17 ± 0.04
6	16.96	Aromandendrene	$C_{15}H_{24}$	1439	1448	——	0.12 ± 0.03
7	17.17	α-Guaiene	$C_{15}H_{24}$	1439	1439	0.21 ± 0.02	0.11 ± 0.02
8	17.39	α-Maaliene	$C_{15}H_{24}$	1480	1479	18.84 ± 0.08	6.64 ± 0.09
9	17.60	Berkheyaradulen	$C_{15}H_{24}$	1492	1493	13.99 ± 0.09	4.82 ± 0.06
10	18.36	β-Muurolene	$C_{15}H_{24}$	1493	1493	8.09 ± 0.05	2.37 ± 0.04
11	18.57	α-Muurolene	$C_{15}H_{24}$	1498	1499	2.20 ± 0.04	3.99 ± 0.07
12	18.98	Bicyclogermacrene	$C_{15}H_{24}$	1500	1501	——	0.66 ± 0.01
13	19.67	*trans*-α-Farnesene	$C_{15}H_{24}$	1508	1507	——	25.03 ± 0.13
14	19.73	α-Bisabolene	$C_{15}H_{24}$	1509	1511	0.58 ± 0.03	——
15	19.84	γ-Cadinene	$C_{15}H_{24}$	1514	1515	1.36 ± 0.04	——
16	20.35	α-Sesquiphellandrene	$C_{15}H_{24}$	1516	1517	0.44 ± 0.01	——
17	20.45	*cis*-Lanceol	$C_{15}H_{24}O$	1525	1527	0.16 ± 0.02	0.09 ± 0.01
18	20.54	Isolongifolene-5-ol	$C_{15}H_{24}O$	1534	1535	0.19 ± 0.01	5.53 ± 0.07
19	20.64	Germacrene D-4-ol	$C_{15}H_{24}$	1574	1576	——	2.35 ± 0.04
20	20.81	Spathulenol	$C_{15}H_{24}O$	1576	1577	0.81 ± 0.03	0.10 ± 0.01
21	20.98	Isoaromadendrene epoxide	$C_{15}H_{24}O$	1580	1579	——	1.50 ± 0.03
22	21.41	Calarenepoxide	$C_{15}H_{24}O$	1592	1592	——	1.07 ± 0.02
23	21.56	Caryophyllene oxide	$C_{15}H_{24}O$	1594	1593	0.86 ± 0.02	0.08 ± 0.01
24	21.66	Salvial-4(14)-en-1-one	$C_{15}H_{24}O$	1595	1595	2.20 ± 0.04	0.38 ± 0.01
25	21.93	Ledene alcohol	$C_{15}H_{24}O$	1729	1731	——	0.97 ± 0.03
26	22.1	Carotol	$C_{15}H_{26}O$	1597	1598	0.78 ± 0.02	——
27	22.42	Humuladienone	$C_{15}H_{24}O$	1607	1605	——	0.32 ± 0.01
28	23.09	Neoclovenoxid	$C_{15}H_{24}O$	1608	1610	0.90 ± 0.02	0.51 ± 0.03
29	23.33	Cubenol	$C_{15}H_{26}O$	1642	1642	——	0.74 ± 0.02
30	23.49	Farnesol	$C_{15}H_{26}O$	1722	1720	1.35 ± 0.05	——
31	23.55	Ledene oxide-(i)	$C_{15}H_{24}O$	1668	1667	——	10.93 ± 0.10
32	23.65	Dendrolasin	$C_{15}H_{22}O$	1574	1575	2.85 ± 0.06	8.37 ± 0.09
33	23.81	Torreyol	$C_{15}H_{26}O$	1645	1644	——	0.55 ± 0.02
34	25	Isospathulenol	$C_{15}H_{24}O$	1625	1627	1.90 ± 0.05	——
35	25.22	tau-Muurolol	$C_{15}H_{26}O$	1646	1646	3.88 ± 0.08	0.51 ± 0.03
36	25.32	Aromadendrene oxide-2	$C_{15}H_{24}O$	1650	1649	8.81 ± 0.11	0.38 ± 0.02
37	25.85	α-Eudesmol	$C_{15}H_{26}O$	1652	1653	6.79 ± 0.08	0.77 ± 0.01
38	26.69	γ-Cadinol	$C_{15}H_{26}O$	1654	1655	0.91 ± 0.04	——
39	26.92	Dehydro-cyclolongifolene oxide	$C_{15}H_{24}O$	1657	1658	10.35 ± 0.12	——
		Diterpenes				——	**2.44%**
40	30.47	Neophytadiene	$C_{20}H_{38}$	1840	1840	——	2.44
		Carotenoids derived compounds				**0.28%**	——
41	27.36	α-Ionone	$C_{13}H_{20}O$	1426	1426	0.28 ± 0.01	——
		Others				**1.33%**	**0.55%**
42	18.66	1-Butanone, 1-(2,3,4,5-tetramethylphenyl)-	$C_{14}H_{20}O$	1660	1661	0.46 ± 0.03	0.32 ± 0.02
43	26.32	Methyl 2,5-octadecadiynoate	$C_{19}H_{30}O_2$	1980	1980	0.64 ± 0.03	0.13 ± 0.01
44	42.28	*n*-Pentacosane	$C_{25}H_{52}$	2500	2500	——	0.10 ± 0.01
45	45.48	*n*-Heptacosane	$C_{27}H_{56}$	2700	2700	0.23±0.01	——
		Total				**96.91%**	**98.21%**

[a] Rt: retention time, [b] Literature Kovats retention index on DB-5 column with reference to *n*-alkanes [36], [c] experimental Kovats retention index; values of each compound are average ± SD from duplicates. The identification of essential oil (EO) components was performed based on the (a) mass spectral data of compounds (MS) and (b) Kovats indices with those of Wiley spectral library collection and NIST (National Institute of Standards and Technology) library database.

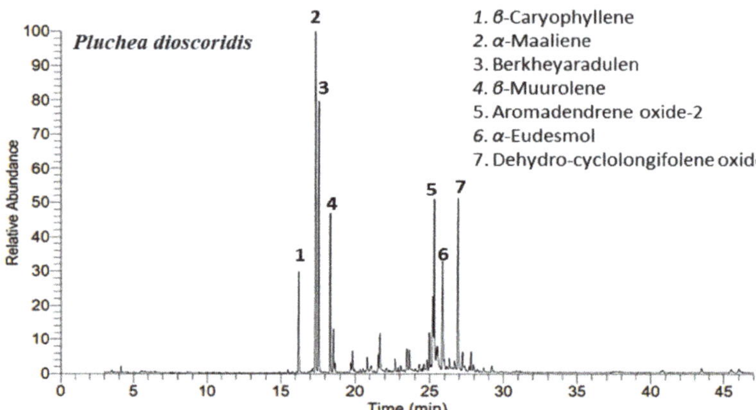

Figure 1. Gas chromatography-mass spectroscopy (GC-MS) chromatograme of the essential oils (EO) of *Pluchea dioscoridis*. The main peaks were numbered (1–7).

The constituents of EO of *P. dioscoridis* were characterized by the presence of four classes of compounds including sesqui- (93.40%), and monoterpenes (0.19%), carotenoid derived compounds (0.28%) in addition to other acyclic compounds (1.33%). The terpenoids were found as abundant compounds with a relative concentration of 93.59% in addition to traces of carotenoids and acyclic compounds with a complete absence of diterpenoids. GC-MS analysis of EO derived from *E. bonariensis*, revealed the presence of four categories of compounds comprising sesqui- (81.06%), and monoterpenes (14.16%), diterpenes (2.44%) in addition to other acyclic compounds (0.55%). Furthermore, the terpenoids were characterized as the main components by a relative concentration of 95.22% with traces of diterpenoids and other compounds. These results deduced the fact of the preponderance of the terpenoids in the different species of *Conyza* genus [32,34,37].

Sesquiterpenoids were found as the main compounds of the EO of *P. dioscoridis* with mixtures of oxygenated and non-oxygenated compounds. The abundance of sesquiterpenes was found in full agreement with previous data of EOs of this plant [32,38]. From all identified sesquiterpenes, α-maaliene (18.84%), berkheyaradulen (13.99%), dehydrocyclolongifolene oxide (10.35%), aromadendrene oxide-2 (8.81%), β-muurolene (8.09%), α-eudesmol (6.79%), β-caryophyllene (4.95%), t-muurolol (3.88%), represented the major compounds.

Berkheyaradulen, muurolene, eudesmol, tau-muurolol, and caryophyllene, were found as marker compounds for this plant in the previous study [32] and this data is in the same line with our results. While the reported data of EO of the leaves of this plant [38] exhibited variations in chemical constituents than those data described previously by our team [32] and also than our results herein. Elshamy, et al. [32] documented that α-cadinol is the main sesquiterpene and this data is different than our results in which γ-cadinol is present as a minor compound. Additionally, eudesmol and tau-muurolol were reported as major sesquiterpenes in EO of the leaves of this plant, and this data agreed with our results.

The results of GC-MS of EO of *P. dioscoridis* revealed that the monoterpenes are traces with only one compound, α-pinene (0.19%). The scarcity of monoterpenes is consistent with the results of Elshamy, et al. [32] and El-Seedi, et al. [38].

In EO derived from *P. dioscoridis*, diterpenes were completely absent and this result is inconsistent with the published data [32,37], while El-Seedi, et al. [38] characterized only one diterpene, phytol, from the leaves of this plant. α-ionone was the only identified carotenoid-derived compound from EO of *P. dioscoridis* that was not reported before from this plant [32].

The other compounds (1.33%) including hydrocarbons were characterized as traces in EO of *P. dioscoridis* that was in agreement with the previous data [32,39]. In contrast,

El-Seedi, et al. [38] documented that the monoterpenoid compounds represented a high concentration (26.6%) of the total mass of EO of the leaves of *P. dioscoridis*.

2.2. Chemical Compositions of EO of E. bonariensis

The hydro-distillation of the above-ground parts of *E. bonariensis* afforded golden yellow EO (0.049%). The chemical characterization of the extracted EO was performed based on the GC-MS analysis. Figure 2 represented the GC-MS chromatogram including the major peaks. Thirty-five components were assigned representing 98.21% of the total oil mass. The characterized constituents as well as retentions times (RIs), molecular formulas (MFs), and literature and calculated Kovats indexes (KIs) were compacted in Table 1.

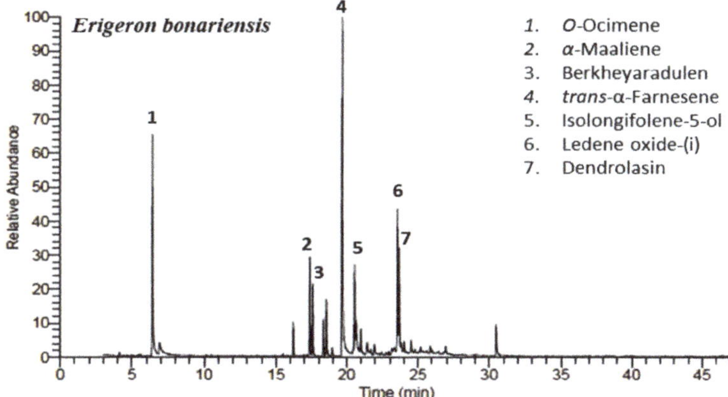

Figure 2. GC-MS chromatogrames of the EO of *Erigeron bonariensis*. The main peaks were numbered (1–7).

In EO of *E. bonariensis*, sesquiterpenes represented also the main constituents including several oxygenated and non-oxygenated metabolites. With a relative concentration of 81.06% of sesquiterpenes, our results are completely agreed with the previous data by Harraz, et al. [34] that reported a relative concentration of 92.50%; trans-α-Farnesene (25.03%), isolongifolene-5-ol (5.53%), α-maaliene (6.64%), berkheyaradulen (4.82%), and α-muurolene (3.99%) were found the main sesquiterpenoid contents. The main sesquiterpene, trans-α-farnesene, was widely distributed in the EOs of *Conyza* species such as *C. bonariensis* (≈*E. bonariensis*) collected from Venezuela and Vietnam [40], *C. canadensis* [39], and *C. sumatrensis* [41]. However, in the only stated study of EO of *E. bonariensis* [34], α-bergamotene was described as the main compound in addition to some farnesene derivatives such as, β-farnesene, and (*E*)-farnesene epoxide. The abundance of α-maaliene (6.64%), berkheyaradulen (4.82%), and α-muurolene (3.99%) were found in perfect harmony with our results of EO of *P. dioscoridis*. The variations of secondary metabolites comprising EOs might be attributed to the plant age and development, plant organs, as well as the environmental factors including such as altitude, seasonality, atmospheric composition and temperature, and water availability [11,42,43].

Monoterpenes represented a remarkable concentration of the EO of *E. bonariensis* with a wealth of *O*-ocimene (12.58%). *O*-Ocimene was reported here for the first time in EO of this plant, contrariwise, Harraz, et al. [34] reported the complete absence of it from EO of the aerial parts of this plant collected from Alexandria, Egypt.

The diterpenoids were represented by a relative concentration of 2.44% from over all mass of the oil of *E. bonariensis*. The total relative concentration of diterpenes was determined in the EO of *E. bonariensis* with only one compound, neophytadiene, which is not reported before from the EO of this plant [34].

Carotenoid derived compounds were not identified from the EO of *E. bonariensis* and this result is in harmony with Mabrouk, et al. [37]; also, hydrocarbons and the other components were represented by traces in EO of *E. bonariensis* (0.55%) that agreed the previous described studies [32,39].

2.3. Chemometric Analysis

The EOs chemical compositions of the major compounds (>3%), reported from different ecospecies of *P. dioscoridis* and *E. bonariensis* were constructed in a matrix. These collected data were subjected to agglomerative hierarchical clustering (AHC) and principal component analysis (PCA). The cluster analysis of *P. dioscoridis* EOs showed that the present studied sample of *P. dioscoridis* is closely correlated to the Egyptian ecospecies collected from El-Sadat City, and little correlated to that collected from Cairo–Suez desert road, Egypt (Figure 3a). However, the present sample was different than those purchased from a commercial source in Cairo, Egypt. This means that the commercial samples are not in pure form or may be mixed with other plants.

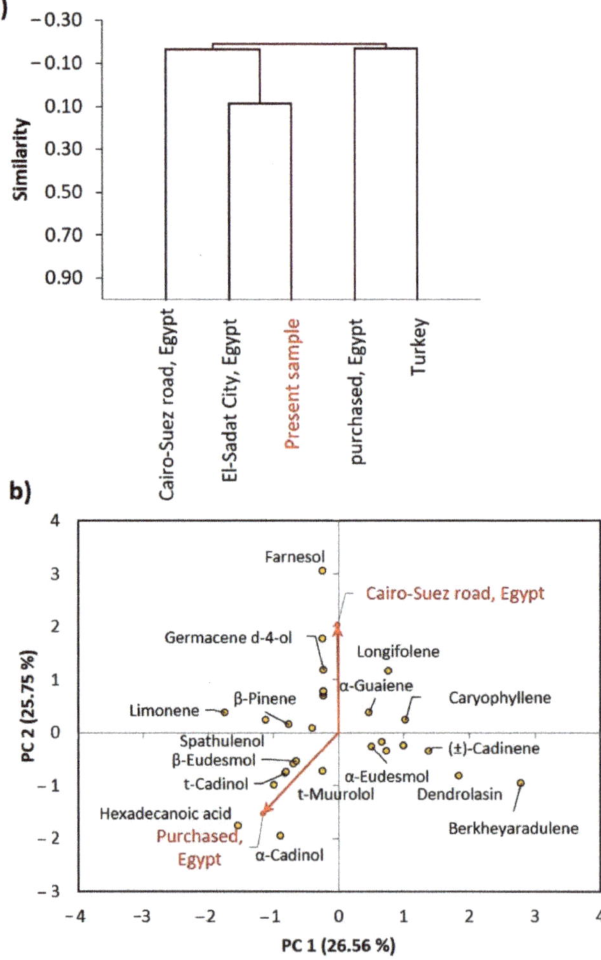

Figure 3. Chemometric analysis of the EOs from the present studied *Pluchea dioscoridis* ecospecies and other reported ecospecies. (**a**) agglomerative hierarchical clustering (AHC) and (**b**) principal component analysis (PCA).

The PCA of the *P. dioscoridis* ecospecies showed that the sample collected from Cairo-Suez desert road, Egypt is mainly characterized by farnesol, germacene d-4-ol, and longifolene (Figure 3b). However, the purchased sample from a commercial source in Cairo, Egypt is characterized by hexadecanoic acid and α-cadinol.

On the other side, the cluster analysis of *E. bonariensis* EOs revealed that the present Egyptian sample is closely related to the Venezuelan ecospecies, while it was different than other ecospecies (Figure 4a).

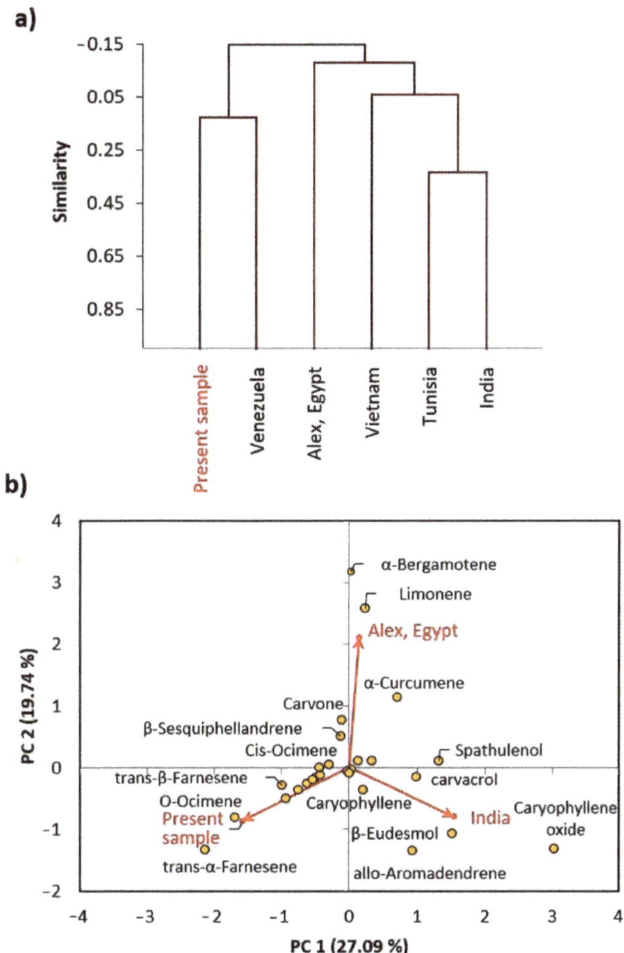

Figure 4. Chemometric analysis of the EOs from the present studied *Erigeron bonariensis* ecospecies and other reported ecospecies. (**a**) agglomerative hierarchical clustering (AHC) and (**b**) principal component analysis (PCA).

While the Indian and Tunisian ecospecies showed a close relation in the composition of the EO. The PCA showed that the present sample of *E. bonariensis* is characterized by *trans*-α-Farnesene, O-ocimene, and *trans*-β-Farnesene (Figure 4b). The sample collected from Alexandria, Egypt, showed a close correlation with α-bergamotene, limonene, and α-curcumene, while the Indian ecospecies is characterized by β-eudesmol, caryophyllene oxide, allo-aromadendrene, and carvacrol.

The observed variation among the present samples and other reported ones revealed the profitable information derived from the EOs analysis, which could be a useful tool in chemotaxonomy [27].

2.4. Anti-Aging Activity

The EOs from *P. dioscoridis*, *E. bonariensis*, and the mixture of the two EOs (1:1) have a strong inhibitory activity of the collagenase, elastase, hyaluronidase, and tyrosinase (Figure 5). All the EO treatments exhibited potent inhibition of collagenase enzyme with IC_{50} of 1.85, 2.90, and 1.73 µg/mL for *P. dioscoridis*, *E. bonariensis*, and the mixture, respectively. Furthermore, the three EO treatments strongly inhibit the elastase enzyme with respective values of IC_{50} of 14.63, 16.52, and 11.01 µg/mL. Furthermore, strong suppression of hyaluronidase was demonstrated via the three EO treatments based upon the respective observed values of IC_{50} of 17.18, 15.16, and 13.54 µg/mL. By the same, the three tested displayed strong tyrosinase enzyme inhibition with IC_{50} values at 19.52, 18.93, and 15.81 µg/mL, respectively. All the results were constructed based upon comparing with the polyphenolic compound, epigallocatechin gallate (EGCG), as a standard ant-aging reference [44] that exhibit inhibition of collagenase, elastase, hyaluronidase, and tyrosinase with IC_{50} of 1.56, 10.29, 12.71, and 14.37 µg/mL.

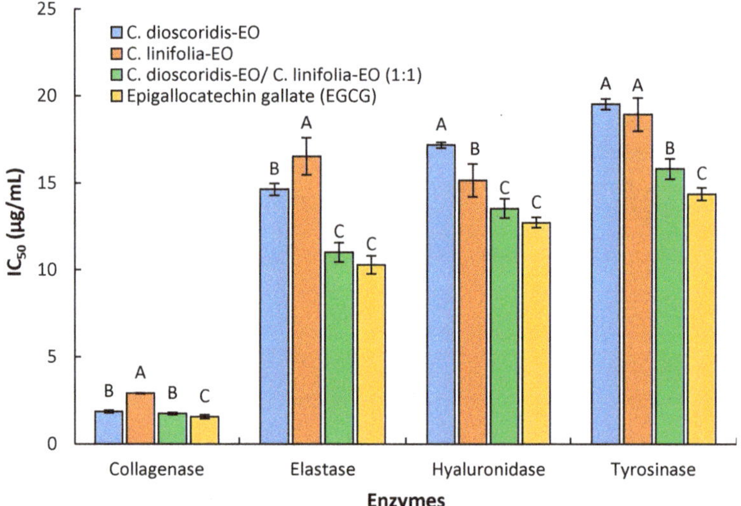

Figure 5. Anti-aging activities of the EOs extracted from *Pluchea dioscoridis* and *Erigeron bonariensis* against the four enzymes: collagenase, elastase, hyaluronidase, and tyrosinase. Values are IC50 (µg/mL) as an average of three replicates and the bars representing the standard deviation. Different letters (A, B, and C) within each enzyme mean values significant at 0.05 probability level after Duncan's test.

In the matrix of extracellular, the elastin and hyaluronan degradation were principally correlated with the two respective proteolytic enzymes, elastase, and hyaluronidase that cause the main reasons for aging of the skin such as wrinkles, sagging. Moreover, tyrosinase caused the regulation of the synthesis of melanin in human melanocytes that lead to skin ailments.

In the present study, the anti-aging of the 1:1 mixture of the two EOs was evaluated to study the synergetic effects of the combination of the two EOs. Results revealed that the EO of *P. dioscoridis*, *E. bonariensis*, and the mixture of the two EOs (1:1) have strong anti-aging activity. These results might be attributed to the chemical components of these oils. The anti-aging activity was directly correlated with antioxidant potentiality [45]. The main constituents in both EOs, sesquiterpenes, were described to play a significant role as antioxidants, anti-inflammatory agents, and thus anti-aging [45]. Tu and Tawata [45] reported that EO of the leaves of *Alpinia zerumbet* exhibit antioxidant and anti-aging activities due to the high concentration of terpenoids, especially sesquiterpenes. Furthermore, the monoterpenes were documented as active anti-aging agents in EO of *Juniperus communis* [46] and *Origanum vulgare* [47]. These reports concluded that the increase of free radical scavenging constituents in EOs lead to an increase in their anti-aging activity. Based upon this fact, the high concentrations of terpenes especially the oxygenated sesqui- and monoterpenes caused increasing in the anti-aging activity of EOs of these two plants. All these reported data deduced the role of synergetic effects between the components of the EOs. This fact of the role of synergetic effect was very clear in our results in which the mixture of the two EOs (1:1) exhibited better activity than the individual EO of each plant. In the mixture of the two EOs, the raising of concentration of oxygenated terpenes as well the synergetic effects between the components caused increasing of the inhibition potentiality.

2.5. Cytotoxic Activity of EOs of P. dioscoridis and E. bonariensis

The cytotoxicity of EOs of the above-ground parts of the two plants, *P. dioscoridis* and *E. bonariensis*, as well as a mixture of the two EOs (1:1) against the three cancer cell lines, breast adenocarcinoma cells (MCF-7), lung cancer cells (A-549), and hepatocellular carcinoma cells (HepG2) are shown in Figure 6. The results exhibited that the EO of *P. dioscoridis* have a significant inhibition of the two cancer cells, MCF-7 and A-549, with IC_{50} of 37.3 and 22.3 µM, respectively (Figure 6A,B), without any activity against HepG2. While, the EO of the *E. bonariensis* showed inhibitory potentiality only against HepG2 with IC_{50} of 25.6 µM (Figure 6C), with negative results against MCF-7 and A-549. The 1:1 mixture of the two EOs did not exhibit any activity against the three cancer cells.

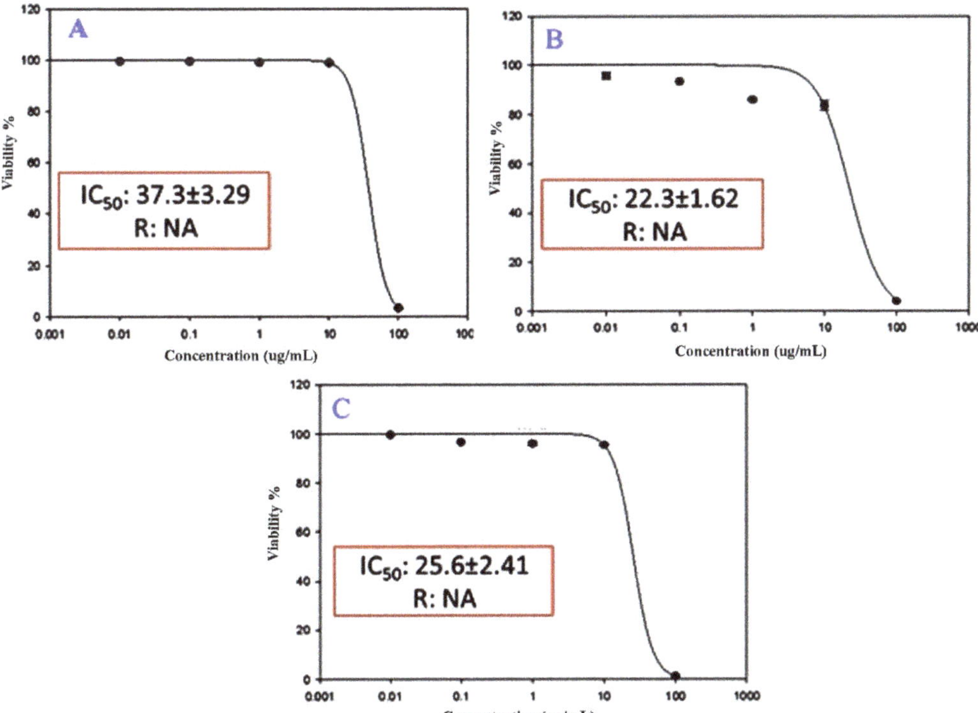

Figure 6. Cytotoxicity of EOs of (**A**) *Pluchea dioscoridis* against MCF-7 cells, (**B**) *P. dioscoridis* against A-549 cells, and (**C**) *Erigeron bonariensis* against HepG2.

The significant activities of the two EOs might be attributed to the chemical composition in which the synergetic effect of the compounds contributes to this activity [48]. The sesquiterpenes in both forms, oxygenated and hydrocarbons, represented very effective compounds as anticancer leaders [49,50]. Several reports deduced that the increasing of sesquiterpene contents in EOs caused increasing in anticancer activity [51,52]. For example, caryophyllene with high concentration in EOs was reported as a known potential cytotoxic agent especially against the growth of breast adenocarcinoma cells (MCF-7) [53,54].

The present data revealed that these two EOs are selective against the tested cancer cells. This selectivity was in full agreement with several documented results of EOs derived from other plants. For example, EO derived from *Sideritis perfoliata, Satureia thymbra, Salvia officinalis, Laurus nobilis,* and *Pistacia palestina* were found to have selective inhibitory effects against, amelanotic melanoma (C32), renal celladenocarcinoma (ACHN), hormone-dependent prostatecarcinoma (LNCaP), and breast cancer (MCF-7) [55]. Moreover, EOs extracted from the three plants, *Satureja montana, Coriandrum sativum,* and *Ocimum basilicum,* were found to have selective cytotoxic activity against HeLa, MDA-MB-453, K562, and MRC-5 [56]. The disappearance of the mixtures of the two EOs (1:1) might be ascribed to the negative synergetic effects of each EO upon the other and this phenomenon was reported in some reports. Haroun and Al-Kayali [57] found that the different extracts of *Thymbra spicata* showed positive synergetic effects via combination with some references antibiotics against some strains of bacteria and a while negative synergetic effects against other strains.

3. Materials and Methods

3.1. Plant Materials Collection and Preparation

The above-ground parts of *P. dioscoridis* and *E. bonariensis*, were collected from two populations along the Cairo–Alexandria desert road, Egypt in November 2019. From each population, the healthy and fresh plant samples were clipped from three individuals and pooled as composite samples (two per each plant; *P. dioscoridis* and *E. bonariensis*). The two plants were authenticated according to Tackholm [58] and Boulos [25]. Voucher specimens (CZ-D-x908-019 & CZ-L-x909-019) have been deposited in the herbarium of the National Research Center, Egypt. The above-ground parts were dried in the shade, ground into a fine powder, and packed in paper bags till further analysis [13].

3.2. Extraction of EOs

The air-dried powder of the above-ground parts of the *P. dioscoridis*, and *E. bonariensis*, (200 gm, each) were subjected separately to Clevenger-type apparatuses using round flask (2.5 L) comtaining water (1.5 L) for hydro-distillation for 3 h. The oily layer of each plant was isolated separately by *n*-hexane, then dried using anhydrous Na_2SO_4 (0.5 g), and finally stored in glass vials in the freezer till further analysis via GC-MS. This extraction of the EO of each plant was repeated as duplicates.

3.3. Gas Chromatography-Mass Spectroscopy (GC-MS) Analysis and Chemical Components Investigations

The four EOs samples (two samples for each plant) were analyzed via Gas Chromatography-Mass Spectroscopy (GC-MS) at National Research Center, Egypt [8]. The adjustment of the GC/MS instrument specifications has occurred as the following conditions: TRACE GC Ultra Gas Chromatographs (THERMO Scientific™ Corporate, Waltham, MA, USA), lined with a Thermo Scientific ISQ™ EC single quadrupole mass spectrometer. The GC-MS system was equipped with a TR-5 MS column with a dimension of 30 m × 0.32 mm i.d., 0.25 μm film thickness. Helium as carrier gas at a flow rate of 1.0 mL/min with a split ratio of 1:10 using the following temperature program: 60 °C for 1 min; rising at 4.0 °C/min to 240 °C and held for 1 min was used for the analyses. Both injector and detector were held at 210 °C. An aliquot of 1 μL of diluted samples in hexane (1:10, *v/v*) was always injected. Mass spectra were recorded by electron ionization (EI) at 70 eV, using a spectral range of *m/z* 40–450.

Chemical constituent of the EOs under investigations was characterized by Automated Mass spectral Deconvolution and Identification (AMDIS) software (www.amdis.net, accessed on 2 January 2020), retention indexes (relative to *n*-alkanes C_8-C_{22}), comparison of the mass spectrum with authentics (if available), and Wiley spectral library collection and NSIT library database (Gaithersburg, MD, USA; Wiley, Hoboken, NJ, USA).

3.4. Anti-Aging Activity of the EOs
3.4.1. Anti-Collagenase Assay

The anti-collagenase assay of the two studied plants EOs as well as the 1:1 mixture were performed according to Thring, et al. [59] with minor modifications for use in a microplate reader. The assay was performed in 50 mM tricine buffer (pH 7.5) with 400 mM NaCl and 10 mM $CaCl_2$. Collagenase from *Clostridium histolyticum* (ChC–EC.3.4.23.3) was dissolved in a buffer for use at an initial concentration of 0.8 units/mL according to the supplier's activity data. The synthetic substrate *N*-[3-(2-furyl) acryloyl]-Leu-Gly-Pro-Ala (FALGPA) was dissolved in tricine buffer to 2 mM. Two studied EOs and the mixture of EOs of the two plants (1:1, *w/w*), separately, were incubated with the enzyme in a buffer for 15 min before adding substrate to start the reaction. Absorbance at 490 nm was measured using a Microplate reader (TECAN, Group Ltd., Männedorf, Switzerland). Epigallocatechin gallate (EGCG) was used as a positive control.

3.4.2. Anti-Elastase Assay

For anti-elastase inhibitory assay the two studied plants EOs as well as the 1:1 mixture, this assay was performed according to Kim, et al. [60] with minor modifications. Briefly; Porcine pancreatic elastase, was dissolved to make a 3.33 mg/mL stock solution in sterile water. The substrate, N-succinyl-Ala-Ala-Ala-*p*-nitroanilide (AAAPVN) was dissolved in buffer at 1.6 mM. The test EOs were incubated with the enzyme for 15 min before adding substrate to begin the reaction. The final reaction mixture (250 µL total volume) contained buffer, 0.8 mM AAAPVN, 1 µg/mL PE and 25 µg test sample. The studied EOs and a mixture of EOs of the two plants (1:1, *w/w*), separately, were incubated. EGCG was used as a positive control. Absorbance values at 400 nm were measured in 96 well microtitre plates using a Microplate reader (TECAN, Inc.). The percentage inhibition for this assay is calculated.

3.4.3. Anti-Tyrosinase Assay

Assays of tyrosinase inhibition of the two plants EOs, as well as the 1:1 mixture, were carried out via measuring of L-DOPA chrome formation according to the described protocol of Batubara, et al. [61]. Briefly, the two EOs and a mixture of them (1:1, *w/w*), separately, were dissolved in a solvent with three certain concentrations (10, 100, and 250 µg/mL). The assays were performed by insertion of the following components: (a) phosphate buffer (120 µL, 20 mM, pH 6.8), (b) 20 µL sample, and (c) 20 µL mushroom tyrosinase (500 U/mL in 20 mM phosphate buffer) in 96-well plates. After 15 min of incubation at 25 °C, the intiation of reaction was occurred by insertion of 20 µL L-tyrosine solution (0.85 mM) for every well and followed by incubation for 10 min at room temperature. The activity of the enzyme was monitored at 475 nm using a Microplate reader (TECAN, Inc.). EGCG was used as a positive control. The calculation of the tyrosinase inhibition % was performed via the following equation:

$$\text{Tyrosinase inhibition (\%)} = [(A - B) - (C - D)]/(A-B) \times 100 \tag{1}$$

where A is the absorbance of the control with the enzyme, B is the absorbance of the control without the enzyme, C is the absorbance of the test sample with the enzyme, and D is the absorbance of the test sample without the enzyme.

3.4.4. Anti-Hyaluronidase Assay

The fluorimetric Morgan–Elson assay method was performed according to Reissig, et al. [62] that modified by Takahashi, et al. [63]. In a brief description, a 5 µL of tested EOs and a mixture of EOs of the two plants (1:1, *w/w*), separately, were incubated for 10 min at 37 °C with bovine hyaluronidase (1.50 U) in 100 µL of 20 mM sodium phosphate buffer solution (pH 7.0), sodium chloride (77 mM), in addition to 0.01% bovine serum albumin (BSA). The assay reaction was initiated via adding the hyaluronic acid sodium salt (100 µL) from rooster comb (0.03% in 300 mM sodium phosphate, pH 5.35) to the incubation mixture, then the mixture was incubated at 37 °C for 45 min. The precipitation of undigested hyaluronic acid was carried out by 1 mL acidic solution of albumin, involving 0.1% BSA in sodium acetate (24 mM) and acetic acid (79 mM, pH 3.75). The mixture was stoped by allowing it for 10 min at room temperature, and fluorescence was detected using a Tecan Infinite microplate reader at 545 nm excitation and 612 nm emission EGCG was used as a positive control.

The percentage of the collagenase, elastase, and hyaluronidase inhibition was calculated via the following equation:

$$\text{Enzyme inhibition (\%)} = [1 - (S/C) \times 100] \tag{2}$$

where S: the corrected absorbance of the samples containing elastase inhibitor (the enzyme activity in the presence of the samples); and C: the corrected absorbance of controls (the enzyme activity in the absence of the samples).

The IC_{50}, the concentration required to inhibit 50% of the enzyme under the assay conditions, was estimated from graphic plots of the dose-response curve for each concentration using Graphpad Prism software (San Diego, CA, USA).

3.5. Cytotoxicity of the Two EOs

Cytotoxic activity of the *P. dioscoridis* and *E. bonariensis* EOs and a mixture of them (1:1, *w/w*), separately were carried out against the three human cancer cells, breast adenocarcinoma cells (MCF-7), lung cancer cells (A-549), and hepatocellular carcinoma cells (HepG2), using sulforhodamine B (SRB) protocol.

3.5.1. Cell Culture

The three cancer cell lines, breast adenocarcinoma cells (MCF-7), lung cancer cells (A-549), and hepatocellular carcinoma cells (HepG2) were obtained from VACCERA, Mokatam, Giza, Egypt. Cells were maintained in DMEM media supplemented with 100 mg/mL of streptomycin, 100 units/mL of penicillin, and 10% of heat-inactivated fetal bovine serum in humidified, 5% (*v/v*) CO_2 atmosphere at 37 °C.

3.5.2. Cytotoxicity Assay

Cell viability was assessed by SRB assay. Aliquots of 100 µL cell suspension (5×10^3 cells) were in 96-well plates and incubated in complete media for 24 h. Cells were treated with another aliquot of 100 µL media containing EOs and a mixture of them (1:1, *w/w*), separately, at various concentrations ranging from (0.01, 0.1, 1, 10, and 100 ug/mL). After 72 h of drug exposure, cells were fixed by replacing media with 150 µL of 10% TCA and incubated at 4 °C for 1 h. The TCA solution was removed, and the cells were washed 5 times with distilled water. Aliquots of 70 µL SRB solution (0.4% *w/v*) were added and incubated in a dark place at room temperature for 10 min. Plates were washed 3 times with 1% acetic acid and allowed to air-dry overnight. Then, 150 µL of TRIS (10 mM) was added to dissolve protein-bound SRB stain; the absorbance was measured at 540 nm using a BMG LABTECH®- FLUOstar Omega microplate reader (Ortenberg, Germany) [64,65].

3.6. Data Treatment

The data of the anti-aging activity of various enzymes were presented in three replications and subjected to one-way ANOVA followed by Duncan's test using CoStat version 6.311 (CoHort, Monterey, CA, USA, http://www.cohort.com).

A matrix of the concentration of a total of 30 major chemical compounds (>3%) identified in the EO of five *P. dioscoridis* ecospecies was constructed, these samples were (1) present sample; (2) purchased from a market in Cairo, Egypt; (3) collected from El-Sadat City, Egypt; (4) collected from Cairo-Suez desert road, Egypt dring April; and (5) collected from Turkey. While for *E. bonariensis*, a matrix of 27 major chemical compounds (>3%) represented six samples was designed, these samples were (1) present sample; (2) collected from Alexandria, Egypt; (3) collected from Monastir, Tunisia; (4) collected from Venezuela; (5) collected from Vietnam; and (6) collected from India. The matrices were subjected to principal component analysis (PCA) and agglomerative hierarchical clustering (AHC) via XLSTAT statistical computer software package (version 2018, Addinsoft Inc., New York, NY, USA).

4. Conclusions

Herein, the GC-MS analysis of EOs of the above-ground parts of *P. dioscoridis* and *E. bonariensis*, revealed the identification of 29 and 35 compounds, respectively. Sesquiterpenes were characterized as the main components of EOs derived from the two plants. The major components of EO of *P. dioscoridis* were α-maaliene, berkheyaradulen, dehydrocyclolongifolene oxide, aromadendrene oxide-2, and β-muurolene. While, *trans*-α-farnesene, *O*-ocimene, and α-maaliene represented the abundant constituents of *E. bonariensis* EO. The observed variation in the EOs composition among the studied ecospecies and that reported support the changing of the taxa names. EO of *P. dioscoridis* exhibited cytotoxicity against the two cancer cells, MCF-7 and A-549, while the EO of the *E. bonariensis* showed activity only against HepG2. The EOs of *P. dioscoridis* and *E. bonariensis* as well as the mixture of them (1:1), exhibited significant anti-aging activity in which the mixture (1:1) > *P. dioscoridis* > *E. bonariensis*. All these data deduced the studied EOs of these two plants may be used as antiaging and anticancer leading agents.

Author Contributions: Conceptualization, A.M.A.-E., A.M.E. and A.I.E.; methodology, A.M.E., R.F.A., A.M.A.-E., A.E.-N.G.E.-G., A.I.E., and M.I.N.; data curation, A.M.A.-E. A.I.E.; software, A.M.A.-E., A.E.-N.G.E.-G. and A.M.E.; validation, A.M.E., R.F.A., A.M.A.-E., A.E.-N.G.E.-G., A.I.E. and M.I.N.; formal analysis, A.M.E., R.F.A., A.M.A.-E., A.E.-N.G.E.G., A.I.E. and M.I.N.; investigation, A.M.E., R.F.A., A.M.A.-E., A.E.-N.G.E.-G., A.I.E. and M.I.N.; funding acquisition, A.M.E.; writing—original draft preparation, A.M.A.-E. and A.I.E.; writing—review and editing, A.M.A.-E., R.F.A., A.M.A.-E., A.E.-N.G.E.-G., M.I.N. and A.I.E. All authors have read and agreed to the published version of the manuscript.

Funding: This research received no external funding.

Institutional Review Board Statement: Not applicable.

Informed Consent Statement: Not applicable.

Data Availability Statement: The data presented in this study are available in the article.

Acknowledgments: The authors gratefully acknowledge the National Research Centre, Cairo, Egypt for the support.

Conflicts of Interest: The authors declare no conflict of interest.

Sample Availability: Samples of the compounds are not available from the authors.

References

1. Cseke, L.J.; Kirakosyan, A.; Kaufman, P.B.; Warber, S.; Duke, J.A.; Brielmann, H.L. *Natural Products from Plants*; CRC Press: Boca Raton, FL, USA, 2016.
2. Abdallah, H.M.I.; Elshamy, A.I.; El Gendy, A.E.-N.G.; Abd El-Gawad, A.M.; Omer, E.A.; De Leo, M.; Pistelli, L. Anti-inflammatory, antipyretic, and antinociceptive effects of a *Cressa cretica* aqueous extract. *Planta Med.* **2017**, *83*, 1313–1320. [CrossRef] [PubMed]
3. Edris, A.E. Pharmaceutical and therapeutic potentials of essential oils and their individual volatile constituents: A review. *Phytother. Res.* **2007**, *21*, 308–323. [CrossRef] [PubMed]
4. Della Pepa, T.; Elshafie, H.S.; Capasso, R.; De Feo, V.; Camele, I.; Nazzaro, F.; Scognamiglio, M.R.; Caputo, L. Antimicrobial and phytotoxic activity of *Origanum heracleoticum* and *O. majorana* essential oils growing in Cilento (Southern Italy). *Molecules* **2019**, *24*, 2576. [CrossRef]
5. Gruľová, D.; Caputo, L.; Elshafie, H.S.; Baranová, B.; De Martino, L.; Sedlák, V.; Gogaľová, Z.; Poráčová, J.; Camele, I.; De Feo, V. Thymol chemotype *Origanum vulgare* L. essential oil as a potential selective bio-based herbicide on monocot plant species. *Molecules* **2020**, *25*, 595. [CrossRef] [PubMed]
6. Al-Rowaily, S.L.; Abd-ElGawad, A.M.; Assaeed, A.M.; Elgamal, A.M.; Gendy, A.E.-N.G.E.; Mohamed, T.A.; Dar, B.A.; Mohamed, T.K.; Elshamy, A.I. Essential oil of *Calotropis procera*: Comparative chemical profiles, antimicrobial activity, and allelopathic potential on weeds. *Molecules* **2020**, *25*, 5203. [CrossRef]
7. Abd El-Gawad, A.; El Gendy, A.; Elshamy, A.; Omer, E. Chemical composition of the essential oil of *Trianthema portulacastrum* L. Aerial parts and potential antimicrobial and phytotoxic activities of its extract. *J. Essent. Oil Bear. Plants* **2016**, *19*, 1684–1692. [CrossRef]
8. Elshamy, A.I.; Ammar, N.M.; Hassan, H.A.; Al-Rowaily, S.L.; Ragab, T.I.; El Gendy, A.E.-N.G.; Abd-ElGawad, A.M. Essential oil and its nanoemulsion of *Araucaria heterophylla* resin: Chemical characterization, anti-inflammatory, and antipyretic activities. *Ind. Crops Prod.* **2020**, *148*, 112272. [CrossRef]
9. Abd-ElGawad, A.M.; Elshamy, A.I.; El-Nasser El Gendy, A.; Al-Rowaily, S.L.; Assaeed, A.M. Preponderance of oxygenated sesquiterpenes and diterpenes in the volatile oil constituents of *Lactuca serriola* L. revealed antioxidant and allelopathic activity. *Chem. Biodivers.* **2019**, *16*, e1900278. [CrossRef]
10. Assaeed, A.; Elshamy, A.; El Gendy, A.E.-N.; Dar, B.; Al-Rowaily, S.; Abd-ElGawad, A. Sesquiterpenes-rich essential oil from above ground parts of *Pulicaria somalensis* exhibited antioxidant activity and allelopathic effect on weeds. *Agronomy* **2020**, *10*, 399. [CrossRef]
11. Elshamy, A.I.; Abd-ElGawad, A.M.; El-Amier, Y.A.; El Gendy, A.E.N.G.; Al-Rowaily, S.L. Interspecific variation, antioxidant and allelopathic activity of the essential oil from three *Launaea* species growing naturally in heterogeneous habitats in Egypt. *Flavour Fragr. J.* **2019**, *34*, 316–328. [CrossRef]
12. Abd El-Gawad, A.M.; El-Amier, Y.A.; Bonanomi, G. Essential oil composition, antioxidant and allelopathic activities of *Cleome droserifolia* (Forssk.) Delile. *Chem. Biodivers.* **2018**, *15*, e1800392. [CrossRef]
13. Abd El-Gawad, A.M. Chemical constituents, antioxidant and potential allelopathic effect of the essential oil from the aerial parts of *Cullen plicata*. *Ind. Crops Prod.* **2016**, *80*, 36–41. [CrossRef]
14. Abd-ElGawad, A.M.; El Gendy, A.E.-N.G.; Assaeed, A.M.; Al-Rowaily, S.L.; Alharthi, A.S.; Mohamed, T.A.; Nassar, M.I.; Dewir, Y.H.; Elshamy, A.I. Phytotoxic effects of plant essential oils: A systematic review and structure-activity relationship based on chemometric analyses. *Plants* **2021**, *10*, 36. [CrossRef] [PubMed]
15. Abd-ElGawad, A.M.; El Gendy, A.E.-N.G.; Assaeed, A.M.; Al-Rowaily, S.L.; Omer, E.A.; Dar, B.A.; Al-Taisan, W.a.A.; Elshamy, A.I. Essential oil enriched with oxygenated constituents from invasive plant *Argemone ochroleuca* exhibited potent phytotoxic effects. *Plants* **2020**, *9*, 998. [CrossRef] [PubMed]
16. Abd El-Gawad, A.M.; El-Amier, Y.A.; Bonanomi, G. Allelopathic activity and chemical composition of *Rhynchosia minima* (L.) DC. essential oil from Egypt. *Chem. Biodivers.* **2018**, *15*, e1700438. [CrossRef]
17. Elshamy, A.I.; Abd-ElGawad, A.M.; El Gendy, A.E.N.G.; Assaeed, A.M. Chemical characterization of *Euphorbia heterophylla* L. essential oils and their antioxidant activity and allelopathic potential on *Cenchrus echinatus* L. *Chem. Biodivers.* **2019**, *16*, e1900051. [CrossRef]
18. Arunachalam, K.; Balogun, S.O.; Pavan, E.; de Almeida, G.V.B.; de Oliveira, R.G.; Wagner, T.; Cechinel Filho, V.; de Oliveira Martins, D.T. Chemical characterization, toxicology and mechanism of gastric antiulcer action of essential oil from *Gallesia integrifolia* (Spreng.) Harms in the in vitro and in vivo experimental models. *Biomed. Pharmacother.* **2017**, *94*, 292–306. [CrossRef] [PubMed]
19. Saleh, I.; Abd-ElGawad, A.; El Gendy, A.E.-N.; Abd El Aty, A.; Mohamed, T.; Kassem, H.; Aldosri, F.; Elshamy, A.; Hegazy, M.-E.F. Phytotoxic and antimicrobial activities of *Teucrium polium* and *Thymus decussatus* essential oils extracted using hydrodistillation and microwave-assisted techniques. *Plants* **2020**, *9*, 716. [CrossRef] [PubMed]
20. Damtie, D.; Braunberger, C.; Conrad, J.; Mekonnen, Y.; Beifuss, U. Composition and hepatoprotective activity of essential oils from Ethiopian thyme species (*Thymus serrulatus* and *Thymus schimperi*). *J. Essent. Oil Res.* **2019**, *31*, 120–128. [CrossRef]
21. Tomaino, A.; Cimino, F.; Zimbalatti, V.; Venuti, V.; Sulfaro, V.; De Pasquale, A.; Saija, A. Influence of heating on antioxidant activity and the chemical composition of some spice essential oils. *Food Chem.* **2005**, *89*, 549–554. [CrossRef]
22. Wang, A.; Wu, H.; Zhu, X.; Lin, J. Species identification of *Conyza bonariensis* assisted by chloroplast genome sequencing. *Front. Genet.* **2018**, *9*, 374. [CrossRef]

23. Peng, L.; Hu, C.; Zhang, C.; Lu, Y.; Man, S.; Ma, L. Anti-cancer activity of *Conyza blinii* saponin against cervical carcinoma through MAPK/TGF-β/Nrf2 signaling pathways. *J. Ethnopharmacol.* **2020**, *251*, 112503. [CrossRef]
24. Ayaz, F.; Küçükboyacı, N.; Demirci, B. Chemical composition and antimicrobial activity of the essential oil of *Conyza canadensis* (L.) Cronquist from Turkey. *J. Essent. Oil Res.* **2017**, *29*, 336–343. [CrossRef]
25. Boulos, L. *Flora of Egypt*; Al Hadara Publishing: Cairo, Egypt, 2002; Volume 3.
26. List, T.P. Version 1.1. Available online: http://www.theplantlist.org/ (accessed on 1 January 2013).
27. Abd El-Gawad, A.; Elshamy, A.; El Gendy, A.E.-N.; Gaara, A.; Assaeed, A. Volatiles profiling, allelopathic activity, and antioxidant potentiality of *Xanthium strumarium* leaves essential oil from Egypt: Evidence from chemometrics analysis. *Molecules* **2019**, *24*, 584. [CrossRef]
28. El-Taher, A.M.; Gendy, A.E.-N.G.E.; Alkahtani, J.; Elshamy, A.I.; Abd-ElGawad, A.M. Taxonomic implication of integrated chemical, morphological, and anatomical attributes of leaves of eight Apocynaceae taxa. *Diversity* **2020**, *12*, 334. [CrossRef]
29. Mohamed, T.A.; Elshamy, A.I.; Abd-ElGawad, A.M.; Hussien, T.A.; El-Toumy, S.A.; Efferth, T.; Hegazy, M.E.F. Cytotoxic and chemotaxonomic study of isolated metabolites from *Centaurea aegyptiaca*. *J. Chin. Chem. Soc.* **2021**, *68*, 159–168. [CrossRef]
30. El Zalabani, S.; Hetta, M.; Ross, S.; Abo Youssef, A.; Zakiand, M.; Ismail, A. Antihyperglycemic and antioxidant activities and chemical composition of *Conyza dioscoridis* (L.) Desf. DC. growing in Egypt. *Aust. J. Basic Appl. Sci.* **2012**, *6*, 257–265.
31. Atta, A.H.; Nasr, S.M.; Mouneir, S.; Alwabel, N.; Essawy, S. Evaluation of the diuretic effect of *Conyza dioscorides* and *Alhagi maurorum*. *Int. J. Pharm. Pharm. Sci.* **2010**, *2*, 162–165.
32. Elshamy, A.I.; El Gendy, A.; Farrag, A.; Nassar, M.I. Antidiabetic and antioxidant activities of phenolic extracts of *Conyza dioscoridis* L. shoots. *Int. J. Pharm. Pharm. Sci.* **2015**, *7*, 65–72.
33. Awaad, A.S.; El-Meligy, R.; Qenawy, S.; Atta, A.; Soliman, G.A. Anti-inflammatory, antinociceptive and antipyretic effects of some desert plants. *J. Saudi Chem. Soc.* **2011**, *15*, 367–373. [CrossRef]
34. Harraz, F.M.; Hammoda, H.M.; El Ghazouly, M.G.; Farag, M.A.; El-Aswad, A.F.; Bassam, S.M. Chemical composition, antimicrobial and insecticidal activities of the essential oils of *Conyza linifolia* and *Chenopodium ambrosioides*. *Nat. Prod. Res.* **2015**, *29*, 879–882. [CrossRef]
35. Atta, A.H.; Mouneir, S.M. Antidiarrhoeal activity of some Egyptian medicinal plant extracts. *J. Ethnopharmacol.* **2004**, *92*, 303–309. [CrossRef] [PubMed]
36. Adams, R.P. *Identification of Essential Oil Components By Gas Chromatography/Mass Spectrometry*; Allured Publishing Corporation: Carol Stream, IL, USA, 2007; Volume 456.
37. Mabrouk, S.; Elaissi, A.; Ben Jannet, H.; Harzallah-Skhiri, F. Chemical composition of essential oils from leaves, stems, flower heads and roots of *Conyza bonariensis* L. from Tunisia. *Nat. Prod. Res.* **2011**, *25*, 77–84. [CrossRef] [PubMed]
38. El-Seedi, H.R.; Azeem, M.; Khalil, N.S.; Sakr, H.H.; Khalifa, S.A.; Awang, K.; Saeed, A.; Farag, M.A.; AlAjmi, M.F.; Pålsson, K. Essential oils of aromatic Egyptian plants repel nymphs of the tick *Ixodes ricinus* (Acari: *Ixodidae*). *Exp. Appl. Acarol.* **2017**, *73*, 139–157. [CrossRef]
39. Lis, A.; Góra, J. Essential oil of *Conyza canadensis* (L.) Cronq. *J. Essent. Oil Res.* **2000**, *12*, 781–783. [CrossRef]
40. Araujo, L.; Moujir, L.M.; Rojas, J.; Rojas, L.; Carmona, J.; Rondón, M. Chemical composition and biological activity of *Conyza bonariensis* essential oil collected in Mérida, Venezuela. *Nat. Prod. Commun.* **2013**, *8*, 1175–1178. [CrossRef]
41. Hoi, T.M.; Huong, L.T.; Chinh, H.V.; Hau, D.V.; Satyal, P.; Tai, T.A.; Dai, D.N.; Hung, N.H.; Hien, V.T.; Setzer, W.N. Essential oil compositions of three invasive *Conyza* species collected in Vietnam and their larvicidal activities against *Aedes aegypti*, *Aedes albopictus*, and *Culex quinquefasciatus*. *Molecules* **2020**, *25*, 4576. [CrossRef]
42. Abd-ElGawad, A.M.; Elshamy, A.I.; El-Amier, Y.A.; El Gendy, A.E.-N.G.; Al-Barati, S.A.; Dar, B.A.; Al-Rowaily, S.L.; Assaeed, A.M. Chemical composition variations, allelopathic, and antioxidant activities of *Symphyotrichum squamatum* (Spreng.) Nesom essential oils growing in heterogeneous habitats. *Arab. J. Chem.* **2020**, *13*, 4237–4245. [CrossRef]
43. Abd-ElGawad, A.M.; Elshamy, A.I.; Al-Rowaily, S.L.; El-Amier, Y.A. Habitat affects the chemical profile, allelopathy, and antioxidant properties of essential oils and phenolic enriched extracts of the invasive plant *Heliotropium curassavicum*. *Plants* **2019**, *8*, 482. [CrossRef] [PubMed]
44. Xiong, L.-G.; Chen, Y.-J.; Tong, J.-W.; Gong, Y.-S.; Huang, J.-A.; Liu, Z.-H. Epigallocatechin-3-gallate promotes healthy lifespan through mitohormesis during early-to-mid adulthood in *Caenorhabditis elegans*. *Redox Biol.* **2018**, *14*, 305–315. [CrossRef]
45. Tu, P.T.B.; Tawata, S. Anti-oxidant, anti-aging, and anti-melanogenic properties of the essential oils from two varieties of *Alpinia zerumbet*. *Molecules* **2015**, *20*, 16723–16740. [CrossRef] [PubMed]
46. Pandey, S.; Tiwari, S.; Kumar, A.; Niranjan, A.; Chand, J.; Lehri, A.; Chauhan, P.S. Antioxidant and anti-aging potential of Juniper berry (*Juniperus communis* L.) essential oil in *Caenorhabditis elegans* model system. *Ind. Crops Prod.* **2018**, *120*, 113–122. [CrossRef]
47. Laothaweerungsawat, N.; Sirithunyalug, J.; Chaiyana, W. Chemical compositions and anti-skin-ageing activities of *Origanum vulgare* L. essential oil from tropical and mediterranean region. *Molecules* **2020**, *25*, 1101. [CrossRef]
48. Pezzani, R.; Salehi, B.; Vitalini, S.; Iriti, M.; Zuñiga, F.A.; Sharifi-Rad, J.; Martorell, M.; Martins, N. Synergistic effects of plant derivatives and conventional chemotherapeutic agents: An update on the cancer perspective. *Medicina* **2019**, *55*, 110. [CrossRef]
49. Mohamed, T.A.; Albadry, H.A.; Elshamy, A.I.; Younes, S.H.; Shahat, A.A.; El-wassimy, M.T.; Moustafa, M.F.; Hegazy, M.E.F. A new Tetrahydrofuran sesquiterpene skeleton from *Artemisia sieberi*. *J. Chin. Chem. Soc.* **2020**, *86*, 1–5. [CrossRef]
50. Xie, Z.Q.; Ding, L.F.; Wang, D.S.; Nie, W.; Liu, J.X.; Qin, J.; Song, L.D.; Wu, X.D.; Zhao, Q.S. Sesquiterpenes from the leaves of *Magnolia delavayi* Franch. and their cytotoxic activities. *Chem. Biodivers.* **2019**, *16*, e1900013. [CrossRef]

51. Blowman, K.; Magalhães, M.; Lemos, M.F.L.; Cabral, C.; Pires, I.M. Anticancer properties of essential oils and other natural products. *J. Evid. Based Compl. Altern. Med.* **2018**, *2018*, 3149362. [CrossRef]
52. Danin, A. *Plants of Desert Dunes*; Springer Science & Business Media: Berlin, Germany, 2012.
53. Wright, B.S.; Bansal, A.; Moriarity, D.M.; Takaku, S.; Setzer, W.N. Cytotoxic leaf essential oils from Neotropical Lauraceae: Synergistic effects of essential oil components. *Nat. Prod. Commun.* **2007**, *2*, 1241–1244. [CrossRef]
54. Ali, N.A.A.; Chhetri, B.K.; Dosoky, N.S.; Shari, K.; Al-Fahad, A.J.; Wessjohann, L.; Setzer, W.N. Antimicrobial, antioxidant, and cytotoxic activities of *Ocimum forskolei* and *Teucrium yemense* (*Lamiaceae*) essential oils. *Medicines* **2017**, *4*, 17. [CrossRef]
55. Loizzo, M.R.; Tundis, R.; Menichini, F.; Saab, A.M.; Statti, G.A.; Menichini, F. Cytotoxic activity of essential oils from Labiatae and Lauraceae families against in vitro human tumor models. *Anticancer Res.* **2007**, *27*, 3293–3299. [PubMed]
56. Elgndi, M.A.; Filip, S.; Pavlić, B.; Vladić, J.; Stanojković, T.; Žižak, Ž.; Zeković, Z. Antioxidative and cytotoxic activity of essential oils and extracts of *Satureja montana* L., *Coriandrum sativum* L. and *Ocimum basilicum* L. obtained by supercritical fluid extraction. *J. Supercrit. Fluids* **2017**, *128*, 128–137. [CrossRef]
57. Haroun, M.F.; Al-Kayali, R.S. Synergistic effect of *Thymbra spicata* L. extracts with antibiotics against multidrug-resistant *Staphylococcus aureus* and *Klebsiella pneumoniae* strains. *Iran. J. Basic Med. Sci.* **2016**, *19*, 1193–1200.
58. Tackholm, V. *Students' Flora of Egypt*, 2nd ed.; Cairo University Press: Cairo, Egypt, 1974.
59. Thring, T.S.A.; Hili, P.; Naughton, D.P. Anti-collagenase, anti-elastase and anti-oxidant activities of extracts from 21 plants. *BMC Compl. Altern. Med.* **2009**, *9*, 27. [CrossRef]
60. Kim, Y.-J.; Uyama, H.; Kobayashi, S. Inhibition effects of (+)-catechin–aldehyde polycondensates on proteinases causing proteolytic degradation of extracellular matrix. *Biochem. Biophys. Res. Commun.* **2004**, *320*, 256–261. [CrossRef] [PubMed]
61. Batubara, I.; Darusman, L.; Mitsunaga, T.; Rahminiwati, M.; Djauhari, E. Potency of Indonesian medicinal plants as tyrosinase inhibitor and antioxidant agent. *J. Biol. Sci.* **2010**, *10*, 138–144. [CrossRef]
62. Reissig, J.L.; Strominger, J.L.; Leloir, L.F. A modified colorimetric method for the estimation of N-acetylamino sugars. *J. Biol. Chem.* **1955**, *217*, 959–966. [CrossRef]
63. Takahashi, T.; Ikegami-Kawai, M.; Okuda, R.; Suzuki, K. A fluorimetric Morgan–Elson assay method for hyaluronidase activity. *Anal. Biochem.* **2003**, *322*, 257–263. [CrossRef] [PubMed]
64. Skehan, P.; Storeng, R.; Scudiero, D.; Monks, A.; McMahon, J.; Vistica, D.; Warren, J.T.; Bokesch, H.; Kenney, S.; Boyd, M.R. New colorimetric cytotoxicity assay for anticancer-drug screening. *J. Natl. Cancer Inst.* **1990**, *82*, 1107–1112. [CrossRef] [PubMed]
65. Allam, R.M.; Al-Abd, A.M.; Khedr, A.; Sharaf, O.A.; Nofal, S.M.; Khalifa, A.E.; Mosli, H.A.; Abdel-Naim, A.B. Fingolimod interrupts the cross talk between estrogen metabolism and sphingolipid metabolism within prostate cancer cells. *Toxicol. Lett.* **2018**, *291*, 77–85. [CrossRef]

Article

Quality Attributes and Storage of Tomato Fruits as Affected by an Eco-Friendly, Essential Oil-Based Product

Panayiota Xylia, Irene Ioannou, Antonios Chrysargyris, Menelaos C. Stavrinides and Nikolaos Tzortzakis *

Department of Agricultural Sciences, Biotechnology and Food Science, Cyprus University of Technology, 3036 Limassol, Cyprus; pa.xylia@edu.cut.ac.cy (P.X.); i.ioannou1996@gmail.com (I.I.); a.chrysargyris@cut.ac.cy (A.C.); m.stavrinides@cut.ac.cy (M.C.S.)
* Correspondence: nikolaos.tzortzakis@cut.ac.cy; Tel.: +357-250-022-80

Abstract: The preservation of fresh produce quality is a major aim in the food industry since consumers demand safe and of high nutritional value products. In recent decades there has been a turn towards the use of eco-friendly, natural products (i.e., essential oils-EOs) in an attempt to reduce chemical-based sanitizing agents (i.e., chlorine and chlorine-based agents). The aim of this study was to evaluate the efficacy of an eco-friendly product (EP—based on rosemary and eucalyptus essential oils) and two different application methods (vapor and dipping) on the quality attributes of tomato fruits throughout storage at 11 °C and 90% relative humidity for 14 days. The results indicated that overall, the EP was able to maintain the quality of tomato fruits. Dipping application was found to affect less the quality attributes of tomato, such as titratable acidity, ripening index and antioxidant activity compared to the vapor application method. Vapor application of 0.4% EP increased fruit's antioxidant activity, whereas tomatoes dipped in EP solution presented decreased damage index (hydrogen peroxide and lipid peroxidation levels), activating enzymes antioxidant capacity (catalases and peroxidases). Moreover, higher EP concentration (up to 0.8%) resulted in a less acceptable product compared to lower concentration (0.4%). Overall, the results from the present study suggest that the investigated EP can be used for the preservation of fresh produce instead of the current commercial sanitizing agent (chlorine); however, the method of application and conditions of application must be further assessed for every commodity tested.

Keywords: tomato; eco-friendly product; essential oils; quality preservation; antioxidants; damage index

Citation: Xylia, P.; Ioannou, I.; Chrysargyris, A.; Stavrinides, M.C.; Tzortzakis, N. Quality Attributes and Storage of Tomato Fruits as Affected by an Eco-Friendly, Essential Oil-Based Product. *Plants* **2021**, *10*, 1125. https://doi.org/10.3390/plants10061125

Academic Editors: Hazem Salaheldin Elshafie, Laura De Martino, Adriano Sofo and Petko Denev

Received: 12 May 2021
Accepted: 28 May 2021
Published: 1 June 2021

Publisher's Note: MDPI stays neutral with regard to jurisdictional claims in published maps and institutional affiliations.

1. Introduction

Fresh produce is considered to be an important source of vitamins (i.e., A, C, niacin, riboflavin, thiamine), minerals (i.e., potassium, calcium, iron) and dietary fibers. Increase consumption of vegetables has been associated with a healthy lifestyle, reducing the risk of vitamin and mineral deficiencies, cancer and other chronic diseases [1]. These benefits derive from the previously mentioned phytonutrients that possess antioxidant, anti-inflammatory, and anti-cancer properties, among others [2].

Vegetables are perishable products and their quality might be affected by various environmental factors throughout the food supply chain [2]. The factors that can influence the quality and storability of vegetables include pre-harvest (i.e., growing temperature and light conditions, irrigation, maturity, pest management, harvesting, cultivation practice) and postharvest (i.e., poor handling, processing, storage temperature, marketing, pathogens) parameters [3–6]. During postharvest handling (including processing, storage, transport and retailing) and under unfavorable conditions (i.e., high temperature, low relative humidity, improper hygiene), vegetables' quality gradually deteriorates resulting in great losses for the food industry [1,7,8].

During storage, fresh produce might exhibit water loss (wilting), degradation of pigments (discoloration, i.e., loss of chlorophylls, carotenoids), and increased susceptibility to diseases and all these result in a less acceptable product by the consumer [2,3]. The main factors that affect vegetables' quality include exposure to undesirable temperature, relative humidity and light [2]. It has been shown that storage at low temperatures and high humidity suppresses the respiration rate of fresh produce, extending their shelf life [9]. Furthermore, the use of sanitizing agents including chlorine and chlorine-based means for fresh produce decontamination, might not be able to sufficiently reduce the microbial load, while at the same time, these products have been associated with the production of harmful, carcinogenic compounds [9,10].

Nowadays there is a turn towards the investigation of natural products in an attempt to reduce the use of chemical sanitizing agents in the food industry and meet consumers' demands for fresh, high nutritional and safe fresh produce [11,12]. Chlorine, the most commonly used sanitizing agent, has been linked with the formation of carcinogenic compounds that can adversely affect human health [10] and its application is of concern. Among the natural products investigated, the essential oils (EOs) from medicinal and aromatic plants gained more attention by researchers due to their antioxidant, anti-inflammatory, antifungal and antibacterial activities, among others [4,11,13–19]. Various EOs have been used in the food industry (as food preservatives) in a variety of foods including meat and meat products, fruits and vegetables, minimally processed products and dairy products [8,19–21].

The application of EOs for the postharvest preservation of fresh produce and the utilization of their properties have been previously reported and the results are promising since they are able to preserve/improve product quality and ensure its safety for consumption [4,8,22]. The use of EOs alone or in combination with other compounds, i.e., chitosan, on fresh and/or minimally processed vegetables (including tomato and cucumber) has been previously studied and the results are encouraging [4,17,23–25]. For instance, EOs from eucalyptus lemon, helichrysum, sage, nutmeg, cinnamon and clove inhibited the growth of *Escherichia coli* in cucumber fruit, preserving fruit's quality and flavor [18]. The use of dittany EO in eggplant fruits decreased gray mold (*Botrytis cinerea*) development and at the same time did not negatively affect fruit's quality attributes [15]. The use of natural products (including sage EO) in vapor phase resulted in suppressed gray mold growth when inoculated on pepper fruits, while sage EO incorporated in *Aloe vera* gel improved (via dipping application) tomato fruit quality attributes alongside with decreased fruit decay throughout storage [4,23]. Moreover, Santoro et al. [26] reported that vapor application of thyme and savory EOs on peaches and nectarines was found to improve fruit's quality attributes (i.e., less weight loss and no significant losses of ascorbic acid and carotenoid content), but at the same time they showed conflicting results on postharvest diseases (brown rot and gray mold). Among EOs, rosemary and eucalyptus have been studied for their many beneficial properties and many uses have been proposed [27,28].

Even though EOs are classified as generally recognized as safe (GRAS) food additives, it is noteworthy that their application might result in phytotoxicity, allergies and undesired alterations in product quality (i.e., appearance, aroma, flavor) if used with inappropriate (high) concentrations and/or food combinations [18,19,29]. Thus, the aim of this study was to evaluate the effects of an eco-product (EP—based on rosemary and eucalyptus essential oils) by two different application methods (vapor and dipping) on the quality attributes of tomato fruits throughout storage at 11 °C and 90% relative humidity for 14 days.

2. Results

2.1. Preliminary Test

The effects of the EP on tomato during the preliminary screening are shown in Figure 1. Both application methods (vapor and dipping) at the highest concentration (0.8% EP) resulted in decreased weight loss compared to the other concentrations tested (Figure 1A,B). Vapor application led to lower scoring on the marketability scale with 0.1%, 0.2% and 0.8% EP, whilst dipping application with 0.4% EP also presented lower scores after two days of storage (Figure 1C,D). Furthermore, all applied concentrations (for both application methods) showed lower scoring values on the aroma scale compare with the control. However, all tested concentrations showed higher scores as compared to the higher applied concentrations (i.e., for vapor: 0.8% and for dipping: 0.4% and 0.8%) (Figure 1E,F).

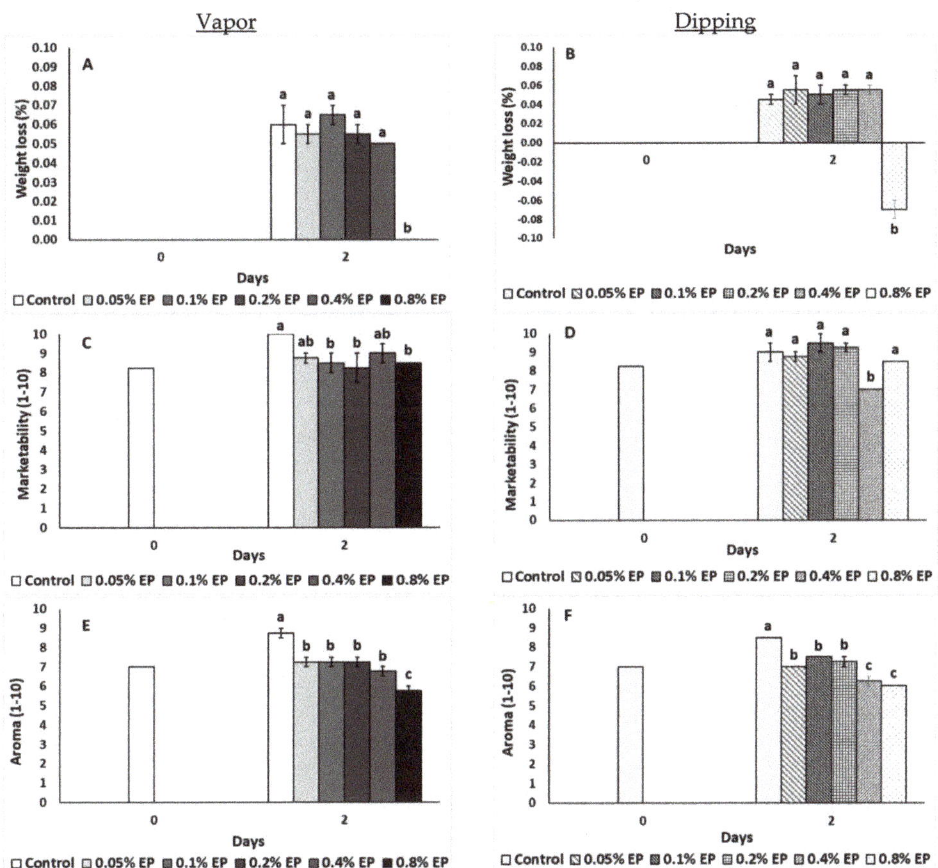

Figure 1. Effects of vapor (**A,C,E**) or dipping (**B,D,F**) application with eco-product (EP) at different concentrations (0%, 0.05%, 0.1%, 0.2%, 0.4% and 0.8%) or control (application of water) on weight loss (%), marketability (scale 1–10) and aroma (scale 1–10) of tomato fruits stored for two days at 11 °C. In each day, means (±SE) followed by different Latin letters significantly differ according to Duncan's MRT ($p = 0.05$).

2.2. Main Experiment

2.2.1. Weight Loss and Decay

Figure 2 illustrates the effects of the EP application (vapor and dipping) on the weight loss and decay of tomato fruits. Dipping application with 0.4% EP resulted in increased weight loss on the 12th day and up to the last day of storage (0.98 and 1.02%, respectively) (Figure 2B), while no differences on weight loss were observed after vapor application (Figure 2A). No significant differences ($p > 0.05$) regarding the decay of tomato fruits were reported throughout storage for both application methods (Figure 2C,D).

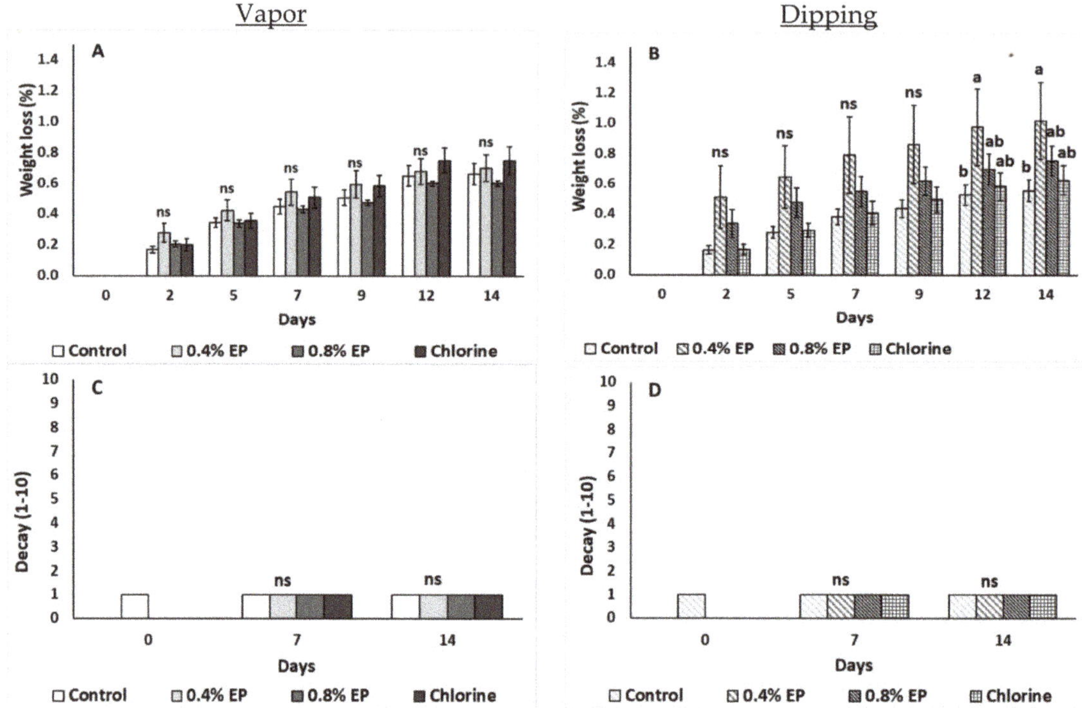

Figure 2. Effects of vapor (**A,C**) or dipping (**B,D**) application with eco-product (EP) at different concentrations (0%, 0.4% and 0.8%), chlorine (0.02%) or control (application of water) on weight loss (%) and decay (scale 1–10) of tomato fruits stored up to 14 days at 11 °C. In each day, means (±SE) followed by different Latin letters significantly differ according to Duncan's MRT (p = 0.05). ns: not significant.

2.2.2. Respiration Rate and Ethylene Production

Figure 3 presents the effects of the EP application (vapor and dipping) on tomato's respiration rate and ethylene production. The vapor application of 0.8% EP and chlorine treatment increased respiration rate on the seventh day of storage (7.14 mL CO_2 kg^{-1} h^{-1}) and this was evident for the chlorine application even at the last day of storage (Figure 3A). Interestingly, dipping application did not significantly affect tomato's respiration rate (p > 0.05) (Figure 3B). Ethylene production was increased on the 7th day of storage with EP (0.4% and 0.8%) vapor application, but this was not persistent after 14 days of storage (Figure 3C). Indeed, dipping application with chlorine increased ethylene production on the 7th day, while both EP and chlorine had increased ethylene levels on the last day of storage, compared to the control (Figure 3D).

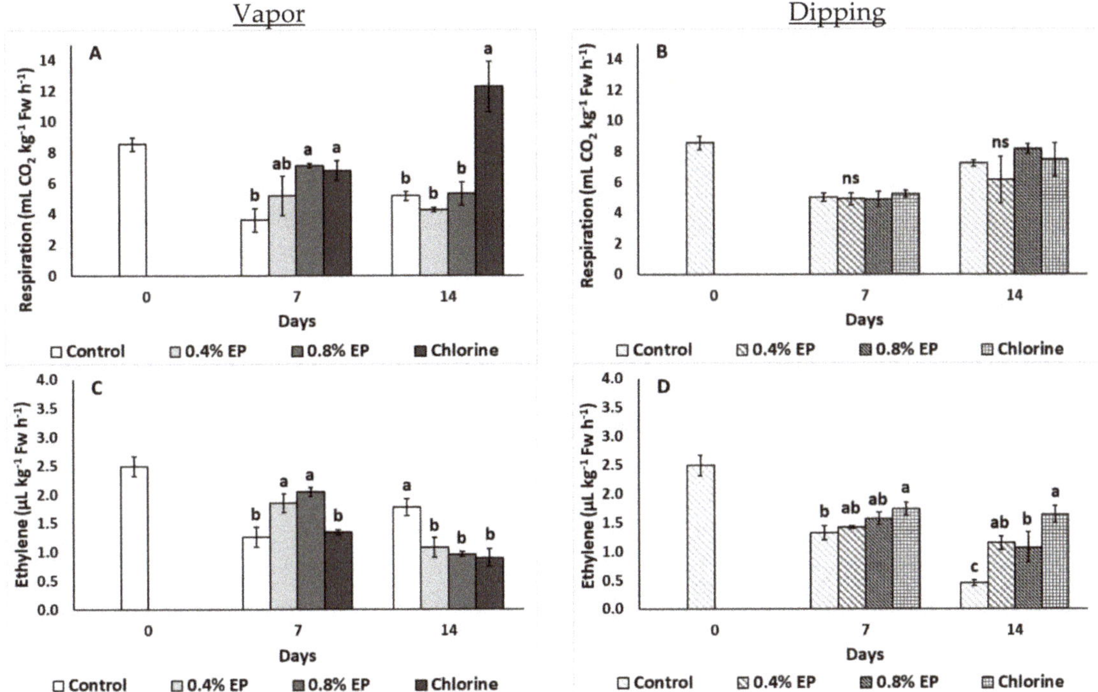

Figure 3. Effects of vapor (**A**,**C**) or dipping (**B**,**D**) application with eco-product (EP) at different concentrations (0%, 0.4% and 0.8%), chlorine (0.02%) or control (application of water) on respiration rate (mL CO_2 kg^{-1} Fw h^{-1}) and ethylene production (μL kg^{-1} Fw h^{-1}) of tomato fruits stored up to 14 days at 11 °C. In each day, means ($\pm$SE) followed by different Latin letters significantly differ according to Duncan's MRT (p = 0.05). ns: not significant.

2.2.3. Firmness, Total Soluble Solids, Titratable Acidity and Ripening Index

The effects of the EP on tomato's quality attributes (firmness, TSS, TA, ripening index) are presented in Figure 4. Fruit firmness was maintained with the EP vapor application compared to control treatment, while chlorine in vapors, decreased firmness compared to 0.4% EP treatment. Fruit firmness was maintained in similar levels after all dipping applications (Figure 4A,B). Interestingly, both methods did not significantly affect TSS of tomato fruit (Figure 4C,D). Vapor treatment of 0.4% EP decreased TA on the 7th day of storage (0.16 g citric acid L^{-1}), while chlorine applied via vapor decreased fruit's ripening index compared to 0.4% EP on the same day (Figure 4E,G). Dipping application resulted in no differences on tomato's TA and ripening index (Figure 4F,H).

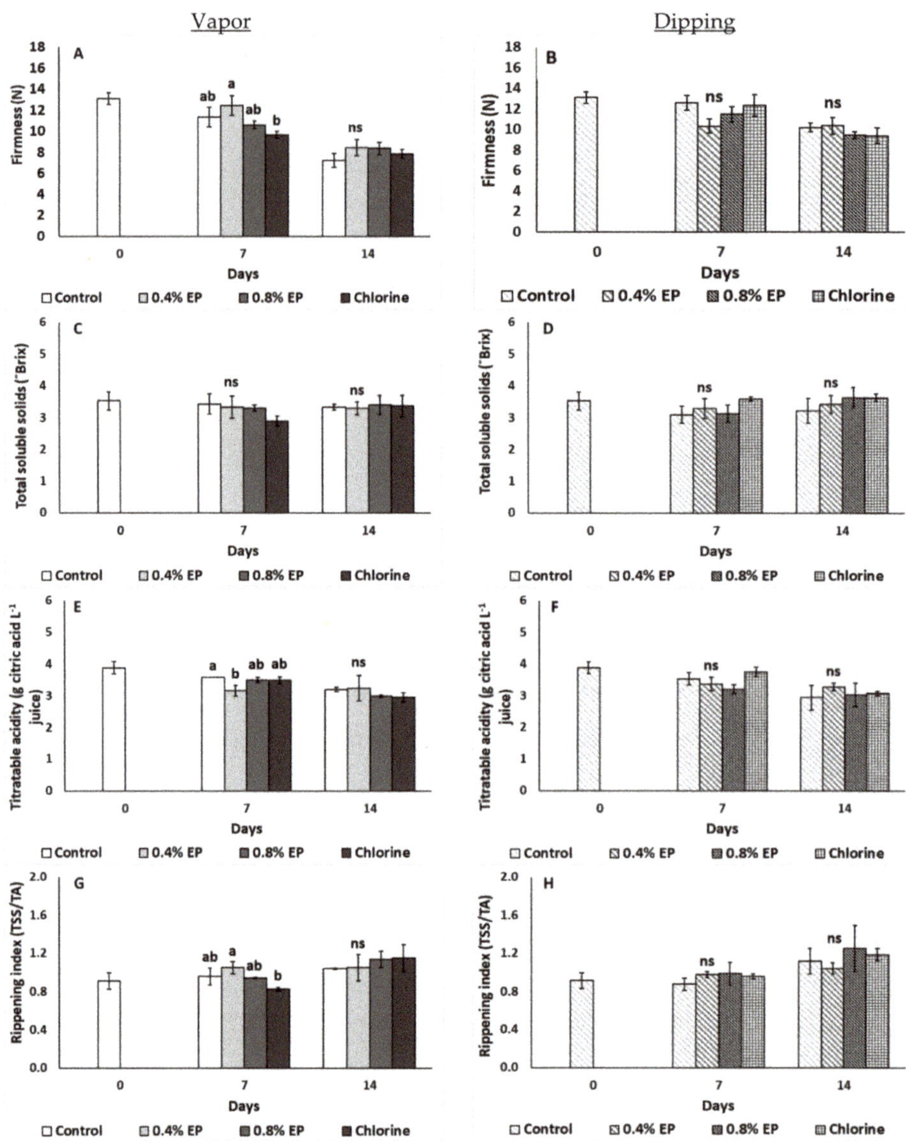

Figure 4. Effects of vapor (**A,C,E,G**) or dipping (**B,D,F,H**) application with eco-product (EP) at different concentrations (0%, 0.4% and 0.8%), chlorine (0.02%) or control (application of water) on firmness (N), total soluble solids (TSS; °Brix), titratable acidity (TA; g citric acid L^{-1} juice) and ripening index (TSS/TA) of tomato fruits stored up to 14 days at 11 °C. In each day, means (±SE) followed by different Latin letters significantly differ according to Duncan's MRT ($p = 0.05$). ns: not significant.

2.2.4. Ascorbic Acid, Lycopene, β-carotene

Figure 5 shows the effects of vapor and dipping application of EP and chlorine on ascorbic acid, lycopene and β-carotene content of tomato fruits. It is noteworthy that tomato's ascorbic acid, lycopene and β-carotene contents did not significantly differ among treatments for both application methods (vapor and dipping) ($p > 0.05$) (Figure 5A–F).

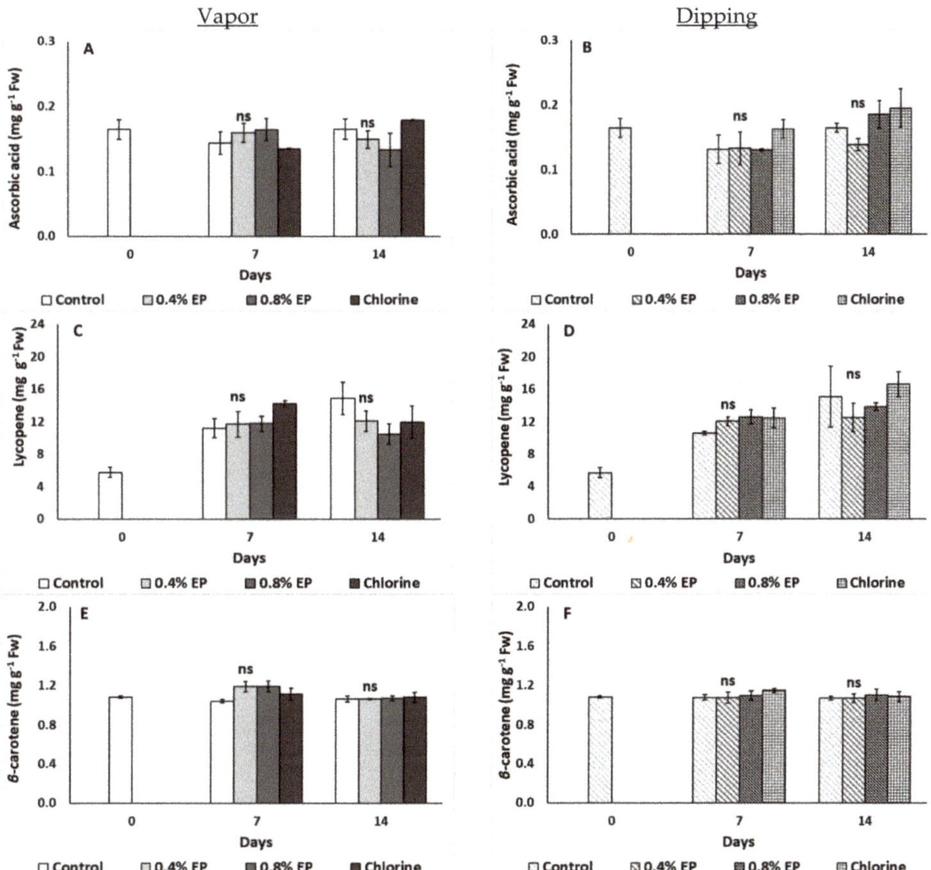

Figure 5. Effects of vapor (**A,C,E**) or dipping (**B,D,F**) application with eco-product (EP) at different concentrations (0%, 0.4% and 0.8%), chlorine (0.02%) or control (application of water) on ascorbic acid (mg g^{-1} Fw), lycopene (mg g^{-1} Fw) and β-carotene (mg g^{-1} Fw) of tomato fruits stored up to 14 days at 11 °C. In each day, means (±SE) followed by different Latin letters significantly differ according to Duncan's MRT ($p = 0.05$). ns: not significant.

2.2.5. Total Phenolic Content and Antioxidant Activity

The effects of vapor and dipping application of EP and chlorine on total phenolic content and antioxidant activity (FRAP, DPPH) of tomato fruit are presented in Figure 6. No significant differences were reported on tomato's total phenolic content among treatments for both application methods (vapor and dipping) ($p > 0.05$) (Figure 6A,B). Vapor application of 0.8% EP decreased antioxidant activity on the last day of storage (FRAP: 0.16 mg trolox g^{-1} Fw) compared to control and chlorine treated fruits (0.22 and 0.21 mg trolox g^{-1} Fw, respectively) (Figure 6C). Antioxidant activity (DPPH) of tomato fruit increased when fruit treated with 0.4% EP on the seventh day (0.60 mg trolox g^{-1} Fw), while vapor application of 0.4% EP and chlorine also increased antioxidants on the last day of storage (0.84 and 0.88 mg trolox g^{-1} Fw, respectively) (Figure 6E). Interestingly, dipping application resulted in no differences on tomato's antioxidant activity (assayed by FRAP and DPPH) (Figure 6D,F).

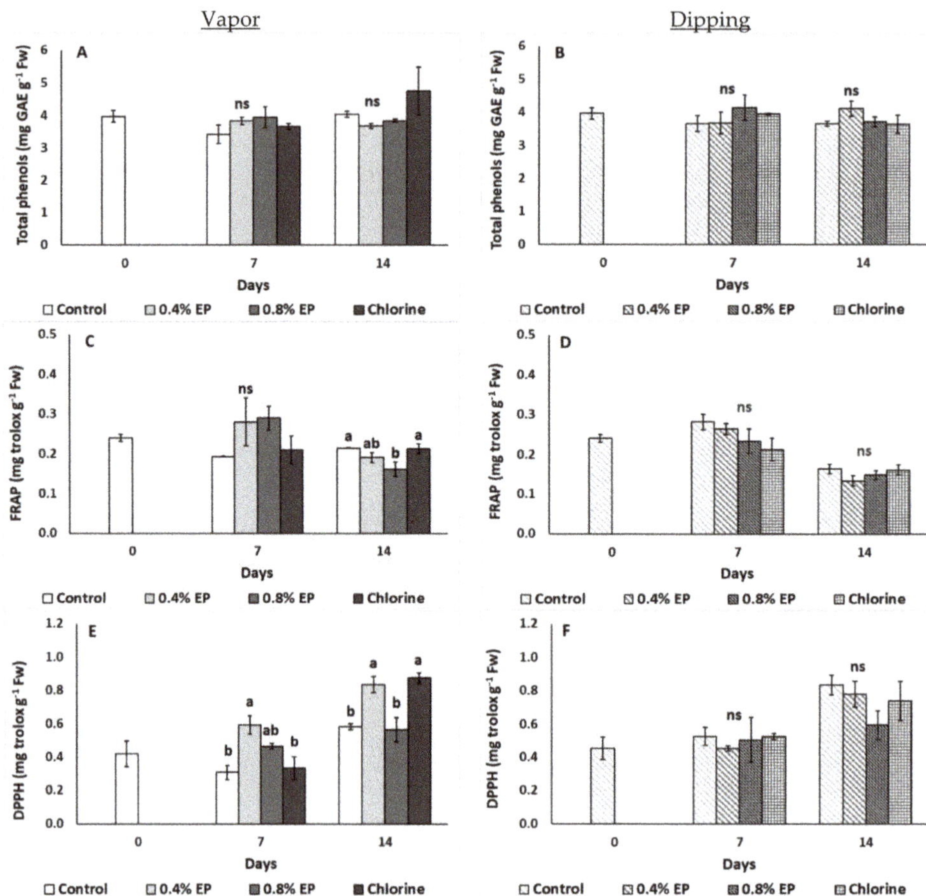

Figure 6. Effects of vapor (**A,C,E**) or dipping (**B,D,F**) application with eco-product (EP) at different concentrations (0%, 0.4% and 0.8%), chlorine (0.02%) or control (application of water) on total phenols content (mg GAE g^{-1} Fw) and antioxidant activity (FRAP, DPPH; mg trolox g^{-1} Fw) of tomato fruits stored up to 14 days at 11 °C. In each day, means ($\pm$SE) followed by different Latin letters significantly differ according to Duncan's MRT ($p = 0.05$). ns: not significant.

2.2.6. Damage Index

The effects of the EP application on the damage index of tomato fruit (H_2O_2 and lipid peroxidation levels) are illustrated in Figure 7. The vapor application of chlorine increased H_2O_2 levels of tomato compared to 0.8% EP-treated and control fruits (0.03 and 0.03 µmol g^{-1} Fw, respectively) on the last day of storage (Figure 7A). On the other hand, all treatments applied via dipping decreased tomato's H_2O_2 levels on the seventh day, compared to control fruits (Figure 7B). All vapor applied treatments resulted in increased lipid peroxidation (increased MDA levels) during the seventh day (compared to untreated fruits). Dipping tomato fruits in 0.8% EP lowered MDA levels in comparison with control fruits (11.34 and 10.90 nmol MDA g^{-1} Fw, respectively) on the 14th day (Figure 7C,D).

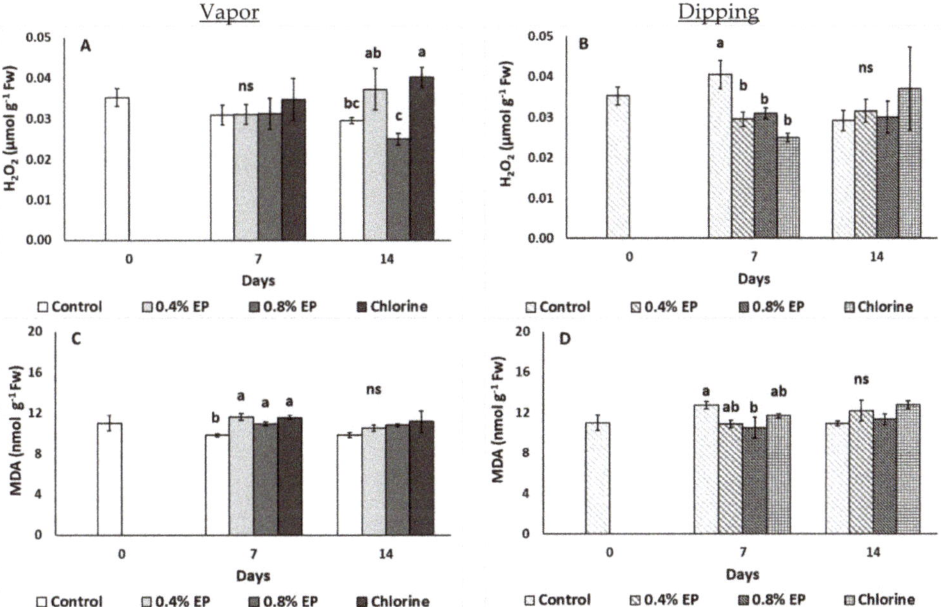

Figure 7. Effects of vapor (**A,C**) or dipping (**B,D**) application with eco-product (EP) at different concentrations (0%, 0.4% and 0.8%), chlorine (0.02%) or control (application of water) on damage index of tomato fruits (H_2O_2 and MDA levels; μmol g^{-1} Fw and nmol g^{-1} Fw, respectively) stored up to 14 days at 11 °C. In each day, means ($\pm$SE) followed by different Latin letters significantly differ according to Duncan's MRT ($p = 0.05$). ns: not significant.

2.2.7. Enzymes Antioxidant Activity

Figure 8 illustrates the effects of the EP application (vapor and dipping) on the enzymatic activity of tomato fruits. During the seventh day of vapor application of 0.8% EP, CAT activity was increased (12.78 units mg^{-1} protein) compared to control and 0.4% EP (7.55 and 7.01 units mg^{-1} protein, respectively), while 0.8% EP also increased CAT activity on the last day of storage in comparison to chlorine treated fruits (Figure 8A). On the other hand, dipping in 0.4% EP decreased CAT activity on the 14th day compared to control (Figure 8B). Interestingly, both EP application methods did not significantly affect SOD activity of tomato fruits (Figure 8C,D). POD activity increased with the EP vapor application (0.4 and 0.8%) compared to control on the seventh day of storage, while 0.8% EP led to decreased enzyme activity on the last day of storage compared to chlorine (2.52 and 3.59 units mg^{-1} protein, respectively) (Figure 8E). In contrast, dipping in 0.4% EP resulted in decreased POD activity on the seventh day of storage compared to chlorine, whilst the same treatment increased enzymes' activity on the last day of storage in comparison to control (Figure 8F).

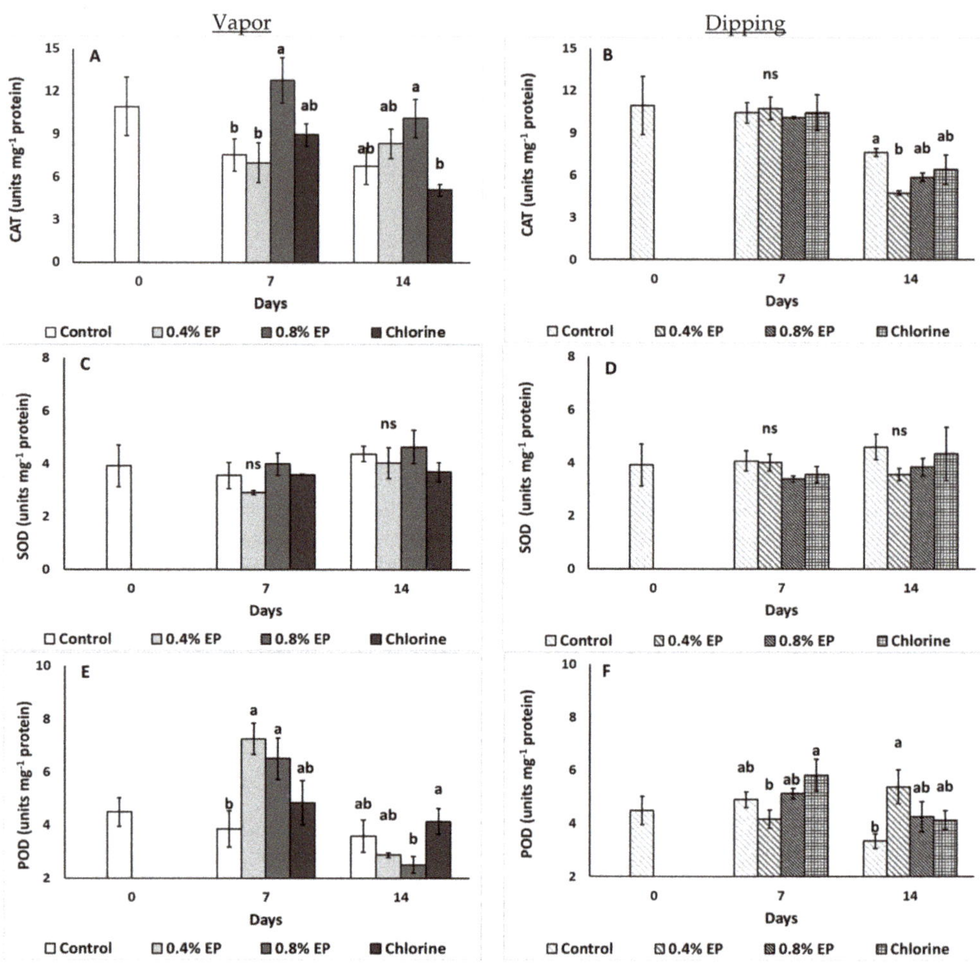

Figure 8. Effects of vapor (**A,C,E**) or dipping (**B,D,F**) application with eco-product (EP) at different concentrations (0%, 0.4% and 0.8%), chlorine (0.02%) or control (application of water) on enzyme activity (catalase-CAT, superoxide dismutase-SOD and peroxidase-POD), expressed as units of enzyme per mg of protein, stored up to 14 days at 11 °C. In each day, means (±SE) followed by different Latin letters significantly differ according to Duncan's MRT (p = 0.05). ns: not significant.

2.2.8. Sensory Evaluation

The effects of the EP's application (via vapor and dipping) on tomato's sensory attributes (marketability, aroma, appearance) are illustrated in Figure 9. Tomato's marketability was not affected during vapor application (Figure 9A), while dipping on 0.4% EP and chlorine decreased marketability (lower scoring values) on the seventh day of storage compared to control (Figure 9B). Vapor treatment with chlorine decreased aroma scoring on the seventh day of storage compared to control, while on the last day of storage all vapor treatments were able to decrease aroma values (Figure 9C). Moreover, dipping in chlorine resulted in lower aroma values on the seventh day compared to other treatments (Figure 9D). During vapor application, all treatments resulted in lower appearance values compared to control on the last day of storage, while dipping method did not affect tomato's appearance (Figure 9E,F).

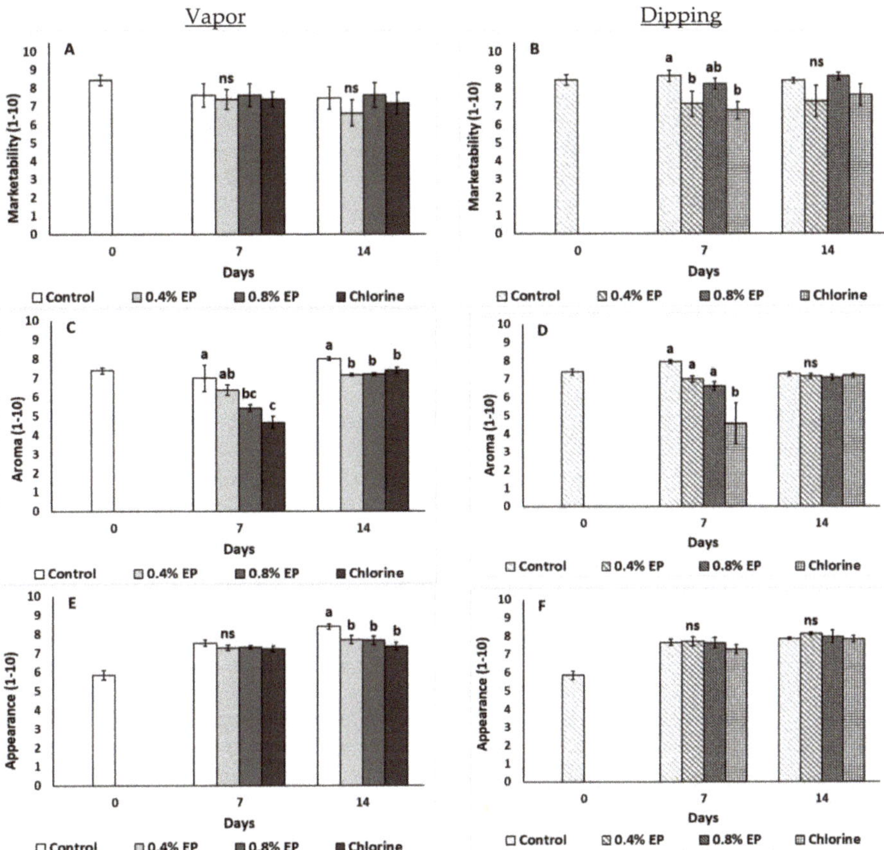

Figure 9. Effects of vapor (**A,C,E**) or dipping (**B,D,F**) application with eco-product (EP) at different concentrations (0%, 0.4% and 0.8%), chlorine (0.02%) or control (application of water) on marketability (scale 1–10), aroma (scale 1–10) and appearance (scale 1–10) of tomato fruits stored up to 14 days at 11 °C. In each day, means (±SE) followed by different Latin letters significantly differ according to Duncan's MRT ($p = 0.05$). ns: not significant.

3. Discussion

Tomato is an important food crop, characterized by high consumption numbers worldwide, and numerous uses and health benefits [30]. As a fresh produce commodity, tomato has a relatively short shelf life and its quality is affected by many pre and postharvest factors. During postharvest handling many parameters can influence tomato's quality attributes resulting in rapid deterioration. In this study, dipping tomato fruits in 0.4% EP increased product's weight loss after the 12th day of storage; however, the weight loss was less than 1.2%, which is not considered of great issue in postharvest storability of fresh produce. Similarly, Tzortzakis et al. [23] reported increased weight loss with dipping tomatoes in 0.5% sage EO compared to 0.1% EO and control fruits after seven and 14 days of storage at 11 °C. In another study, the vapor application of oregano EO did not result in any significant differences in tomato's weight loss [31]. These effects can be attributed to the similar EO composition of the main components of rosemary (isoborneol, α-pinene, α terpineol, 1.8-cineole), eucalyptus (1.8-cineole, α-pinene and δ-3 carene) and sage (α-thujone, camphor, 1.8-cineole, camphene and α-pinene) in comparison to oregano EO (carvacrol, *p*-cymene, γ-terpinene) [4,23,31,32].

Tomato, as a typical climacteric fruit, is characterized by increased respiration rate followed by increased ethylene production while ripening [33]. In that sense, vapor application increased respiration rates mainly at high EP levels and chlorine after seven days and at chlorine application at 14 days. Such changes in respiration were not evidenced in dipping application, indicating a non-lasting impact of the dipping, compared to the vapor method. We speculate that any boost in respiration rates of dipped tomatoes could possibly happen before the seven days, as indicated by the increased ethylene levels at days 7 and 14. However, considering the effects of EP product, vapor application of EP (0.4%, 0.8%) increased ethylene emission on the seventh day, while respiration increase was observed only at the 0.8% EP vapor application, on the same day. Both vapor and dipping application methods retarded ethylene emission on the last day of storage compared to day zero (fruits were at room temperature), and this is evidenced by the low temperatures used in the present study in comparison to room temperature, by retarding the ripening process and slowing down metabolic changes of the fruit. In a previous study, it has been shown that respiration rate and ethylene emission of tomato fruit were increased after dipping the fruits in 0.5% sage EO, compared to control and 0.1% EO, after the seventh day and up to the last day (14th) of storage at 11 °C [23]. The differences in respiration rate and ethylene emission of these products can be attributed to disturbance of gas exchange and cell wall degradation that can be caused by the duration of the EO application and/or even the method of application [34,35]. The increased respiration and/or ethylene production are processes related to increased fruit metabolism and ripening.

Tomato fruit firmness decreases through storage and maturation as fruit gets softer, while it has been previously suggested that EO treated fruits maintained higher firmness values compared to non-treated fruits [35]. Oregano EO when applied to tomato fruits was found not to affect fruit's firmness in comparison with control fruits while at the same time it decreased fruit decay after 14 days of storage [31], highlighting a prolonged postharvest storage period. Vapor application of EP (0.4%) resulted in firmness maintenance compared to vapor chlorine on the seventh day of storage. Moreover, in the present study, no symptoms of decay were observed on the examined tomatoes in all treatments and methods of applications (vapor vs. dipping). Similarly, in another study, eucalyptus EO vapor-treated tomatoes maintained their firmness [36]. No significant differences regarding firmness were reported with the application (via dipping) of sage EO (0.5% and 0.1%) on tomato fruits even up to 14 days of storage at 11 °C [23]. These results are in accordance with the observations of dipping application of the EP (0.4% and 0.8%) in the present study. Indeed, EO effectiveness from the previous mentioned EOs on fruit firmness is attributed to the common main components of the tested EOs, such as 1,8-cineole and α-pinene; however, that statement needs further substantiation by testing individual chemical components and/or mixtures of them in ratios similar to the examined EOs. The Eos' effectiveness is not related only to the main component of the oil, but to the synergistic action of the major components, usually numbering 3–5 components in each EO. Nevertheless, EOs effectiveness can vary, and that can be attributed to the different species, even varieties, to the cultivation practices, and components' composition.

Acidity (organic acids) is one of the main taste characteristics of tomatoes that influence fruit quality and decreases throughout storage, as fruit ripens [37,38]. Tomato's TA was decreased on the seventh day by the vapor application of 0.4% EP. The use of sage EO decreased tomato's TA, while increasing its sweetness (ripening index) after 14 days of storage at 11 °C [23]. Similarly, Adams et al. [35] mentioned a decrease in TA in EO-treated (ginger EO) tomatoes compared to control. Aminifarda and Mohammadi [39] reported that tomato fruits treated (dipping method) with ammi and anise EOs (concentration range: 200–800 μL L^{-1}) presented higher TSS compared to the control ones. During the ripening of tomato fruits, sugars accumulate above the required levels for respiration purposes and this is reflected in an increased TSS value [37]. In the present study, both application methods (dipping and vapor) did not affect tomato's TSS throughout storage. However, it has been previously mentioned that EO-treated tomatoes seem to have a slightly decreased TSS due to respiratory metabolism [35].

Color is one of the most important quality attributes of tomato fruit. The development of pigments changes during tomato fruit maturation, with the production of carotenoids and the breakdown of chlorophylls. In the present study, both application methods (dipping and vapor) did not affect tomato's AA, β-carotene and lycopene content throughout storage, which is important for maintaining the fruit's high nutritive value. These results are of agreement with another study in which no significant differences were reported on the carotenoid content of peaches and nectarines with the application of thyme and savory EO (1% and 10% in vapor phase) [26]. Carotenoid content decreases during storage and exposure of fruits to light, however the use of EOs has been proven to prevent oxidation processes (scavenging free radicals) and even preserve carotenoid levels due to their antioxidant activities [8]. On the other hand, the application of 0.1% sage EO resulted in higher AA levels of tomato fruits compared to control and 0.5% EO treated fruits [23]. Interestingly, AA and carotenoid content (β-carotene and lycopene) were found to be increased with the dipping of tomato fruits in ammi and anise EOs (concentrations range: 200–800 μL L^{-1}) compared to control fruits, initiating the antioxidative metabolism of the fruit [39]. Moreover, application of oregano EO (0.40 mL L^{-1}) resulted in tomato's increased lycopene and AA content (compared to control fruits) after 14 days of storage [31]; however, oregano EO levels were 10-fold lower than the ones used in the present study (i.e., 0.4%). Santoro et al. [26] reported that the reduced loss of the nutritional value macromolecules such as AA and carotenoids might be attributed to the EOs and their components (i.e., 1,8-cineole, α-pinene and camphor) that possess antioxidant activities, protecting sensitive nutrients from oxidation.

The results of this study indicate that both application methods (dipping and vapor) did not cause significant changes in total phenolic content. Considering that AA, carotenoids (β-carotene and lycopene), and total phenolics remained unchanged in general, this indicates that non-enzymatic antioxidants compounds were not influenced by the examined treatments. Therefore, any possible stress, as indicated by MDA and H_2O_2 increases, could alter the enzymatic antioxidant capacity of the fruit (for example, the increased CAT and POD values). Tzortzakis et al. [23] also reported no significant differences in phenolic content of tomato fruits treated with sage EO (0.1% and 0.5% via dipping). Additionally, in the same study no significant differences in antioxidant levels (DPPH, FRAP) were observed on the seventh day of storage, while increased antioxidants were reported on the last day of storage when tomato fruits were dipped in 0.1% sage EO (compared to 0.5% EO and control) [23], being in agreement with our results of 0.4% EP (vapor method) for DPPH activity at seven days of storage. The application of EOs on fresh produce can enhance the content as well as the production of antioxidant compounds when applied to fresh produce due to their own antioxidant capacity [23]. However, the increase in antioxidants during fresh produce storage is also attributed to other extrinsic postharvest factors such as chilling or adverse storage temperatures and low RH, among others.

Considering damage indices, dipping in EP product decreased tomato's H_2O_2 levels on the seventh day of storage, whereas dipping in 0.8% EP decreased MDA levels on the same day. The decreased MDA levels observed in EO-treated fruits can be attributed to the induced activation of defense-related enzymes, towards the oxidative stress challenge [40]. When citrus fruits were treated with 0.5% clove EO, increased H_2O_2 levels were revealed up to four days of storage at 25 °C [40]. At the same study, clove EO-treated fruits presented lower MDA levels compared to non-treated fruits from 12 h after the application and up to four days [40]. Another study mentioned that kiwifruits treated with 0.6 μL mL^{-1} citral presented decreased H_2O_2 and MDA levels when stored at 1 °C for up to 90 days [41]. On the other hand, our findings indicate that vapor application of the EP increased MDA levels (increased lipid peroxidation) during the seventh day of storage. Shao et al. [42] reported that vapor application of tea tree EO (0.9 g L^{-1}) increased H_2O_2 levels of strawberry fruits stored at 20 °C for three days. Increased oxidative stress—production of reactive oxygen species (ROS) such as H_2O_2—can be reported throughout storage of fresh produce and this might be attributed to the ripening process, the time and conditions of storage as well as the applied treatments (including the application of EOs). EOs when applied in fruits seem to directly and indirectly initiate the activation of fruit defense mechanisms, including the activity of antioxidant enzymes [42].

It has been previously mentioned that the application of clove EO (0.4%) on citrus fruits enhanced the activity of plant defense-related enzymes including POD, phenylalanine ammonia-lyase (PAL), polyphenol oxidase (PPO), and lipoxygenase (LOX) [40]. In the present study, the application of EP (based on rosemary and eucalyptus essential oil components) revealed an increase in the CAT and POD activity of tomato's fruits during the seventh day of storage, highlighting the induction of enzymatic antioxidant mechanisms of the fruit. Similarly, tea tree EO vapor treated (0.9 g L^{-1}) strawberries showed higher SOD, PAL and POD activities during a three-day storage at 20 °C [42]. In our study, a decrease in POD activity was reported on the seventh day with dipping application of 0.4% EP, followed by an increase in POD activity on the 14th day. These results differ from a previous study where increased POD activity was recorded throughout storage of citral-treated (0.6 µL mL^{-1}) kiwifruits [41], and such EO variation in its effectiveness is related to the EO composition, application practices and/or tested commodities. However, both tomatoes and kiwifruits are climacteric fruits and a speed-up of the ripening metabolism is taking place during storage, as indicated by the increased respiration and ethylene emission rates, mainly after seven days of storage. EO-treated fruits tend to present increased activity of antioxidant enzymes in order to suppress the accumulation of ROS during ripening and senescence of fruits during storage conditions and processing. Moreover, the activity of POD has also been correlated with fresh produce quality attributes (including color) [8]. This might also partially explain the lower appearance scoring values of tomato fruits vapor treated with the EP.

Sensory evaluation (aroma, optical/visual appearance) of fresh produce is one of the main factors affecting consumers' buying decisions. In the present study, dipping tomato fruits in 0.4% EP decreased marketability on the seventh day compared to control fruits. In contrast, the application of ginger EO (application time 20 and 30 min) increased acceptance of tomato fruit compared to control fruits [35]. Furthermore, the application of 0.8% EP (vapor method) in our study resulted in a less tomato-like aroma on the seventh day, while all applied vapor treatments also decreased aroma (less tomato-like aroma) on the last day of storage. When fruits were treated with the vapor method (all treatments), the appearance of the fruits decreased on the last day of storage indicating a less red product (less acceptable by consumers). Tzortzakis et al. [23] mentioned that tomato fruits treated with 0.1% sage EO presented greater appearance, aroma, texture and marketability compared to control fruits. On the other hand, the same study reports that increased sage EO concentration (0.5%) resulted in a less acceptable product [23]. The time of application can also affect the quality of the end product. In a previous study, the dipping application of ginger EO for 20 min resulted in a less acceptable ("sour") product compared to 30 min application which resulted in a more acceptable product [35]. Further investigation of EO and EP applied to postharvest fresh produce is needed, as it is essential that fresh produce and applied EOs are being combined in a harmonious manner enhancing fresh produce quality attributes.

4. Materials and Methods

4.1. Plant Material and Experimental Set Up

The present study took place at the laboratory facilities of Food Science and Technology of the Cyprus University of Technology, Limassol, Cyprus. Tomato (*Solanum lycopersicum* cv. F179) fresh produce were purchased from a commercial greenhouse. Crops were grown under common cultivation practices in a clay sandy-loam soil and drip irrigation and fertigation were applied according to crop needs. The cultivation took place during winter–spring months and the temperature ranged between 19 °C and 31 °C.

An eco-product (EP; named as "Agriculture Green-tech E", Meydan Solution Ltd., Larnaca, Cyprus) based on rosemary (*Rosmarinus officinalis* L.) and eucalyptus (*Eucalyptus crabra* L.) essential oil was used. Individual EOs were analyzed by gas chromatography-mass spectrometry (GC/MS-Shimadzu GC2010 gas chromatograph interfaced Shimadzu GC/MS QP2010 plus mass spectrometer) and constituents were determined [43] and presented in Table S1. *Rosemarinus officinalis* essential oils used in the tested formula were found rich in isoborneol (30.29%), α pinene (25.71%), α terpineol (14.89%) and 1.8-cineole (10.81%), while the dominant compounds of the essential oils from *Eucalyptus crabra* were 1.8-cineole (26.51%), α pinene (24.12%) and δ-3 carene (20.10%), being in accordance with previous studies [44,45]. This product was a mixture of these two essential oils (eucalyptus: rosemary in approximately 2:1 v/v ratio) and it also contained vinegar < 5% w/w as well as emulsifier treated water (<80%). Chlorine was used as a commercial sanitizer at 0.02% (v/v).

Fresh produce was selected based on uniform size, appearance, and absence of physical defects and used immediately in the different experiments. The fruits were disinfected with a chlorine (0.05% v/v) solution for 4 min and washed four times with distilled water before use (to avoid any microbial load). A preliminary screening and main experiments were implemented.

4.2. Preliminary Test

A preliminary experimental set-up has been conducted in order to determine the possible phytotoxicity or negative effects of the EP on the examined produce quality. Tomatoes were placed in 1 L capacity polystyrene (2 fruits per container), snap-on lid containers.

A total of five concentrations of the EP were examined (0.05%–0.1%–0.2%–0.4%–0.8% v/v), while purified water (0.0% EP) was used as control treatment and applied either as vapor or dipping. Three replications were used for each concentration and for each application. Containers were placed in a chamber at 11 °C and 90% relatively humidity (RH), in the dark. To maintain high RH during storage period, wet filter paper was displaced inside each container, and was remoistened every second day [46]. Container lids were open every 48 h and aerated in order to avoid air composition abnormalities (i.e., decreased O_2 and increased CO_2 levels). Fresh produce was monitored for phytotoxicity (marked spots), marketability, aroma and weight loss (as described at the main experiment) after 2 days of storage. Based on the results, the concertation of 0.4% was selected for further investigation and it was compared with a double level of the eco-product (0.8% EP) and common postharvest sanitizer (i.e., chlorine) in vapor or dipping applications.

4.3. Main Experiment

The following four treatments were examined: (i) purified water (control), (ii) 0.4% EP, (iii) 0.8% EP (iv) chlorine (0.02%). The treatments were applied either in vapor or dipping. For vapor application, tomatoes were placed in the container, together with the vaporized solution of the selected concentrations (in a 2 mL Eppendorf tube). For dipping application, based on the preliminary tests and previous studies [8], fresh produce was immersed for 10 min in the examined solution and then fruits were left to dry for 20 min at room temperature. Then, each two tomatoes were placed into the polystyrene, snap-on lid containers (1 L capacity). In total, six biological replications per treatment, for each storage period of 7 and 14 days, were used. Containers were placed in a refrigerated chamber at 11 °C and 90% relatively humidity (RH), in the dark. In summary, the experimental set up consisted of four treatments × six replications × two storage periods (plus day 0) and two application (dipping and vapor) methods. Fresh produce enclosed in containers was kept at room temperature for 2–3 h to allow EP vapor-activation, and then transferred to chilled conditions. To maintain high RH during storage period, wet filter paper was displaced inside each container, and was remoistened every second day, as described above. Container lids were open every 48 h and aerated in order to avoid air composition abnormalities.

4.3.1. Decay Evaluation

Fruit decay was visually evaluated at days 7 and 14, after storage at 11 °C. All fresh produce from each container were used for the evaluation. A commodity was considered as decayed when the symptoms of mycelia or bacteria development were present. A scale from 1 to 10 showing the surface infection percentage as 1: 0–10% infection; 2: 11–20% infection; 3: 21–30% infection; 4: 31–40% infection; 5: 41–50% infection; 6: 51–60% infection; 7: 61–70% infection; 8: 71–80% infection; 9: 81–90% infection; and 10: 91–100% infection was assessed to estimate the degree of produce infection.

4.3.2. Respiration Rate and Ethylene Production

The carbon dioxide (CO_2) and ethylene production were assessed by placing each fruit in a 1 L plastic container sealed for 1 h, at room temperature. Each fruit was weighed and its volume was measured. Moreover, CO_2 and ethylene of room air were tested and subtracted from the measurements, by equipment zeroing, prior to and during measurement. For the determination of the respiration rate, container gas atmosphere was sucked by a dual gas analyzer (GCS 250 Analyzer, International Control Analyser Ltd., Kent, UK) for 40 s and results were expressed as milliliter of CO_2 per kilogram per hour (mL CO_2 kg^{-1} h^{-1}). Ethylene was measured with an ethylene analyzer (ICA 56 Analyzer, International Control Analyser Ltd., Kent, UK) whereas the container air sample was sucked for 15 s. Results were expressed as microliter of ethylene per kilogram per h (μL ethylene kg^{-1} h^{-1}) (three replications per treatment and storage period; $n = 3$).

4.3.3. Weight Loss and Fruit Firmness

Fruit weight was recorded on the harvesting day (day 0) for each fresh produce ($n = 6$) per treatment and every other day, up to the last day of storage (day 14). Fruit firmness was measured at two points on the shoulder of each tomato fruit by applying a plunger of 3 mm in diameter at a speed of 2 mm s^{-1} and the penetration depth was 12 mm, using a texture analyzer (TA.XT plus, Stable Micro Systems, Surrey, UK). The amount of force (in Newtons; N) required to break the radial pericarp (i.e., surface) of each commodity ($n = 6$) was recorded at ambient temperature (22–24 °C).

4.3.4. Soluble Solids, Titratable Acidity, Ascorbic Acid and Carotenoids

Total soluble solids (TSS) concentration was determined in triplicate from the juice obtained from two pooled fresh fruits for each replication ($n = 3$) with a temperature-compensated digital refractometer (model Sper Scientific 300017, Scottsdale, AZ, USA) at 20 °C, and results were expressed in °Brix. Titratable acidity (TA) was measured via potentiometric titration (Mettler Toledo DL22, Columbus, OH, USA) of 5 mL of fruit juice diluted to 50 mL with distilled water, using 0.1 N NaOH up to pH 8.1. Results were expressed as g of citric acid per 1 L juice (g citric acid L^{-1} juice). The fruit sweetness/ripening index was calculated using TSS/TA ratio.

Ascorbic acid (AA) was determined by the 2,6-Dichloroindophenol titrimetric method as described previously [47]. An aliquot of 5 mL of pooled tomato juice was diluted with 45 mL of oxalic acid 0.1% and was titrated by the dye solution until the color changed. Data were expressed as mg of AA per gram of fresh weight (mg AA g^{-1} Fw).

Carotenoids (lycopene and β-carotene) for tomatoes were determined according to the method described by Nagata and Yamashita [48]. Briefly, 1 g of blended tomatoes was homogenized with 20 mL of acetone:hexane 4:6 (*v:v*) and after sonication and vigorous vortex the two phases were separated automatically. An aliquot from the upper phase was used for absorbance measurement at 663, 645, 505 and 453 nm in a spectrophotometer, using a reference of acetone:hexane (4:6) ratio. Lycopene and β-carotene contents were calculated according to the Nagata and Yamashita [48] equations:

$$\text{Lycopene (mg 100 mL}^{-1}\text{ of extract)} = -0.0458 \times A_{663} + 0.204 \times A_{645} + 0.372 \times A_{505} - 0.0806 \times A_{453}. \tag{1}$$

$$\beta\text{-Carotene (mg 100 mL}^{-1} \text{ of extract)} = 0.216 \times A_{663} - 1.22 \times A_{645} - 0.304 \times A_{505} + 0.452 \times A_{453}. \tag{2}$$

Results were expressed as mg per gram of fresh weight (mg g^{-1} Fw).

4.3.5. Total Phenolics and Antioxidant Activity

Total phenolic content was measured in blended fruit tissue (1 g) extracted with 10 mL of 50% (*v*/*v*) methanol, as reported previously [43]. Results were expressed as mg gallic acid equivalents (GAE) per gram of fresh weight (mg GAE g^{-1} Fw). The antioxidant activity was determined using the ferric-reducing antioxidant power (FRAP) and 2,2-diphenyl-1-picrylhydrazyl (DPPH) radical-scavenging activity assays (at 593 and 517 nm, respectively) as described by Chrysargyris et al. [43]. The results were expressed in mg trolox per gram of fresh weight (mg trolox g^{-1} Fw). All biological samples were analyzed in triplicate.

4.3.6. Damage Index (Hydrogen Peroxide and Lipid Peroxidation)

Hydrogen peroxide (H_2O_2) levels were estimated using the procedure previously described by Loreto and Velikova [49]. After measuring the optical density (OD) at 390 nm, results were expressed as μmol of H_2O_2 per gram of fresh weight (μmol g^{-1} Fw). Lipid peroxidation was determined with the 2-thiobarbituric acid reactive substances (TBARS) assay according to de Azevedo Neto et al. [50]. The absorbance was measured at 352 nm (discarding the non-specific absorbance at 600 nm) and results were expressed as nmol malondialdehyde (MDA) per gram of fresh weight (nmol g^{-1} Fw).

4.3.7. Enzymatic Antioxidant Activity

Fresh fruit tissue, powdered with liquid nitrogen (0.5 g) was homogenized with 4 mL of 50 mM potassium phosphate buffer (pH 7.0), which contained 1 mM phenylmethylsulfonyl fluoride (PMSF), 1 mM ethylenediaminetetraacetic acid (EDTA), 1% *w*/*v* polyvinylpolypyrrolidone (PVPP) and 0.05% Triton X-100. Samples were then centrifuged at 16000 g for 20 min, at 4 °C, and the supernatant was used for the determination of the antioxidant enzyme activity. Catalase (CAT, EC 1.11.1.6) activity was determined by following the consumption of H_2O_2 at 240 nm for 3 min. Results were expressed as CAT units per milligram of protein (units mg^{-1} protein), using the extinction coefficient of 39.4 mM cm^{-1} [51].

Superoxide dismutase (SOD, EC 1.15.1.1) was assayed as described by Chrysargyris et al. [52]. The reaction of 13 mM methionine, 75 μM nitro blue tetrazolium (NBT), 0.1 mM EDTA, 2 μM riboflavin and extract, was started after the addition of riboflavin. Reaction was then incubated under a light source of two 15-watt fluorescent lamps, for 15 min. The absorbance was determined at 560 nm and the activity was expressed as SOD units per mg of protein (units mg^{-1} protein). One unit of SOD activity was defined as the amount of enzyme required to cause 50% inhibition of the NBT photoreduction rate.

Peroxidase activity (POD, EC 1.11.1.7) was determined according to the method used by Tarchoune et al. [53], using pyrogallol as substrate. The increase in absorbance at 430 nm was measured on a kinetic cycle for 3 min, and results were expressed as units of POD per milligram of protein (units mg^{-1} protein).

An aliquot of the extract was used to determine the protein content by the Bradford method [54], with bovine serum albumin (BSA) as the protein standard.

4.3.8. Sensory Evaluation

Fresh produce marketability, aroma and appearance were recorded by at least six panelists to compare the external visual aspect and marketability of treated and control fresh produce after 7 and 14 days of storage at 11 °C. Aroma was evaluated by using a 1–10 scale, with 1: bad aroma but not EP odor; 3: EP odor with some unpleasant smell; 5: EP smell but it is pleasant; 8: less tomato-like; 10: intense tomato-like. Appearance was evaluated by using a 1–10 scale, with 1: green color of 50%; 3: yellow-green; 5: orange; 8: red; 10: deep red. Marketability was evaluated by using a 1–10 scale, with 1: not marketable quality (i.e., malformation, wounds, infection); 3: low marketable with malformation; 5:

marketable with few defects i.e., small size, decolorization (medium quality); 8: marketable (good quality); 10: marketable with no defects (extra quality).

4.4. Statistical Analysis

Statistical analysis was performed using IBM SPSS version 22 comparing data means ($\pm$ standard error-SE) with one-way ANOVA, and Duncan's multiple range tests were calculated for the significant data at $p = 0.05$. Measurements were done in three (preliminary test) or three or six (main study) biological replications/treatment (each replication consisted of a pool of two to three individual measures/sample) on different assays.

5. Conclusions

The EP based on the mixture of EOs (eucalyptus and rosemary) had different impacts on tomato fruits. Tomato fruit's quality attributes including TA, ripening index as well as antioxidants were not significantly affected by EP dipping application, whilst the same application method resulted in decreased damage index (H_2O_2 and MDA levels). Exposure of tomato fruits to vapor EP low concentration (0.4%) increased fruit's antioxidants. It is important to consider the application method of the EP, as it seems from this study that vapor applications had more profound effects than dipping application. The nutritional value (mostly AA and carotenoid content) remained unaffected in different EP levels and/or applications vapor and dipping). The findings of this study indicate that the investigated EP presented similar and/or better results on tomato fruits compared to those treated with chlorine (a commonly used sanitizing agent of the food industry); thus, this product could possibly be considered as an alternative and environmentally friendly agent used during postharvest handling of fresh produce. The use of natural products in fresh produce preservation should be further investigated to determine the optimum conditions of application (i.e., method, time, concentration) for each commodity examined in each case. However, caution should be taken when applying EOs and products based on their components to fresh produce commodities, since products' sensory attributes might be adversely impacted (undesirable/intense aroma and flavor) or even allergy issues might arise during consumption.

Supplementary Materials: The following are available online at https://www.mdpi.com/article/10.3390/plants10061125/s1, Table S1: Chemical composition (%) of essential oils of rosemary (*Rosmarinus officinalis* L.) and eucalyptus (*Eucalyptus crabra* L.).

Author Contributions: Conceptualization, P.X. and N.T.; methodology, P.X., I.I. and A.C.; software, P.X. and M.C.S.; validation, P.X., A.C., M.C.S. and N.T.; formal analysis, P.X.; investigation, P.X., I.I., and A.C.; resources, M.C.S. and N.T.; data curation, P.X. and A.C.; writing—original draft preparation, P.X. and A.C.; writing—review and editing, A.C., M.C.S. and N.T.; visualization, P.X. and M.C.S.; supervision, P.X., A.C. and N.T.; project administration, M.C.S. and N.T.; funding acquisition, M.C.S. and N.T. All authors have read and agreed to the published version of the manuscript.

Funding: This research was funded by Cyprus Research and Innovation Foundation programmes for research, technological development and innovation "RESTART 2016–2020", grant number ENTERPRISES/0916/0025; project PlantSafe. The project is co-funded by the European Regional Development Fund and the Republic of Cyprus.

Institutional Review Board Statement: Not applicable.

Informed Consent Statement: Not applicable.

Data Availability Statement: The data presented in this study are available in the article.

Conflicts of Interest: The authors declare no conflict of interest.

References

1. Jahan, S.E.; Hassan, M.K.; Roy, S.; Ahmed, Q.M.; Hasan, G.N.; Muna, A.Y.; Sarkar, M.N. Effects of different postharvest treatments on nutritional quality and shelf life of cucumber. *Asian J. Crop Soil Sci. Plant Nutr.* **2020**, *2*, 51–61. [CrossRef]
2. Cortbaoui, P.E.; Ngadi, M.O. A New Method to Quantify Postharvest Quality Loss of Cucumber using the Taguchi Approach. *Food Sci. Qual. Manag.* **2015**, *44*, 13–22.
3. Kang, H.M.; Park, K.W.; Saltveit, M.E. Elevated growing temperatures during the day improve the postharvest chilling tolerance of greenhouse-grown cucumber (*Cucumis sativus*) fruit. *Postharvest Biol. Technol.* **2002**, *24*, 49–57. [CrossRef]
4. Tzortzakis, N.; Chrysargyris, A.; Sivakumar, D.; Loulakakis, K. Vapour or dipping applications of methyl jasmonate, vinegar and sage oil for pepper fruit sanitation towards grey mould. *Postharvest Biol. Technol.* **2016**, *118*, 120–127. [CrossRef]
5. Tzortzakis, N. Ozone: A powerful tool for the fresh produce preservation. In *Postharvest Management Approaches for Maintaining Quality of Fresh Produce*; Siddiqui, M., Zavala, J., Hang, C.-A., Eds.; Springer International Publishing: Cham, Switzerland, 2016; pp. 175–208.
6. El-Garhy, H.A.S.; Abdel-Rahman, F.A.; Shams, A.S.; Osman, G.H.; Moustafa, M.M.A. Comparative analyses of four chemicals used to control black mold disease in tomato and its effects on defense signaling pathways, productivity and quality traits. *Plants* **2020**, *9*, 808. [CrossRef]
7. Xylia, P.; Botsaris, G.; Chrysargyris, A.; Skandamis, P.; Tzortzakis, N. Variation of microbial load and biochemical activity of ready-to-eat salads in Cyprus as affected by vegetable type, season, and producer. *Food Microbiol.* **2019**, *83*, 200–210. [CrossRef]
8. Xylia, P.; Clark, A.; Chrysargyris, A.; Romanazzi, G.; Tzortzakis, N. Quality and safety attributes on shredded carrots by using *Origanum majorana* and ascorbic acid. *Postharvest Biol. Technol.* **2019**, *155*, 120–129. [CrossRef]
9. Imaizumi, T.; Yamauchi, M.; Sekiya, M.; Shimonishi, Y.; Tanaka, F. Responses of phytonutrients and tissue condition in persimmon and cucumber to postharvest UV-C irradiation. *Postharvest Biol. Technol.* **2018**, *145*, 33–40. [CrossRef]
10. Tzortzakis, N.; Singleton, I.; Barnes, J. Deployment of low-level ozone-enrichment for the preservation of chilled fresh produce. *Postharvest Biol. Technol.* **2007**, *43*, 261–270. [CrossRef]
11. Kawhena, T.G.; Tsige, A.A.; Opara, U.L.; Fawole, O.A. Application of gum arabic and methyl cellulose coatings enriched with thyme oil to maintain quality and extend shelf life of "acco" pomegranate arils. *Plants* **2020**, *9*, 1690. [CrossRef]
12. Panahirad, S.; Naghshiband-Hassani, R.; Bergin, S.; Katam, R.; Mahna, N. Improvement of postharvest quality of plum (*Prunus domestica* L.) using polysaccharide-based edible coatings. *Plants* **2020**, *9*, 1148. [CrossRef] [PubMed]
13. Burt, S. Essential oils: Their antibacterial properties and potential applications in foods—A review. *Int. J. Food Microbiol.* **2004**, *94*, 223–253. [CrossRef]
14. Scollard, J.; Francis, G.A.; O'Beirne, D. Some conventional and latent anti-listerial effects of essential oils, herbs, carrot and cabbage in fresh-cut vegetable systems. *Postharvest Biol. Technol.* **2013**, *77*, 87–93. [CrossRef]
15. Stavropoulou, A.; Loulakakis, K.; Magan, N.; Tzortzakis, N. Origanum dictamnus Oil Vapour Suppresses the Development of Grey Mould in Eggplant Fruit in Vitro. *BioMed Res. Int.* **2014**, *2014*, 562679. [CrossRef] [PubMed]
16. Teixeira, B.; Marques, A.; Pires, C.; Ramos, C.; Batista, I.; Saraiva, J.A.; Nunes, M.L. Characterization of fish protein films incorporated with essential oils of clove, garlic and origanum: Physical, antioxidant and antibacterial properties. *LWT Food Sci. Technol.* **2014**, *59*, 533–539. [CrossRef]
17. Istúriz-Zapata, M.A.; Hernández-López, M.; Correa-Pacheco, Z.N.; Barrera-Necha, L.L. Quality of cold-stored cucumber as affected by nanostructured coatings of chitosan with cinnamon essential oil and cinnamaldehyde. *LWT* **2020**, *123*, 109089. [CrossRef]
18. Cui, H.; Zhao, C.; Li, C.; Lin, L. Essential Oils-Based Antibacterial Agent Against *Escherichia coli* O157:H7 Biofilm on Cucumber. *J. Food Process. Preserv.* **2017**, *41*, e13140. [CrossRef]
19. Falleh, H.; Ben Jemaa, M.; Saada, M.; Ksouri, R. Essential oils: A promising eco-friendly food preservative. *Food Chem.* **2020**, *330*, 127268. [CrossRef] [PubMed]
20. Alharbi, A.A. Effectiveness of Acetic Acid, Essential Oils and Trichoderma spp. in Controlling Gray Mold Disease on Cucumber. *Alexandria J. Agric. Sci.* **2018**, *63*, 275–281. [CrossRef]
21. Xylia, P.; Chrysargyris, A.; Tzortzakis, N. The Combined and Single Effect of Marjoram Essential Oil, Ascorbic Acid, and Chitosan on Fresh-Cut Lettuce Preservation. *Foods* **2021**, *10*, 575. [CrossRef]
22. Xylia, P.; Chrysargyris, A.; Botsaris, G.; Tzortzakis, N. Potential application of spearmint and lavender essential oils for assuring endive quality and safety. *Crop Prot.* **2017**, *102*, 94–103. [CrossRef]
23. Tzortzakis, N.; Xylia, P.; Chrysargyris, A. Sage essential oil improves the effectiveness of *Aloe vera* gel on postharvest quality of tomato fruit. *Agronomy* **2019**, *9*, 635. [CrossRef]
24. Olufunmilayo, S.O.; Uzoma, O. Postharvest physicochemical properties of cucumber fruits (*Cucumber sativus* L.) treated with chitosan-lemon grass extracts under different storage durations. *African J. Biotechnol.* **2016**, *15*, 2758–2766. [CrossRef]
25. Poimenidou, S.V.; Bikouli, V.C.; Gardeli, C.; Mitsi, C.; Tarantilis, P.A.; Nychas, G.J.; Skandamis, P.N. Effect of single or combined chemical and natural antimicrobial interventions on *Escherichia coli* O157: H7, total microbiota and color of packaged spinach and lettuce. *Int. J. Food Microbiol.* **2016**, *220*, 6–18. [CrossRef]
26. Santoro, K.; Maghenzani, M.; Chiabrando, V.; Bosio, P.; Gullino, M.L.; Spadaro, D.; Giacalone, G. Thyme and savory essential oil vapor treatments control brown rot and improve the storage quality of peaches and nectarines, but could favor gray mold. *Foods* **2018**, *7*, 7. [CrossRef]

27. De Medeiros Barbosa, I.; da Costa Medeiros, J.A.; de Oliveira, K.Á.R.; Gomes-Neto, N.J.; Tavares, J.F.; Magnani, M.; de Souza, E.L. Efficacy of the combined application of oregano and rosemary essential oils for the control of *Escherichia coli*, *Listeria monocytogenes* and *Salmonella Enteritidis* in leafy vegetables. *Food Control* **2016**, *59*, 468–477. [CrossRef]

28. Ponce, A.G.; Del Valle, C.; Roula, S.I. Shelf Life of Leafy Vegetables Treated with Natuaral Essential Oils. *J. Food Sci.* **2004**, *69*, 50–56.

29. Werrie, P.Y.; Durenne, B.; Delaplace, P.; Fauconnier, M.L. Phytotoxicity of essential oils: Opportunities and constraints for the development of biopesticides: A review. *Foods* **2020**, *9*, 1291. [CrossRef]

30. Pék, Z.; Helyes, L.; Lugasi, A. Color changes and antioxidant content of vine and postharvest-ripened tomato fruits. *HortScience* **2010**, *45*, 466–468. [CrossRef]

31. Tzortzakis, N.G.; Tzanakaki, K.; Economakis, C.D.C.D. Effect of origanum oil and vinegar on the maintenance of postharvest quality of tomato. *Food Nutr. Sci.* **2011**, *2*, 974–982. [CrossRef]

32. Chrysargyris, A.; Charalambous, S.; Xylia, P.; Litskas, V.; Stavrinides, M.; Tzortzakis, N. Assessing the biostimulant effects of a novel plant-based formulation on tomato crop. *Sustainability* **2020**, *12*, 8432. [CrossRef]

33. Serrano, M.; Martínez-Romero, D.; Guillén, F.; Valverde, J.M.; Zapata, P.J.; Castillo, S.; Valero, D. The addition of essential oils to MAP as a tool to maintain the overall quality of fruits. *Trends Food Sci. Technol.* **2008**, *19*, 464–471. [CrossRef]

34. Rabiei, V.; Shirzadeh, E.; Rabbiangourani, H.; Sharafi, Y. Effect of thyme and lavender essential oils on the qualitative and quantitative traits and storage life of apple "Jonagold" cultivar. *J. Med. Plant Res.* **2011**, *5*, 5522–5527.

35. Adams, A.-R.; Adama, A.-R.; Kuunuori Thadius, T.; Barau, B. Ginger Essential Oil for Postharvest Quality of Datterino Tomato: Effect of Immersion Duration and Storage Temperature. *J. Postharvest Technol.* **2018**, *6*, 109–121.

36. Tzortzakis, N.G. Maintaining postharvest quality of fresh produce with volatile compounds. *Innov. Food Sci. Emerg. Technol.* **2007**, *8*, 111–116. [CrossRef]

37. Oms-Oliu, G.; Hertog, M.L.A.T.M.; Van de Poel, B.; Ampofo-Asiama, J.; Geeraerd, A.H.; Nicolai, B.M. Metabolic characterization of tomato fruit during preharvest development, ripening, and postharvest shelf-life. *Postharvest Biol. Technol.* **2011**, *62*, 7–16. [CrossRef]

38. Tabarestani, H.S.; Sedaghat, N.; Alipour, A. Shelf life Improvement and Postharvest Quality of Cherry Tomato (*Solanum lycopersicum* L.) Fruit Using Basil Mucilage Edible Coating and Cumin Essential Oil. *Int. J. Agron. Plant Prod.* **2013**, *4*, 2346–2353.

39. Aminifard, M.H.; Mohammadi, S. Effect of essential oils on postharvest decay and quality factors of tomato in vitro and in vivo conditions. *Arch. Phytopathol. Plant Prot.* **2012**, *45*, 1280–1285. [CrossRef]

40. Chen, C.; Cai, N.; Chen, J.; Wan, C. Clove essential oil as an alternative approach to control postharvest blue mold caused by *Penicillium italicum* in citrus fruit. *Biomolecules* **2019**, *9*, 197. [CrossRef]

41. Wei, L.; Chen, C.; Wan, C.; Chen, M.; Chen, J. Citral Delays Postharvest Senescence of Kiwifruit by Enhancing Antioxidant Capacity under Cold Storage. *J. Food Qual.* **2021**, *2021*, 6684172. [CrossRef]

42. Shao, X.; Wang, H.; Xu, F.; Cheng, S. Effects and possible mechanisms of tea tree oil vapor treatment on the main disease in postharvest strawberry fruit. *Postharvest Biol. Technol.* **2013**, *77*, 94–101. [CrossRef]

43. Chrysargyris, A.; Panayiotou, C.; Tzortzakis, N. Nitrogen and phosphorus levels affected plant growth, essential oil composition and antioxidant status of lavender plant (*Lavandula angustifolia* Mill.). *Ind. Crops Prod.* **2016**, *83*, 577–586. [CrossRef]

44. Bajalan, I.; Rouzbahani, R.; Pirbalouti, A.G.; Maggi, F. Antioxidant and antibacterial activities of the essential oils obtained from seven Iranian populations of *Rosmarinus officinalis*. *Ind. Crops Prod.* **2017**, *107*, 305–311. [CrossRef]

45. Bhatti, H.; Iqbal, Z.; Chatha, S.; Bukhari, I. Variations in oil potential and chemical composition of *Eucalyptus crebra* among different districts of Punjab-Pakistan. *Int. J. Agric. Biol.* **2007**, *9*, 136–138.

46. Chrysargyris, A.; Nikou, A.; Tzortzakis, N. Effectiveness of *Aloe vera* gel coating for maintaining tomato fruit quality. *N. Z. J. Crop Hortic. Sci.* **2016**, *44*, 203–217. [CrossRef]

47. AOAC. *Official Methods of Analysis*, 18th ed.; AOAC: Gaithersburg, MD, USA, 2007.

48. Nagata, M.; Yamashita, I. Simple Method for Simultaneous Determination of Chlorophyll and Carotenoids in Tomato Fruit. *Nippon Shokuhin Kogyo Gakkaishi* **1992**, *39*, 925–928. [CrossRef]

49. Loreto, F.; Velikova, V. Isoprene produced by leaves protects the photosynthetic apparatus against ozone damage, quenches ozone products, and reduces lipid peroxidation of cellular membranes. *Plant Physiol.* **2001**, *127*, 1781–1787. [CrossRef]

50. De Azevedo Neto, A.D.; Prisco, J.T.; Enéas-Filho, J.; De Abreu, C.E.B.; Gomes-Filho, E. Effect of salt stress on antioxidative enzymes and lipid peroxidation in leaves and roots of salt-tolerant and salt-sensitive maize genotypes. *Environ. Exp. Bot.* **2006**, *56*, 87–94. [CrossRef]

51. Jiang, M.; Zhang, J. Involvement of plasma-membrane NADPH oxidase in abscisic acid- and water stress-induced antioxidant defense in leaves of maize seedlings. *Planta* **2002**, *215*, 1022–1030. [CrossRef] [PubMed]

52. Chrysargyris, A.; Michailidi, E.; Tzortzakis, N. Physiological and biochemical responses of *Lavandula angustifolia* to salinity under mineral foliar application. *Front. Plant Sci.* **2018**, *9*, 1–23. [CrossRef]

53. Tarchoune, I.; Sgherri, C.; Izzo, R.; Lachaâl, M.; Navari-Izzo, F.; Ouerghi, Z. Changes in the antioxidative systems of *Ocimum basilicum* L. (cv. Fine) under different sodium salts. *Acta Physiol. Plant.* **2012**, *34*, 1873–1881. [CrossRef]

54. Bradford, M.M. A rapid and sensitive method for the quantitation of microgram quantities of protein utilizing the principle of protein-dye binding. *Anal. Biochem.* **1976**, *72*, 248–254. [CrossRef]

Article

Vapour Application of Sage Essential Oil Maintain Tomato Fruit Quality in Breaker and Red Ripening Stages

Antonios Chrysargyris [1,2], Charalampos Rousos [1], Panayiota Xylia [1] and Nikolaos Tzortzakis [1,*]

[1] Department of Agricultural Sciences, Biotechnology and Food Science, Cyprus University of Technology, Limassol 3036, Cyprus; a.chrysargyris@cut.ac.cy (A.C.); rousos9@hotmail.com (C.R.); pa.xylia@edu.cut.ac.cy (P.X.)

[2] Department of Life Sciences, School of Sciences, European University of Cyprus, Nicosia 1516, Cyprus

* Correspondence: nikolaos.tzortzakis@cut.ac.cy; Tel.: +357-25-002-280

Citation: Chrysargyris, A.; Rousos, C.; Xylia, P.; Tzortzakis, N. Vapour Application of Sage Essential Oil Maintain Tomato Fruit Quality in Breaker and Red Ripening Stages. *Plants* **2021**, *10*, 2645. https://doi.org/10.3390/plants10122645

Academic Editors: Adriano Sofo, Laura De Martino and Hazem Salaheldin Elshafie

Received: 15 November 2021
Accepted: 30 November 2021
Published: 1 December 2021

Publisher's Note: MDPI stays neutral with regard to jurisdictional claims in published maps and institutional affiliations.

Abstract: Consumers seek safe, high-nutritional-value products, and therefore maintaining fresh produce quality is a fundamental goal in the food industry. In an effort to eliminate chemical-based sanitizing agents, there has been a shift in recent decades toward the usage of eco-friendly, natural solutions (e.g., essential oils-EOs). In the present study, tomato fruits (*Solanum lycopersicum* L. cv. Dafni) at breaker and red ripening stage were exposed to sage essential oils (EO: 50 μL L^{-1} or 500 μL L^{-1}) for 2, 7 and 14 days, at 11 °C and 90% relative humidity (RH). Quality-related attributes were examined during (sustain effect—SE) and following (vapour-induced memory effect—ME; seven days vapours + seven days storage) vapour treatment. In breaker tomatoes, EO-enrichment (sustained effect) retained fruit firmness, respiration rates, and ethylene emission in low EO levels (50 μL L^{-1}). In contrast, breaker fruit metabolism sped up in high EO levels of 500 μL L^{-1}, with decreased firmness, increased rates of respiration and ethylene, and effects on antioxidant metabolism. The effects were more pronounced during the storage period of 14 days, comparing to the fruit exposed to common storage-transit practice. In red fruits, the EOs impacts were evidenced earlier (at two and seven days of storage) with increased rates of respiration and ethylene, increased β-carotene, and decreased lycopene content. In both breaker and red ripening fruit, EO application decreased weight losses. Considering the fruits pre-exposed to EOs, quality attributes were more affected in green fruits and affected to a lesser level in the red ones. Furthermore, based on appearance, color, and texture evaluations, organoleptic trials demonstrated an overwhelming preference for EO-treated red fruit during choice tests. EOs had lower effects on total phenolics, acidity, total soluble solids, and fruit chroma, with no specific trend for both breaker and red tomatoes. Natural volatiles may aid to retain fruit quality in parallel with their antimicrobial protection offered during storage and transportation of fresh produce. These effects may persist after the EO is removed from the storage conditions.

Keywords: fruit storage; natural products; quality-related attributes; tomato; volatiles

1. Introduction

The increased demands on fresh produce, fruits, vegetables, and herbs is challenged nowadays, with efforts focusing on the increasing yields and quality during the crop production. Moreover, efforts have also been targeted to decrease produce losses during the postharvest storage. As a consequence, the increased consumption of fresh produce has driven commercial desire for better storage and transportation conditions. There is an increased interest on effective sanitation means a decrease in postharvest losses due to decay, while maintaining fruit quality, including flavor, color, nutritional value, texture, and storability [1,2]. Non-single preservation means are efficient enough to be applied in a wide range of fresh produce, microorganisms, and environmental conditions, for each crop. Despite the fact that chemical applications in postharvest are of high effectiveness, there are significant challenges including current sanitation procedures and health and environmental concerns due to the possible generation of toxic by-products and residues [3–5]. Moreover, the use of chemicals as fungicides for postharvest sanitation is partially restricted in many countries [6]. Therefore, alternative, safe, eco-friendly but effective sanitizing agents are explored for the fresh produce preservation [7–11].

One candidate is the essential oils (EOs) derived from medicinal and aromatic plants (MAP) due to a wide range of biocidal activities, including antifungal [12–14], antibacterial [15–17], antioxidant [15,17], cytotoxic [18], and anti-inflammatory [19] activities, to name a few. Essential oils from a variety of plant species, including sage (*Salvia* spp.), have demonstrated bioactivity against a variety of plant diseases [7,20,21]. However, there has been not much research on the beneficial effects of the EOs application on the fruit quality of pears, tomatoes, eggplants, strawberries, and cherries [2,6,12,22,23], processed fresh produce [24] and cut flowers [25,26]. Although the strong aroma of EO can restrict the final product's organoleptic acceptability, it is known to have strong antioxidant and antimicrobial properties [7,27,28]. Recent research has revealed that EO (i.e., *Thymus capitatus*; thyme oil), can act as signaling compound. Therefore, EO application is triggering a signal to induce a defense mechanism in vegetables by increasing the activity of defense-related enzymes and increasing antioxidant ability [29]. Essentials oils can be both applied in aqueous and in vaporized phase, with the latter being an advantage for some commodities (i.e., strawberries and grapes) where aqueous sanitation is a limitation. The EO's antimicrobial activity is linked to its hydrophobic properties, which allow it to penetrate into microbial cells' phospholipid membranes, causing structural disorder and permeability [30]. However, the use of EOs in high levels is restricted due to probable unfavorable sensory effects, and as a result of that, the concentrations used need to be optimized for each commodity.

Tomatoes (*Solanum lycopersicum* Mill.) are harvested at different stages of ripeness to meet various consumption needs (e.g., fresh and processed). For red-fleshed tomatoes, six ripeness stages (i.e., green, breaker, turning, pink, light red, and red) are described based on the surface color [31]. Breaker-turning ripeness stage is used for longer fruit storage and transportation. Tomato is a climacteric fruit with a limited postharvest life due to the elevated levels of respiration, transpiration, ethylene emission and postharvest decay, resulting in an increased ripening process and senescence [32,33]. Tomato ripening is accompanied with chlorophyll break down, lycopene accumulation, tissue softening, and shifts in aroma and other compositional properties [34]. Following harvest, the fruit continues to have several biochemical changes on quality and deteriorates rapidly. In some cases, fruit deterioration can be during or after transport and marketing. Tomatoes are stored at comparably high temperatures (10–12.5 °C) depending on the maturity stage to prevent chilling injury which is evidenced at lower temperatures, below 7–10 °C [35].

Only fresh produce that meets the consumer's standards is suitable at the market interface. As a result, it is vital to assess the impact of potentially revolutionary applications on the sensory and organoleptic features of fruits and vegetables. Sage EO has been effective in fruit quality and observed antimicrobial activity [23,36]. The goal of this study was to examine if the vapor phase of essential oils obtained from sage (*Salvia trilova* L.) had any effect on tomato fruit quality attributes including: (i) physiological parameters (including weight loss, fruit firmness, and rates of respiration and ethylene production); (ii) fruit chemical composition (for example, vitamin C content, antioxidant capability, organic acid content (citrate), total soluble solids, carotenoids (lycopene, β-carotene) and total phenolic content) and damage index; and (iii) sensory qualities as determined by a consumer panel under controlled settings.

2. Results

The experimental lay out is presented in Figure 1, with tomato fruits exposed to sage EO (50 or 500 µL L^{-1}) during storage for up to 2, 7, and 14 days or exposed to EO for 7 days and then stored to chilled conditions for an additional 7 days.

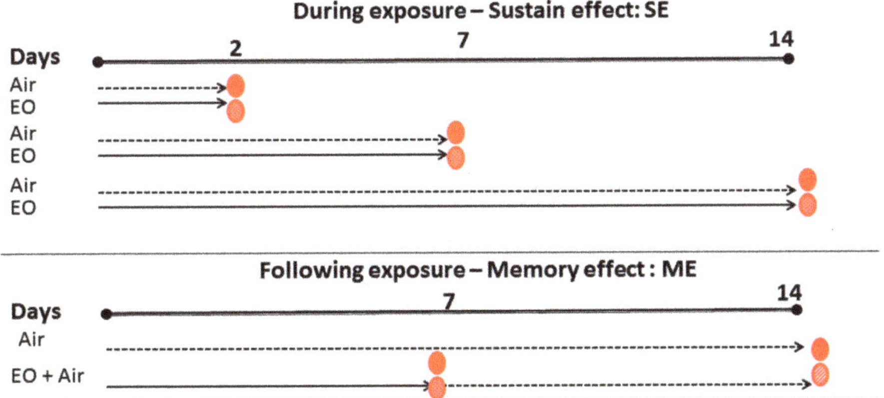

Figure 1. Layout of experiments: Tomato fruit were exposed to ambient air or essential oil (EO: 50 or 500 μL L^{-1}) in the dark at 11 °C and 90% RH. Experiment 1: Tomato fruit were exposed to air or EO for 2, 7, and 14 days and sampling took place during exposure (sustain effect—SE) to air or EO. Experiment 2: Tomato fruit were exposed to air or EO for seven days, and then transferred for additional seven days to air. Sampling took place following EO exposure (memory effect—ME) at 14 days of storage. Air (⋯⋯▶), EO exposure (→). Tomato exposed to air ⬤, tomato exposed to EOs ◉.

2.1. Fruit Decay

Neither the EO-treated nor the control fruit showed signs of degradation until day 7 of the storage period. At the end of the trial (day 14), control fruit showed evidence of deterioration (assessed as 2.05 and 2.75 on a 1–5 scale, for breaker and red fruit, respectively) [principally symptoms of anthracnose rot (caused by *Colletotrichum coccodes*) and secondary symptoms of black spot] (caused by *Alternaria alternata*)] as shown in Table 1.

Table 1. Effect of sage essential oil (EO) of tomato fruit decay at breaker and red ripening stage, exposed to ambient air (control) or EO (50 or 500 μL L^{-1}) for 14 days (Sustain effect—S) or up to 7 days and then transferred to ambient air for an additional 7 days (memory effect—M). Fruit were maintained throughout at 11 °C and 90% RH. The degree of infection on fruit was rated using a scale of 1 to 5 (1-clean, no infection; 2-trace infection; 3-slight infection; 4-moderate infection; 5-severe infection).

		Fruit Decay					
	Treatments	Breaker Stage			Red Stage		
		0 Days	7 Days	14 Days	0 Days	7 Days	14 Days
S/M	Control	1.00 ± 0.00 *	1.17 ± 0.12 a	2.05 ± 0.15 a	1.00 ± 0.00 *	1.16 ± 0.10 a	2.75 ± 0.22 a
S	EO-50 μL L^{-1}		1.03 ± 0.04 a	1.10 ± 0.15 b		1.06 ± 0.07 a	1.25 ± 0.10 b
S	EO-500 μL L^{-1}		1.00 ± 0.00 a	1.00 ± 0.00 b		1.00 ± 0.00 a	1.00 ± 0.00 b
M	EO-50 μL L^{-1}			1.15 ± 0.10 b			1.40 ± 0.20 b
M	EO-500 μL L^{-1}			1.30 ± 0.25 b			1.95 ± 0.35 ab

Values represent the mean (±SE) of evaluation made on eight independent fruit per treatment per storage period. Values followed by the same letter in each column do not differ significantly ($p < 0.05$). Symbols of * indicating significance among controls through storage period.

2.2. Fruit Weight Loss, Firmness and Colour

Fruit weight loss increased when storage time was extended, reaching 1.65% in control and in 1.32% in 500 μL L^{-1} EO-treated breaker fruit after 14 days at 11 °C (Figure 2A) while the relevant values in red fruits were 1.61% and 1.17%, respectively (Figure 2B). Fruit weight loss (%) was at similar levels for both breaker and red tomatoes during 2 days of storage. However, fruit weight loss was significantly decreased (up to 45%) in EO-treated tomatoes after 7 and 14 days, comparing with fruits maintained throughout in ambient air at 11 °C (Figure 2A,B). Interestingly, the effects were persisted when fruit removed from 50 µL L^{-1} of EO exposure (including 500 µL L^{-1} of EO for red tomatoes), and stored additionally for seven days (memory effect).

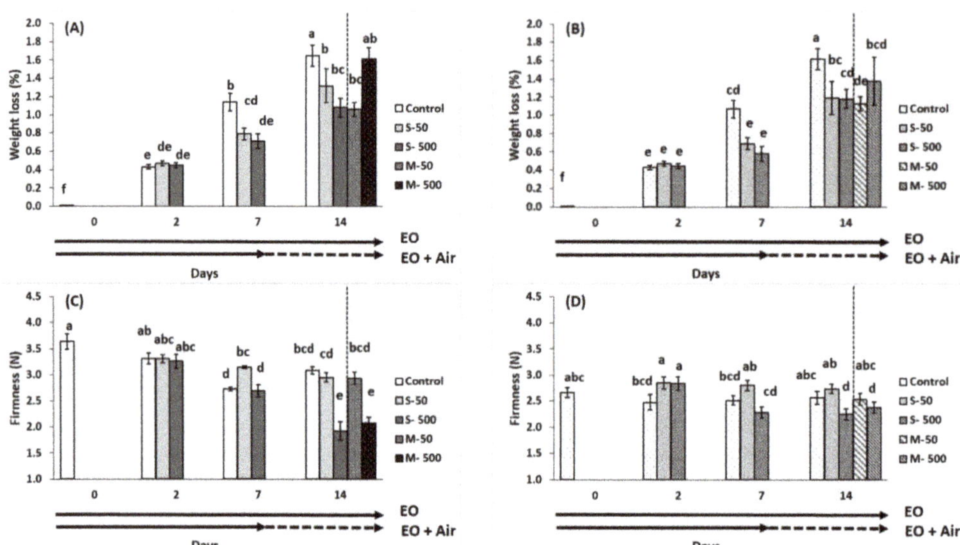

Figure 2. Impacts of sage essential oil (EO) on weight loss (%) and firmness (expressed in Newtons) of tomato fruit at breaker (**A**,**C**) and red (**B**,**D**) ripening stage, exposed to ambient air (control) or EO (50 or 500 µL L^{-1}) for 2, 7, and 14 days (sustain effect—S) or up to 7 days and then transferred to ambient air for an additional 7 days (memory effect—M). Fruits were maintained throughout at 11 °C and 90% RH. Values represent mean (±SE) of measurements made on eight independent fruit per treatment and storage period. Means followed by different Latin letters significantly differ according to Duncan's MRT (p = 0.05).

The effect of EO and fruit ripening stage on the tomato firmness is presented in Figure 2C,D. Two days' storage in an EO-enriched atmosphere revealed no changes in the tomato firmness for breaker stage fruits, but maintained in red tomatoes. Tomatoes-enriched with 50 µL L^{-1} EOs maintained fruit firmness up to 14 days comparing with higher concentration (500 µL L^{-1}) in both red and breaker fruits. However, when treated fruit was transferred to ambient air, breaker and red fruit previously exposed to 50 µL L^{-1} EO remained substantially (p = 0.01) firmer than fruit subjected to 500 mg L^{-1} EO storage conditions throughout.

Fruit colour was mainly affected by the storage and ripening stage of tomatoes rather than the EO application (Figure S1). At breaker stage, L^* value was greater in 500 µL L^{-1} EO application at 14 days of storage for both sustain and memory treatments (Figure S1A). Moreover, breaker-tomatoes revealed decreased chroma and a^* value but increased b^* value in 500 µL L^{-1} EO application at 14 days of storage (Figure S1C,G). Red tomatoes revealed increased L^* value at 2 and 7 days but decreased Chroma and a^* value at high EO (500 µL L^{-1}) application at 14 days of storage (Figure S1B,D,H).

2.3. Soluble Solids, Organic Acid and Ripening Index

Total soluble solids (TSS) levels attained a maximum after 14 days of storage for breaker and red tomatoes (Figure 3A,B). Two days' storage in an EO-enriched atmosphere revealed in increased levels for TSS content in red tomatoes. However, soluble sugar levels were found reduced in breaker tomatoes either exposed to EO (500 µL L^{-1}) for 14 days or pre-exposed to EO for 7 days and then transferred to ambient air (Figure 3A). Citric acid content measured by titratable acidity (TA), was declined (p < 0.05) as fruit ripened

(Figure 3C,D), whereas EO application in general resulted in no changes in the citric acid content. Ripening index indicated by the ratio of TSS/TA did not differ among the tested applications and/or storage period (data not presented).

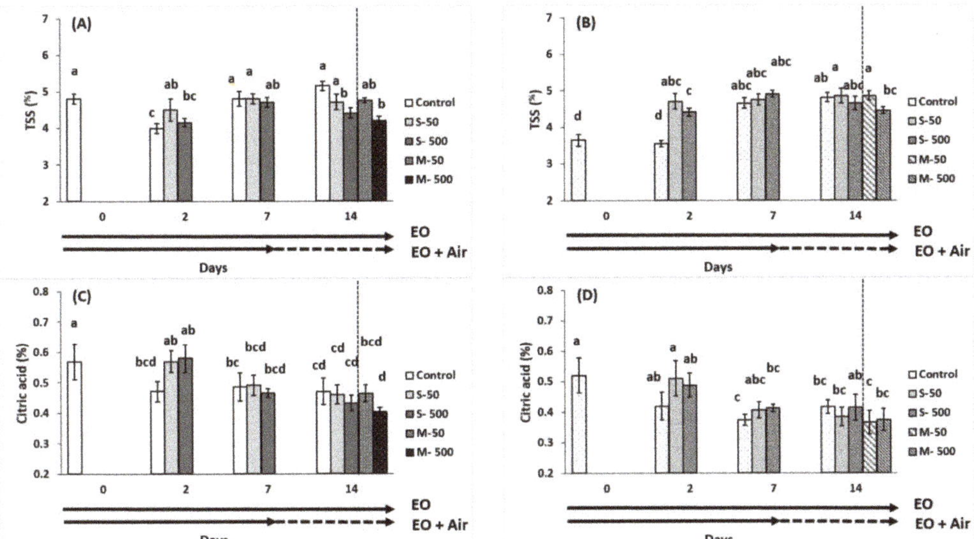

Figure 3. Impacts of sage essential oil (EO) on total soluble solids (TSS; in %) and acidity (% citric acid) of tomato fruit at breaker (**A,C**) and red (**B,D**) ripening stage, exposed to ambient air (control) or EO (50 or 500 µL L^{-1}) for 2, 7, and 14 days (sustain effect—S) or up to 7 days and then transferred to ambient air for an additional 7 days (memory effect—M). Fruits were maintained throughout at 11 °C and 90% RH. Values represent mean (±SE) of measurements made on eight independent fruit per treatment and storage period. Means followed by different Latin letters significantly differ according to Duncan's MRT ($p = 0.05$).

2.4. Respiration Rate and Ethylene Emission

Fruit treated with volatiles (500 µL L^{-1}) revealed an increased respiration rate after 2 and 7 days at breaker tomatoes and after 7 and 14 days of storage at red tomatoes (Figure 4A,B). Low level (50 µL L^{-1}) EO-treated tomatoes at breaker stage respired greater than the relevant control (fruits stored in ambient air) following 14 days of storage. Respiration rate was increased in pre-exposed fruit to EO (500 µL L^{-1}) for red and for breaker (including the 50 µL L^{-1} EO) tomatoes. Indeed, red tomatoes pre-exposed to 50 µL L^{-1} EO and followed 7 days of storage revealed the lowest respiration rate (Figure 4B).

Similar trend to respiration rates was observed in fruit ethylene production (Figure 4C,D). Therefore, the high EO (500 µL L^{-1}) concentration increased the ethylene emission in both exposed and pre-exposed tomatoes to EO, for both breaker and red maturation tomato stages.

2.5. Carotenoid Composition and Ascorbic Acid

In breaker fruits, EO application at 500 µL L^{-1} increased β-carotene content at 7 days of storage but this effect did not persist after 14 days of storage (Figure 5A). Fruits pre-exposed to 500 µL L^{-1} of EO and stored for additional seven days in ambient air marked an increase in β-carotene content, almost twice more than the control fruits. A steady increase in β-carotene content marked at two days of volatiles application in red tomatoes (Figure 5B).

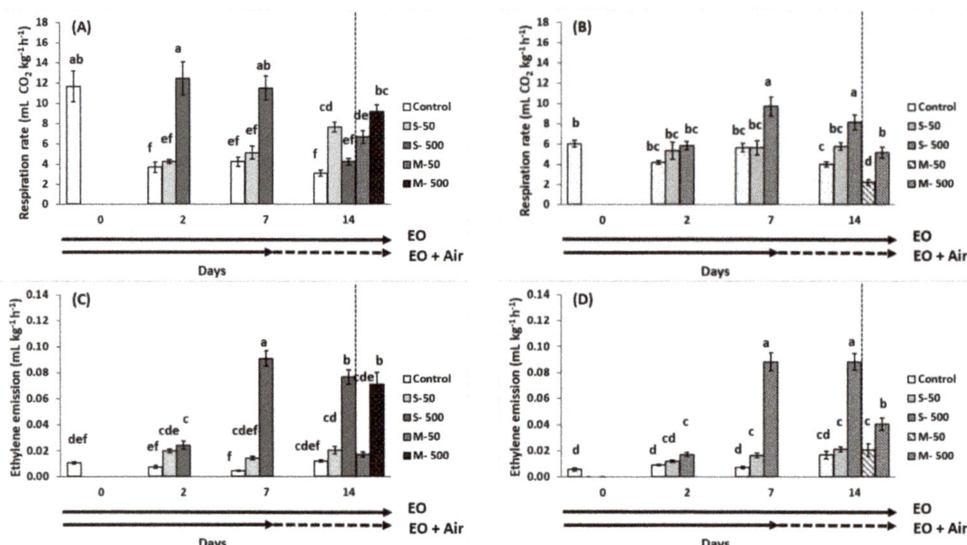

Figure 4. Impacts of sage essential oil (EO) on respiration rate (mL CO_2 kg^{-1} h^{-1}) and ethylene emission (mL kg^{-1} h^{-1}) of tomato fruit at breaker (**A,C**) and red (**B,D**) ripening stage, exposed to ambient air (control) or EO (50 or 500 μL L^{-1}) for 2, 7, and 14 days (sustain effect—S) or up to 7 days and then transferred to ambient air for an additional 7 days (memory effect—M). Fruits were maintained throughout at 11 °C and 90% RH. Values represent mean ($\pm$SE) of measurements made on eight independent fruit per treatment and storage period. Means followed by different Latin letters significantly differ according to Duncan's MRT (p = 0.05).

Sage oil-treated fruit at breaker and red ripening stage with 50 μL L^{-1} maintained or increased lycopene content at 7 and 14 days of exposure, while lycopene content in red tomatoes reduced due to EO application at 2 days of storage (Figure 5C,D). Lycopene content was significantly (p < 0.01) reduced in pre-exposed breaker tomatoes to 500 μL L^{-1} EO, following storage of seven days, but such effects were not found in the relevant red tomatoes, tomatoes pre-exposed to 500 μL L^{-1} EO, and storage of seven days.

EO-enrichment resulted in increased ascorbic acid (AA) content in breaker fruits at 2 days and this effect was persisted in 50 μL L^{-1} EO-treated fruits at 14 days and in the pre-exposed fruits with 50 μL L^{-1} EO and stored for 7 days in ambient air (Figure 5E). Indeed, the relevant pre-exposed fruits with 500 μL L^{-1} EO had decreased AA content compared with the control at 14 days of storage. The non-treated fruits with EOs, revealed increased levels of AA during the storage period. Similarly, in red tomatoes, EO-treated fruits had increased AA content at 2 days but this effect was not persisted thereafter (Figure 5F).

2.6. Total Phenols Content and Antioxidant Activity

EO-treatment tended to decrease or not to affect fruit total phenols content, and the effects did not attain statistical significance (Figure 6A,B). However, non-treated fruits with EOs, revealed increased levels of total phenolics up to seven days of storage, but this effect did not persist thereafter. Antioxidant activity measured by ferric-reducing antioxidant power (FRAP) and 2,2-diphenyl-1-picrylhydrazyl (DPPH) radical scavenging activity methods was reduced in breaker tomatoes exposed to EO for two days. However, at seven days of storage DPPH activity was increased in EO-treated breaker tomatoes (Figure 6C,E). Moreover, pre-exposed fruit to 500 μL L^{-1} of EO following storage of additional 7 days in ambient air revealed decreased DPPH levels, compared to the relevant control at 14 days of storage. In red tomatoes, DPPH antioxidant capacity of fruit was increased with the 500 μL L^{-1} of EO application, and this effect was persisted in pre-exposed fruits as well (Figure 6D). FRAP activity was not changed during storage and/or EO application (Figure 6F).

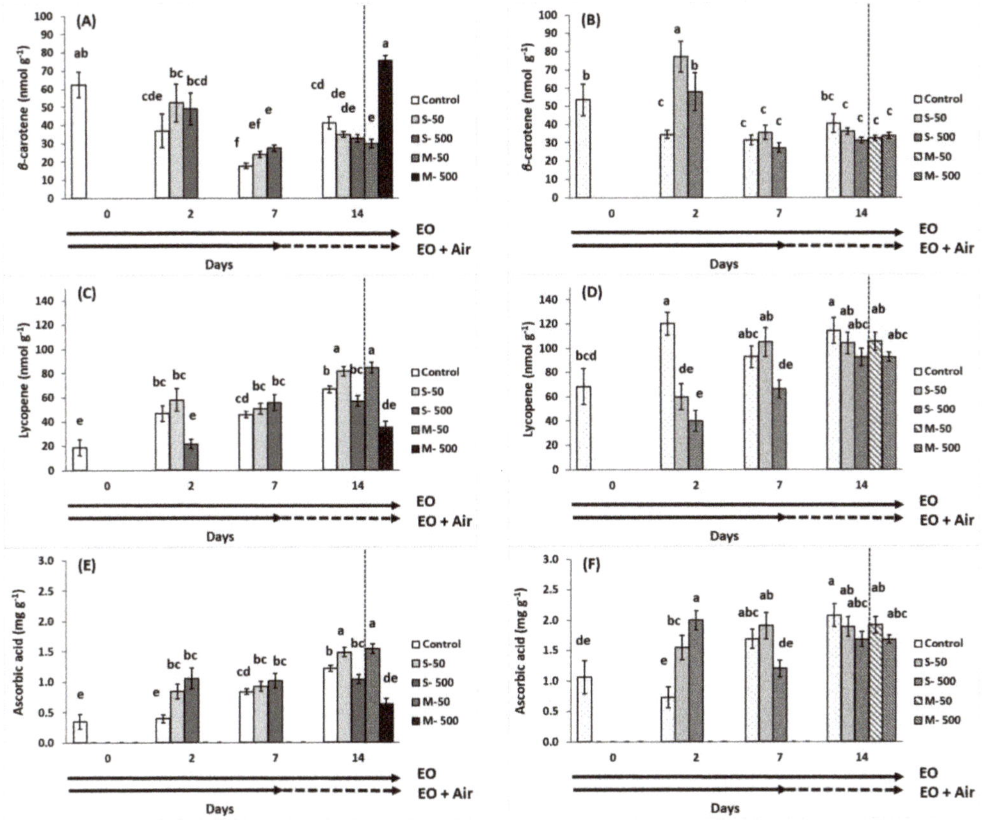

Figure 5. Impacts of sage essential oil (EO) on β-carotene (nmol g^{-1}), lycopene (nmol g^{-1}) and ascorbic acid (mg g^{-1}) in tomato fruit at breaker (**A,C,E**) and red (**B,D,F**) ripening stage, exposed to ambient air (control) or EO (50 or 500 µL L^{-1}) for 2, 7, and 14 days (sustain effect—S) or up to 7 days and then transferred to ambient air for an additional 7 days (memory effect—M). Fruits were maintained throughout at 11 °C and 90% RH. Values represent mean (±SE) of measurements made on eight independent fruit per treatment and storage period. Means followed by different Latin letters significantly differ according to Duncan's MRT ($p = 0.05$).

2.7. Plant Stress Indicators

In breaker and red fruits, hydrogen peroxide (H_2O_2) fluctuated among the treatments without a specific trend (Figure 7A,B). In breaker stage fruits, lipid peroxidation as assessed in terms of malondialdehyde (MDA) content, was not changed in EO-treated fruits and/or during storage duration of 14 days (Figure 7C). However, in red tomatoes MDA decreased at 50 µL L^{-1} of EO application after 7 and 14 days of storage (Figure 7D).

2.8. Sensory Evaluation

Sensory evaluation revealed significant changes between treatments that the jurors were able to detect and marked as 86% and 50% for breaker and red tomatoes. In breaker tomatoes, jurors preferred un-treated fruit with 57% and followed by 50 µL L^{-1} treated fruits with 43%. In the case of the red tomatoes, jurors preferred low EO-treated fruits and then control fruits with 79% and 21%, respectively. Interestingly, no preference was noticed for high-EO treated tomatoes, revealing the lowest sensory scores (Table 2). Those who preferred low EO-treated fruits did so due to the improved appearance and texture, which was comparable to that of the untreated fruits (Table 2). In red tomatoes, taste panelists preferred fruit subjected to low-level EO-enrichment (50 μL L^{-1}), and this effect was mirrored by the increased scores in appearance, color, aroma, texture, and sweetness.

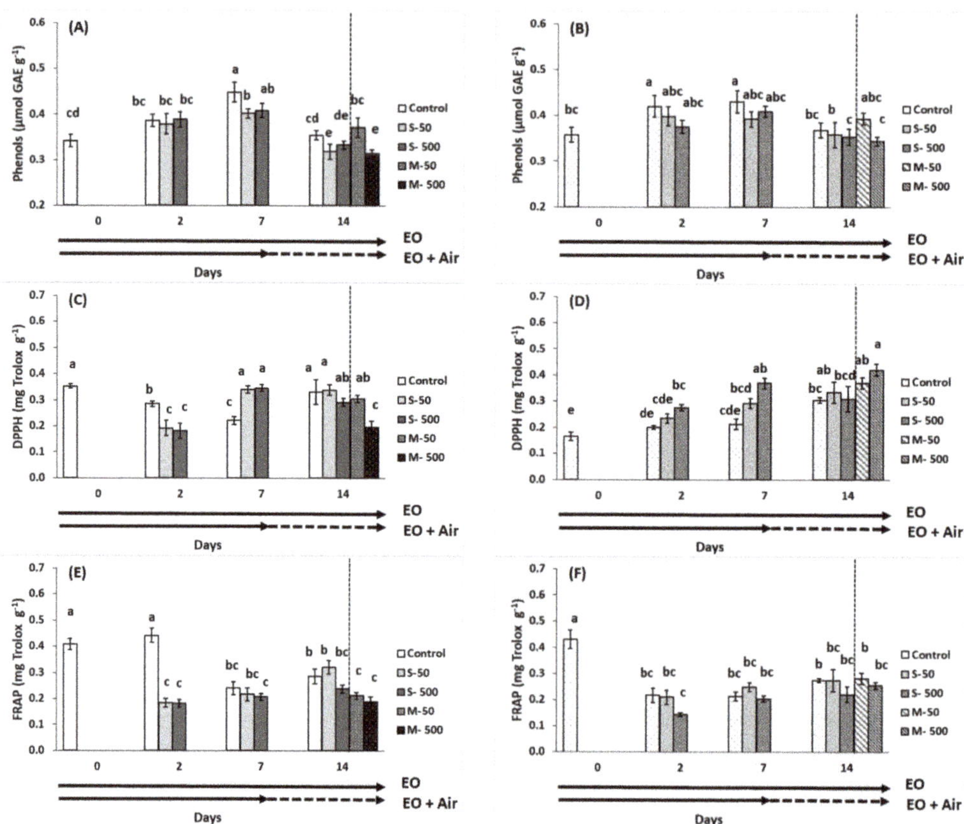

Figure 6. Impacts of sage essential oil (EO) on total phenolics (GAE μmol g^{-1}) and antioxidant activity (mg Trolox g^{-1}) in tomato fruit at breaker (**A,C,E**) and red (**B,D,F**) ripening stage, exposed to ambient air (control) or EO (50 or 500 μL L^{-1}) for 2, 7, and 14 days (sustain effect—S) or up to 7 days and then transferred to ambient air for an additional 7 days (memory effect—M). Fruit were maintained throughout at 11 °C and 90% RH. Values represent mean (±SE) of measurements made on eight independent fruit per treatment and storage period. Means followed by different Latin letters significantly differ according to Duncan's MRT (*p* = 0.05).

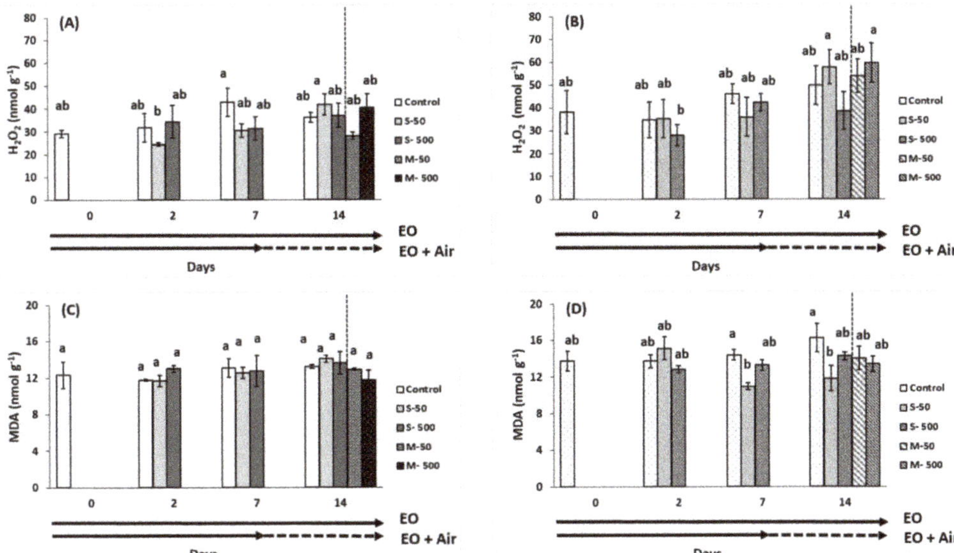

Figure 7. Impacts of sage essential oil (EO) on the fruit content of hydrogen peroxide (H_2O_2) and malondialdehyde (MDA) in tomato fruit at breaker (**A,C**) and red (**B,D**) ripening stage, exposed to ambient air (control) or EO (50 or 500 µL L^{-1}) for 2, 7, and 14 days (sustain effect—S) or up to 7 days and then transferred to ambient air for an additional 7 days (memory effect—M). Fruits were maintained throughout at 11 °C and 90% RH. Values represent mean (±SE) of measurements made on eight independent fruit per treatment and storage period. Means followed by different Latin letters significantly differ according to Duncan's MRT (*p* = 0.05).

Table 2. Quantitative analysis of the impacts of sage essential oil (EO) on sensory attributes of tomato fruit at breaker and red ripening stage, exposed to ambient air (control) or EO (50 or 500 µL L^{-1}) for 14 days. Fruits were maintained throughout at 11 °C and 90% RH. Values represent mean (±SE) of assessments made by 14 panelists per treatment.

	Breaker Tomatoes			Red Tomatoes		
	Control	EO-50 µL L^{-1}	EO-500 µL L^{-1}	Control	EO-50 µL L^{-1}	EO-500 µL L^{-1}
Appearance	61.8 ± 4.6 a	64.4 ± 4.9 a	40.0 ± 4.2 b	71.8 ± 3.4 a	78.8 ± 3.9 a	38.5 ± 4.4 b
Color	70.1 ± 5.8 a	52.8 ± 5.4 b	50.1 ± 5.8 b	74.2 ± 4.9 a	79.5 ± 3.3 a	51.4 ± 5.0 b
Aroma	70.0 ± 5.7 a	57.0 ± 4.5 b	24.2 ± 2.2 c	68.5 ± 4.1 a	67.1 ± 4.5 a	27.1 ± 2.6 b
Texture	65.8 ± 5.7 a	72.7 ± 6.4 a	34.2 ± 3.8 b	65.2 ± 3.8 a	74.0 ± 3.5 a	30.0 ± 3.4 b
Sweetness	45.7 ± 3.8 a	38.5 ± 3.9 ab	30.0 ± 4.1 b	67.1 ± 5.7 a	59.2 ± 3.8 a	32.8 ± 4.5 b
Satisfaction	61.4 ± 3.9 a	49.7 ± 5.3 b	22.8 ± 1.9 c	68.5 ± 4.0 a	55.7 ± 5.2 b	21.4 ± 1.4 c
Marketability	67.1 ± 5.7 a	54.7 ± 5.8 b	21.4 ± 1.4 c	80.0 ± 5.5 a	64.2 ± 7.6 b	20.0 ± 0.0 c

In each row, mean values in percentage followed by the same small (breaker stage) are not significantly different, *p* ≤ 0.05.

3. Discussion

Only fresh produce that meets the consumer's standards is suitable at the market interface. As a result, evaluating the impact of possible novel procedures on the sensory and organoleptic features of fruits and vegetables is critical. Weight loss, color, firmness, total soluble solids, total acidity, and antioxidants are only a few attributes that affect postharvest fruit quality. Moreover, the postharvest performance of the tomato ripening stage and understanding the physiological changes taken place during storage are of high research interest [37]. In the present study, mature green tomato fruit when were subjected to EO-enrichment (sustained effect) were perceptibly retained their firmness in low EO levels (50 µL L^{-1}). However, the rates of respiration and ethylene as well as the antioxidant metabolism were increased in high EO levels of 500 µL L^{-1}, and the effects were more pronounced during the storage period of 14 days, in comparison to the control fruits (subjected to typical storage and transportation methods). When red tomatoes (being in higher maturation stage compared to mature green fruit) were subjected to EOs, effects on quality attributes were appeared even earlier, after two days of EOs exposure, with increase of TSS, and β-carotene and decrease on lycopene content. Considering the pre-exposed fruits to EOs, quality attributes were more affected in mature green fruits and to a lesser level in the red fruits. Furthermore, based on appearance, color, and texture evaluations, taste panel trials demonstrated an overwhelming preference for EO-treated red fruits during choice testing.

The relationship between increased ethylene production and tomato ripening is well understood [33] and effects are related to the fruit ripening stage, by altering signaling genes related to ethylene metabolic pathway [37]. In addition to the ripening stage, biotic and abiotic stresses have an impact on ethylene production [38]. In tomato fruit, the increase in respiration occurs either concurrently or shortly after the increase in ethylene production [39,40] and this was evidenced in both mature-green and red tomatoes, starting from the 2nd day up to 14th day of storage for the high EO concentration of 500 μL L^{-1}. Interestingly, mature green tomatoes stimulated more the respiration rates compared to the relevant red tomatoes with the EO of 500 μL L^{-1}. When ethylene is added to mature-green tomato fruit, ethylene hastens the climacteric and ripening process [37], meaning induced respiratory climacteric due to increased endogenous ethylene output in tomato [41]. That could be the case in our study, but further research is needed to that direction before final conclusions. As a result, a comprehensive investigation at the molecular level is required to investigate the effects of EOs on gene and/or protein expression in metabolic pathways, such as the ethylene biosynthesis pathway, which is linked to fruit ripening (particularly in climacteric fruits such as tomatoes).

Fresh commodities loss weight mainly by vapor pressure at different locations [42] but also through the respiration process [43]. A loss of more than 5%, on the other hand, is a limiting factor for the fruit marketing and consumption [44]. However, weight loss in the present study was <1.8%. Fruit treated with EO lost less weight during storage than fruit that had not been treated with EOs, and weight loss increased progressively over time. This decrease in weight loss could be attributed to the ability of the EOs to decrease water exchange and solute movement due to EOs hydrophobic properties [45]. The ability of the essential oil to act as a barrier and the antioxidant activity of the essential oil coatings were responsible for the reduced weight loss rate in coated fruits during storage [46,47].

During fruit maturation and storage time, titratable acidity is decreased and TSS is increased in general, as this trend was observed in our study. TA was decreased up to 17% from day 0 to day 14, and the values were ranged from 0.2 to 0.6%, being in agreement with previous records [40].

EO application as preservative means is well documented due to their antimicrobial and antioxidant activities [12,13,48,49]. Moreover, sage (*Salvia officinalis* L.) antibacterial and antifungal properties have been reported previously [50]. In the present study, sage EO maintained their antimicrobial efficiency up to 14 days, with reduced decay symptoms, being in accordance with previous applications of cinnamon and eucalyptus EO on tomatoes and strawberries [22]. In the present study, the main component of sage EO was eucalyptol, as described at the Section 4, with proven antimicrobial activity [51,52]. Additionally, secondary components of sage EO, such as camphor and α-pinene have also antimicrobial activity [53,54]. Both primary and secondary components of an EO contribute to the antimicrobial activity of the oil, affecting the quality of the fresh produce. Fruit decay causes metabolic alterations that are responsible for unpleasant smell and flavor [40]. Based on the findings of this study, it is hypothesized that the active component in sage EO continues to be released throughout storage, extending the fruit's shelf-life. Additionally, the effects were persisted even when fruits removed from the EO and were stored in ambient air for seven days, indicating a residual effect. Similarly, sage EO revealed residual effects in pepper fruits [36].

The pigment content of the fruit changes during development, whereas the chlorophyll level falls during ripening, prompting the synthesis of carotenoids, including the red pigment lycopene as well as β-carotene. In red tomatoes 500 µL L^{-1} EO-treated fruits had lighter (higher L^* value) color than the untreated ones during two and seven days of storage at 11 °C, suggesting delayed color development by EO treatment [55]. This was evidenced by the decreased lycopene levels for the EO-treated fruits up to seven days of storage. However, this effect did not persist after 14 days of storage. Noticeably, delay in color development was evidenced in pre-exposed mature-green tomatoes to 500 µL L^{-1} EO-treated after seven days of EO exposure and additional seven days of storage in clean air (as "7 + 7 days" treatment).

The mechanisms underlying the effects of EOs on fruit firmness are unknown. However, it is known that during fruit ripening, cell wall matrices, particularly pectins, are disrupted, and these modifications are thought to be responsible for the decrease in tissue firmness that occurs with ripening [56,57]. In the present study, fruit firmness was maintained in tomatoes-enriched with 50 µL L^{-1} EOs for up to 14 days, compared with higher concentration (500 µL L^{-1}) in both red and breaker fruits. The effect of EO was even persisted in fruits pre-exposed to EO (50 µL L^{-1}) and stored for an additional seven days in ambient air.

Depending on the species, cultivar, temperature, and climatic and environmental conditions during the growing period, the evolution of total phenolics in fruit during storage could be different [58]. The key contributors to the soluble antioxidant activity in tomato fruit, ascorbic acid, and soluble phenolics increased with storage, resulting in an increase in antioxidant activity in tomato fruit [59]. According to one study, ascorbic acid comprises 28–38% of soluble antioxidant activity, with soluble phenolics accounting for the rest [60]. In our study, increased AA levels were found at two days of EO-treated fruits reflecting the increased DPPH levels found at two and seven days in red tomatoes and at seven days in breaker fruits. Antioxidants help to avoid the build-up of potentially harmful reactive oxygen species (ROS), which are produced as a by-product of cellular metabolism and serve as secondary messengers in hormone signaling [61]. Since tomato fruit is known to be particularly rich in antioxidants [62], such as vitamin C, carotenoids (especially lycopene; [62]), and vitamin A, the antioxidative characteristics of tomato fruit and tomato products are affected by storage procedures, which is a source of worry [63]. Indeed, tomato fruit is an essential nutritional source of several of these compounds, which are vital in the prevention of chronic diseases including heart disease and cancer [64]. The temporary rise in AA content in breaker and red tomatoes (including β-carotene in red tomatoes) after two days of EO-enriched atmosphere is noteworthy in this regard. Moreover, EO of 50 µL L^{-1} in red tomatoes kept MDA levels down indicating less stress on the fruits.

During choice testing, panel trials demonstrated a clear preference for EO of 50 µL L^{-1} treated fruits in red tomatoes compared to the untreated fruits, while the opposite was evidenced in breaker tomatoes. Appearance and texture were the main indicators for breaker fruits, while for red tomatoes, not only appearance and texture but also color, aroma, and sweetness were scored to similar levels in low EO-treated fruits and in the control.

4. Materials and Methods

4.1. Plant Material and Experimental Design

Tomato fruit (*Solanum lycopersicum* L. cv. Dafni F1) was collected from a local field Limassol, Cyprus (crop cultivated for six months under commercial conditions and standard cultural practices in a clay loam soil [65], frequently irrigated by drippers according to crop needs, during spring with temperatures ranging from 18 °C to 28 °C). Fruits were collected by the third inflorescences of the plants. At the laboratory, fruits were selected to obtain homogeneous batches based on color, size, ripeness [breaker stage-mature green (two and three ripening stage)—and light red and red (five and six ripening stage)] and free from defect or injury and then were utilized for experimental purposes. To avoid microbial contamination, the fruits were submerged in a diluted chlorine solution for 3 min before being washed four times with distilled water.

Organic essential oils were extracted by hydrodistillation from sage [*Salvia trilova* L. (Lamiaceae)] gathered in a hilly area of Crete, Greece (without any human inputs) (Clevenger apparatus for 3 h). The composition of the EO was analysed by Gas Chromatography-Mass Spectroscopy (GC-MS), and the main (>2.0%) components were: α-Pinene (3.1%), Camphene (2.3%), β-Pinene (4.1%), Eucalyptol (53.5%), *cis*-Thujone (6.7%), *trans*-Thujone (3.3%) and Camphor (7.9%), as described previously [36].

Breaker and fully ripe tomato fruits were placed in 1 L polystyrene containers with snap-on lids for each treatment. Two tomatoes were placed in each container, resulting in eight containers (biological replications) per treatment for each of the storage periods. Sage EO used in this study (concentrations based on previous research [22]) were 50 µL L^{-1} and 500 µL L^{-1}. Aliquot of each EO solution was placed into individual Eppendorf (1.5 mL) tubes, which were subsequently placed inside the plastic containers shortly before the lids were covered. Filter paper dampened with water was inserted in each container to maintain high relative humidity level during storage, as described in Tzortzakis [22]. The EO volatile components were allowed to spontaneously evaporate inside the containers at 20 °C for 2 h. The containers were then moved to a cold room for storage. Tomato fruit exposed to control (ambient air) or EO (sustainable effect—SE) for 2, 7, and 14 days at 11 °C and 90% relatively humidity (RH~90%) in darkness. Following 1-week exposure, a second batch of fruits were transferred to ambient air and stored at 11 °C for an additional one week (memory effect-ME) named as "7 + 7 d" treatment. To summarize, the experimental set up consisted of 3 treatments × 2 ripening stages × 8 replications (2 fruits per replication) × 4 storage periods (plus day 0) with total of 400 fruits used (Figure 1). Sixteen samples of treated and control fruits were taken after 2, 7, and 14 days and 7 + 7 days for immediate analysis for each ripening stage. For day zero measurements, washed fruits (eight containers) with chlorine were used. Containers were aerated every 72 h avoiding air concentration abnormalities. Volatiles exposure did not cause any phytotoxic effect on the tomato fruit.

4.2. Decay Evaluation

After 2, 7, and 14 days of storage at 11 °C, the severity of fruit degradation (in individual fruits in each container; total 16 fruits per treatment per storage period) was visually assessed. Tomato fruit showing surface mycelia growth was considered decayed. On a scale of one to five, the degree of infection on fruit was rated: 1-clean, no infection, 2-trace infection, 3-slight infection, 4-moderate infection, and 5-severe infection. Rots were distinguished by tomato tissue subculture onto Potato Dextrose Agar (PDA) media as described previously [66].

4.3. Respiration Rate and Ethylene Emission

The carbon dioxide (CO_2) and ethylene production were measured by placing each tomato in a 1 L glass jar hermetically sealed with a rubber stopper for 1 h at ambient room temperature. Fruits were weighed and volume was measured. Additionally, CO_2 and ethylene of room air were tested and subtracted from the measurements, by equipment zeroing, prior to and during experimentation. For respiration rate determination, the holder atmosphere was sucked by a dual gas analyzer (International Control Analyser Ltd., UK) for 30 s. Results were the mean of two determinations for each jar (eight jars per treatment and storage period; $n = 8$) and expressed as milliliter of CO_2 per kilogram per hour. Ethylene was quantified by using an ethylene analyzer (ICA 56 Analyser, International Control Analyser Ltd., UK) whereas container air sample was sucked for 30 s. Results were the mean of two determinations for each jar and expressed as microliter of ethylene per kilogram per hour (eight jars per treatment and storage period; $n = 8$). CO_2 and ethylene evolution were calculated according to the following Equation: rate of evolution = M × $(V_1 - V_2) \times (1/w) \times (1/t)$; where, M represents the measurement; V_1, V_2 represent jar and fruit volume (mL), respectively; w represents fruit weight (g); and t represents incubation time (h).

4.4. Weight Loss, Colour and Fruit Firmness

Individual tomato weights were measured on the day of harvesting (day 0) and after the different sampling dates. Weight loss was calculated for each fruit ($n = 8$) per treatment and storage time as follows: weight loss % = 100 ($W_o - W_f$)/W_o, with W_o being the initial weight and W_f the final weight of the fruit.

Color was determined using the Hunter Lab System and a Minolta colorimeter model CR400 (Konica Minolta, Osaka, Japan). Following the recording of individual L^*, a^*, and b^* parameters, and chroma value (C) was calculated by the following equations C = $(a^{*2} + b^{*2})^{1/2}$ as described previously [24]. Results were the mean of determinations made on four points for each fruit ($n = 8$) along the equatorial axis, for each treatment and storage time.

Fruit firmness was measured at two points on the shoulder of each tomato fruit (1 cm^2 of skin removed), respectively for each treatment by applying a plunger of 8 mm in diameter, using a texturometer FT 011 (TR Scientific Instruments, Forli, Italy). The amount of force (in Newtons; N) required to break the radial pericarp (i.e., surface) of each tomato ($n = 8$) was recorded at ambient (21–23 °C) temperature for each treatment and storage time.

4.5. Soluble Solids, Titratable Acidity, Ripening Index, Ascorbic Acid and Carotenoids

Total soluble solids concentration was determined in triplicate from the juice obtained from two pooled tomatoes for each replication ($n = 8$) with a temperature-compensated digital refractometer (model Atago PR-101, Atago Co. Ltd., Tokyo, Japan) at 20 °C, and results were expressed in percentage (%). The titratable acidity was measured via potentiometric titration (Mettler Toledo DL22, Columbus, OH, USA) of 5 mL juice diluted to 50 mL with distilled water using 0.1 N NaOH up to pH 8.1. The results were expressed as percentage of citric acid. The ratio of TSS/TA was used to evaluate the sweetness/ripening index of the fruit.

Ascorbic acid (being the major part in Vitamin C) in eight independent pools of tomato juice was determined by the 2,6-Dichloroindophenol titrimetric method [67]. An aliquot of 5 mL of pooled tomato juice was diluted with 5 mL of water and was titrated by the dye solution until the color changed. Data were expressed as mg of ascorbic acid per gram of fresh weight.

Carotenoids (lycopene and β-carotene) were determined according to the Nagata and Yamashita [68] method following modification [69]. Eight individual samples (each sample pooled of two fruits) were examined per treatment and storage period. Thus, 1 g of blended tomatoes were placed in 50 mL falcons and stored in −20 °C till analysis (within 48 h). A volume of 16 mL of acetone:hexane 4:6 (*v:v*) were added to each sample, the samples were shaken vigorously and the two phases were separated automatically. An aliquot was taken from the upper solution for measurement of optical density at 663, 645, 505, and 453 nm in a spectrophotometer, using a reference acetone:hexane (4:6) ratio. Lycopene and β-carotene contents were calculated according to the Nagata and Yamashita [68] equations:

$$\text{Lycopene (mg 100 mL}^{-1} \text{ of extract)} = -0.0458 \times A663 + 0.204 \times A645 + 0.372 \times A505 - 0.0806 \times A453.$$

$$\beta\text{-carotene (mg 100 mL}^{-1} \text{ of extract)} = 0.216 \times A663 - 1.22 \times A645 - 0.304 \times A505 + 0.452 \times A453.$$

Results were expressed as nmol per gram of fresh weight.

4.6. Total Phenols and Antioxidant Activity

Eight individual samples (each sample pooled of two fruits) were examined per treatment and storage period. Samples of 5 g were milled in an Ultraturrax (T25 digital ultra-turrax, IKA, Germany) with 10 mL methanol (50% *v/v*) for 30 s, and polyphenol extraction was assisted with ultrasound (Ultrasonic cleaning baths-150, Raypa, Spain) for

5 min. The slurry was centrifuged for 30 min on $5000\times g$ at 4 °C (Sigma 3–18 K, Sigma Laboratory Centrifuge, Germany). The supernatant was transferred to a 15 mL falcon tube, and was stored at 4 °C until analysis (within 48 h) for evaluation of total phenolic content and total antioxidant activity.

The total phenols content of the methanolic extracts was determined by using Folin–Ciocalteu reagent (Merck), according to the procedure described by Tzortzakis et al. [70]. Briefly, 125 μL of plant extract was mixed with 125 μL of Folin reagent. The mixture was shaken, before addition of 1.25 mL of 7% Na_2CO_3, adjusting with distilled water to a final volume of 3 mL, and thorough mixing. After incubation in the dark for 90 min, the absorbance at 755 nm was measured versus the prepared blank. Total phenolic content was expressed as μmol of gallic acid equivalents (GAE) per gram of fresh weight, through a calibration curve with gallic acid. All samples were analyzed in triplicate.

A sample of 3 mL of freshly prepared ferric-reducing antioxidant power solution (0.3 mol L^{-1} acetate buffer, pH 3.6), containing 10 mmol L^{-1} TPTZ (Tripyridil-s-triazine) and 40 mmol L^{-1} $FeCl_3 \cdot 10H_2O$ and 20 μL of extract (50 mg mL^{-1}) were incubated at 37 °C for 4 min and the absorbance was measured at 593 nm. The absorbance change was converted into a FRAP value, by relating the change of absorbance at 593 nm of the test sample to that of the standard solution of trolox $((\pm)$-6-Hydroxy-2,5,7,8-tetramethylchromane-2-carboxylic acid). Standard curve was prepared using different concentrations of trolox, and the results were expressed as mg trolox per gram of fresh weight [69]. All samples were analysed in triplicate.

Radical-scavenging activity was determined according to Wojdylo et al. [71] with some modifications. The 2,2-diphenyl-1-picrylhydrazyl radical scavenging activity of the plant extracts was measured from the bleaching of the purple-colored 0.3 mM solution of DPPH. One milliliter of the DPPH solution in ethanol, 1.98 mL (50% *v/v*) methanol and 0.02 mL of plant extract were mixed. After shaking, the mixture was incubated at room temperature in the dark for 30 min, and then the absorbance was measured at 517 nm. The results were expressed in mg trolox per gram of fresh weight. All samples were analyzed in triplicate.

4.7. Plant Stress Indicators

Cell damage index of lipid peroxidation in leaves was assessed in terms of malondialdehyde content, which was determined by the thiobarbituric acid reaction [72]. Hydrogen peroxide content was measured according to the method of Loreto and Velikova [73]. The results were expressed as nmol MDA or μmol H_2O_2 per g FW. Four replicates were analyzed for each treatment and sampling date.

4.8. Sensory Evaluation

For the sensory evaluation, 14 panelists of similar ratio of males and females (aged from 22 to 44 years old) were employed to assess fruit of the two ripening stages and subject to storage for 14 days in ambient air or EO-enriched air (50 μL L^{-1} or 500 μL L^{-1}). All panelists had at least some training in the sensory evaluation of tomato fruit. To ensure representative results, the panel was initially asked to assess treatment preferences, with each panelist being given more than one fruit from each sample. Panelists were subsequently challenged with fresh fruit from each treatment and asked to rate appearance, colour, aroma, sweetness, texture, and marketability using scales (values of acceptance) with anchor points 1: 'Poor/unsweet/soft' and 5: 'excellent/very sweet/firm'. Scales were converted to percentage values. Individual panelists were given two sets of fruit (representing the two stages of ripening) and each set had three plates (one for each treatment) containing three whole tomato fruits and three halved tomato fruits for sensory analysis, all tests being conducted under the same conditions and with no time limit. To avoid intermixing of panel members, panel testing was conducted in isolation in booths in the same room.

4.9. Statistical Analysis

The data were checked for normality before being subjected to an analysis of variance (ANOVA). The time of storage and the treatments were the sources of variation. Following one-way ANOVA, significant differences between mean values were detected using Tukey's HSD test ($p = 0.05$). SPSS was used to conduct statistical analysis (SPSS Inc., Chicago, IL, USA).

5. Conclusions

The current study emphasizes the possibility of employing natural volatiles obtained from sage essential oils to preserve tomato fruit during storage and/or transit at 11 °C and high RH levels of 90%. In breaker tomatoes, EO-enrichment (sustained effect) retained fruit firmness, respiration rates, and ethylene emission in low EO levels (50 μL L^{-1}), while fruit metabolism was sped up in high EO levels of 500 μL L^{-1}, with decreased firmness and increased rates of respiration and ethylene and effects on antioxidant capacity. The effects were more pronounced during the storage period of 14 days, in comparison with fruit subject to traditional storage/transit practice. In red fruits, the EOs impacts were evidenced earlier (at two and seven days of storage) with increased rates of respiration and ethylene, increased TSS and β-carotene, and decreased lycopene content. Considering the pre-exposed fruits to EOs, quality attributes were more affected in mature green fruits and to a lesser extent in red fruits. Furthermore, based on appearance, color, and texture evaluations, taste panel trials demonstrated an overwhelming preference for EO-treated red fruit during choice testing. Additional investigation is needed to encapsulate the EOs and to examine the application of EOs mixtures, based on their active ingredients, for the preservation of tomato fruits. The use of natural products to preserve fresh commodities should be researched further to determine the best application conditions (i.e., method, duration, and concentration) for each commodity in each case.

Supplementary Materials: The following are available online at https://www.mdpi.com/article/10.3390/plants10122645/s1. Figure S1: Impacts of sage essential oil (EO) on *L**, *a**, *b** and chroma values in tomato fruit at breaker and red ripening stage, exposed to ambient air (control) or EO (50 or 500 μL L^{-1}).

Author Contributions: P.X.: investigation, data curation; A.C.: investigation, methodology, and writing—original draft preparation, review and editing; C.R.: investigation; N.T.: conceptualization, supervision, formal analysis, writing—review and editing, funding acquisition, and project administration. All authors have read and agreed to the published version of the manuscript.

Funding: This research was funded by PRIMA StopMedWaste project, which is funded by PRIMA, a programme supported by the European Union with co-funding by the Funding Agencies RIF–Cyprus.

Conflicts of Interest: The authors declare no conflict of interest.

References

1. Brummell, D.A.; Harpster, M.H. Cell wall metabolism in fruit softening and quality and its manipulation in transgenic plants. *Plant Mol. Biol.* **2001**, *47*, 311–340. [CrossRef] [PubMed]
2. Ju, Z.; Duan, Y.; Ju, Z. Plant oil emulsion modifies internal atmosphere, delays fruit ripening, and inhibits internal browning in Chinese pears. *Postharvest Biol. Technol.* **2000**, *20*, 243–250. [CrossRef]
3. Spotts, R.A.; Peters, B.B. Chlorine and Chlorine Dioxide for Control of d' Anjou Pear Decay. *Plant Dis.* **1980**, *64*, 1095. [CrossRef]
4. Tzortzakis, N.; Singleton, I.; Barnes, J. Deployment of low-level ozone-enrichment for the preservation of chilled fresh produce. *Postharvest Biol. Technol.* **2007**, *43*, 261–270. [CrossRef]
5. Tzortzakis, N. Ozone: A powerful tool for the fresh produce preservation. In *Postharvest Management Approaches for Maintaining Quality of Fresh Produce*; Siddiqui, M., Zavala, J., Hang, C.-A., Eds.; Springer: Cham, Switzerland, 2016; pp. 175–208.
6. Serrano, M.; Martínez-Romero, D.; Castillo, S.; Guillén, F.; Valero, D. The use of natural antifungal compounds improves the beneficial effect of MAP in sweet cherry storage. *Innov. Food Sci. Emerg. Technol.* **2005**, *6*, 115–123. [CrossRef]
7. Tzortzakis, N. Essential oil: Innovative tool to improve the preservation of fresh produce—A review. *Fresh Prod.* **2009**, *3*, 87–97.
8. Abdollahi, M.; Rezaei, M.; Farzi, G. Improvement of active chitosan film properties with rosemary essential oil for food packaging. *Int. J. Food Sci. Technol.* **2012**, *47*, 847–853. [CrossRef]

9.   Castillo, S.; Navarro, D.; Zapata, P.J.; Guillén, F.; Valero, D.; Serrano, M.; Martínez-Romero, D. Antifungal efficacy of *Aloe vera* in vitro and its use as a preharvest treatment to maintain postharvest table grape quality. *Postharvest Biol. Technol.* **2010**, *57*, 183–188. [CrossRef]

10. Camele, I.; Elshafie, H.S.; Caputo, L.; Sakr, S.H.; De Feo, V. *Bacillus mojavensis*: Biofilm formation and biochemical investigation of its bioactive metabolites. *J. Biol. Res.* **2019**, *92*, 39–45. [CrossRef]

11. Elshafie, H.S.; Sakr, S.; Bufo, S.A.; Camele, I. An attempt of biocontrol the tomato-wilt disease caused by *Verticillium dahliae* using *Burkholderia gladioli* pv. *Agaricicola* and its bioactive secondary metabolites. *Int. J. Plant Biol.* **2017**, *8*, 57–60.

12. Stavropoulou, A.; Loulakakis, K.; Magan, N.; Tzortzakis, N. *Origanum dictamnus* Oil Vapour Suppresses the Development of Grey Mould in Eggplant Fruit in Vitro. *Biomed Res. Int.* **2014**, *2014*, 562679. [CrossRef]

13. Tzortzakis, N.G.; Economakis, C.D. Antifungal activity of lemongrass (*Cympopogon citratus* L.) essential oil against key postharvest pathogens. *Innov. Food Sci. Emerg. Technol.* **2007**, *8*, 253–258. [CrossRef]

14. Tzortzakis, N.G. Ethanol, vinegar and *Origanum vulgare* oil vapour suppress the development of anthracnose rot in tomato fruit. *Int. J. Food Microbiol.* **2010**, *142*, 14–18. [CrossRef]

15. Chrysargyris, A.; Xylia, P.; Botsaris, G.; Tzortzakis, N. Antioxidant and antibacterial activities, mineral and essential oil composition of spearmint (*Mentha spicata* L.) affected by the potassium levels. *Ind. Crops Prod.* **2017**, *103*, 202–212. [CrossRef]

16. Basile, A.; Senatore, F.; Gargano, R.; Sorbo, S.; Del Pezzo, M.; Lavitola, A.; Ritieni, A.; Bruno, M.; Spatuzzi, D.; Rigano, D.; et al. Antibacterial and antioxidant activities in *Sideritis italica* (Miller) Greuter et Burdet essential oils. *J. Ethnopharmacol.* **2006**, *107*, 240–248. [CrossRef]

17. Teixeira, B.; Marques, A.; Ramos, C.; Neng, N.R.; Nogueira, J.M.F.; Saraiva, J.A.; Nunes, M.L. Chemical composition and antibacterial and antioxidant properties of commercial essential oils. *Ind. Crops Prod.* **2013**, *43*, 587–595. [CrossRef]

18. Loizzo, M.R.; Tundis, R.; Menichini, F.; Saab, A.M.; Statti, G.A.; Menichini, F. Cytotoxic activity of essential oils from Labiatae and Lauraceae families against in vitro human tumor models. *Anticancer Res.* **2007**, *27*, 3293–3299. [PubMed]

19. Borges, R.S.; Ortiz, B.L.S.; Pereira, A.C.M.; Keita, H.; Carvalho, J.C.T. *Rosmarinus officinalis* essential oil: A review of its phytochemistry, anti-inflammatory activity, and mechanisms of action involved. *J. Ethnopharmacol.* **2019**, *229*, 29–45. [CrossRef]

20. Lopez-Reyes, J.G.; Spadaro, D.; Prelle, A.; Garibaldi, A.; Gullino, M.L. Efficacy of plant essential oils on postharvest control of rots caused by fungi on different stone fruits in vivo. *J. Food Prot.* **2013**, *76*, 631–639. [CrossRef] [PubMed]

21. Ben Farhat, M.; Jordán, M.J.; Chaouech-Hamada, R.; Landoulsi, A.; Sotomayor, J.A. Variations in essential oil, phenolic compounds, and antioxidant activity of tunisian cultivated *Salvia officinalis* L. *J. Agric. Food Chem.* **2009**, *57*, 10349–10356. [CrossRef]

22. Tzortzakis, N.G. Maintaining postharvest quality of fresh produce with volatile compounds. *Innov. Food Sci. Emerg. Technol.* **2007**, *8*, 111–116. [CrossRef]

23. Tzortzakis, N.; Xylia, P.; Chrysargyris, A. Sage essential oil improves the effectiveness of *Aloe vera* gel on postharvest quality of tomato fruit. *Agronomy* **2019**, *9*, 635. [CrossRef]

24. Xylia, P.; Clark, A.; Chrysargyris, A.; Romanazzi, G.; Tzortzakis, N. Quality and safety attributes on shredded carrots by using *Origanum majorana* and ascorbic acid. *Postharvest Biol. Technol.* **2019**, *155*, 120–129. [CrossRef]

25. Teerarak, M.; Laosinwattana, C. Essential oil from ginger as a novel agent in delaying senescence of cut fronds of the fern (*Davallia solida* (G. Forst.) Sw.). *Postharvest Biol. Technol.* **2019**, *156*, 110927. [CrossRef]

26. Hassani, R.N.; Haghi, D.Z.; Pouya, Z. The effect of clove oil as preservative solution on vase life of cut rose flower. *Int. J. Adv. Sci. Eng. Technol.* **2017**, *5*, 28–30.

27. Cindi, M.D.; Soundy, P.; Romanazzi, G.; Sivakumar, D. Different defense responses and brown rot control in two *Prunus persica* cultivars to essential oil vapours after storage. *Postharvest Biol. Technol.* **2016**, *119*, 9–17. [CrossRef]

28. Elshafie, H.S.; Caputo, L.; De Martino, L.; Gruľová, D.; Zheljazkov, V.Z.; De Feo, V.; Camele, I. Biological investigations of essential oils extracted from three Juniperus species and evaluation of their antimicrobial, antioxidant and cytotoxic activities. *J. Appl. Microbiol.* **2020**, *129*, 1261–1271. [CrossRef]

29. Ben-Jabeur, M.; Ghabri, E.; Myriam, M.; Hamada, W. Thyme essential oil as a defense inducer of tomato against gray mold and Fusarium wilt. *Plant Physiol. Biochem.* **2015**, *94*, 35–40. [CrossRef]

30. Nikkhah, M.; Hashemi, M. Boosting antifungal effect of essential oils using combination approach as an efficient strategy to control postharvest spoilage and preserving the jujube fruit quality. *Postharvest Biol. Technol.* **2020**, *164*, 111159. [CrossRef]

31. United States Department of Agriculture (USDA). *USDA, Agricultural Marketing Service*; USDA: Washington, DC, USA, 1991.

32. Zapata, P.; Guillen, F.; Martinez-Romero, D.; Castillo, S.; Valero, D.; Serrano, M. Use of alginate or zein as edible coatings to delay postharvest ripening process and to maintain tomato (*Solanum lycopersicon* Mill) quality. *J. Sci. Food Agric.* **2008**, *88*, 1287–1293. [CrossRef]

33. Cara, B.; Giovannoni, J.J. Molecular biology of ethylene during tomato fruit development and maturation. *Plant Sci.* **2008**, *175*, 106–113. [CrossRef]

34. Tiecher, A.; de Paula, L.A.; Chaves, F.C.; Rombaldi, C.V. UV-C effect on ethylene, polyamines and the regulation of tomato fruit ripening. *Postharvest Biol. Technol.* **2013**, *86*, 230–239. [CrossRef]

35. Chomchalow, S.; El Assi, N.M.; Sargent, S.A.; Brecht, J.K. Fruit maturity and timing of ethylene treatment affect storage performance of green tomatoes at chilling and nonchilling temperatures. *Horttechnology* **2002**, *12*, 104–114. [CrossRef]

36. Tzortzakis, N.; Chrysargyris, A.; Sivakumar, D.; Loulakakis, K. Vapour or dipping applications of methyl jasmonate, vinegar and sage oil for pepper fruit sanitation towards grey mould. *Postharvest Biol. Technol.* **2016**, *118*, 120–127. [CrossRef]

37. Suzuki, Y.; Nagata, Y. Postharvest ethanol vapor treatment of tomato fruit stimulates gene expression of ethylene biosynthetic enzymes and ripening related transcription factors, although it suppresses ripening. *Postharvest Biol. Technol.* **2019**, *152*, 118–126. [CrossRef]

38. Wang, K.L.C.; Li, H.; Ecker, J.R. Ethylene biosynthesis and signaling networks. *Plant Cell* **2002**, *14*, 131–151. [CrossRef]

39. Colombié, S.; Beauvoit, B.; Nazaret, C.; Bénard, C.; Vercambre, G.; Le Gall, S.; Biais, B.; Cabasson, C.; Maucourt, M.; Bernillon, S.; et al. Respiration climacteric in tomato fruits elucidated by constraint-based modelling. *New Phytol.* **2017**, *213*, 1726–1739. [CrossRef]

40. de Jesús Salas-Méndez, E.; Vicente, A.; Pinheiro, A.C.; Ballesteros, L.F.; Silva, P.; Rodríguez-García, R.; Hernández-Castillo, F.D.; de Lourdes Virginia Díaz-Jiménez, M.; Flores-López, M.L.; Villarreal-Quintanilla, J.Á.; et al. Application of edible nanolaminate coatings with antimicrobial extract of *Flourensia cernua* to extend the shelf-life of tomato (*Solanum lycopersicum* L.) fruit. *Postharvest Biol. Technol.* **2019**, *150*, 19–27.

41. Grierson, D.; Kader, A.A. Fruit ripening and quality. In *The Tomato Crop*; Chapman and Hall: London, UK, 1986; pp. 241–280.

42. Yaman, Ö.; Bayoindirli, L. Effects of an edible coating and cold storage on shelf-life and quality of cherries. *LWT-Food Sci. Technol.* **2002**, *35*, 146–150. [CrossRef]

43. Pan, J.C.; Bhowmilk, S.R. Shelf-life of mature green tomatoes stored in controlled atmosphere and high humidity. *J. Food Sci.* **1992**, *57*, 948–953.

44. Aktas, H.; Bayindir, D.; Dilmaçünal, T.; Koyuncu, M.A. The effects of minerals, ascorbic acid, and salicylic acid on the bunch quality of tomatoes (*Solanum lycopersicum*) at high and low temperatures. *HortScience* **2012**, *47*, 1478–1483. [CrossRef]

45. Shehata, S.A.; Abdeldaym, E.A.; Ali, M.R.; Mohamed, R.M.; Bob, R.I.; Abdelgawad, K.F. Effect of some citrus essential oils on post-harvest shelf life and physicochemical Quality of Strawberries during Cold Storage. *Agronomy* **2020**, *10*, 1466. [CrossRef]

46. Martínez, K.; Ortiz, M.; Albis, A.; Castañeda, C.G.G.; Valencia, M.E.; Tovar, C.D.G. The effect of edible chitosan coatings incorporated with thymus capitatus essential oil on the shelf-life of strawberry (*Fragaria* x *ananassa*) during cold storage. *Biomolecules* **2018**, *8*, 155. [CrossRef] [PubMed]

47. Dhital, R.; Mora, N.B.; Watson, D.G.; Kohli, P.; Choudhary, R. Efficacy of limonene nano coatings on post-harvest shelf life of strawberries. *LWT* **2018**, *97*, 124–134.

48. Camele, I.; Elshafie, H.S.; Caputo, L.; De Feo, V. Anti-quorum Sensing and Antimicrobial Effect of Mediterranean Plant Essential Oils Against Phytopathogenic Bacteria. *Front. Microbiol.* **2019**, *10*, 2619.

49. Elshafie, H.S.; Camele, I. An overview of the biological effects of some mediterranean essential oils on human health. *Biomed Res. Int.* **2017**, *2017*, 9268468. [CrossRef]

50. Elshafie, H.S.; Sakr, S.; Mang, S.M.; De Feo, V.; Camele, I. Antimicrobial activity and chemical composition of three essential oils extracted from Mediterranean aromatic plants. *J. Med. Food.* **2016**, *19*, 1096–1103. [CrossRef]

51. Morcia, C.; Malnati, M.; Terzi, V. In vitro antifungal activity of terpinen-4-ol, eugenol, carvone, 1,8-cineole (eucalyptol) and thymol against mycotoxigenic plant pathogens. *Food Addit. Contam. Part A* **2012**, *29*, 415–422.

52. Xylia, P.; Chrysargyris, A.; Ahmed, Z.F.; Tzortzakis, N. Application of Rosemary and Eucalyptus Essential Oils and Their Main Component on the Preservation of Apple and Pear Fruits. *Horticulture* **2021**, *7*, 479. [CrossRef]

53. Mokbel, A.A.; Alharbi, A.A. Antifungal effects of basil and camphor essential oils against *Aspergillus flavus* and *A. parasiticus*. *Aust. J. Crop Sci.* **2015**, *9*, 532–537.

54. Da Silva, A.C.R.; Lopes, P.M.; de Azevedo, M.M.B.; Costa, D.C.M.; Alviano, C.S.; Alviano, D.S. Biological Activities of α-Pinene and β-Pinene Enantiomers. *Molecules* **2012**, *17*, 6305–6316. [CrossRef] [PubMed]

55. Jobling, J.; Pradhan, R.; Morris, S.C.; Mitchell, L.; Rath, A.C. The effect of ReTain plant growth regulator [aminoethoxyvinylglycine (AVG)] on the postharvest storage life of 'Tegan Blue' plums. *Aust. J. Exp. Agric.* **2003**, *43*, 515–518. [CrossRef]

56. Onelli, E.; Ghiani, A.; Gentili, R.; Serra, S.; Musacchi, S.; Citterio, S. Specific changes of exocarp and mesocarp occurring during softening differently affect firmness in melting (MF) and non melting flesh (NMF) fruits. *PLoS ONE* **2015**, *10*, e0145341. [CrossRef]

57. Saladié, M.; Matas, A.J.; Isaacson, T.; Jenks, M.A.; Goodwin, S.M.; Niklas, K.J.; Xiaolin, R.; Labavitch, J.M.; Shackel, K.A.; Fernie, A.R.; et al. A reevaluation of the key factors that influence tomato fruit softening and integrity. *Plant Physiol.* **2007**, *144*, 1012–1028. [CrossRef] [PubMed]

58. Kalt, W.R. Concise Reviews/Hypotheses in Food Science Effects of Production and Processing Factors on Major Fruit and Vegetable Antioxidants. *J. Food Sci.* **2005**, *70*, R11–R19. [CrossRef]

59. Jagadeesh, S.L.; Charles, M.T.; Gariepy, Y.; Goyette, B.; Raghavan, G.S.V.; Vigneault, C. Influence of Postharvest UV-C Hormesis on the Bioactive Components of Tomato during Post-treatment Handling. *Food Bioprocess Technol.* **2011**, *4*, 1463–1472. [CrossRef]

60. Toor, R.K.; Savage, G.P. Antioxidant activity in different fractions of tomatoes. *Food Res. Int.* **2005**, *38*, 487–494. [CrossRef]

61. Andrews, P.K.; Fahy, D.A.; Foyer, C.H. Relationships between fruit exocarp antioxidants in the tomato (*Lycopersicon esculentum*) high pigment-1 mutant during development. *Physiol. Plant.* **2004**, *120*, 519–528. [CrossRef]

62. Beecher, G. Nutrient content of tomatoes and tomato products. *Proc. Soc. Exp. Biol. Med.* **1998**, *218*, 98–100. [CrossRef]

63. Scalfi, L.; Fogliano, V.; Pentangelo, A.; Graziani, G.; Giordano, I.; Ritieni, A. Antioxidant activity and general fruit characteristics in different ecotypes of Corbarini small tomatoes. *J. Agric. Food Chem.* **2000**, *48*, 1363–1366. [CrossRef]

64. Arai, Y.; Watanabe, S.; Kimira, M.; Shimoi, K.; Mochizuki, R.; Kinae, N. Dietary intakes of flavonols, flavones and isoflavones by Japanese women and the inverse correlation between quercetin intake and plasma LDL cholesterol concentration. *J. Nutr.* **2000**, *130*, 2243–2250. [CrossRef] [PubMed]

65. U.S. Department of Agriculture (USDA). *USDA Agricultural Handbook No. 18*; U.S. Government Printing Office: Washington, DC, USA, 1951.
66. Tzortzakis, N.; Singleton, I.; Barnes, J. Impact of low-level atmospheric ozone-enrichment on black spot and anthracnose rot of tomato fruit. *Postharvest Biol. Technol.* **2008**, *47*, 1–9. [CrossRef]
67. Horwitz, W.; Latimer, G.W.; AOAC International. *Official Methods of Analysis of AOAC International*, 18th ed.; AOAC International: Gaithersburg, MD, USA, 2007.
68. Nagata, M.; Yamashita, I. Simple Method for Simultaneous Determination of Chlorophyll and Carotenoids in Tomato Fruit. *Nippon Shokuhin Kogyo Gakkaishi* **1992**, *39*, 925–928. [CrossRef]
69. Chrysargyris, A.; Nikou, A.; Tzortzakis, N. Effectiveness of *Aloe vera* gel coating for maintaining tomato fruit quality. *N. Z. J. Crop Hortic. Sci.* **2016**, *44*, 203–217. [CrossRef]
70. Tzortzakis, N.G.; Tzanakaki, K.; Economakis, C.D. Effect of origanum oil and vinegar on the maintenance of postharvest quality of tomato. *Food Nutr. Sci.* **2011**, *2*, 974–982. [CrossRef]
71. Wojdyło, A.; Oszmiański, J.; Czemerys, R. Antioxidant activity and phenolic compounds in 32 selected herbs. *Food Chem.* **2007**, *105*, 940–949. [CrossRef]
72. De Azevedo Neto, A.D.; Prisco, J.T.; Enéas-Filho, J.; De Abreu, C.E.B.; Gomes-Filho, E. Effect of salt stress on antioxidative enzymes and lipid peroxidation in leaves and roots of salt-tolerant and salt-sensitive maize genotypes. *Environ. Exp. Bot.* **2006**, *56*, 87–94. [CrossRef]
73. Loreto, F.; Velikova, V. Isoprene produced by leaves protects the photosynthetic apparatus against ozone damage, quenches ozone products, and reduces lipid peroxidation of cellular membranes. *Plant Physiol.* **2001**, *127*, 1781–1787. [CrossRef]

 *plants*

Article

A New Sesquiterpene Essential Oil from the Native Andean Species *Jungia rugosa* Less (Asteraceae): Chemical Analysis, Enantiomeric Evaluation, and Cholinergic Activity

Karyna Calvopiña [1,2], Omar Malagón [1], Francesca Capetti [3], Barbara Sgorbini [3], Verónica Verdugo [1,4] and Gianluca Gilardoni [1,*]

[1] Departamento de Química, Universidad Técnica Particular de Loja, Calle M. Champagnat s/n, Loja 110107, Ecuador; kmcalvopina1@utpl.edu.ec (K.C.); omalagon@utpl.edu.ec (O.M.); vmverdugo@utpl.edu.ec (V.V.)
[2] Carrera de Ingeniería Química, Facultad de Ingenierías, Universidad Técnica "Luis Vargas Torres" de Esmeraldas, Ciudadela Nuevos Horizontes s/n, Esmeraldas 179619, Ecuador
[3] Dipartimento di Scienza e Tecnologia del Farmaco, Università degli Studi di Torino, 10125 Torino, Italy; francesca.capetti@unito.it (F.C.); barbara.sgorbini@unito.it (B.S.)
[4] Unidad Educativa Ambrosio Andrade Palacios-Suscal, Vía Durán Tambo Eloy Alfaro, Suscal 030206, Ecuador
* Correspondence: ggilardoni@utpl.edu.ec or gianluca.gilardoni@gmail.com

Citation: Calvopiña, K.; Malagón, O.; Capetti, F.; Sgorbini, B.; Verdugo, V.; Gilardoni, G. A New Sesquiterpene Essential Oil from the Native Andean Species *Jungia rugosa* Less (Asteraceae): Chemical Analysis, Enantiomeric Evaluation, and Cholinergic Activity. *Plants* **2021**, *10*, 2102. https://doi.org/10.3390/plants10102102

Academic Editors: Hazem Salaheldin Elshafie, Laura De Martino and Adriano Sofo

Received: 6 September 2021
Accepted: 23 September 2021
Published: 4 October 2021

Publisher's Note: MDPI stays neutral with regard to jurisdictional claims in published maps and institutional affiliations.

Abstract: As part of a project devoted to the phytochemical study of Ecuadorian biodiversity, new essential oils are systematically distilled and analysed. In the present work, *Jungia rugosa* Less (Asteraceae) has been selected and some wild specimens collected to investigate the volatile fraction. The essential oil, obtained from fresh leaves, was analysed for the first time in the present study. The chemical composition was determined by gas chromatography, coupled to mass spectrometry (GC-MS) for qualitative analysis, and to flame ionization detector (GC-FID) for quantitation. The calculation of relative response factors (RRF), based on combustion enthalpy, was carried out for each quantified component. Fifty-six compounds were identified and quantified in a 5% phenyl-polydimethylsiloxane non-polar column and 53 compounds in a polyethylene glycol polar column, including four undetermined compounds. The main feature of this essential oil was the exclusive sesquiterpenes content, both hydrocarbons (74.7% and 80.4%) and oxygenated (8.3% and 9.6%). Major constituents were: γ-curcumene (47.1% and 49.7%) and β-sesquiphellandrene (17.0% and 17.9%), together with two abundant undetermined oxygenated sesquiterpenes, whose abundance was 6.7–7.2% and 4.7–3.3%, respectively. In addition, the essential oil was submitted to enantioselective evaluation in two β-cyclodextrin-based enantioselective columns, determining the enantiomeric purity of a minor component (1*S*,2*R*,6*R*,7*R*,8*R*)-(+)-α-copaene. Finally, the AChE inhibition activity of the EO was evaluated in vitro. In conclusion, this volatile fraction is suitable for further investigation, according to two main lines: (a) the purification and structure elucidation of the major undetermined compounds, (b) a bio-guided fractionation, intended to investigate the presence of new sesquiterpene AChE inhibitors among the minor components.

Keywords: *Jungia rugosa*; *Jungia bullata*; *Jungia jelskii*; *Jungia malvifolia*; Asteraceae; essential oil; enantiomers; sesquiterpenes; Ecuador

1. Introduction

Ecuador, due to multiple combinations of factors, has been configured as a megadiverse country, with a high rate of plant endemism per surface area, which makes it one of the richest countries in biodiversity and endemism of the world [1,2]. Some figures presented in the Fifth and Sixth National Report for the Convention on Biological Diversity regarding the emergence of new plant species illustrate this peculiarity: between 1999 and 2012, 2443 new species were reported for the country, of which 1663 were also new to the science. In 2013, 18,198 species of vascular plants were registered, which meant 1140 more

Plants **2021**, *10*, 2102. https://doi.org/10.3390/plants10102102 — https://www.mdpi.com/journal/plants

species than those reported in 2010 and representing about 7.6% of the vascular plants registered worldwide. It is estimated that the total number of vascular plants could reach 25,000 [3,4].

Along with the above, indigenous cultures possess a strong tradition about plants as a means of treating diseases, which has allowed ancestral knowledge to be transferred through generations from ancient times to the present, promoting the abundant use of medicinal plants. For all these reasons, Ecuador is an invaluable source of natural products and unprecedented knowledge about plant applications. In contrast, the number of high-impact scientific studies in this area is relatively low, given the potential that the country's biodiversity offers [5]. In this respect, to the best of the authors' knowledge, the essential oil (EO), distilled from the leaves of *Jungia rugosa* Less, has never been described.

Within the Asteraceae, the *Jungia* genus corresponds to flowering plants that mostly develop at high altitudes and cold climates, being characteristic of the Andean regions of Ecuador, Peru, and Argentina. Despite many articles describing the phytochemistry of genus *Jungia*, only three deal with EOs. In fact, only the volatile fractions of *Jungia paniculata* and *Jungia polita* have been described so far, the first one being very popular in the Andes and known with the traditional name "matico" [6–8]. Concerning *J. rugosa*, two phytochemical studies have been published. However, they are devoted to non-volatile fractions and their biological activities [9,10].

Jungia rugosa Less (Asteraceae) is a native Andean species, growing at altitudes between 1500 and 4000 m above sea level [11]. It is characterised by great resistance to frost and low temperatures, which is why it prevails in cold and humid climates. This plant grows up to 5 m in height, presenting a thin, woody, smooth, hard, and green stem. Its intense green leaves with a pale green underside, measure between 5 and 12 centimetres and are covered with villi; they are also petiolate, presenting an anti-parallel rib. Its main root divides, giving rise to an abundant root system. Its flower is whitish in colour, presented in a green capsule, which generates small black seeds. In some localities located in the Andean region of Ecuador, it is better known as "carne humana". Based on the indigenous heritage of the central Ecuadorian region (Cotopaxi), this species is used as an anti-inflammatory remedy, for instance, in treating bruises, and for other unspecified healing purposes [12]. The anti-inflammatory activity is probably the most important medicinal property of this plant, since it has also been confirmed by two scientific studies, together with the closely related antioxidant capacity [13,14]. Some sources also report that leaf decoctions are applied to treat wounds and skin ulcerations, gastric problems, and kidney disorders, among others [15,16]. In addition to medicinal applications, this species is also used to prepare ropes in the Chimborazo region of Ecuador [12]. Furthermore, *J. rugosa* is also known with three botanical synonyms: *Jungia bullata* Turcz., *Jungia jelskii* Hieron., and *Jungia malvifolia* Muschl [17]. None of these synonyms corresponds to any chemical literature.

So far, many plant species from Ecuador have been described for producing new EOs, often characterised by important biological activities such as analgesic, antioxidant, antibacterial, anticancer, and sedative, among others [5,18–20]. In particular, EOs rich in sesquiterpenes have been presented as promising anti-proliferative agents, whose constituents are able to easily reach certain organs, such as heart, liver, and kidneys [20]. Among all the biological properties of EOs and their constituents, we are particularly interested in the inhibition of acetylcholinesterase (AChE) and butyrylcholinesterase (BChE), due to the serious implications that neurodegenerative diseases are ever-more producing in western countries [18,21–23].

In accordance with the above, the objectives of this research were to investigate the chemical and enantiomeric composition of *J. rugosa* EO and to evaluate the presence of cholinergic molecules in this volatile fraction. All this information will provide a contribution to the phytochemical and phytopharmacological knowledge of the Ecuadorian flora.

2. Results

2.1. Distillation and Physical Properties

The essential oil of the fresh aerial parts of *Jungia rugosa* was obtained by steam distillation for 4 h, yielding an average of 0.09 (*w/w*). The physical properties, chemical composition and enantiomeric analysis are discussed below.

Two physical properties were determined: relative density (d = 0.898 ± 0.012 g/mL) and refractive index (η = 1.505 ± 0.002). These properties are notoriously determined by genetic characteristics, geographical location, and phenological stage of the plant [24].

2.2. Chemical Analysis of the EO

In the chemical analysis of the EO of *J. rugosa*, all components identified were sesquiterpenes corresponding to a total of 56 and 53 compounds, respectively, with DB-5ms and HP-INNOWax columns (see Section 4.4). Most of the constituents (52) were identified by comparing the electron impact mass spectrum (EIMS) and the linear retention index (LRI) with literature, whereas four remained unidentified. According to their molecular weight, the unknown components are consistent with one sesquiterpene (204 amu) and three oxygenated sesquiterpenoids (220, 262 and 280 amu). The significant difference between calculated and reference LRIs is within the experimental error.

Concerning the quantitative analysis (see Section 4.5), 50 compounds were quantified on at least one column, with a detection threshold of 0.1%, whereas six compounds (β-cubebene, α-chamipinene, δ-amorphene, *allo*-aromadendrene epoxide, *cis*-thujopsenal, and 8-α-acetoxyelemol) appeared as traces (<0.1%) in both columns. Quantified components corresponded to 98.3% and 99.6% of the EO total mass, on the non-polar and polar column, respectively; the sesquiterpene hydrocarbons, corresponding to 75.8% and 80.8% and oxygenated sesquiterpenes, corresponding to 22.3–18.8%, respectively. The major components, with an average amount ≥3% over the two columns, were γ-curcumene (47.1%, 49.7%), β-sesquiphellandrene (17.0%, 17.9%), *ar*-curcumene (3.4%, 4.2%) and two undetermined oxygenated sesquiterpenes with molecular weight 220 (6.7%, 7.2%) and 262 (4.7%, 3.3%). A standard deviation of less than 5% was obtained between the percentages of each analyte with both columns. The GC-MS chromatograms on both columns are reported in Figures 1 and 2. Table 1 shows the identified components together with their relative percent abundance, calculated vs. *n*-nonane as the internal standard.

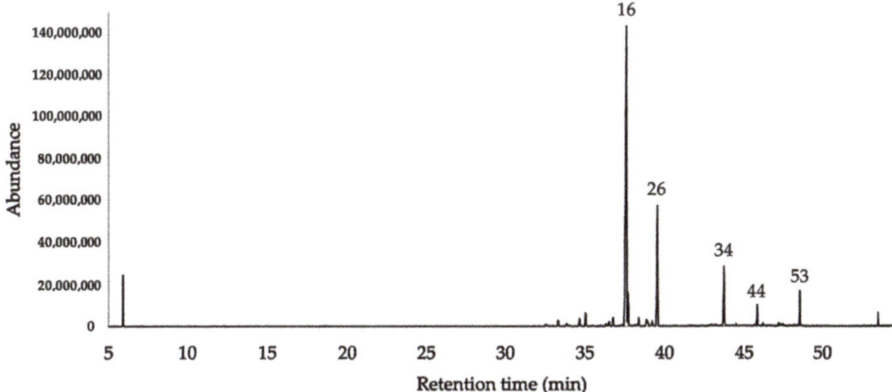

Figure 1. GC-MS chromatogram of *J. rugosa* EO on DB-5ms column.

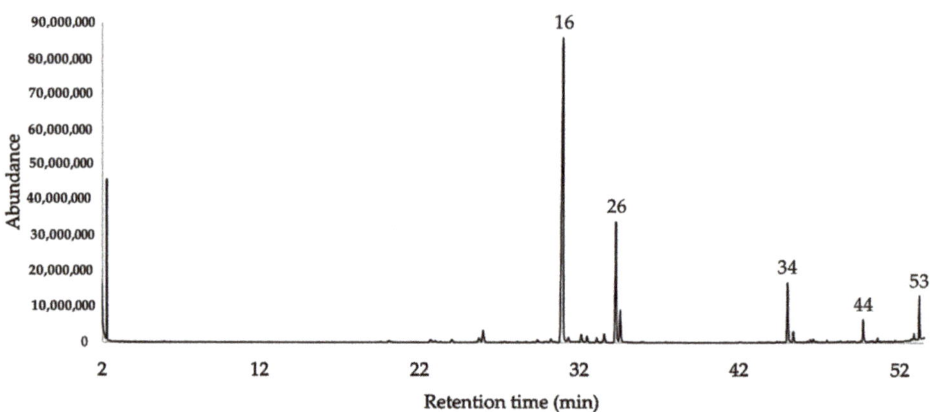

Figure 2. GC-MS chromatogram of *J. rugosa* EO on HP-INNOWax column.

Table 1. Qualitative and quantitative chemical analyses of the EO from *J. rugosa* fresh leaves.

N.	Compounds	DB-5ms			HP-INNOWax			DB-5ms		HP-INNOWax	
		LRI [1]	LRI	Ref.	LRI [1]	LRI	Ref.	(%) [2]	σ	(%) [2]	σ
1	α-copaene	1370	1374	[25]	1464	1489	[26]	0.2	0.01	0.3	0.01
2	β-cubebene	1378	1387	[25]	1536	1542	[27]	trace	0.02	trace	-
3	7-epi-sesquithujene	1385	1390	[25]	1576	-		0.9	0.02	trace	-
4	Italicene	1395	1405	[25]	1525	1536	[28]	0.3	0.08	0.5	0.02
5	α-chamipinene	1397	1396	[25]	1552	-		trace	0.05	trace	-
6	undetermined (MW 204)	1412	-		1549	-		1.1	0.02	0.4	0.01
7	α-cis-bergamotene	1420	1411	[25]	1584	1577	[28]	1.6	0.03	1.7	0.05
8	α-trans-bergamotene	1430	1432	[25]	1530	1560	[29]	0.1	0.01	0.3	0.02
9	seychellene	1442	1444	[25]	1663	-		0.1	-	trace	-
10	α-humulene	1448	1452	[25]	1652	1667	[27]	1.1	0.01	0.5	0.01
11	allo-aromadendrene	1451	1458	[25]	1626	1637	[30]	0.1	0.02	trace	-
12	(E)-β-farnesene	1453	1454	[25]	1669	1664	[27]	trace	-	0.8	0.02
13	6-demethoxy ageratochromene	1458	1461	[25]	2083	2075	[31]	0.9	0.02	1.4	0.02
14	2-epi-(E)-caryophyllene	1466	1465	[32]	1674	1669	[32]	0.1	0.01	trace	-
15	ishwarane	1471	1465	[25]	1609	1636	[33]	0.1	0.07	trace	-
16	γ-curcumene	1476	1481	[25]	1685	1692	[27]	47.1	0.70	49.7	0.40
17	ar-curcumene	1479	1479	[25]	1768	1774	[27]	3.4	0.34	4.2	0.32
18	γ-muurolene	1486	1478	[25]	1692	1690	[27]	0.1	0.07	0.8	0.02
19	β-selinene	1489	1489	[25]	1710	1717	[27]	0.3	0.04	1.3	0.02
20	α-zingiberene	1491	1493	[25]	1698	1713	[34]	0.1	0.11	trace	-
21	epi-cubebol	1493	1493	[25]	1943	1928	[35]	1.0	0.13	trace	-
22	β-bisabolene	1505	1505	[25]	1719	1728	[27]	0.9	0.02	1.0	0.02
23	α-cuprenene	1506	1505	[25]	1733	1759	[36]	0.5	0.01	0.6	0.01
24	δ-amorphene	1511	1511	[25]	1704	1710	[37]	trace	-	trace	-
25	γ-cadinene	1513	1513	[25]	1744	1763	[27]	0.7	0.01	0.8	0.01
26	β-sesquiphellandrene	1521	1521	[25]	1762	1771	[27]	17.0	0.20	17.9	0.16
27	8,14-cedranoxide	1549	1541	[25]	1842	1858	[38]	0.1	-	trace	-
28	cis-muurol-5-en-4-α-ol	1569	1559	[25]	2210	2221	[39]	0.1	0.02	trace	-
29	spathulenol	1575	1577	[25]	2141	2140	[40]	0.1	-	0.5	0.01
30	allo-cedrol	1588	1589	[25]	2261	-		0.1	0.03	trace	-
31	sesquithuriferol	1605	1604	[25]	2125	2113	[41]	0.3	0.01	0.3	0.01
32	isolongifolan-7-α-ol	1609	1618	[25]	2117	-		0.3	0.01	0.2	0.11
33	cis-isolongifolanone	1613	1612	[25]	2168	-		0.2	0.01	0.3	0.01
34	undetermined (MW 220)	1627	-		2070	-		6.7	0.10	7.2	0.14
35	3-iso-thujopsanone	1632	1641	[25]	2106	-		0.2	0.13	trace	-

Table 1. *Cont.*

N.	Compounds	DB-5ms			HP-INNOWax			DB-5ms		HP-INNOWax	
		LRI [1]	LRI	Ref.	LRI [1]	LRI	Ref.	(%) [2]	σ	(%) [2]	σ
36	allo-aromadendrene epoxide	1634	1639	[25]	2096	2095	[42]	trace	-	trace	-
37	epi-α-muurolol	1642	1640	[25]	2194	2186	[27]	0.1	0.01	0.1	0.01
38	3-thujopsanone	1650	1653	[25]	2265	-		0.3	-	0.2	0.01
39	α-cadinol	1653	1652	[25]	2244	2227	[27]	0.2	0.02	trace	-
40	14-hydroxy-9-epi-(E)-caryophyllene	1658	1668	[25]	2110	-		0.1	0.04	trace	-
41	7-epi-α-eudesmol	1664	1662	[25]	2209	2205	[43]	0.2	0.01	trace	-
42	bulnesol	1667	1670	[25]	2204	2200	[44]	0.2	0.01	trace	-
43	8-cedren-13-ol	1685	1688	[25]	2335	2359	[45]	0.4	0.03	trace	-
44	cyperotundone	1690	1695	[25]	2474	-		2.5	0.04	2.5	0.28
45	zizanal	1701	1697	[25]	2450	-		0.6	0.01	1.4	0.07
46	cis-thujopsenal	1705	1708	[25]	2294	-		trace	-	trace	-
47	14-hydroxy-α-humulene	1713	1713	[25]	-	-		0.1	-	trace	-
48	vetiselinenol	1718	1730	[25]	2445	-		0.2	-	trace	-
49	γ-costol	1742	1745	[25]	2337	-		0.3	0.01	trace	-
50	xanthorrhizol	1749	1751	[25]	2657	2674	[42]	0.6	0.02	1.1	0.27
51	cedryl acetate	1776	1767	[25]	2132	2150	[46]	0.2	0.02	0.3	0.01
52	8-cedren-13-ol acetate	1782	1788	[25]	2248	-		0.3	0.09	trace	-
53	undetermined (MW 262)	1789	-		2272	-		4.7	0.18	3.3	0.07
54	8-α-acetoxyelemol	1793	1792	[25]	-	-		trace	-	trace	-
55	undetermined (MW 280)	2029	-		-	-		1.3	0.30	trace	-
56	n-tricosane	2301	2300	[25]	2300	2300	[47]	0.2	0.07	trace	-
	monoterpene hydrocarbons							-		-	
	oxygenated monoterpenes							-		-	
	sesquiterpene hydrocarbons							75.8%		80.8%	
	oxygenated sesquiterpenes							22.3%		18.8%	
	others							0.2%		trace	
	total							98.3%		99.6%	

[1] Calculated linear retention index (LRI) according to van den Dool and Kratz [48]; [2] Trace for % < 0.1.

2.3. Enantioselective Evaluation of the EO

The enantioselective analysis of the EO was carried out on a 2,3-diethyl-6-*tert*-butyldimethylsilyl-β-cyclodextrin based capillary column. Only the very minor compound (1*S*,2*R*,6*R*,7*R*,8*R*)-(+)-α-copaene could be certainly identified, appearing enantiomerically pure in the EO. No more enantiomeric pairs or enantiomerically pure compounds could be identified in the enantioselective chromatogram.

2.4. AChE Inhibition Activity

The AChE inhibitory activity of the investigated essential oils was measured in vitro, using a spectrophotometric assay based on Elman's method. Galanthamine and *Laurus nobilis* EO were used as positive controls, the latter being considered an active EO in literature (see Section 3. Discussion). All results are summarised in Table 2.

Table 2. Percent inhibition of AChE by *J. rugosa* EO compared to *L. nobilis* EO and galantamine as positive controls.

Sample	Enzymatic Inhibition (%)	σ
Galanthamine 1.0 μg/mL	49.2	5.2
Laurus nobilis EO 38 μg/mL	38.8	4.2
Jungia rugosa EO 38 μg/mL	25.9	13.9

3. Discussion

3.1. The Chemical Composition

About the chemical composition of the EO, the hydrocarbon sesquiterpene fraction was predominant, corresponding to 74.7% and 80.4% with a non-polar and a polar column respectively. Furthermore, an oxygenated sesquiterpene fraction was present between 9.6% and 8.3% of the whole amount. No monoterpenes were detected in the EO. Major components of this volatile fraction were γ-curcumene and β-sesquiphellandrene, together with two undetermined oxygenated sesquiterpenes (molecular weight 220 amu and 262 amu, respectively). If we compare these results with the only two partial analyses, known so far for EOs of genus *Jungia* (*J. paniculata* and *J. polita*), the prevalence of sesquiterpenes is confirmed [7,8]. However, unlike our case, (*E*)-β-caryophyllene and caryophyllene oxide are there the main components. Regarding γ-curcumene, it derives its name from *Curcuma longa* L. (turmeric), but we must look at *Helichrysum italicum* (Roth) G. Don (Asteraceae) to find an important and widely studied botanical species where γ-curcumene is often a major constituent. Other *Helichrysum* species are also familiar with similar sesquiterpene compositions [49]. On the one hand, despite γ-curcumene being quite common and known for a long time, no exhaustive studies on its pharmacology can be found. On the other hand, the EOs where it is an important component are widely described, with all the typical biological activities known for volatile fractions. In regards to β-sesquiphellandrene, it is also a typical hydrocarbon sesquiterpene of *Curcuma longa*. The most important study on its pharmacological properties is probably a recent publication, where β-sesquiphellandrene has been described as a potent anticancer agent. Its activity is comparable with the one of curcumin. According to that investigation, β-sesquiphellandrene would exert an antiproliferative activity, by inhibiting the formation of cancer cell colonies and inducing apoptosis. The neoplastic formations, that appeared to be more sensitive to this metabolite, were leukaemia, multiple myeloma, and colorectal cancer. Furthermore, cancer cells expressing p-53 protein resulted in being more sensitive to β-sesquiphellandrene than those lacking it [50]. Finally, we must mention the presence of two important undetermined compounds, contributing to the mass of the EO with the non-negligible amounts of 6.7–7.2% and 4.7–3.3%, respectively (see peaks 34 and 53 in Figures 1 and 2). These constituents showed a molecular ion of 220 and 262 amu—the first one being characteristic of sesquiterpenoids with molecular formula $C_{15}H_{24}O$, whereas the second one is consistent with the rare sesquiterpene derivatives of formula $C_{18}H_{30}O$ (e.g., sesquiterpenes acetones) [25].

3.2. The Enantiomeric Evaluation

For what concerns the enantiomeric evaluation, the EO was submitted to enantioselective GC, in a classical β-cyclodextrin-based capillary column. However, the only chiral terpene that could be identified was (1*S*,2*R*,6*R*,7*R*,8*R*)-(+)-α-copaene, present as a pure enantiomer. No other sesquiterpene could be detected, both as a pure enantiomer or enantiomeric pair. This result is not surprising. In fact, most of the enantiomerically pure available standards are indeed monoterpenes, whose use is necessary to determine the elution order of the enantiomers from an enantioselective column. Since the present EO is exclusively constituted by sesquiterpenes, the corresponding enantioselective GC information is, for most of them, unavailable. Furthermore, the similarity among the spectra for this class of metabolites excluded the possibility to certainly identify enantiomeric pairs within the peaks. The only exception, although a minor component, was (1*S*,2*R*,6*R*,7*R*,8*R*)-(+)-α-copaene, since the enantiomerically pure standards of this compound are available.

3.3. The Cholinergic Activity

Finally, the inhibition activity of this EO against AChE can be discussed. Observing our results, shown in Table 2, the inhibition capacity of *J. rugosa* EO was compared to the ones of galanthamine and *L. nobilis* EO. However, whereas the biological activity of galanthamine is clearly extremely high, mainly because it is a pure compound, the biological activity of *L. nobilis* EO is decidedly lower. Nevertheless, *L. nobilis* EO is considered as an active mixture in this kind of assay, and it can be subsequently used as a better positive control while working with EOs [51]. In our case, the inhibition power of *J. rugosa* EO is about 68% compared to that of *L. nobilis* EO, clearly resulting in less activity but not inactive. This fact could be explained by the presence of at least one active minor sesquiterpene in the mixture. If that is the case, the EO may be considered useless as it is, but suitable to be studied, through a bio-guided fractionation, in search of new sesquiterpene inhibitors. The interest in this aspect resides in that, to the best of the authors' knowledge, the most active EOs are characterised by an important monoterpene fraction (except for the case where the EO is dominated by (*E*)-β-caryophyllene) [18,22]. However, due to their toxicity, hydrocarbon monoterpenes can hardly be used as pharmaceutical active principles, which cannot be assumed for sesquiterpenes. Therefore, the discovery of new sesquiterpene inhibitors of AChE is a matter of some pharmaceutical interest. Consequently, this volatile fraction is suitable for further investigation, according to two main lines: a) the purification and structure elucidation of the major undetermined compounds, by mean of preparative chromatography and NMR spectroscopy; b) a bio-guided preparative fractionation, intended to investigate the presence of new sesquiterpene AChE inhibitors among the minor components. Due to the low distillation yield of this EO, a non-classical approach should be applied. On the one hand, a tentative method for purification and structure elucidation could be the use of preparative thin-layer chromatography (TLC) and microprobe NMR spectroscopy. In this way, about 1 mg of a pure compound would be enough to be submitted to a complete series of NMR experiments. On the other hand, the bio-guided investigation could be faced through a bioautographic method. Based on a TLC analysis, a bioautographic assay can be carried out with few micrograms of EO. Since the active compounds possibly are known sesquiterpenes, the combined use of bioautography, preparative TLC and GC-MS should afford the desired information.

In regards to the traditional use of *J. rugosa*, some previous studies have described the antioxidant and anti-inflammatory activities of the non-volatile fraction, mainly attributed to flavonoids [13,14]. Since these properties are fully consistent with the ethnobotanical use, the EO could probably be exempted to be considered the active fraction.

4. Materials and Methods

4.1. General Information

The chemical and enantioselective analyses of the *J. rugosa* EO were performed with a gas chromatography-mass spectrometry (GC-MS) system, consisting of a 6890 N Agilent Technologies gas chromatograph with an autoinjector model 7683. The instrument was coupled to an Agilent Technologies simple quadrupole mass spectrometry detector (MSD) model 5973 INERT (Santa Clara, CA, USA), and a common flame ionization detector (FID). The MSD operated in SCAN mode (scan range 35–350 *m/z*), with an electron ionization (EI) source at 70 eV. The qualitative and quantitative analyses were carried out with both non-polar and polar stationary phase capillary columns from Agilent Technologies. The non-polar column was based on 5% phenyl-methylpolysiloxane phase (DB-5ms, 30 m long, 0.25 mm internal diameter, and 0.25 μm film thickness), while the polar column was provided with a polyethylene glycol stationary phase (HP-INNOWax, 30 m × 0.25 mm × 0.25 μm). The enantioselective analysis was run with an enantioselective capillary column, based on 30% diethyl-*tert*-butyldimethylsilyl-β-cyclodextrin in PS-086 as chiral stationary phases as a chiral selector (25 m × 250 μm internal diameter × 0.25 μm phase thickness, purchased from Mega, MI, Italy). For all the analyses, GC purity grade helium (Indura, Guayaquil, Ecuador) was used as the carrier gas, set at the constant flow, with a rate of 1 mL/min. For the biological assays, a Spectronic Genesys 6 spectrophotometer was used, purchased from Thermo-Fisher Scientific (Waltham, MA, USA).

All solvents for GC analysis, the mixture of *n*-alkanes C_9–C_{25} for linear retention indices (LRI), internal standard (*n*-nonane), and reagents for the inhibition activity assays were purchased from Sigma-Aldrich. The calibration standard was isopropyl caproate, obtained by synthesis in the authors' laboratory and purified to 98.8% (GC-FID).

4.2. Plant Material

The leaves of *J. rugosa* were collected in February 2020 in the south-central area of the Ecuadorian highlands (sector Citar, province Cañar). The specimens grew at an altitude of 3445 m above sea level, with coordinates 02°35′387″ S and 78°56′309″ W. The collection was carried out under governmental permission (MAAE-ARSFC-2020-0638), issued by the Ministry of Environment of Ecuador (MAE). The identification was achieved at the Universidad Técnica Particular de Loja, by the botanist Dr. Itziar Arnelas, and a botanical specimen was deposited at the UTPL herbarium, with voucher number D-HUTPL-2020-6. The fresh leaves of *J. rugosa* were distilled the day after collection.

4.3. Distillation of the EO and GC Sample Preparation

Fresh leaves (4.8 kg) of *J. rugosa* were steam distilled in a Clevenger-type apparatus for 4 h. The EO, which spontaneously separated from the aqueous layer, was immediately dried on anhydrous sodium sulphate and stored in the dark at $-15\,^{\circ}$C until use. The analytical samples for each GC injection were prepared as described in previous studies [18–22].

4.4. GC-MS Qualitative Analyses

The EO was analysed by injecting 1 µL of analytical sample into the GC instrument that operated in split mode (40:1). The injector temperature was set at 220 °C. The elution with the DB-5ms column was conducted according to the following oven temperature program: 60 °C was kept for 5 min, followed by a first thermal gradient to 100 °C at a rate of 2 °C/min, a second gradient to 150 °C at a rate of 3 °C/min, then a third gradient to 200 °C at a rate of 5 °C/min; in the end, the oven temperature was maintained at 250 °C for 15 min at a rate of 15 °C/min. With the HP-INNOWax column, the same conditions and oven program were applied, except for the final temperature, which was set at 230 °C.

In order to identify the components of the EO, a homologous series of *n*-alkanes, from *n*-nonane to *n*-pentacosane, was injected in each column. The linear retention index (LRI) of each constituent was calculated according to Van Den Dool and Kratz [48]. This way each volatile metabolite was identified by comparing the corresponding LRI value and EI-MS spectrum with tabulated data for DB-5ms [25] and HP-INNOWax [26–47].

4.5. GC-FID Quantitative Analyses

The quantitative analyses were performed, with both columns, under the same conditions and configurations described for the qualitative ones. All the samples were injected in quadrupled and the percentage of each analyte in the EO was calculated as the average value over the four injections. The quantification was achieved by external calibration and use a process internal standard, calculating the relative response factor (RRF) of each EO constituent, based on its combustion enthalpy [52]. The original method was modified, isopropyl caproate instead of the methyl octanoate reported in the literature, was chosen as the calibration standard for this analysis. This approach is based on the principle that the RRF of an organic compound only depends on the molecular formula and number of aromatic rings, being the same for isomers.

Two external calibration curves were build-up, according to what is described in previous articles [18–22]. All calibration curves achieved an $R^2 > 0.995$.

4.6. Enantioselective Analysis of the EO

The enantioselective analysis of the EO was carried out by GC-MS on the same samples of the qualitative and quantitative analyses. The injector temperature was the same as for the EO qualitative analysis, whereas the injector operated in split mode, with a ratio of 50:1. The following oven thermal program was applied: The initial temperature was 60 °C for 2 min, followed by a thermal gradient of 2 °C/min until 220 °C, maintained for 2 min. In addition, a mixture of *n*-alkanes (C_9–C_{25}) was injected under the same conditions as for conventional analysis to determine LRIs. The enantiomeric pairs of chiral sesquiterpenes were identified based on the EI-MS spectra and elution order, determined according to literature data for the same chiral selector [53,54].

4.7. AChE Inhibition Spectrophotometric Assay

The protocol followed in this study was that of Rhee et al., with slight modifications [55]. Enzyme solution 0.22 U/mL was prepared in 50 mM Tris-HCl, pH 8, containing 0.1% bovine serum albumin (BSA). Acetyltiocholine (ATCI) solution 1.5 mM was prepared in Millipore water. Ellman's reagent solution 3mM was prepared in 50 mM Tris–HCl, pH 8, containing 0.1 M NaCl and 0.02 M $MgCl_2$ hexahydrate. Essential oils stock solutions 15 mg/mL and 30 mg/mL were prepared in dimethyl sulfoxide (DMSO). *Laurus nobilis* essential oil solution 15 mg/mL and galanthamine solution 0.4 mg/mL in DMSO were used as a positive control.

The reagents were placed in the cuvette in the following order: 150 µL of ATCI solution, 900 µL of buffer B, 5 µL of the essential oil/galanthamine solution and finally 150 µL of enzyme solution. BSA 1%, Tween-20 0.1 and 0.5% and Tween-80 0.1 and 0.5% were tested as buffer B detergents to increase the essential oil solubility in the aqueous reaction mixture. The reaction mixture was incubated for 6 minutes at room temperature (25 °C). Absorbance values were collected after 6 minutes of incubation at 412 nm. The absorbance corresponding to 100% of AChE activity was measured by replacing the EOs/galanthamine solution with 5 µL of pure DMSO. Control and sample blank solutions were prepared by replacing the 150 µL of enzyme solution with 150 µL of buffer B. The percentage of AChE inhibition was measured according to the equation below:

$$\% \text{ Inhibition} = \Delta A \text{ (Control)} - \Delta A \text{ (Sample)}/\Delta A \text{ (Control)} \times 100$$

$$\Delta A \text{ (Control) or (Sample)} = A_{412} \text{ (Control) or (Sample)} - A_{412} \text{ (Control Blank) or (Sample Blank)}$$

5. Conclusions

The fresh leaves of *Jungia rugosa* Less afforded, by steam distillation, an essential oil in quite a low yield (0.09% by weight). This volatile fraction was composed exclusively of sesquiterpenes, whose major constituents were γ-curcumene (more than 45%) and β-sesquiphellandrene (about 17%). The other two unknown oxygenated sesquiterpenoids were detected among the main constituents (about 7% and 5% of the whole mixture). The EO also manifested a weak inhibition activity against AChE.

Author Contributions: Conceptualization, O.M. and G.G.; investigation, K.C., F.C. and V.V.; data curation, O.M., B.S. and G.G.; writing—original draft preparation, K.C.; writing—review and editing, O.M. and G.G.; and supervision, O.M., B.S. and G.G. All authors have read and agreed to the published version of the manuscript.

Funding: This research received no external funding.

Institutional Review Board Statement: Not applicable.

Informed Consent Statement: Not applicable.

Data Availability Statement: Raw data are available from the authors (K.C.).

Acknowledgments: We are grateful to the Universidad Técnica Particular de Loja (UTPL) for supporting this investigation (2nd call funding TFT, Apr–Aug 2020) and open access publication.

Conflicts of Interest: The authors declare no conflict of interest.

References

1. Mittermeier, R.A.; Myers, N.; Thomsen, J.B.; Da Fonseca, G.A.; Olivieri, S. Biodiversity hotspots and major tropical wilderness areas: Approaches to setting conservation priorities. *Conserv. Biol.* **1998**, *12*, 516–520. [CrossRef]

2. Araujo-Baptista, L.; Vimos-Sisa, K.; Cruz-Tenempaguay, R.; Falconí-Ontaneda, F.; Rojas-Fermín, L.; González-Romero, A. Componentes químicos y actividad antimicrobiana del aceite esencial de *Lasiocephalus ovatus* (Asteraceae) que crece en Ecuador. *Acta Biol. Colomb.* **2020**, *25*, 22–28. [CrossRef]

3. Ministerio del Ambiente del Ecuador. Quinto Informe Nacional para el Convenio sobre la Diversidad Biológica. Available online: https://www.ambiente.gob.ec/wp-content/uploads/downloads/2015/06/QUINTO-INFORME-BAJA-FINAL-19.06.2015.pdf (accessed on 27 September 2021).

4. Convention on Biological Diversity. 6th National Report for the Convention on Biological Diversity. Available online: https://www.cbd.int/nr6/ (accessed on 27 September 2021).

5. Malagón, O.; Ramírez, J.; Andrade, J.M.; Morocho, V.; Armijos, C.; Gilardoni, G. Phytochemistry and ethnopharmacology of the Ecuadorian flora. A review. *Nat. Prod. Commun.* **2016**, *11*, 297–314. [CrossRef]

6. Zapata, B.; Duran, C.; Stashenko, E.; Betancur-Galvis, L.; Mesa-Arango, A.C. Actividad antimicótica y citotóxica de aceites esenciales de plantas de la familia Asteraceae. *Rev. Iberoam. Micol.* **2010**, *27*, 101–103. [CrossRef]

7. Ruiz, C.; Díaz, C.; Rojas, R. Composición Química de Aceites Esenciales de 10 Plantas aromáticas peruanas. *Rev. Soc. Quím. Perú* **2015**, *81*, 81–94. [CrossRef]

8. Duschatzky, C.B.; Possetto, M.L.; Talarico, L.B.; García, C.C.; Michis, F.; Almeida, N.V.; de Lampasona, M.P.; Schuff, C.; Damonte, E.B. Evaluation of chemical and antiviral properties of essential oils from South American plants. *Antivir. Chem. Chemother.* **2005**, *16*, 247–251. [CrossRef]

9. Wilches, I.; Tobar, V.; Peñaherrera, E.; Cuzco, N.; Jerves, L.; Vander-Heyden, Y.; León-Tamariz, F.; Vila, E. Evaluation of anti-inflammatory activity of the methanolic extract from *Jungia rugosa* leaves in rodents. *J. Ethnopharmacol.* **2015**, *173*, 166–171. [CrossRef]

10. Acosta-León, K.L.; Tacuamán-Jácome, S.E.; Vinueza-Tapia, D.R.; Vélez-Correal, F.X.; Pilco-Bonilla, G.A.; García-Veloz, M.J.; Abdo-López, S.P. Hypoglycemic activity of *Jungia rugosa* on induced diabetic mice (*Mus musculus*). *Pharmacologyonline* **2019**, *1*, 239–245.

11. Jorgensen, P.; Leon-Yanez, S. *Catalogue of the Vascular Plants of Ecuador*; Monogram St. Louis: St. Louis, MO, USA, 1999; Volume 75, p. 291.

12. De la Torre, L.; Navarrete, H.; Muriel, P.M.; Macía, M.J.; Balslev, H. *Enciclopedia de las Plantas Útiles del Ecuador*; Herbario QCA de la Escuela de Ciencias Biológicas de la Pontificia Universidad Católica del Ecuador: Quito, Ecuador; Herbario AAU del Departamento de Ciencias Biológicas de la Universidad de Aarhus: Aarhus, Denmark, 2008; p. 228.

13. Graf, B.L.; Rojas-Silva, P.; Baldeón, M.E. Discovering the Pharmacological Potential of Ecuadorian Market Plants using a Screens-to-Nature Participatory Approach. *J. Biodivers. Biopros. Dev.* **2016**, *3*, 1000156. [CrossRef]

14. Enciso, E.; Arroyo, J. Efecto antiinflamatorio y antioxidante de los flavonoides de las hojas de *Jungia rugosa* Less (matico de puna) en modelo experimental en ratas. *Fac. Med.* **2011**, *72*, 231. [CrossRef]

15. Criollo, K.; Molina, N. Evaluación de la Estabilidad de Extractos Obtenidos a Partir de Distintos Procesos de Secado de *Jungia Rugosa*. Bachelor's Thesis, Universidad de Cuenca, Cuenca, Ecuador, 2008.

16. Campoverde, J.; Verdugo, M. Determinación del Efecto Cicatrizante de las Hojas de Carne Humana (*Jungia cf. rugosa*). Bachelor's Thesis, Universidad de Cuenca, Cuenca, Ecuador, 2008.

17. Tropicos.org. Missouri Botanical Garden. Available online: https://www.tropicos.org (accessed on 31 August 2021).

18. Espinosa, S.; Bec, N.; Larroque, C.; Ramírez, J.; Sgorbini, B.; Bicchi, C.; Gilardoni, G. Chemical, enantioselective, and sensory analysis of a cholinesterase inhibitor essential oil from *Coreopsis triloba* SF Blake (Asteraceae). *Plants* **2019**, *8*, 448. [CrossRef]

19. Alvarado, B. Actividad Antioxidante y Citotóxica de 35 Plantas Medicinales de la Cordillera Negra. Master's Thesis, Universidad Nacional Mayor de San Marcos, Lima, Peru, 2017.

20. Pant, P.; Sut, S.; Castagliuolo, I.; Gandin, V.; Maggi, F.; Gyawali, R.; Dall'Acqua, S. Sesquiterpene rich essential oil from Nepalese Bael tree (*Aegle marmelos* (L.) Correa) as potential antiproliferative agent. *Fitoterapia* **2019**, *138*, 104266. [CrossRef]

21. Montalván, M.; Peñafiel, M.A.; Ramírez, J.; Cumbicus, N.; Bec, N.; Larroque, C.; Bicchi, C.; Gilardoni, G. Chemical Composition, Enantiomeric Distribution, and Sensory Evaluation of the Essential Oils Distilled from the Ecuadorian Species *Myrcianthes myrsinoides* (Kunth) Grifo and *Myrcia mollis* (Kunth) DC. (Myrtaceae). *Plants* **2019**, *8*, 511. [CrossRef]

22. Espinosa, S.; Bec, N.; Larroque, C.; Ramírez, J.; Sgorbini, B.; Bicchi, C.; Cumbicus, N.; Gilardoni, G. A novel chemical profile of a selective in vitro cholinergic essential oil from *Clinopodium taxifolium* (Kunth) Govaerts (Lamiaceae), a native Andean species of Ecuador. *Molecules* **2021**, *26*, 45. [CrossRef]

23. Armijos, C.; Gilardoni, G.; Amay, L.; Lozano, A.; Bracco, F.; Ramirez, J.; Bec, N.; Larroque, C.; Vita Finzi, P.; Vidari, G. Phytochemical and ethnomedicinal study of *Huperzia* species used in the traditional medicine of Saraguros in Southern Ecuador; AChE and MAO inhibitory activity. *J. Ethnopharmacol.* **2016**, *193*, 546–554. [CrossRef] [PubMed]

24. Benyelles, B.; Allali, H.; Dib, M.E.A.; Djabou, N.; Paolini, J.; Costa, J. Chemical Composition Variability of Essential Oils of *Daucus gracilis* Steinh. from Algeria. *Chem. Biodivers.* **2017**, *14*, e1600490. [CrossRef]

25. Adams, R.P. *Identification of Essential Oil Components by Gas Chromatography/Mass Spectrometry*, 4th ed.; Allured Publishing Corporation: Carol Stream, IL, USA, 2007; ISBN 10-1932633219.

26. Saroglou, V.; Marin, P.; Rancic, A.; Veljic, M.; Skaltsa, H. Composition and antimicrobial activity of the essential oil of six *Hypericum* species from Serbia. *Biochem. Syst. Ecol.* **2007**, *35*, 146–152. [CrossRef]

27. Babushok, V.I.; Linstrom, P.J.; Zenkevich, I.G. Retention Indices for Frequently Reported Compounds of Plant Essential Oils. *J. Phys. Chem. Ref. Data* **2011**, *40*, 043101. [CrossRef]

28. Bianchini, A.; Tomi, P.; Bernardini, A.; Morelli, I.; Flamini, G.; Cioni, P.; Usaï, M.; Marchetti, M. A comparative study of volatile constituents of two *Helichrysum italicum* (Roth) Guss. Don Fil subspecies growing in Corsica (France), Tuscany and Sardinia (Italy). *Flavour Fragr. J.* **2003**, *18*, 487–491. [CrossRef]

29. Smadja, J.; Rondeau, P.; Shum Cheong Sing, A. Volatile constituents of five *Citrus Petitgrain* essential oils from Reunion. *Flavour Fragr. J.* **2005**, *20*, 399–402. [CrossRef]

30. Skaltsa, H.; Demetzos, C.; Lazari, D.; Sokovic, M. Essential oil analysis and antimicrobial activity of eight *Stachys* species from Greece. *Phytochemistry* **2003**, *64*, 743–752. [CrossRef]

31. Boti, J.; Muselli, A.; Tomi, F.; Koukoua, G.; N'Guessan, T.; Costa, J.; Casanova, J. Combined analysis of *Cymbopogon giganteus* Chiov. leaf oil from Ivory Coast by GC/RI, GC/MS and ^{13}C-NMR. *Comptes Rendus Chimie* **2006**, *9*, 164–168. [CrossRef]

32. Duquesnoy, E.; Castola, V.; Casanova, J. Composition and chemical variability of the twig oil of *Abies alba* Miller from Corsica. *Flavour Fragr. J.* **2007**, *22*, 293–299. [CrossRef]

33. Cavalli, J.; Tomi, F.; Bernardini, A.; Casanova, J. Composition and chemical variability of the bark oil of *Cedrelopsis grevei* H. Baillon from Madagascar. *Flavour Fragr. J.* **2003**, *18*, 532–538. [CrossRef]

34. Grujic-Jovanovic, S.; Skaltsa, H.; Marin, P.; Sokovic, M. Composition and antibacterial activity of the essential oil of six *Stachys* species from Serbia. *Flavour Fragr. J.* **2004**, *19*, 139–144. [CrossRef]

35. Hachicha, S.; Skanji, T.; Barrek, S.; Ghrabi, Z.; Zarrouk, H. Composition of the essential oil of *Teucrium ramosissimum* Desf. (Lamiaceae) from Tunisia. *Flavour Fragr. J.* **2007**, *22*, 101–104. [CrossRef]

36. Wedge, D.; Klun, J.; Tabanca, N.; Demirci, B.; Ozek, T.; Husnu Can Baser, K.; Liu, Z.; Zhang, S.; Cantrell, C.; Zhang, J. Bioactivity-Guided Fractionation and GC/MS Fingerprinting of *Angelica sinensis* and *Angelica archangelica* Root Components for Antifungal and Mosquito Deterrent Activity. *J. Agric. Food Chem.* **2009**, *57*, 464–470. [CrossRef]

37. Martinez, J.; Rosa, P.; Menut, C.; Leydet, A.; Brat, J.; Pallet, D.; Meireles, M. Valorization of Brazilian Vetiver (*Vetiveria zizanioides* (L.) Nash ex Small) Oil. *J. Agric. Food Chem.* **2004**, *52*, 6578–6584. [CrossRef]

38. Maksimovic, M.; Vidic, D.; Milos, M.; Solic, M.; Abadzic, S.; Siljak-Yakovlev, S. Effect of the environmental conditions on essential oil profile in two Dinaric *Salvia* species: *S. brachyodon* Vandas and *S. officinalis* L. *Biochem. Syst. Ecol.* **2007**, *35*, 473–478. [CrossRef]

39. Chéraif, I.; Jannet, H.; Hammami, M.; Khouja, M.L.; Mighri, Z. Chemical composition and antimicrobial activity of essential oils of *Cupressus arizonica* Greene. *Biochem. Syst. Ecol.* **2007**, *35*, 813–820. [CrossRef]

40. Kundakovic, T.; Fokialakis, N.; Kovacevic, N.; Chinou, I. Essential oil composition of *Achillea lingulata* and *A. umbellata*. *Flavour Fragr. J.* **2007**, *22*, 184–187. [CrossRef]

41. Duquesnoy, E.; Dinh, N.; Castola, V.; Casanova, J. Composition of a Pyrolytic oil from *Cupressus funebris* Endl. of Vietnamese origin. *Flavour Fragr. J.* **2006**, *21*, 453–457. [CrossRef]

42. Paolini, J.; Tomi, P.; Bernardini, A.; Bradesi, P.; Casanova, J.; Kaloustian, J. Detailed analysis of the essential oil from Cistus albidus L. by combination of GC/RI, GC/MS and ^{13}CNMR spectroscopy. *Nat. Prod. Res. Former. Nat. Prod. Lett.* **2008**, *22*, 1270–1278.

43. Cozzani, S.; Muselli, A.; Desjobert, J.; Bernardini, A.; Tomi, F.; Casanova, J. Chemical composition of essential oil of *Teucrium polium* subsp. *capitatum* (L.) from Corsica.. *Flavour Fragr. J.* **2005**, *20*, 436–441. [CrossRef]

44. Boti, J.; Koukoua, G.; N'Guessan, T.; Muselli, A.; Bernardini, A.; Casanova, J. Composition of the leaf, stem bark and root bark oils of *Isolona cooperi* investigated by GC (retention index), GC-MS and ^{13}C-NMR spectroscopy. *Phytochem. Anal.* **2005**, *16*, 357–363. [CrossRef]

45. Tunalier, Z.; Kirimer, J.; Husnu Can Baser, K. Wood Essential Oils of *Juniperus foetidissima* Willd. *Holzforschung* **2003**, *57*, 140–144. [CrossRef]

46. Choi, H.; Sawamura, M. Composition of the Essential Oil of *Citrus tamurana* Hort. ex Tanaka (Hyuganatsu). *J. Agric. Food Chem.* **2000**, *48*, 4868–4873. [CrossRef]

47. Khaoukha, G.; Jemia, M.B.; Amira, S.; Laouer, H.; Bruno, M.; Scandolera, E.; Senatore, F. Characterisation and antimicrobial activity of the volatile components of the flowers of Magydaris tomentosa (Desf.) DC. collected in Sicily and Algeria. *Nat. Prod. Res. Former. Nat. Prod. Lett.* **2014**, *28*, 1152–1158. [CrossRef]

48. Van Den Dool, H.; Kratz, P.D. A generalization of the retention index system including linear temperature programmed gas—Liquid partition chromatography. *J. Chromatogr.* **1963**, *11*, 463–471. [CrossRef]

49. Aćimović, M.; Ljujić, J.; Vulić, J.; Zheljazkov, V.D.; Pezo, L.; Varga, A.; Tumbas Šaponjac, V. *Helichrysum italicum* (Roth) G. Don Essential Oil from Serbia: Chemical Composition, Classification and Biological Activity—May It Be a Suitable New Crop for Serbia? *Agronomy* **2021**, *11*, 1282. [CrossRef]

50. Tyagi, A.K.; Prasad, P.; Yuan, W.; Li, S.; Aggarwal, B.B. Identification of a novel compound (β-sesquiphellandrene) from turmeric (*Curcuma longa*) with anticancer potential: Comparison with curcumin. *Investig. New Drugs* **2015**, *33*, 1175–1186. [CrossRef] [PubMed]

51. Yakoubi, R.; Megateli, S.; Sadok, T.H.; Bensouici, C.; Bağci, E. A synergistic interaction of Algerian essential oils of *Laurus nobilis* L., *Lavandula stoechas* L. and *Mentha pulegium* L. on anticholinesterase and antioxidant activities. *Biocatal. Agric. Biotechnol.* **2021**, *31*, 101891. [CrossRef]

52. Tissot, E.; Rochat, S.; Debonneville, C.; Chaintreau, A. Rapid GC-FID quantification technique without authentic samples using predicted response factors. *Flavour. Fragr. J.* **2012**, *27*, 290–296. [CrossRef]

53. Gilardoni, G.; Matute, Y.; Ramírez, J. Chemical and Enantioselective Analysis of the Leaf Essential Oil from *Piper coruscans* Kunth (Piperaceae), a Costal and Amazonian Native Species of Ecuador. *Plants* **2020**, *9*, 791. [CrossRef]
54. Ramírez, J.; Andrade, M.D.; Vidari, G.; Gilardoni, G. Essential Oil and Major Non-Volatile Secondary Metabolites from the Leaves of Amazonian *Piper subscutatum*. *Plants* **2021**, *10*, 1168. [CrossRef]
55. Rhee, I.K.; van de Meent, M.; Ingkaninan, K.; Verpoorte, R. Screening for acetylcholinesterase inhibitors from Amaryllidaceae using silica gel thin-layer chromatography in combination with bioactivity staining. *J. Chromatogr. A* **2001**, *915*, 217–223. [CrossRef]

plants

Article

Biochemical Characterization, Antifungal Activity, and Relative Gene Expression of Two *Mentha* Essential Oils Controlling *Fusarium oxysporum*, the Causal Agent of *Lycopersicon esculentum* Root Rot

Seham A. Soliman [1], Elsayed E. Hafez [1], Abdu M. G. Al-Kolaibe [2], El-Sayed S. Abdel Razik [1], Sawsan Abd-Ellatif [3], Amira A. Ibrahim [1], Sanaa S. A. Kabeil [4] and Hazem S. Elshafie [5,*]

[1] Plant Protection and Biomolecular Diagnosis Department, Arid Lands Cultivation Research Institute, City of Scientific Research and Technology Applications, Borg EL-Arab, Alexandria 21934, Egypt; sehamsoliman50@yahoo.com (S.A.S.); elsayed_hafez@yahoo.com (E.E.H.); eshabaan@srtacity.sci.eg (E.-S.S.A.R.); amiranasreldeen@yahoo.com (A.A.I.)

[2] Microbiology Department, Faculty of Science, Taiz University, Taiz 6803, Yemen; abdu_71@yahoo.com

[3] Bioprocess Development Department, Genetic Engineering and Biotechnology Research Institute, City of Scientific Research and Technology Applications, Borg EL-Arab, Alexandria 21934, Egypt; sabdellatif@srtacity.sci.eg

[4] Protein Research Department, Genetic Engineering and Biotechnology Research Institute, City of Scientific Research and Technology Applications, Borg EL-Arab, Alexandria 21934, Egypt; sanaa.protein@gmail.com

[5] School of Agricultural, Forestry, Food and Environmental Sciences, University of Basilicata, Viale dell'Ateneo Lucano 10, 85100 Potenza, Italy

* Correspondence: hs.elshafie@gmail.com; Tel.: +39-0971-205522; Fax: +39-0971-205503

Citation: Soliman, S.A.; Hafez, E.E.; Al-Kolaibe, A.M.G.; Abdel Razik, E.-S.S.; Abd-Ellatif, S.; Ibrahim, A.A.; Kabeil, S.S.A.; Elshafie, H.S. Biochemical Characterization, Antifungal Activity, and Relative Gene Expression of Two *Mentha* Essential Oils Controlling *Fusarium oxysporum*, the Causal Agent of *Lycopersicon esculentum* Root Rot. *Plants* 2022, 11, 189. https://doi.org/10.3390/plants11020189

Academic Editor: Franklin Gregory

Received: 14 December 2021
Accepted: 5 January 2022
Published: 11 January 2022

Publisher's Note: MDPI stays neutral with regard to jurisdictional claims in published maps and institutional affiliations.

Abstract: Tomato (*Lycopersicon esculentum* Mill.) is important food in daily human diets. Root rot disease by *Fusarium oxysporum* caused huge losses in tomato quality and yield annually. The extensive use of synthetic and chemical fungicides has environmental risks and health problems. Recent studies have pointed out the use of medicinal plant essential oils (EOs) and extracts for controlling fungal diseases. In the current research, *Mentha spicata* and *Mentha longifolia* EOs were used in different concentrations to control *F. oxysporum*. Many active compounds are present in these two EOs such as: thymol, adapic acid, menthol and menthyl acetate. These compounds possess antifungal effect through malformation and degradation of the fungal cell wall. The relative expression levels of distinctly upregulated defense-related WRKY genes (WRKY1, WRKY4, WRKY33 and WRKY53) in seedling root were evaluated as a plant-specific transcription factor (TF) group in different response pathways of abiotic stress. Results showed significant expression levels of WRKY, WRKY53, WRKY33, WRKY1 and WRKY4 genes. An upregulation was observed in defense-related genes such as chitinase and defensin in roots by application EOs under pathogen condition. In conclusion, *M. spicata* and *M. longifolia* EOs can be used effectively to control this plant pathogen as sustainable and eco-friendly botanical fungicides.

Keywords: WRKY transcription factor; essential oils; Fusarium root rot; *Mentha spicata*; *Mentha longifolia* GC–MS; antioxidant enzymes; antifungal activity

1. Introduction

The tomato plant (*Lycopersicon esculentum* Mill.) is considered one of the third most important vegetable plants worldwide. Moreover, it is one of the most widespread vegetable crops grown across the globe. The tomato plant is highly sensitive to various biotic and abiotic stresses, which results in high economic losses [1,2]. The biotic stress which affects tomato plant growth and production is *Fusarium oxysporum* f. sp. *lycopersici* (*Fol*) [3]. *Fusarium oxysporum* is a soil-borne pathogen that targets the plant by attacking the tomato roots, resulting in wilt disease [4]. Wilt disease frequency in tomato crops is very high in some countries where it reaches up to 25 ± 5% [5,6].

Additionally, in the presence of suitable conditions for the fungus, especially in developing countries, economic losses may increase up to 80% [7]. Consequently, the fungus gains its capability in tomato infection through secretion of mycotoxins [8], which have hazardous effects on animal and human health [9]. Mint essential oil (EO) has been reported to have a strong antimicrobial activity against several pathogenic microorganisms [10,11]. Many researchers have studied the biological activity of different EOs from Mentha against different pathogenic fungi, especially *Fusarium* species [11–15]. In particular, the two studied *Mentha* EOs have illustrated a strong antifungal activity against potato pathogens in addition to different soil-borne diseases in tomatoes [16–18].

The most active chemical compounds in EOs of *Mentha* species are piperitone oxide, pulegone, and 3-cyclopenten-1-one, 2-hydroxy-3-(3-methyl-2-butenyl) [19,20]. These compounds play an important role in defense against pests, pathogens, and fungi [21–23]. Sharma et al. [24] studied the effect of mint, clove, lemongrass, and eucalyptus EOs on wilt-causing fungus *F. oxysporum*. Plants' EOs have a vital role in enhancing plant defense systems by increasing the production of phytochemicals as phenolic compounds and peroxidases enzymes which lead to strengthening of the cell wall and increasing lignification against phytopathogens [25].

When any pathogen infects plants, it is well known that they induce a plant's defense system which works to resist both the pathogen attack and development of disease [26]. The plant defense system works once the plant is exposed to any stress; plant transcription factors belonging to multiple families play a critical role in stress mitigation or other adjustment mechanisms by modulating the gene expression patterns [27]. There is a large gene family, "WRKY", which is considered the transcriptional factors distributed in all plant parts [28]. In addition, the WRKY genes were previously discovered in non-photosynthetic eukaryotes [29], and consequently have been identified and characterized in different plant species [30,31].

The main role of WRKY genes is defense; these genes work in the plant acquired resistance by using different pathways, including different enzymes [32]. Researchers have reported that they play a role in the defense mechanism of the *Arabidopsis* plant infected with necrotrophic fungal pathogens *Botrytis cinerea* and *Alternaria brassicicola* [33]. Several studies revealed that WRKY genes might bind with the promoter of phytoalexin deficient 3 and 1-aminocyclopropane-1-carboxylic acid synthase 2 when the plant is attacked by *Botrytis cinerea* [34].

The aims of the present study are (i) investigating the potential antifungal activities of different concentrations of *M. spicata* and *M. longifolia* EOs against Fusarium root rot disease caused by *F. oxysporum* in tomato plant; (ii) demonstrating the possible alterations in seedling germination, total phenols, and the activity of several antioxidant enzymes; and (iii) discovering the mode of action between the fungus and plant through analyzing the expression levels of defense-related genes as chitinase (PR3) and defensin (PR12) and *WRKY* transcriptional factors (TFs) (such as WRKY1, WRKY4, WRKY33, and WRKY53) by investigating the upregulation or downregulation profile of studied defense and WRKY genes against *Fusarium* attack.

2. Results

2.1. Screening of Antifungal Activity of Studied EOs

Antifungal activity of *M. spicata* and *M. longifolia* EOs at various concentrations (0.25%, 0.5%, 0.75%, 1.0%, and 1.25%) was investigated against *F. oxysporum*. The average reductions in *F. oxysporum* radial growth in response to *M. spicata* and *M. longifolia* EOs colloid treatment are shown in the graph (Figure 1).

All tested concentrations exhibited varied inhibitory activity compared with positive control fungicide (nystatin 0.005%) and the untreated experimental control. Inhibition of mycelium growth increased with time and complete growth reduction was achieved after 7 days of incubation at $28 \pm 2\,°C$ in the case of *M. longifolia* EOs at 1.0% and 1.25%. Whereas, the highest significant growth inhibition (92.55** $\pm$ 0.08% and 90.63** $\pm$ 0.04) was achieved in the case of *M. spicata* EO (1.25%) and nystatin (0.005%), respectively, compared with control. The lowest insignificant growth inhibition, 14.33 ns%, was obtained with *M. spicata* EO at 0.25%.

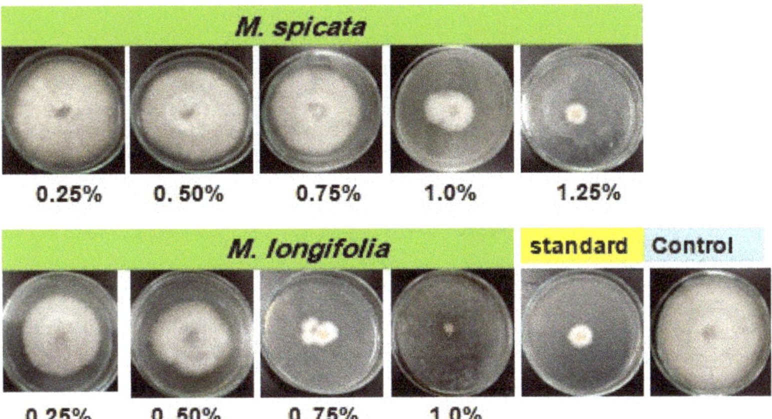

Figure 1. Effects of the two studied EOs at different concentrations on mycelial growth reduction percentages of *F. oxysporum* after 7 days of incubation at $28 \pm 2\,°C$.

2.2. Chemical Composition of Mentha EOs

Selected GC–MS analysis chromatograms of the two studied EOs are shown in Figures 2 and 3. The component relative concentrations (calculated as GC peak area percentages and retention index) are shown in Tables 1 and 2. GC–MS results identified that the principal bioactive components belonged to different chemical groups. The major constituents in *M. spicata* EO included thymol (28.19%), adipic acid (25.82%), piperitone (24.76%), and menthol (24.18%), whereas, the major constituents in *M. longifolia* EO included menthol (51.4%), menthyl acetate (20.5%), and d-limonene (11.15%).

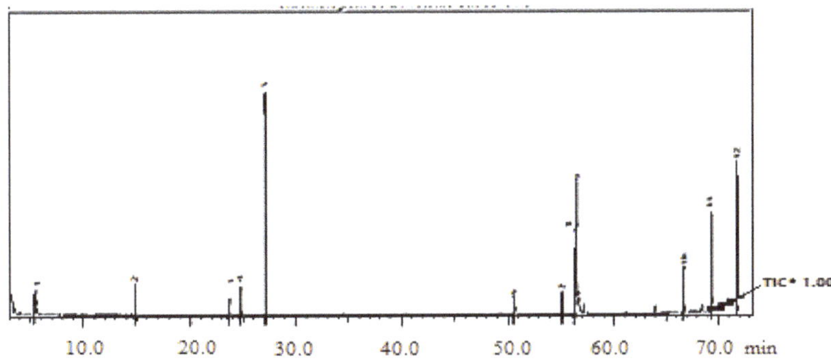

Figure 2. GC–MS chromatogram of *M. spicata* EO. Where (*) is the Total Ion Current.

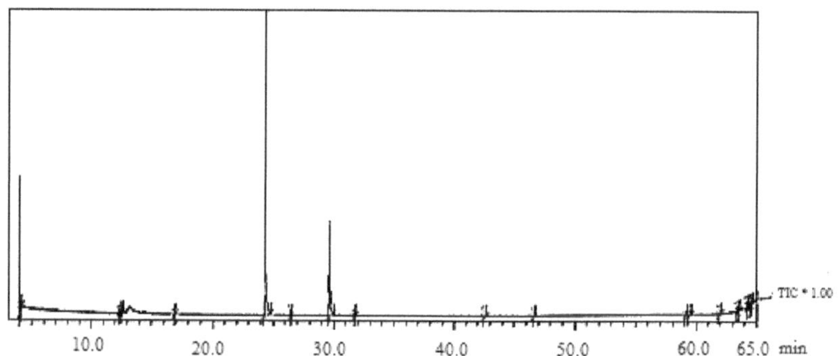

Figure 3. GC–MS chromatogram of *M. longifolia* EO. Where (*) is the Total Ion Current.

Table 1. The relative percentage of the area of peak of *M. spicata* EO constituents.

Quantitative ID	Component Identified	Retention Time (min)	Retention Index (RI)	Area (%)	Identification
1	Piperitone	5.20	1245	12.34	RI, MS
2	IR-alpha-pinene	7.82	933	0.76	RI, MS
3	D-limonene	11.83	1031	0.77	RI, MS
4	Cis-P-menthan	17.63	984	0.99	RI, MS
5	Menthone	18.13	1150	0.67	RI, MS
6	P-menthan-3-ol alcohol	18.55	1164	2.27	RI, MS
7	Alpha-terpin	19.49	1187	6.11	RI, MS
8	Gamma-terpineol	19.73	1185	0.82	RI, MS
9	β-caryophyllene	27.19	1325	18.7	RI, MS
10	Butanedioic acid	32.28	1580	6.26	RI, MS
11	Ethyl 4-heptyl ester	36.54	1378	8.83	RI, MS
12	Adipic acid	56.70	1507	13.24	RI, MS
13	Menthol	66.70	1182	12.08	RI, MS
14	Thymol	72.10	1290	14.62	RI, MS
15	Others	-	-	1.54	RI, MS

MS: Mass spectrometry (GC–MS).

Table 2. The relative percentage of the area of peak *M. longifolia* EO constituents.

Quantitative ID	Component Identified	Retention Time (min)	Retention Index (RI)	Area (%)	Identification
1	Alpha piene	4.05	933	6.87	RI, MS
2	β-pinene	12.21	964	0.54	RI, MS
3	D-limonene	12.43	1031	10.33	RI, MS
4	Borneol	12.72	1165	0.38	RI, MS
5	β-terpinyl acetate	16.89	1267	0.47	RI, MS
6	Menthol	24.32	1182	44.17	RI, MS
7	Menthone	26.48	1150	0.55	RI, MS
8	Menthyl acetate	29.63	1294	18.36	RI, MS
9	Linalool	31.74	1083	0.43	RI, MS
10	Eugenol	43.6	1209	0.64	RI, MS
11	Carvone	46.89	1242	0.33	RI, MS
12	Thymol	5.08	1290	11.23	RI, MS
13	Cavacrol	59.55	1298	0.36	RI, MS
14	Cis-jasmone	63.41	1394	0.56	RI, MS
15	Cinerolone	63.52	1641	0.53	RI, MS
16	Caryophyllene	64.32	1418	0.74	RI, MS
17	β-farnesene	64.38	1452	0.35	RI, MS
18	β-cubebene	64.57	1389	0.36	RI, MS
19	Alpha-cadinol	64.78	1627	0.82	RI, MS
20	Others	-	-	1.98	RI, MS

2.3. Effects of EOs on Tomato Growth Parameters

Tomato seedlings, treated with the highest concentrations (1.0% and 1.25%) of *M. spicata* and *M. longifolia*, respectively showed in vitro complete and maximum reduction in mycelial growth in *F. oxysporum* colonies and disease incidence of Fusarium root rot compared with untreated control (Figure 4). The magnitude of root rot symptoms and disease severity (86.39 ± 0.025) was observed in *Fusarium* inoculated experiments, whereas a significant reduction in root rot disease severity by 5.6 ± 0.01% and 3.5 ±.02% was shown in treated plants with 1.25% *M. spicata* and 1.0% *M. longifolia* EOs treatments.

In addition, the application of 1.25% *M. spicata* and 1.0% *M. longifolia* EOs exhibited a significant increase in the following growth parameters: plant height, shoot and root fresh, and dry weights, compared with treated-infected, infected, and negative control EOs (Table 3). A remarkable maximum plant height (32.42 ± 0.02 cm) was observed in the case of 1.0% *M. longifolia*-treated plants followed by (28.9 ± 0.01 cm) with 1.25% *M. spicata* treatment compared with control (24.32 ± 0.02). In contrast, the lowest plant height (16.54 ± 0.02) was observed in the case of pathogen-inoculated plants. The treatment with 1.0% *M. spicata* and *M. longifolia* resulted in a significant increase in all other measured germination features such as radicle and root lengths, and fresh and dry weights of radicle and root compared with control (Figure 4), while germination features were greatly decreased due to fungal infestation, where a clear root rot severity was observed. The total chlorophyll content and electrical leakage results are shown in Figures 5 and 6 where the maximum chlorophyll content (74.16 ± 0.01 and 62.33 ± 0.02) were measured in 1.0% *M. longifolia* and 1.25% *M. spicata*, respectively, compared with *Fusarium* inoculated plants (42.06 ± 0.02). Whereas, the maximum electrical leakage (EL %) (127.18 ± 0.01) was recorded in pathogen-treated plants with 1.25% *M. spicata* compared with (6.97 ± 0.01) in case of control.

Figure 4. Effects of highest concentrations of *M. spicata* and *M. longifolia* EOs formulations on *L. esculentum* seedling growth potential under Fusarium root rot disease infection. C = control, P = pathogen, T1 = 1.25% *M. spicata*, T2 = 1% *M. longifolia*, P + T1 = pathogen + 1.25% *M. spicata*, and P + T2 = pathogen + 1% *M. longifolia*.

2.4. Protein, Total Phenols, Flavonoids, Malondialdehyde Contents, and Antioxidant Enzymes

Data presented in Figures 7 and 8 represent the effects of *M. spicata* and *M. longifolia* application on *L. esculentum* seedling on lipid peroxidation level, protein content (PC), total phenols content (TPC), total flavonoids content (TFC), and different antioxidant enzymes of *L. esculentum* seedling in all treatments of 35 DAS and 1.25% *M. spicata* and 1.0% *M. longifolia* application under Fusarium root rot infection. Overall, the pathogen-infected plants showed significant increase in MDA level, TPC, and TFC contents compared with the untreated control, whereas *M. spicata* and *M. longifolia* applications showed a slight increase in MDA, TPC, and TFC. In addition, PC of pathogen-infected plants treated with 1.25% *M. spicata* and 1.0% *M. longifolia* had the highest content at 19.09 μmol/g and 18.54 μmol/g of FW, respectively, compared with control and pathogen-infected plants.

Assessed results of antioxidant enzymes in Figure 8 show the effect of *Fusarium* infection and 1.25% *M. spicata* and 1.0% *M. longifolia* applications for two weeks of seedling transplantation on SOD, CAT, and APOX enzymes activities in *L. esculentum* plant leaves. Our results reveal a significant increase in SOD, CAT, and APOX enzymes activities in all treatments and *Fusarium*-infected experiments compared with untreated control.

2.5. qRT-PCR of the Plant Defence System

qRT-PCR was carried out with mRNA to assess the expression levels of different WRKY transcription factors WRKY1, WRKY4, WRKY33, and WRKY53 (Figures S1–S4) which play an important role in biotic and abiotic tolerance. In addition, qRT-PCR represented the relative expression levels of defense-related proteins such as chitinase (PR3) and defensin (PR12) genes (Figures S5 and S6) in tomato plant roots after two weeks from *Fusarium* inoculation and EOs application, respectively.

Table 3. Effects of *M. spicata* and *M. longifolia* EOs on *L. esculentum* seedling growth under Fusarium root rot disease infection.

	EOs Treatment	PH (cm)	SFW (g)	SDW (g)	RFW (g)	RDW (g)
Control	Control Plants	24.32 ± 0.02 c	15.44 ± 0.02 c	1.98 ± 0.01 c	1.52 ± 0.01 c	0.16 ± 0.01 c
	1.25% *M. spicata*	28.9 ± 0.01 b	15.96 ± 0.02 b	2.09 ± 0.01 b	1.64 ± 0.01b	0.18 ± 0.01 b
	1% *M. longifolia*	32.42 ± 0.02 a	16.33 ± 0.01 a	2.34 ± 0.01 a	1.87 ± 0.01 a	0.24 ± 0.01 a
	Infected Plants	16.54 ± 0.02 f	9.52 ± 0.03 f	0.97 ± 0.01 f	1.21 ± 0.02 f	0.1 ± 0.01 f
Pathogen	1.25% *M. spicata*	18.76 ± 0.01 e	12.97 ± 0.01 e	1.67 ± 0.01 e	1.34 ± 0.15 e	0.12 ± 0.01 e
	1% *M. longifolia*	19.86 ± 0.02 d	13.64 ± 0.01 d	1.85 ± 0.01 d	1.41 ± 0.02 d	0.14 ± 0.01 d

PH = plant height, SFW = shoot fresh weight, SDW = shoot dry weight, RFW = root fresh weight, RDW = root dry weight, and EOs = essential oils. Different letters indicate significant differences between different treatments at $p \leq 0.05$. Data are expressed as the mean of three replicates ± SDs.

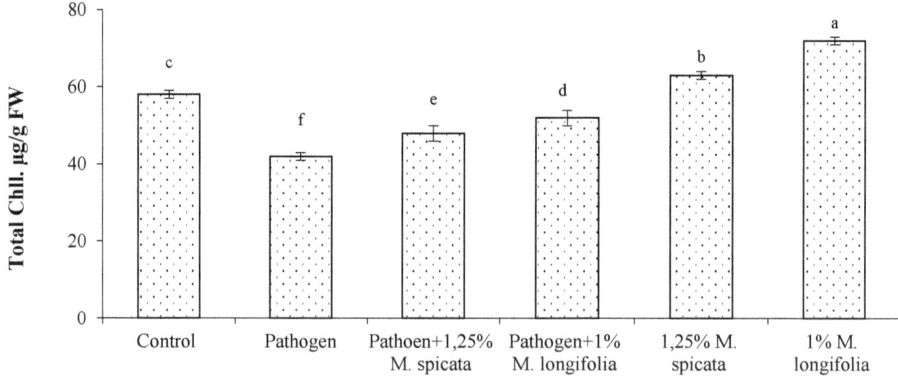

Figure 5. Effects of highest concentrations of *M. spicata* and *M. longifolia* EOs on total chlorophyll content (Chll) in *L. esculentum* seedling under Fusarium root rot disease infection. Bars with different letters indicate significant differences between treatments at $p \leq 0.05$. Data are expressed as the mean of three replicates ± SDs.

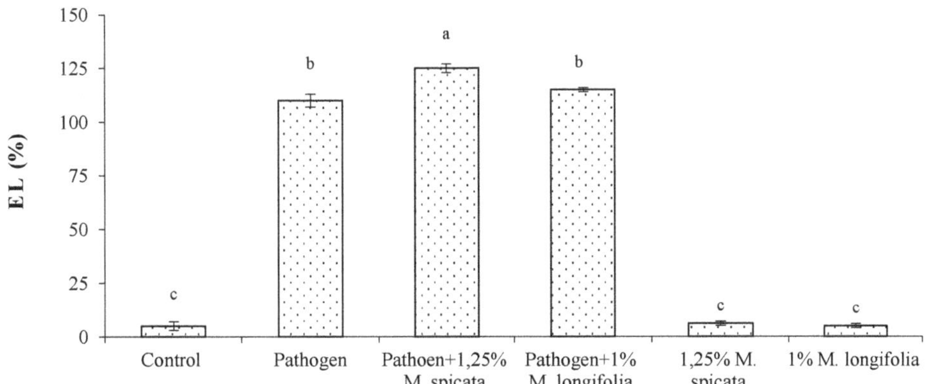

Figure 6. Effects of highest concentrations of *M. spicata* and *M. longifolia* EOs on electrical leakage percentage (EL%) in *L. esculentum* seedling leaves under Fusarium root rot disease conditions. Bars with different letters indicate significant differences between treatments at $p \leq 0.05$. Data are expressed as the mean of three replicates ± SDs.

In our study, the highest expression mRNA level (57.24 ± 0.01) was recorded for chitinase gene in *L. esculentum* plant roots at 1.0% *M. longifolia* under Fusarium infection, followed with (57.16 ± 0.02, 56.4 ± 0.02) of pathogen and *M. spicata* application treatments, respectively, compared with (1.0 ± 0.0) in untreated control treatments. Overall, the WRKY TFs genes (WRKY1, WRKY4, WRKY33, and WRKY53) in tomato seedling roots showed positive expression levels (upregulation) under *M. spicata* and *M. longifolia* treatments and positive control, compared with the untreated control. The highest expression mRNA levels of WRKY transcriptional factors WRKY53 gene (39.233 ± 0.03) represented in tomato roots at 1.0% *M. longifolia* treatment under Fusarium infection condition, followed with (38.12 ± 0.02) in pathogen treatment. Figure 9 shows a hierarchical clustering heat map and the correlation among different treatments and their gene expression. The red color demonstrates the highest correlation, and the blue color demonstrates the lowest correlation.

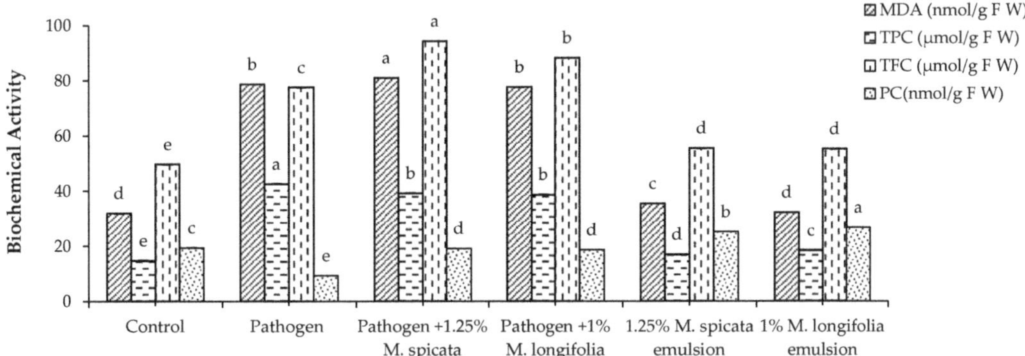

Figure 7. Effects of highest concentrations of *M. spicata* and *M. longifolia* EOs on MDA, TPC, TFC, and PC level in *L. esculentum* seedling under Fusarium root rot disease infection. Bars with different letters indicate significant differences between treatments at $p \leq 0.05$. Data are expressed as the mean of three replicates ± SDs.

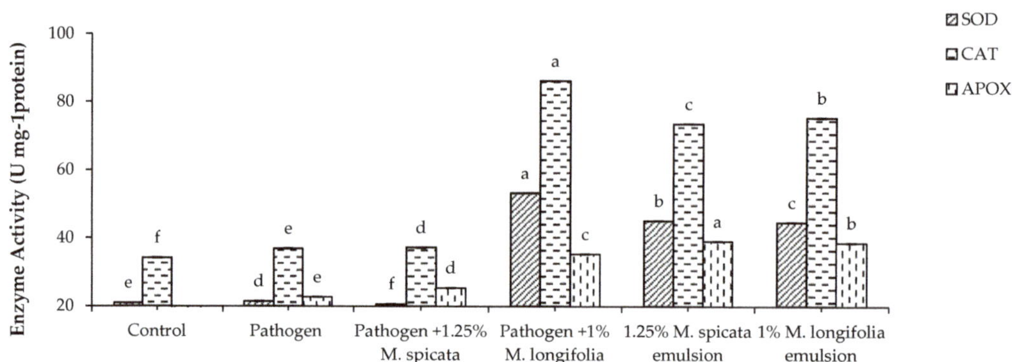

Figure 8. Effects of *M. spicata* and *M. longifolia* EOs on SOD, CAT, and APOX activity in *L. esculentum* seedling under Fusarium root rot disease infection. Bars with different letters indicate significant differences between treatments at $p \leq 0.05$. Data are expressed as the mean of three replicates $\pm$ SDs.

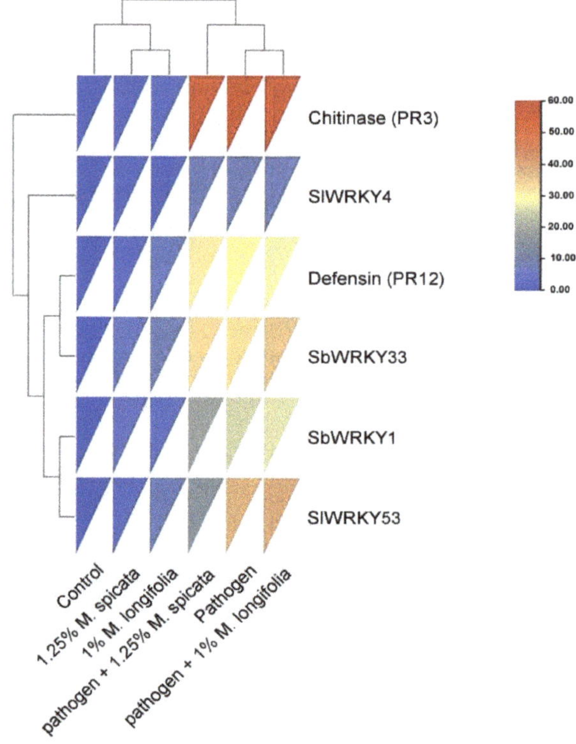

Figure 9. Hierarchical clustering heat map of relative expression levels of chitinase, defensin genes, and WRKY1, WRKY4, WRKY33, and WRKY53 transcripts in *L. esculentum* seedling under Fusarium injection and application of 1.25% *M. spicata* and 1.0% *M. longifolia* EOs treatments.

3. Discussion

F. oxysporum f. sp. *radicis-lycopersici* is a wide-spread fungus in the plant rhizosphere which causes Fusarium crown and root rot (FCRR) disease and leads to losses of tomato production even in greenhouses and systems of soil production [35]. There are different management methods for root rot disease of tomato crop using chemical and biological controls [36]. Biological control for pathogenic fungi is the new management trend for reducing the harmful effects of chemicals (fungicides) [37,38]. There are four types of biocontrol management: microorganisms, semi-chemical products, plant-based natural products, or living microorganisms [39–43]. Furthermore, plant EOs are effective biocontrol agents against a variety of pathogenic fungi and bacteria [44–47].

The present investigation revealed the effect of *Mentha spicata* and *M. longifolia* on EOs on root rot disease of tomato infected with *F. oxysporum* both in vitro and in vivo. *M. spicata* and *M. longifolia* EOs had potentiality against *Fusarium*. The order of efficient EOs against *Fusarium* pathogen was: *M. longifolia* > *M. spicata*. The highest antifungal activity was observed in the case of all concentrations of *M. longifolia* EO in agreement with previous studies [48]. The capability of two *Mentha* EOs against fusarium is due to the ability of bioactive chemical molecules to penetrate the fungal cell wall and cytoplasmic membrane and destroy mitochondrial membranes [49]. Plant EOs contributed to loss of rigidity of the hyphal cell wall as well as damaging the cellular enzyme system, resulting in cell death [50,51]. Many other studies reported that the antifungal activity of *M. spicata* EO against *F. oxysporum* and *Aspergillus niger* depends on the chemical constituents: menthol, thymol, and piperitone, individually or in synergic effect [52,53]. Thymol compound activity presented in the malformation of the cellular membrane in addition to inhibition for ATPase activity [43,54]; regarding the effect of thymol and eugenol, they correlated to the ability of thymol compounds' lysis of the external membrane of microorganisms which facilitated the entrance of eugenol to cytoplasm and interacted with protein [44,53]. Krishna Kishore et al. [55] demonstrated the antifungal activity of carvacrol, -terpineol, terpinen-4-ol, and linalool against *Rhizoctonia solani*, *F. oxysporum*, *Penicillium digitatum*, *A. niger*, *Alternaria alternate*, and *A. flavus*; these produce an effect against different microbial cells due to the ability of these compounds to penetrate the cell membrane, inactivate the enzyme pathway, and disturb their active transport [56,57].

The biological effect of the studied EOs on physiological parameters of tomato seedling is due to the presence of terpenes, alcohols, and phenolic compounds [55,58–60]. Moreover, the infected plants treated with *M. longifolia* EO showed the highest values of plant height (19.86 cm), shoot fresh weight (13.64 g), shoot dry weight (1.85 g), root fresh weight (1.41 g), and root dry weight (0.14 g). The main components in *M. spicata*, adipic acid 25.82% and piperitone 24.76%, are different than their percentage in *M. spicata* EO as reported by Bayan and Küsek [61]. Chemical constituents for *M. longifolia* EO were d-limonene, menthol, menthyl acetate, linalool, and eugenol, with percentages that differ from those reported from GC–MS analysis conducted by Desam et al. [62]. This difference in the percentage of single constituents for *Mentha* EOs may be due to differences in extraction methods or genetic diversity of these plants [63].

Menthol, menthyl acetate, linalool, and eugenol constituents alter cell permeability for *Fusarium* fungi and cause plasmolysis and cell death [64]. This study recorded collapsing of mycelium hyphae for *F. oxysporum* f. sp. *lycopersici* treated with EOs of *M. spicata and M. longifolia* with potential activity as biological control and therapeutic effect against root rot disease.

To analyze the role of chitinase, defensin, WRKY1, WRKY4, WRKY33, and WRKY53 transcripts in *L. esculentum* plant defense against *F. oxysporum* fungal pathogen, we analyzed their expression after pathogen infection, and pathogen infection and application of 1.25% *M. spicata and* 1.0% *M. longifolia* EOs treatments. The expression results showed over-expression by 57.24 fold and changes in the level of WRKY33 transcription factor were recorded in infected plants treated with 1.0% *M. longfolia*, followed with 57.16 and 56.4 fold changes in pathogen, and 1.25% *M. spicata* treatments, respectively, while minimum expression of WRKY33 2.43 fold was observed with 1.25% *M. spicata* EO compared with control. The expression profile of WRKY35 TF revealed a significant upregulation of 39.23 and 38.12 fold changes at the pathogen-infected plant under 1.0% *M. longfolia* and pathogen treatments, respectively, and the minimum fold change of 3.52 was at 1.25%

M. spicata EO compared with control. Data obtained in this study showed upregulation in the WRKY4 TF expression of 37.65 and 32.95 change folds at the pathogen-infected plant under 1.0% *M. longfolia* and pathogen treatments, respectively. In comparison, WRKY1 TF expression patterns were 26.35 and 25.17 change folds at the pathogen-infected plant under 1.0% *M. longfolia* and pathogen treatments, respectively, as compared with minimum expression level of 4.23 folds with 1.0% *M. longfolia*. Our results were in agreement with previous studies which reported that *WRKY3* and *WRKY4* encode two structurally similar WRKY proteins, and their expression was responsive to stress conditions. Stress-induced expression of *WRKY4* but not *WRKY3* was further enhanced by pathogen infection. These results strongly suggest that WRKY4 regulates crosstalk between SA and JA/ET-mediated signaling pathways and, as a result, plays opposite roles in resistance to the two different types of microbial pathogens. Interestingly, WRKY proteins such as WRKY4, WRKY33, and redundant WRKY18, WRKY40, and WRKY60 play a positive role in plant resistance to necrotrophic pathogens.

The expression of defense-related genes showed over-expression under pathogen infection conditions and with the pathogen under 1.0% of *M. longfolia* and 1.25% of *M. spicata*. In contrast, chitinase gene was upregulated with 31.25, 29.6, and 27.83 fold changes at pathogen with 1.25% *M. spicata*, pathogenated plants, and then pathogenated with 1.0% *M. longfolia* treatments, respectively. In addition, defensin gene expressed as 18.65, 16.16, and 15.76 fold changes at pathogen with 1.0% *M. longfolia*, pathogen with 1.25% *M. spicata*, and pathogen treatments compared with minimum 2.44 fold expression was recorded at 1.25% *M. spicata* EO treatment.

These results collectively indicate that overexpression of chitinase, defensin, and WRKY transcripts play a positive role in inducing plant resistance against *F. oxysporum* and working toward reducing disease severity. Additionally, WRKY transcripts and PR3, PR12 genes were further enhanced by pathogen infection, and they are already considered as a marker for the plant–microbe interaction.

WRKY family members have diverse regulatory mechanisms; their protein can be effectively combined with W-box elements and bind to acting elements to activate or inhibit the transcription of downstream target genes through the cis-acting mechanism [65]. Thus, WRKY as a transcription factor plays an important role in plant defense in response to attacks by several pathogens. The response works by activating the expression of resistance genes directly or indirectly. It has been reported that WRKY DNA binding proteins bind to the promoter region of Arabidopsis natriuretic peptide receptor 1 (NPR1), which activated the plant defense system [66]. Moreover, WRKY33 activates the plant resistance system against necrotrophic fungi *Alternaria brassicicola* and *Botrytis cinerea* [29], and it can regulate the SAR system in the infected plants and also the PR genes [67,68]. Moreover, the high expression of such transcription factors could regulate the plant pathogen sensitivity to mutants of AtWRKY4, AtWRKY3, and AtWRKY3 WRKY4, increasing the plant suscep-tibility toward the fungus *B. cinerea*. In contrast, the high expression of the non-mutated AtWRKY4 enhanced the plant's resistance toward the *Pseudomonas syringae* [69]. Many plant WRKY genes are induced by biotrophic and necrotrophic pathogens, including fungi and viruses, through the induction of SA-dependent SAR and PR genes [70,71].

4. Materials and Methods

4.1. Sample Collection, Identification, and Preparation

4.1.1. Fungal Isolate and Tomato Variety

The studied *Fusarium oxysporum* isolate was investigated as an aggressive fungal pathogen in tomato and other plants and was deposited in GenBank under accession number (KJ831189) and obtained from the Department of Plant Protection and Biomolecular Diagnosis, Arid Lands Cultivation Research Institute, Alexandria (Egypt). The fungal isolate was maintained on potato dextrose agar (PDA) slants and stored at 4 °C until further bioassay. Tomato seed (super strain B) variety was obtained from the Egyptian Ministry of Agriculture.

4.1.2. Medicinal Plant Materials

Leaves of *M. spicata* and *M. longifolia* were collected from unrestricted habitats, Alexandria (Egypt), in June 2020. The plants' identification was performed in the Botany Department, Faculty of Science, Mansoura University, Mansoura (Egypt).

4.2. Extraction of Essential Oils

Previously collected *M. spicata* and *M. longifolia* healthy fresh leaves at age one month were prewashed with tap water, disinfected with 2% sodium hypochlorite for 30 min, then rinsed with sterile distilled water. Air-dried plant leaves were homogenized to a fine powder in a mill, stored in airtight dark bottles, and kept until use. Leaf powders were subjected to hydrodistillation for 3 h using a Clevenger-type apparatus, Shiva Scientific Glass Pvt. Ltd., New Delhi, India [72].

4.3. GC–MS of Essential Oils

The chemical composition of the volatile content of two studied EOs was determined using GC–MS- QP2010 Ultra analysis system (Shimadzu, Tokyo, Japan). Compounds were separated on an Inc DB-5 60 m × 0.25 mm/0.25 micron column (Agilent Technologies, Santa Clara, CA, USA). The oven temperature program was initiated at 50 °C, held for 3 min, then increased at rate of 8 °C to 250 °C min^{-1} and held for 10 min. The spectrophotometer was operated in electron impact mode. The injector, interface, and ion source were kept at 250 °C, 250 °C, and 220 °C, respectively. Split injection (1 μL diluted sample in n-hexane (1:1, *v/v*) injected) was conducted with a split ratio of 1:20 and column flow of 1.5 mL/min, and helium was the carrier gas.

Identification of the components of the sample was based on a comparison of their relative indices and mass spectra by computer matching with WILEY and National Institute of Standards and Technology (NIST08) libraries data (http://webbook.nist.gov accessed on 20 November 2021) provided with the computer controlling GC–MS system. Individual isolated compound identifications were also performed by comparing their mass spectra and retention times with authentic compounds and literature data [73].

4.4. Preparation of EOs

The *M. spicata* and *M. longifolia* EOs colloid solutions were prepared by slowly adding 20 mL of *M. spicata* and *M. longifolia* EOs to 1 mL of non-ionic surfactant Tween 80 (1%), and the dispersion was performed under gentle stirring. Then, 80 mL of distilled water was added to reach the final mixture of 100% with continuous stirring using a magnetic stirrer for 30 min. The mixture was fed into a liquefied potato dextrose medium at different concentrations for further in vitro antifungal activity assay and greenhouse experiments.

4.5. In Vitro Antifungal Activity of EOs

Assessment of the antifungal activity of *M. spicata* and *M. longifolia* EOs were conducted in vitro and evaluated against *F. oxysporum* radial mycelial growth using the agar plate technique according to Tatsadjieu et al. [74]. The *M. spicata* and *M. longifolia* EOs were liquefied in sterilized PDA media to obtain a final concentration of 0.25%, 0.5%, 0.75%, 1.0% and 1.25%. Twenty mL of broth medium was poured into Petri dishes (90 mm diameter). Plates supplemented with 0.05% of fungicide (nystatin at 0.5 μL/mL) were used as control. Sterile distilled water was used in the bioassays instead of EO as a negative control. All plates were inoculated with mycelial disc (5 mm diameter) of *F. oxysporum* from the PDA plate margins (5–7 days old). Three replicate plates were used for each treatment. Then, the Petri-dishes were incubated at 25 °C and the fungal colony diameter was measured at 7 days.

4.6. Preparation of Fungal Suspension

The fungal suspension was prepared as follows: five discs (5 mm diameter) of mycelia agar plugs (7 days old) were added to 1 kg of sterilized maize grains, sand (2:1 *v/v*). Ten mL of sterile water was added to the last mixture in a 2 L flask and incubated at $25 \pm 2\,^\circ\text{C}$ for two weeks. After that, the mixture was put into plastic pots (20 cm diameter). Five sterile discs of PDA medium were inoculated into a control flask [75].

4.7. Greenhouse Experiments

Seeds of tomato were surface sterilized in sodium hypochlorite for 30 min, washed five times in sterile water, and germinated in peat moss for three weeks (21DAS). The experiment was irrigated regularly and subsequently moved to experimental pots. Four weeks later, tomato seedlings were removed and their roots were washed and transplanted into the 20 cm diameter pots filled with pasteurized sandy clay soil at 0.9 kg per pot. The seedlings were treated with *M. spicata* and *M. longifolia* EOs at 1.25% and 1.0%, respectively, in the rhizosphere soil. Pots were arranged in a randomized complete block design with three replications. In the first experiment, the pots were divided into two main groups: untreated plants as negative control (C) and plants inoculated with *F. oxysporum* fungal suspension as positive control (P). In the second experiment, after 2 weeks from inoculation, negative control was treated with 50 mL *M. spicata* EOs (1.25%) (T1) and 50 mL *M. longifolia* EOs (1.0%) (T2). In addition, the positive control was treated with 50 mL *M. spicata* EOs (1.25%) (P + T1) and 50 mL *M. longifolia* EOs (1.0%) (P+T2). All the plants continued growth after transplantation with regular irrigation every 3 days for 2 weeks in a greenhouse at $22/16\,^\circ\text{C}$, 65–70% humidity. We then evaluated all pots for the incidence of *F. oxysporum* root rot.

4.7.1. Disease Assessments

The disease severity (DS) was evaluated using the 0–5 scale described by Filion et al. [76]:

$$\text{Disease severity (\%)} = \left(\sum ab/AK\right) \times 100$$

where a = number of diseased plants with the same infection degree, b = infection degree, A = total number of the evaluated plants, and K = the greatest infection degree.

Whereas the disease incidence (DI) was calculated according to the following equation:

$$\text{Disease incidence (\%)} = (a/A) \times 100$$

where a = number of diseased plants, and A = total number of evaluated plants.

4.7.2. Analysis of Plant Growth Parameters

Tomato seedlings of 21-day samples were collected to measure morphological traits. Three plants of each experiment were harvested and transferred to the laboratory and carefully uprooted, washed using tap water for measuring plant height and shoot and root fresh weight. Shoot and root dry weight were measured after oven drying at $40\,^\circ\text{C}$ for 48 h.

According to Lichtenthaler et al [77] method, chlorophyll content was determined after 35 days using spectrophotometry. The photosynthetic pigments were ground and extracted from 0.5 g of a third of the fully expanded plant leaves between 8:00 and 10:00 a.m., and suspended in 10 mL of 80% (*v/v*) acetone in the dark using a pestle and mortar. Extracts were filtrated and the content of total chlorophyll was determined at 645 nm and 663 nm.

4.8. Electrolytes Leakage

Determination of electrolytes leakage was conducted by adding 200 mg of fresh tomato leaves to a test tube containing 4 mL of de-mineralized water and shacked for 30 min. It was then rinsed 3 times to eliminate surface electrolytes [78]. Malondialdehyde (MDA) content was examined in the fresh tomato leaves using the method described by Heath and Packer [79]. Briefly, the MDA contents were determined after centrifugation ($12,000 \times g$) for 10 min; the absorbance reading was 600 nm and 532 nm using a UV–VIS spectrometer (Jenway, Tokyo, Japan).

4.9. Determination of Total Phenolic and Flavonoid Contents

Total phenolic content (TPC) of tomato was measured by dissolving 5 mg of air-dried leaf powder in 10 mL methanol according to Slinkard and Singleton [80] using Folin–Ciocalteu reagent protocol. Total flavonoid content (TFC) of tomato leaves was evaluated using the aluminum chloride colorimetry method described by Chavan et al. [81]. A standard calibration curve was constructed using quercetin in different concentrations (0.05–1 mg/mL). Tomato extract or quercetin (2 mL) was mixed with 500 µL of 10% aluminum chloride solution and 500 µL of 0.1 mM sodium nitrate solution. The absorbance of the reaction mixture was measured after incubation at room temperature for 30 min at wavelength 430 nm using a UV–VIS spectrometer (Jenway, Tokyo, Japan). Soluble protein content (PC) was estimated in both control and treated plants following Bradford [82] using Coomassie Brilliant Blue G-250 dye and the absorbance was recorded at 595 nm using bovine serum albumin as standard.

4.10. Assay of Antioxidant Enzymes

Antioxidant enzymes were extracted by homogenizing 1 gm fresh tomato leaf tissue in chilled 50 mM phosphate buffer (pH 7.0) supplemented with 1% polyvinyl pyrolidine and 1 mM EDTA using prechilled pestle and mortar. After centrifuging at $18,000 \times g$ for 30 min at 40 °C, the supernatant was used for enzyme assay. Determination of the activity of superoxide dismutase (SOD, EC 1.15.1.1) and NBT photochemical reductions was recorded at 560 nm using the Bayer and Fridovich [83] method in a 1.5 mL assay mixture containing sodium phosphate buffer (50 mM, pH 7.5), 100 µL EDTA, L-methionine, 75 µM NBT, riboflavin, and 100 µL enzyme extract.

The catalase assay (CAT, EC1.11.1.6) activity method of Luck [84] was used and monitored the change in absorbance at 240 nm for 2 min. For the calculation, an extinction coefficient of 39.4 mM^{-1} cm^{-1} was used. Ascorbate peroxidase (APX, EC 1.11.1.11) activity was tested by monitoring absorption change at 290 nm for 3 min in a 1 mL reaction mixture containing potassium phosphate buffer (pH 7.0), 0.5 mM ascorbic acid, hydrogen peroxide, and enzyme extract. The calculation of the extinction coefficient of 2.8 mM^{-1} cm^{-1} was used [85].

4.11. Gene Expression

According to the manufacturer's protocol, total mRNA was isolated from 0.5 g tomato plant root of control and all treatments using the Plant RNA Kit (Sigma-Aldrich, St. Louis, MO, USA). The purified RNA was quantitated using SPECTROstar Nano (BMG LABTECH, Ortenberg, Germany). For each sample, 10 µg total RNA was treated with DNAse RNAse-free (Fermentas, Waltham, MA, USA), 5 µg of which was reverse transcribed in a reaction mixture consisting of oligo dT primer (10 pmL/µL), 2.5 µL 5X buffer, 2.5 µL MgCl2, 2.5 µL 2.5 mM dNTPs, 4 µL from oligo (dT), 0.2 µL (5 Unit/µL) reverse transcriptase (Promega, Walldorf, Germany), and 2.5 µL RNA. RT-PCR amplification was performed in a thermal cycler PCR, programmed at 42 °C for 1 h and 72 °C for 20 min. Quantitative real-time PCR was carried out on 1 µL 1:10 diluted cDNA templates by triplicate using the real-time analysis (Rotor-Gene 6000, QIAGEN GmbH, Hilden, Germany) system. The primer sequences used in qRT-PCR are given in Table 4. Primers of three PRs (PR3, PR12) genes, three WRKY transcriptional factors (WRKY1, WRKY4, WRKY33, and WRKY53 TFs genes), and housekeeping gene (reference gene) were used for gene expression analysis using SYBR® Green-based method. The reaction mixture consists of 1 µL of template, 10 µL of SYBR Green Master Mix, 2 µL of reverse primer, 2 µL of forwarding primer, and sterile distilled water for a total reaction volume of 20 µL. PCR assays were performed using the following conditions: 950 °C for 15 min followed by 40 cycles of 950 °C for 30s and 600 °C for 30 s. The CT of each sample was used to calculate ∆CT values (target gene CT subtracted from β-Actin gene CT [86]). The relative gene expression was determined using the 2-∆∆Ct method [87].

Table 4. Sequences of primers used in qRT-PCR analysis.

Gene Name		Sequence	ID & Reference
Chitinase (PR3)	F	5′- ATGGCGGAAACTGTCCTAGTGGAA -3′	Medeiros et al. [88]
	R	5′ ACATGGTCTACCATCAGCTTGCCA -3′	
Defensin (PR12)	F	5′- TCACCAAACTATTGGATTTCAA -3′	Hafez et al. [89]
	R	5′- GACTCAATTTTTGACTTCTTAATCC -3′	
WRKY1	F	5′- CGCAACTCAAAGAGACGGAAG-3′	Solyc07g047960.2.1
	R	5′- CATTGACTACATCCACTTCACTGC-3′	
WRKY4	F	5′- CGTTGCACATACCCTGGATG -3′	Solyc05g012770.2.1
	R	5′- GGCCTCCAAGTTGCAATCTC -3′	
WRKY33	R	5′- CCACCTCCTTCACTTCCATT -3′	Solyc09g014990.2.1
	F	5′- GATGGAAAACTCCCAGTCGT -3′	
WRKY53	F	5′- CACATACCGAGGCTCCCATAA -3′	Solyc08g008280.2.1
	R	5′- CCTGTTGGATAAACGGCTTGG -3′	
β-Actin	F	5′- TCCTTCTTGGGTATGGAATCCT-3′	NM_007393.5
	R	5′- CAGCACTGTGTTGGCATAGA-3′	

4.12. Statistical Analyses

All the experiments were performed in triplicates. Obtained data and results were expressed as mean $\pm$ standard deviation ($\pm$SD). Some experiments were arranged in a completely randomized blocks design and data were statistically analyzed by one-way ANOVA test using SPSS 16. The probability values $p \leq 0.05$ were considered statistically significant based on Duncan's least significant difference test. The heatmap was constructed to study the similarity and dissimilarity among studied taxa based on essential and non-essential amino acids using the TBtools package [90].

5. Conclusions

We can conclude that EOs of two *Mentha* aromatic plants (*M. spicata* and *M. longifolia*) have the highest potential against Fusarium root rot disease for tomatoes. The chemical essential oils of these EOs have a lethal effect against *F. oxysporum* fungi. In addition, these EOs enhance the growth of tomato plants by increasing their physiological activities. Due to the pathogen attack, this activity could be induced by the plant defense system. This activity was regulated by different families of genes such as WRKY and PR proteins. On the other hand, *L. esculentum* seed priming or seedling root treatments with *M. spicata* and *M. longifolia* EOs could be used at lower concentrations (1.0–1.25%) to enhance seedling growth and alleviate the adverse effects of the fungal disease by supporting antioxidant enzymes and total phenols accumulation which could help make seeds or plants tolerate oxidative stress conditions further. Finally, the EOS of *Mentha* medicinal plants can be used as a safe alternative to fungicides that improve the growth of the infected plant without defects in plant and human health.

Supplementary Materials: The following supporting information can be downloaded at: https://www.mdpi.com/article/10.3390/plants11020189/s1, Figure S1. Relative expression level of WRKY1 gene in *S. lycopersicum* seedling under *Fusarium* inoculation and application of *M. spicata* and *M. longifolia* EOs. Different letters indicate significant differences between different treatments at $p \leq 0.05$. Figure S2. Relative expression level of WRKY4 gene in *S. lycopersicum* seedling under *Fusarium* inoculation and application of *M. spicata* and *M. longifolia* EOs. Different letters indicate significant differences between different treatments at $p \leq 0.05$. Figure S3. Relative expression level of WRKY33 gene in *S. lycopersicum* seedling under *Fusarium* inoculation and application of *M. spicata* and *M. longifolia* EOs. Different letters indicate significant differences between different treatments at $p \leq 0.05$. Figure S4. Relative expression level of WRKY53 gene in *S. lycopersicum* seedling under *Fusarium* inoculation and application of *M. spicata* and *M. longifolia* EOs. Different letters indicate significant differences between different treatments at $p \leq 0.05$. Figure S5. Relative expression level of Chitinase gene in *S. lycopersicum* seedling under *Fusarium* inoculation and application of *M. spicata* and *M. longifolia* EOs. Different letters indicate significant differences between different treatments at $p \leq 0.05$.

Plants **2022**, *11*, 189

Figure S6. Relative expression level of defensin gene in *S. lycopersicum* seedling under *Fusarium* inoculation and application of *M. spicata* and *M. longifolia* EOs. Different letters indicate significant differences between different treatments at $p \leq 0.05$.

Author Contributions: Conceptualization, S.A.S., E.E.H. and E.-S.S.A.R. and A.A.I.; data curation, E.-S.S.A.R., S.S.A.K., A.A.I. and H.S.E.; formal analysis, E.E.H., S.A.-E. and A.A.I.; investigation, S.A.S.; methodology, E.-S.S.A.R., S.A.-E. and A.A.I.; resources, E.E.H., E.-S.S.A.R. and A.A.I.; software, A.M.G.A.-K. and E.-S.S.A.R.; supervision, E.E.H. and H.S.E.; validation, E.E.H., S.S.A.K. and H.S.E.; visualization, S.A.S., A.M.G.A.-K., E.-S.S.A.R., S.A.-E. and A.A.I.; and writing—review and editing, A.A.I. All authors have read and agreed to the published version of the manuscript.

Funding: This research received no external funding.

Institutional Review Board Statement: Not applicable.

Informed Consent Statement: Not applicable.

Data Availability Statement: Relevant data applicable to this research are within the paper.

Conflicts of Interest: The authors declare no conflict of interest.

References

1. Gupta, S.; Rashotte, A.M. Expression patterns and regulation of SlCRF3 and SlCRF5 in response to cytokinin and abiotic stresses in tomato (*Solanum lycopersicum*). *J. Plant Physiol.* **2014**, *171*, 349–358. [CrossRef] [PubMed]
2. Bergougnoux, V. The history of tomato: From domestication to biopharming. *Biotechnol. Adv.* **2014**, *32*, 170–189. [CrossRef] [PubMed]
3. Panthee, D.R.; Chen, F. Genomics of fungal disease resistance in tomato. *Curr. Genom.* **2010**, *11*, 30–39. [CrossRef] [PubMed]
4. Rongai, D.; Pulcini, P.; Pesce, B.; Milano, F. Antifungal activity of pomegranate peel extract against fusarium wilt of tomato. *Eur. J. Plant Pathol.* **2017**, *147*, 229–238. [CrossRef]
5. Asha, B.B.; Nayaka, C.S.; Shankar, U.A.; Srinivas, C.; Niranjana, S.R. Biological control of *F. oxysporum* f. sp. *lycopersici* causing wilt of tomato by *Pseudomonas fluorescens*. *Int. J. Microbiol. Res.* **2011**, *3*, 79–84.
6. Pandey, K.K.; Gupta, R.C. Pathogenic and cultural variability among Indian isolates of *Fusarium oxysporum* f. sp. lycopersici causing wilt in tomato. *Indian Phytopathol.* **2014**, *67*, 383–387.
7. Kirankumar, R.; Jagadeesh, K.S.; Krishnaraj, P.U.; Patil, M.S. Enhanced growth promotion of tomato and nutrient uptake by plant growth promoting rhizobacterial isolates in presence of tobacco mosaic virus pathogen. *Karnataka J. Agric. Sci.* **2010**, *21*, 309–311.
8. Singh, V.K.; Singh, H.B.; Upadhyay, R.S. Role of fusaric acid in the development of 'Fusarium wilt' symptoms in tomato: Physiological, biochemical and proteomic perspectives. *Plant Physiol. Biochem.* **2017**, *118*, 320–332. [CrossRef]
9. Bhat, R.; Rai, R.V.; Karim, A.A. Mycotoxins in food and feed: Present status and future concerns. *Compr. Rev. Food Sci. Saf.* **2010**, *9*, 57–81. [CrossRef]
10. Gulluce, M. Antimicrobial and antioxidant properties of the essential oils and methanol extract from *Mentha longifolia* L. ssp. *Longifolia*. *Food Chem.* **2007**, *103*, 1449–1456. [CrossRef]
11. Moghaddam, M.; Pourbaige, M.; Tabar, H.K.; Farhadi, N.; Hosseini, S.M.A. Composition and antifungal activity of peppermint (*Mentha piperita*) essential oil from Iran. *J. Essent. Oil Bear. Plants* **2013**, *16*, 506–512. [CrossRef]
12. Hussain, A.I.; Anwar, F.; Nigam, P.S.; Ashraf, M.; Gilani, A.H. Seasonal variation in content, chemical composition and antimicrobial and cytotoxic activities of essential oils from four *Mentha* species. *J. Sci. Food Agric.* **2010**, *90*, 1827–1836. [CrossRef]
13. Della Pepa, T.; Elshafie, H.S.; Capasso, R.; De Feo, V.; Camele, I.; Nazzaro, F.; Scognamiglio, M.R.; Caputo, L. Antimicrobial and phytotoxic activity of *Origanum heracleoticum* and *O. majorana* essential oils growing in cilento (Southern Italy). *Molecules* **2019**, *24*, 2576. [CrossRef]
14. Sonam, K.A.; Kumar, V.; Guleria, I.; Sharma, M.; Kumar, A.; Alruways, M.W.; Khan, N.; Raina, R. Antimicrobial Potential and Chemical Profiling of Leaves Essential Oil of *Mentha* Species Growing under North-West Himalaya Conditions. *J. Pure Appl. Microbiol.* **2021**, *15*, 2229–2243. [CrossRef]
15. Rizwana, H.; Alwhibi, M.S. Biosynthesis of silver nanoparticles using leaves of *Mentha pulegium*, their characterization, and antifungal properties. *Green Process Synth.* **2021**, *10*, 824–834. [CrossRef]
16. Hajlaoui, H.; Trabels, N.; Noumi, E. Biological activities of the essential oils and methanol extract of two cultivated mint species (*Mentha logifolia* and *Mentha pulegium*) used in the Tunisian folkloric medicine. *World J. Microbiol. Biotechnol.* **2009**, *25*, 2227–2238. [CrossRef]
17. Singh, Y.R.; Sandeep, K.; Anupam, D. Antifungal properties of essential oil *Mentha spicata* l. var. MSS-5. *Indian J. Crop Sci.* **2006**, *1*, 197–200.
18. Kadoglidou, K.; Chatzopoulou, P.; Maloupa, E.; Kalaitzidis, A.; Ghoghoberidze, S.; Katsantonis, D. Mentha and Oregano Soil Amendment Induces Enhancement of Tomato Tolerance against Soilborne Diseases, Yield and Quality. *Agronomy* **2020**, *10*, 23. [CrossRef]

19. Brahmi, F.; Adjaoud, A.; Marongiu, B.; Porcedda, S.; Piras, A.; Falconieri, D.; Yalaoui-Guellal, D.; Elsebai, M.F.; Madani, K.; Chibane, M. Chemical composition and in vitro antimicrobial, insecticidal and antioxidant activities of the essential oils of *Mentha pulegium* L. and *Mentha rotundifolia* (L.) Huds growing in Algeria. *Ind. Crops Prod.* **2016**, *88*, 96–105. [CrossRef]

20. Silva, C.L.; Câmara, J.S. Profiling of volatiles in the leaves of Lamiaceae species based on headspace solid phase microextraction and mass spectrometry. *Food Res. Int.* **2013**, *51*, 378–387. [CrossRef]

21. Bayan, Y.; Aksit, H. Antifungal activity of volatile oils and plant extracts from *Sideritis germanicopolitana* BORNM growing in Turkey. *Egypt J. Biol. Pest Control* **2016**, *26*, 333–337.

22. Kumar, P.; Mishra, S.; Malik, A.; Satya, S. Insecticidal properties of *Mentha* species: A review. *Ind. Crops Prod.* **2011**, *34*, 802–817. [CrossRef]

23. Al-Harbi, N.A.; Al Attar, N.M.; Hikal, D.M.; Mohamed, S.E.; Abdel Latef, A.A.H.; Ibrahim, A.A.; Abdein, M.A. Evaluation of insecticidal effects of plants essential oils extracted from basil, black seeds and lavender against *Sitophilus oryzae*. *Plants* **2021**, *10*, 829. [CrossRef]

24. Sharma, A.; Rajendran, S.; Srivastava, A.; Sharma, S.; Kundu, B. Antifungal activities of selected essential oils against *Fusarium oxysporum* f. sp. *lycopersici* 1322, with emphasis on *Syzygium aromaticum* essential oil. *J. Biosci. Bioeng.* **2017**, *123*, 308–313. [CrossRef]

25. Ben-Jabeur, M.; Ghabri, E.; Myriam, M.; Hamada, W. Thyme essential oil as a defense inducer of tomato against gray mold and Fusarium wilt. *Plant Physiol. Biochem.* **2015**, *94*, 35–40. [CrossRef]

26. De Vos, M.; Van Oosten, V.R.; Van Poecke, R.M.; Van Pelt, J.A.; Pozo, M.J.; Mueller, M.J.; Pieterse, C.M. Signal signature and transcriptome changes of Arabidopsis during pathogen and insect attack. *Mol. Plant Microbe Interact.* **2005**, *18*, 923–937. [CrossRef]

27. Zhang, Z.; Liu, W.; Qi, X.; Liu, Z.; Xie, W.; Wang, Y. Genome-wide identification, expression profiling, and SSR marker development of the bZIP transcription factor family in *Medicago truncatula*. *Biochem. Syst. Ecol.* **2015**, *61*, 218–228. [CrossRef]

28. Nan, H.; Li, W.; Lin, Y.; Gao, L. Genome-Wide Analysis of WRKY genes and their response to salt stress in the wild progenitor of Asian cultivated rice, *Oryza rufipogon*. *Front. Genet.* **2020**, *11*, 359. [CrossRef]

29. Zhang, Y.; Wang, L. The WRKY transcription factor superfamily: Its origin in eukaryotes and expansion in plants. *BMC Evol. Biol.* **2005**, *5*, 1. [CrossRef]

30. Wen, F.; Zhu, H.; Li, P.; Jiang, M.; Mao, W.; Ong, C.; Chu, Z. Genome-wide evolutionary characterization and expression analyses of WRKY family genes in *Brachypodium distachyon*. *DNA Res.* **2014**, *21*, 327–339. [CrossRef]

31. Guo, C.; Guo, R.; Xu, X.; Gao, M.; Li, X.; Song, J.; Zheng, Y.; Wang, X. Evolution and expression analysis of the grape (*Vitis vinifera* L.) WRKY gene family. *J. Exp. Bot.* **2014**, *65*, 1513–1528. [CrossRef] [PubMed]

32. Meng, X.; Zhang, S. MAPK cascades in plant disease resistance signaling. *Ann. Rev. Phytopathol.* **2013**, *51*, 245–266. [CrossRef] [PubMed]

33. Zheng, Z.; Qamar, S.A.; Chen, Z.; Mengiste, T. Arabidopsis WRKY33 transcription factor is required for resistance to necrotrophic fungal pathogens. *Plant J.* **2006**, *48*, 592–605. [CrossRef] [PubMed]

34. Li, G.; Meng, X.; Wang, R.; Mao, G.; Han, L.; Liu, Y.; Zhang, S. Dual-level regulation of ACC synthase activity by MPK3/MPK6 cascade and its downstream WRKY transcription factor during ethylene induction in Arabidopsis. *PLoS Genet.* **2012**, *8*, 1002767. [CrossRef]

35. Szczechura, W.; Staniaszek, M.; Habdas, H. *Fusarium oxysporum* F. sp. *Radicis lycopersici*—The cause of fusarium crown and root rot in tomato cultivation. *J. Plant Protec. Res.* **2013**, *53*, 172–176. [CrossRef]

36. Ozbay, N.; Newman, S.E. Fusarium crown and root rot of tomato and control methods. *Plant Pathol. J.* **2004**, *3*, 9–18. [CrossRef]

37. Cosentino, C.; Labella, C.; Elshafie, H.S.; Camele, I.; Musto, M.; Paolino, R.; Freschi, P. Effects of different heat treatments on lysozyme quantity and antimicrobial activity of jenny milk. *J. Dairy Sci.* **2016**, *99*, 5173–5179. [CrossRef]

38. Elshafie, H.S.; Viggiani, L.; Mostafa, M.S.; El-Hashash, M.A.; Bufo, S.A.; Camele, I. Biological activity and chemical identification of ornithine lipid produced by *Burkholderia gladioli* pv. *agaricicola* ICMP 11096 using LC-MS and NMR analyses. *J. Biol. Res.* **2017**, *90*, 96–103. [CrossRef]

39. Ravensberg, W. Crop protection in 2030: Towards a natural, efficient, safe and sustainable approach. In Proceedings of the IBMA International Symposium, Swansea University, Swansea, UK, 7–9 September 2015.

40. Camele, I.; Elshafie, H.S.; Caputo, L.; Sakr, S.H.; De Feo, V. *Bacillus mojavensis*: Biofilm formation and biochemical investigation of its bioactive metabolites. *J. Biol. Res.* **2019**, *92*, 39–45. [CrossRef]

41. Elshafie, H.S.; Devescovi, G.; Venturi, V.; Camele, I.; Bufo, S.A. Study of the regulatory role of N-acyl homoserine lactones mediated quorum sensing in the biological activity of *Burkholderia gladioli* pv. *agaricicola* causing soft rot of *Agaricus* spp. *Front. Microbiol.* **2019**, *10*, 2695. [CrossRef]

42. Sofo, A.; Elshafie, H.S.; Scopa, A.; Mang, S.M.; Camele, I. Impact of airborne zinc pollution on the antimicrobial activity of olive oil and the microbial metabolic profiles of Zn-contaminated soils in an Italian olive orchard. *J. Trace Elem. Med. Biol.* **2018**, *49*, 276–284. [CrossRef]

43. Elshafie, H.S.; Racioppi, R.; Bufo, S.A.; Camele, I. In vitro study of biological activity of four strains of *Burkholderia gladioli* pv. *agaricicola* and identification of their bioactive metabolites using GC–MS. *Saudi J. Biol. Sci.* **2017**, *24*, 295–301. [CrossRef]

44. Raveau, R.; Fontaine, J.; Lounès-Hadj Sahraoui, A. Essential oils as potential alternative biocontrol products against plant pathogens and weeds: A Review. *Foods* **2020**, *9*, 365. [CrossRef]

45. Elshafie, H.S.; Caputo, L.; De Martino, L.; Gruľová, D.; Zheljazkov, V.D.; De Feo, V.; Camele, I. Biological investigations of essential oils extracted from three Juniperus species and evaluation of their antimicrobial, antioxidant and cytotoxic activities. *J. Appl. Microbiol.* **2020**, *129*, 1261–1271. [CrossRef]
46. Camele, I.; Gruľová, D.; Elshafie, H.S. Chemical composition and antimicrobial properties of *Mentha piperita* cv. 'Kristinka' essential oil. *Plants* **2021**, *10*, 1567. [CrossRef]
47. Ali, H.M.; Elgat, W.; El-Hefny, M.; Salem, M.Z.M.; Taha, A.S.; Al Farraj, D.A.; Elshikh, M.S.; Hatamleh, A.A.; Abdel-Salam, E.M. New Approach for Using of *Mentha longifolia* L. and *Citrus reticulata* L. Essential Oils as Wood-Biofungicides: GC-MS, SEM, and MNDO Quantum Chemical Studies. *Materials* **2021**, *14*, 18. [CrossRef]
48. Ragab, M.M.M.; Ashour, A.M.A.; Abdel-Kader, M.M.; ElMohamady, R.; Abdel-Aziz, A. In vitro evaluation of some fungicides alternatives against *Fusarium oxysporum* the causal of wilt disease of pepper (*Capsicum annum* L.). *Int. J. Agric. For.* **2012**, *2*, 70–77.
49. Arnal-Schnebelen, B.; Hadji-Minaglou, F.; Peroteau, J.F.; Ribeyre, F.; de Billerbeck, V.G. Essential oils in infectious gynaecological disease: A statistical study of 658 cases. *Int. J. Aromatherapy.* **2004**, *14*, 192–197. [CrossRef]
50. Sharma, N.; Tripathi, A. Effects of *Citrus sinensis* (L.) *Osbeck epicarp* essential oil on growth and morphogenesis of *Aspergillus niger* Van Tieghem. *Microbiol. Res.* **2008**, *163*, 337–344. [CrossRef]
51. Camele, I.; Elshafie, H.S.; De Feo, V.; Caputo, L. Anti-quorum sensing and antimicrobial effect of Mediterranean plant essential oils against phytopathogenic bacteria. *Front. Microbiol.* **2019**, *10*, 2619. [CrossRef]
52. Regmi, S.; Jha, S.K. Antifungal activity of plant essential oils against *Fusarium oxysporum* schlecht. and *Aspergillus niger* van tiegh. from papaya. *Inter. J. Curr. Trends Sci. Tech.* **2017**, *8*, 20196–20204.
53. Gruľová, D.; Caputo, L.; Elshafie, H.S.; Baranová, B.; De Martino, L.; Sedlák, V.; Camele, I.; De Feo, V. *Thymol chemotype Origanum vulgare* L. essential oil as a potential selective bio-based herbicide on monocot plant species. *Molecules* **2020**, *25*, 595. [CrossRef]
54. Yakhlef, G.; Hambaba, L.; Pinto, D.; Silva, A.M.S. Chemical composition and insecticidal, repellent and antifungal activities of essential oil of *Mentha rotundifolia* (L.) from Algeria. *Ind. Crops Prod.* **2020**, *158*, 8. [CrossRef]
55. Krishna Kishore, G.; Pande, S.; Harish, S. Evaluation of essential oils and their components for broad-spectrum antifungal activity and control of late leaf spot and crown rot diseases in peanut. *Plant Dis.* **2007**, *91*, 375–379. [CrossRef]
56. Oussalah, M.; Caillet, S.; Saucier, L.; Lacroix, M. Antimicrobial effects of selected plant essential oils on the growth of a *Pseudomonas putida* strain isolated from meat. *Meat Sci.* **2006**, *73*, 236–244. [CrossRef]
57. Chen, C.J.; Li, Q.Q.; Zeng, Z.Y.; Duan, S.S.; Wang, W.; Xu, F.R.; Cheng, Y.X.; Dong, X. Efficacy and mechanism of *Mentha haplocalyx* and *Schizonepeta tenuifolia* essential oils on the inhibition of *Panax notoginseng* pathogens. *Ind. Crops Prod.* **2020**, *145*, 12. [CrossRef]
58. Elshafie, H.S.; Ghanney, N.; Mang, S.M.; Ferchichi, A.; Camele, I. An in vitro attempt for controlling severe phytopathogens and human pathogens using essential oils from Mediterranean plants of genus *Schinus*. *J. Med. Food* **2016**, *19*, 266–273. [CrossRef]
59. Elshafie, H.S.; Sakr, S.; Mang, S.M.; De Feo, V.; Camele, I. Antimicrobial activity and chemical composition of three essential oils extracted from Mediterranean aromatic plants. *J. Med. Food.* **2016**, *19*, 1096–1103. [CrossRef] [PubMed]
60. Burt, S. Essential oils: Their antibacterial properties and potential applications in foods a review. *Int. J. Food Microbiol.* **2004**, *94*, 223–253. [CrossRef] [PubMed]
61. Bayan, Y.; Küsek, M. Chemical composition and antifungal and antibacterial activity of *Mentha spicata* L. volatile oil. *Cienc. Investig. Agrar.* **2018**, *45*, 64–69. [CrossRef]
62. Desam, N.R.; Al-Rajab, A.J.; Sharma, M.; Mylabathula, M.M.; Gowkanapalli, R.R.; Albratty, M. Chemical composition, antibacterial and antifungal activities of Saudi Arabian *Mentha longifolia* L. essential oil. *J. Coast. Life Med.* **2017**, *5*, 441–446. [CrossRef]
63. Kumar, P.; Mishra, S.; Malik, A.; Satya, S. Compositional analysis and insecticidal activity of *Eucalyptus globulus* (family: Myrtaceae) essential oil against housefly (*Musca domestica*). *Acta Trop.* **2012**, *122*, 212–218. [CrossRef]
64. Devi, K.P.; Nisha, S.A.; Sakthivel, R.; Pandian, S.K. Eugenol (an essential oil of clove) acts as an antibacterial agent against *Salmonella typhi* by disrupting the cellular membrane. *J. Ethnopharmacol.* **2010**, *130*, 107–115. [CrossRef]
65. Phukan, U.J.; Jeena, G.S.; Shukla, R.K. WRKY transcription factors: Molecular regulation and stress responses in plants. *Front. Plant Sci.* **2016**, *7*, 760. [CrossRef]
66. Després, C.; Chubak, C.; Rochon, A.; Clark, R.; Bethune, T.; Desveaux, D.; Fobert, P. The Arabidopsis NPR1 disease resistance protein is a novel cofactor that confers redox regulation of DNA binding activity to the basic domain/leucine zipper transcription factor TGA1. *Plant Cell* **2003**, *15*, 2181–2191. [CrossRef]
67. Wang, D.; Amornsiripanitch, N.; Dong, X. A genomic approach to identify regulatory nodes in the transcriptional network of systemic acquired resistance in plants. *PLoS Pathog.* **2006**, *2*, 1024–1050. [CrossRef]
68. Koo, S.C.; Moon, B.C.; Kim, J.K.; Kim, C.Y.; Sung, S.J.; Kim, M.C.; Cho, M.; Cheong, Y. OsBWMK1 mediates SA-dependent defense responses by activating the transcription factor OsWRKY33. *Biochem. Biophys. Res. Commun.* **2009**, *387*, 365–370. [CrossRef]
69. Lai, Z.; Vinod, K.M.; Zheng, Z.; Fan, B.; Chen, Z. Roles of Arabidopsis WRKY3 and WRKY4 transcription factors in plant responses to pathogens. *BMC Plant Biol.* **2008**, *8*, 68. [CrossRef]
70. Dong, J.; Chen, C.; Chen, Z. Expression profile of the *Arabidopsis* WRKY gene superfamily during plant defense response. *Plant Mol. Biol.* **2003**, *51*, 21–37. [CrossRef]
71. Xu, X.; Chen, C.; Fan, B.; Chen, Z. Physical and functional interactions between pathogen-induced *Arabidopsis* WRKY18, WRKY40, and WRKY60 transcription factors. *Plant Cell* **2006**, *18*, 1310–1326. [CrossRef]

72. Périno, S.; Chemat-Djenni, Z.; Petitcolas, E.; Giniès, C.; Chemat, F. Downscaling of industrial turbo-distillation to laboratory turbo-clevenger for extraction of essential oils. Application of concepts of green analytical chemistry. *Molecules* **2019**, *24*, 2734. [CrossRef]

73. Elshafie, H.S.; Mancini, E.; Sakr, S.; De Martino, L.; Mattia, C.A.; De Feo, V.; Camele, I. Antifungal activity of some constituents of *Origanum vulgare* L. essential oil against postharvest disease of peach fruit. *J. Med. Food* **2015**, *18*, 929–934. [CrossRef]

74. Tatsadjieu, N.L.; Dongmo, P.J.; Ngassoum, M.B.; Etoa, F.X.; Mbofung, C.M.F. Investigations on the essential oil of *Lippia rugosa* from Cameroon for its potential use as antifungal agent against *Aspergillus flavus* Link ex. Fries. *Food Control* **2009**, *20*, 161–166. [CrossRef]

75. Abd-El-Moity, T.H. Effect of single and mixture of *Trichoderma harzianum* isolates on controlling three different soil borne pathogens. *Egyptian J. Microbiol.* **1985**, 111–120.

76. Filion, M.; St-Arnaud, M.; Jabaji-Hare, S.H. Quantification of *Fusarium solani* f. sp. phaseoli in mycorrhizal bean plants and surrounding mycorrhizosphere soil using real-time polymerase chain reaction and direct isolations on selective media. *Phytopathology* **2003**, *93*, 229–235. [CrossRef]

77. Lichtenthaler, H.; Wellburn, A. Determinations of total carotenoids and chlorophylls a and b of leaf extracts in different solvents. *Biochem. Soc. Trans.* **1983**, *11*, 591–592. [CrossRef]

78. Blum, A.; Ebercon, A. Cell membrane stability as a measure of drought and heat tolerance in wheat. *Crop Sci.* **1981**, *21*, 43–47. [CrossRef]

79. Heath, R.L.; Packer, L. Photoperoxidation in isolated chloroplasts. I. Kinetics and stoichiometry of fatty acid peroxidation. *Arch. Biochem. Biophisics* **1968**, *125*, 189–198. [CrossRef]

80. Slinkard, K.; Singleton, V.L. Total phenol analysis: Automation and comparison with manual methods. *Am. J. Enol. Vitic.* **1977**, *28*, 49–55.

81. Chavan, J.J.; Gaikwad, N.B.; Kshirsagar, P.R.; Dixit, G.B. Total phenolics, flavonoids and antioxidant properties of three *Ceropegia* species from Western Ghats of India. *South Afr. J. Bot.* **2013**, *88*, 273–277. [CrossRef]

82. Bradford, M.M. A rapid and sensitive method for the quantitation of microgram quantities of protein utilizing the principle of protein-dye binding. *Anal. Biochem.* **1976**, *72*, 248–254. [CrossRef]

83. Beyer, W.F., Jr.; Fridovich, I. Assaying for superoxide dismutase activity: Some large consequences of minor changes in conditions. *Anal. Biochem.* **1987**, *161*, 559–566. [CrossRef]

84. Bergmeyer, H.U. *Methods of Enzymatic Analysis*, 2nd ed.; Academic Press: New York, NY, USA, 1974; p. 855.

85. Nakano, Y.; Asada, K. Hydrogen peroxide is scavenged by ascorbate-specific peroxidase in spinach chloroplasts. *Plant Cell Physiol.* **1981**, *22*, 867–880.

86. Rebouças, E.D.L.; Costa, J.J.N.; Passos, M.J.; Passos, J.R.S.; Hurk, R.; Silva, J.R.V. Real time PCR and importance of housekeepings genes for normalization and quantification of mRNA expression in different tissues. *Braz. Arch. Biol. Technol.* **2013**, *56*, 143–154. [CrossRef]

87. Livak, K.J.; Schmittgen, T.D. Analysis of relative gene expression data using real-time quantitative PCR and the 2-DDCT Method. *Methods* **2001**, *25*, 402–408. [CrossRef] [PubMed]

88. Medeiros, F.C.L.; Resende, M.L.V.; Medeiros, F.H.V.; Zhang, H.M.; Pare, P.W. Defense gene expression induced by a coffee-leaf extract formulation in tomato. *Physiol. Mol. Plant Pathol.* **2009**, *74*, 175–183. [CrossRef]

89. Hafez, E.E.; Hashem, M.; Balbaa, M.M.; El-Saadani, M.A.; Ahmed, S.A. Induction of new defensin genes in tomato plants via pathogens-biocontrol agent interaction. *J. Plant Pathol. Microbiol.* **2013**, *4*, 167.

90. Chen, C.; Chen, H.; Zhang, Y.; Thomas, H.R.; Frank, M.H.; He, Y.; Xia, R. TBtools: An integrative toolkit developed for interactive analyses of big biological data. *Mol. Plant* **2020**, *13*, 1194–1202. [CrossRef]

Article

Yield and In Vitro Antioxidant Potential of Essential Oil from *Aerva javanica* (Burm. f.) Juss. ex Schul. Flower with Special Emphasis on Seasonal Changes

Suzan Marwan Shahin [1,2], **Abdul Jaleel** [1] **and Mohammed Abdul Muhsen Alyafei** [1,*]

[1] Department of Integrative Agriculture, College of Agriculture and Veterinary Medicine, United Arab Emirates University, Al Ain 15551, United Arab Emirates; drsuzan.s@uaqu.ac.ae (S.M.S.); abdul.jaleel@uaeu.ac.ae (A.J.)

[2] Research and Development Head, Umm Al Quwain University, Umm Al Quwain 536, United Arab Emirates

* Correspondence: mohammed.s@uaeu.ac.ae

Abstract: The essential oil (EO) of the desert cotton (*Aerva javanica* (Burm. f.) Juss. ex Schul.) was extracted by hydrodistillation, from *A. javanica* flowers growing in the sandy soils of the United Arab Emirates (UAE) wild desert. The influence of seasonal variation on flowers' EO yield was studied. The flowers' EO yield obtained from spring samples (0.011%) was significantly the highest followed by early summer (0.009%), winter (0.007%), and autumn samples (0.006%), respectively. The flowers' EO antioxidant analysis were tested by DPPH, FRAP and ABTS assays (in vitro). Results proved that *A. javanica* flowers' EO, isolated during the four seasons, is a good source of natural bioactive antioxidants. Based on the three tested assays, the highest antioxidant activity was recorded in the spring. Testing of the chemical composition of the flowers' EO was conducted for the season with the highest yield and the best antioxidant performance, recorded in spring, by a combination of gas chromatograph (GC) and gas chromatograph-mass spectrometer (GC-MS). This led to the identification of 29 volatile components, in which the flowers' oil was characterized by angustione as a major compound. Photos by scanning electron microscope (SEM) showed prominent availability of star-shaped trichomes in the epidermis of the flowers.

Keywords: *Aerva javanica*; sandy soil; hydrodistillation; antioxidant activity; seasonal variation; GC-MS; angustione; trichomes

Citation: Shahin, S.M.; Jaleel, A.; Alyafei, M.A.M. Yield and In Vitro Antioxidant Potential of Essential Oil from *Aerva javanica* (Burm. f.) Juss. ex Schul. Flower with Special Emphasis on Seasonal Changes. *Plants* 2021, 10, 2618. https://doi.org/10.3390/plants10122618

Academic Editors: Hazem Salaheldin Elshafie and Laura De Martino

Received: 24 September 2021
Accepted: 21 October 2021
Published: 29 November 2021

Publisher's Note: MDPI stays neutral with regard to jurisdictional claims in published maps and institutional affiliations.

1. Introduction

Aerva javanica (Burm. f.) Juss. ex Schul. (English names: desert cotton, snow bush) (Arabic names: Al ara', twaim, efhe, tirf) [1], a perennial xerophyte belonging to the Amaranthaceae family (common name: cockscomb). This shrub grows up to 100 cm. It has a woody base, erect and branched stems, covered with fine hairs. The leaves are alternate with greyish-green color on very short stalks, lance-shaped to oblong (1–1.5 × 4–5 cm) with clear veins and midrib on the underside covered with hairs. The flowers are aromatic, soft like cotton and generally available throughout the year [2]. It has five petals on a long spike (5–10 cm) from the leaf nodes. The open flowers are white and the buds are pinkish. The bracts (leaf-like structure) just below the flowers are covered with long woolly white hairs that become more intense as the season progresses. The inside of the fruits have a woolly covering, including one small black or brown shiny seed (0.1 × 0.15 cm) [1].

The genus *Aerva* is well-known as an important medicinal genus, including many species with proven biological activities, such as, antioxidant, hypoglycemic, analgesic, antivenin, antimalarial, and anthelmintic activities [2,3]. *A. javanica* is an important medicinal and commercial plant. It is recommended by Ayurvedic medicine as one of the best sources of natural remedies (e.g., to treat the bladder and kidney stones) [3]. *A. javanica* is an essential oil-bearing shrub (or small tree), common in tropical and subtropical dry areas, with various folk applications related to its flowers in traditional herbal practices. For example, the flowers were mixed with water as a paste to stop wounds bleeding, and to pack suppurating wounds after cleaning. Furthermore, flowers were used for curing kidney and rheumatism problems [1,2].

Consequently, it is of great interest to investigate the flowers' EO yield, which is the first objective of this work, in order to recommend the best harvesting season with the highest yield. Furthermore, for the sake of testing the quality, the antioxidant activity of *A. javanica* flowers' EO will be tested for the first time (in vitro) and during the four seasons, which is the second objective of this research, seeking for a scientific justification for the rich ethnomedicinal applications of this plant's flowers. Besides, the volatiles of flowers' EO will be identified, to the best of our knowledge, for the first time in the literature. The same would be identified for the best flowers' EO yield obtained. In addition, the microscopic photos under scanning electron microscope are included in this study; providing a better understanding of the finer morphology of the trichomes of the flower epidermis.

2. Materials and Methods

2.1. Plant Material Identification

The mature plants of *A. javanica* were identified by Ali El-Keblawy, Professor of Plant Ecology (College of Sciences, University of Sharjah) and Tamer Mahmoud, Researcher Botanist (Sharjah Seed Bank and Herbarium, Sharjah Research Academy). Furthermore, fresh *A. javanica* flowers were collected from the study location in Al Ain on 5th of November (temperature: 22–33 °C) and identified by Mohamed Taher Mousa from the Biology Department of the UAEU; the voucher specimen of the plant was deposited in the Herbarium of the Biology Department, UAEU (voucher No. 14668).

2.2. Study Location and Plant Collection Schedules

The natural communities of *A. javanica* were growing wild in the Al Ain area (latitude: 24°11′ N, longitude: 55°39′ E). The mature flowers of *A. javanica* were randomly collected and separated in the early morning according to the following time schedule and details:

The flowers were harvested once at around the middle of each season, except for the summer batch of flowers, which was collected on the first of June instead of July. The same was done, due to the unavailability of flowers, after the first of June; since they will both complete and end their life cycles. The collected flower batch permitted examining the seasonal variation influence on the flowers' EO yields. The details of the harvesting schedules are as follows:

Spring: Mid-April (21–35 °C, average: 28 °C).
Summer: First of June (23–47.2 °C, average: 35.1 °C)
Autumn: End of October (24–35 °C, average: 29.5 °C).
Winter: Mid-January (11–22 °C, average: 16.5 °C).

All weather data were obtained from the iPhone application, "Weather", which collects data from the Weather Channel: Aqbiyah Weather Station, UAE.

2.3. Soil Physical and Chemical Analysis

The analysis of soil physical properties (done in triplicate) showed that the soil has a sandy soil texture, measured by dry sieve method (sand% "2 to 0.053 mm": 97.30 ± 1.02; silt% "0.032 mm": 2.09 ± 1.01; clay% "<0.032 mm": 0.61 ± 0.37).

The analysis of soil chemical properties (done in triplicate) showed that the soil electric conductivity (EC) is 0.971 ± 0.064 mS measured by EC meter with pH 7.6 ± 0.035 measured at 21 °C by pH meter; both tests were done for disturbed samples. The soil samples consisted of 0.520 ± 0.047% organic matter measured by the Walkley–Black method. The utilized soil consisted of 0.0761 ± 0.015% nitrogen, 6.074 ± 0.021 ppm phosphorus and 102.54 ± 0.468 ppm potassium. The total nitrogen was measured by Vario MACRO cube CHNS and manufactured by Elementar Co. (Langenselbold, Germany), while the total phosphorus and potassium were measured by inductively coupled plasma atomic emission spectroscopy (ICP-OES), model 710-ES. The total soil calcium carbonate ($CaCO_3$) was 27.841 $\pm$ 0.844% measured by calcimeter method.

2.4. Statistics and Experimental Design

The data were subjected to a statistical analysis using SPSS statistical software version 21. One-way analysis of variance (ANOVA) followed by Tukey (honestly significant differences, HSD) multiple range test were employed, and the differences between the individual means were deemed to be significant at $p \leq 0.05$ significance level.

The figures and tables illustrate the significant differences between the individual groups by letters, in which the use of either different letters in the same treatment group, the use of the symbol (*), or both, mean that there is a significant difference at $p \leq 0.05$ (unless a different significance level is specified), while the use of similar letters or absence of letters means that the difference between the same individual group is not significant.

The experimental design of all the conducted experiments is completely randomized. The number of samples and their replications used are illustrated separately in the description of each experiment.

2.5. EO Extraction by Hydrodistillation

The EO isolation was done by hydrodistillation method using Clevenger type apparatus. The seasonal variation influence on flowers' EO yields were examined, in which three (3) replicates were considered.

The seasonal variations' influence on *A. javanica* flowers' EO was tested. The summary of the tested treatments (mentioned as points) is represented in Table 1.

Table 1. Summary of *A. javanica* experimental treatments.

Plant Part	Drying Methods	Particle Size	Harvesting Season
Flowers	Air-Drying: Shaded at 25 °C	2 mm	Spring (Mid-April) Summer (First of June) Autumn (End of October) Winter (Mid-January)

2.5.1. Sample Preparation

The flowers collected throughout the year in each of the four seasons (as described in Table 1) were washed thoroughly with distilled water three times, and left in a well ventilate shaded area (27 °C) for two weeks (until completely dried). The flowers were then weighed (dry weight), labelled, and stored in clean and closed paper bags until use for further analysis.

On the distillation day, the flowers were cut into homogenous sizes (2 mm), measured by U.S. metric sieves (SOILTEST, ASTM), and directly subjected to hydrodistillation. It is worth mentioning that the size of *A. javanica* flowers, which is 2 mm, was obtained by separating the flowers from the mother plant by hand. The dry matter contents were measured using the following formula:

$$\text{DM}(\%) = \frac{\text{DW}}{\text{FW}} \times 100$$

where DM is the dry matter, FW is the fresh weight, and DW is the dry weight.

2.5.2. Extraction Conditions and Yield Determination

The hydrodistillation process was performed for around 100 g of plant matrix (dry weight) at 85 °C, for a period of 4 h, until no further EO was extracted. At the end of the distillation, the EO was accumulated as a waxy extract collected and easily separated from water. Thus, drying with anhydrous sodium sulfate (anhydrous Na_2SO_4) was not needed. The pure collected EO was weighed and stored in a sealed glass amber vial at 4 °C until analyzed. After the distillation, the Clevenger apparatus was subjected to a process using water, soap, methanol, and ethanol.

The EO yield was calculated based on the plant dry weight and the herbal extract weight using the following equation:

$$\text{EO yield } (\%) = \frac{W_{EO}}{W_{plant}} \times 100$$

where, W_{plant} is the dry weight of the plant (in grams) and W_{EO} is the weight of the extracted EO (in grams).

2.6. Antioxidant Activity

The antioxidant activity was determined in microplate to evaluate the antioxidant activity (in vitro) by spectrophotometer (Microplate reader, BioTek, EPOCH 2) using a 96-well plate. The tested antioxidant assays were: DPPH (2,2-diphenyl-1-picryl hydrazyl), FRAP (ferric reducing antioxidant power), and ABTS (2, 2-azinobis (3-ethylbenzothiazoline-6-sulfonate)).

2.6.1. Sample Preparation

EO waxy samples of *A. javanica* (flowers) for different seasons (in triplicate) were dissolved in 500 µL of dimethyl sulfoxide (DMSO) and vortexed very well.

2.6.2. Standard Curves

The standard curves were prepared using Trolox as a strong antioxidant reagent. The absorbance of Trolox at different concentrations was measured at 517 nm, 593 nm, and 734 nm for DPPH, FRAP, and ABTS assays, respectively. The results of all conducted assays were calculated and expressed as mg Trolox equivalents/g EO extract. The standard curves are illustrated in Figures 1–3.

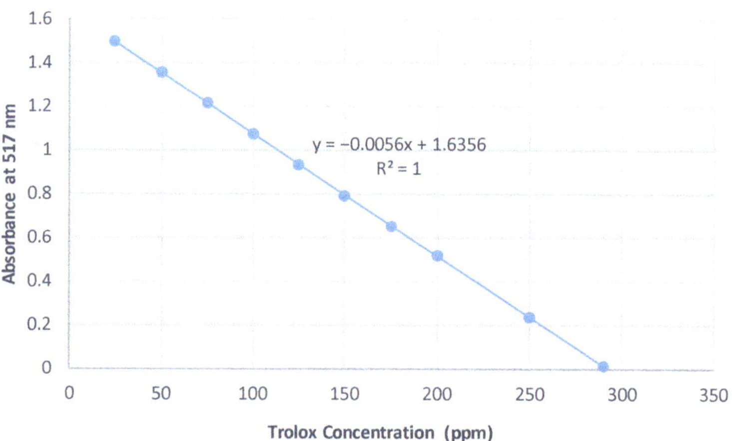

Figure 1. Standard curve of DPPH assay.

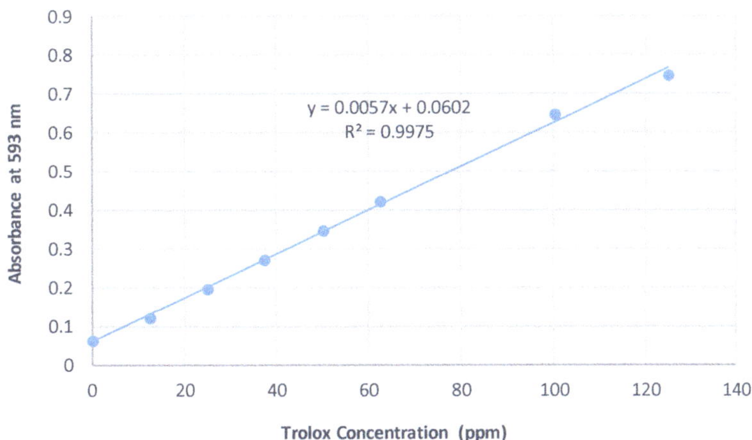

Figure 2. Standard curve of FRAP assay.

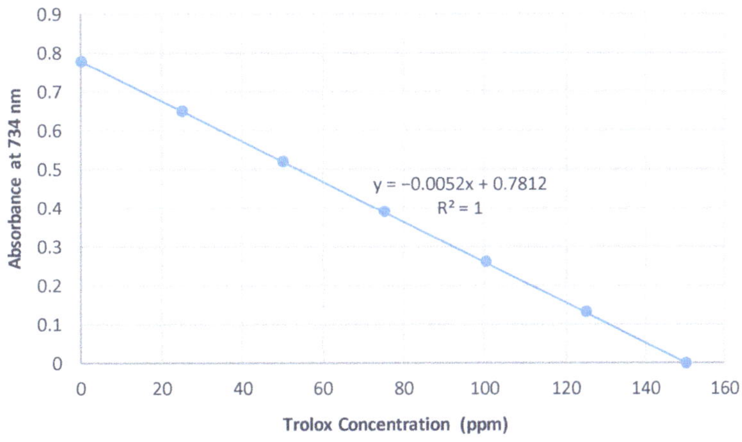

Figure 3. Standard curve of ABTS assay.

2.7. Chromatographic Analysis

EO sample analysis was performed on an Agilent gas chromatograph (GC) equipped with flame ionization detection (FID) and an HP-5MS capillary column, 30 m length × 0.25 mm internal diameter, 0.25 μm film thickness, and temperature programmed as follows: 70–300 °C at 10 °C/min. The carrier gas was helium at a flow of 3.0 mL/min; injector port and detector temperature were 250 °C and 300 °C, respectively. One microliter of the sample (diluted with acetone, 1:10 ratio) was injected by splitting and the split ratio was 1:100.

GC-MS Analysis

The gas chromatography-mass spectrometry (GC-MS) analysis was performed on an Agilent C 5975 GC-MS system with an HP-5MS capillary column (30 m × 0.25 mm internal diameter × 0.25 μm film thickness). The operating conditions were the same conditions as described above for the chromatographic analysis. The mass spectra were taken at 70 eV. The scan mass range was from 40 to 700 m/z at a sampling rate of 1.0 scan/s. The quantitative data were obtained from the electronic integration of the FID peak areas.

The components of the EO were identified by their retention time (RT), retention indices (RI) and mass spectra, relative to C9–C28 n-alkanes, computer matching with the library of NIST Mass Spectrometry Data Center (NIST11.L), as well as by comparison of their retention indices with those of the authentic samples or with the data already available in the literature [4]. The percentage of composition of the identified compounds was computed from the GC peak areas without any correction factors and was calculated relatively.

2.8. Plant Morphology by SEM Photos

The finer morphological details of the *A. javanica* flower (collected from the experimental location in mid-April) were examined using a scanning electron microscope (SEM) XL Series, Netherlands. The process was done at the Electron Microscope Unit, College of Medicine, UAEU.

2.8.1. Chemicals

Chemicals used were as follows: fixative, phosphate buffer, osmium tetroxide buffer, and ethanol.

2.8.2. Sample Preparation Protocol

First, the specimens were cut into 4×4 mm pieces, and transferred into fixative at pH 7.2 in a 7 mL glass vial with attached lids and stand overnight at 4 °C. The next day, specimens were washed in 0.1 M phosphate buffer (3 times, 20 min each), and stored at 4 °C. Furthermore, 1 mL of buffered 1% osmium tetroxide was then added to each vial, and mixed for 1 h at room temperature on a Rota mixer. After that, the osmium tetroxide was removed and samples were washed three times in distilled water (for 2 min each). The samples were dehydrated through a series of ethanol washes. The next day, the specimens were washed again (twice) with 100% ethanol. The final flower sample was kept inside a desiccator for four days to allow gradual drying.

3. Results and Discussion

3.1. EO Extraction by Hydrodistillation

EO physical characteristics: The flowers' EO has a yellowish color with waxy nature, and has a characteristic floral odor similar to the fresh flowers.

3.2. Effect of Seasonal Variation

3.2.1. Quantitative Analysis (by Yield)

The influence of seasonal variations on the EO yield (%, v/w of dry weight) of *A. javanica* air-dried flowers is illustrated in Figure 4. The mean results are represented with their standard deviations (SDs) and standard deviation error bars.

The analysis of variance, by one-way ANOVA, showed significant effect for the seasonal variation on flowers' EO yield (quantitatively) at 0.042 significance level ($p \leq 0.05$). The analysis of multiple comparisons between groups by Tukey HSD test showed significant variations of EO yield obtained in spring at 0.036 significance level with results obtained in autumn. Our results are similar to other studies found in the literature, which report that a seasonal variation has significant influence on EO yield [5–7]. According to Hussain et al. [5], the highest EO yield was obtained in spring, which was reduced afterward. Therefore, our findings are in agreement with their obtained results.

Annually, *A. javanica*'s flowering stage started approximately in November (end of autumn season), continuing to grow in the winter, reaching maturity in the spring, and completing the life cycle by the beginning of summer. This corresponds with the trend of our obtained EO yields, which shows the lowest results in the autumn, followed by an increased EO yield in winter, reaching the maximum yield in spring, which then starts decreasing by the beginning of summer.

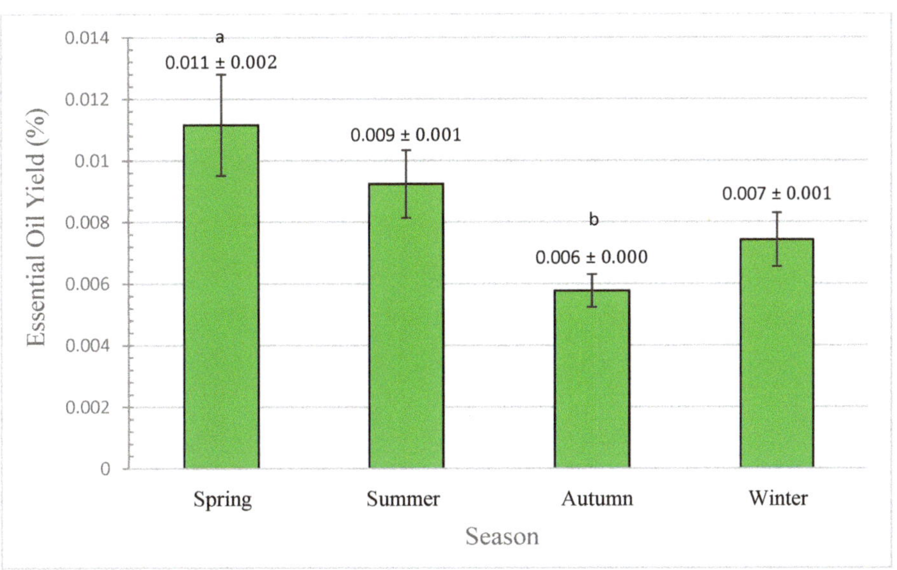

Figure 4. Effect of seasonal variation on flowers of *A. javanica* EO yield. The different letters are indicators of the significant variation between groups.

Our results are in agreement with the results obtained by Hussain et al. [5], Omer et al. [6], and Villa-Ruano et al. [7], which report that the seasonal variation has a significant influence on the quantitative EO yield. According to Hussain et al. [5], high EO yield was recorded in spring and reduced afterward, which is similar to our results.

During spring, the flowers reach the maturity stage, thus the EO yield from the glandular cells will reach maximum levels, which will play a major role in the pollination process. However, with the arrival of the summer, which is characterized by high temperature and high variations between daily minimum and maximum temperatures, the proline content of the leaves will reach maximum levels, while flowers will end their life cycle, and accordingly the glandular cells will collapse, thus flowers' EO content will be reduced.

Based on our results, the best season to extract the highest quantitative EO yield ($0.011 \pm 0.002\%$) obtained from *A. javanica* flowers is spring, followed by early summer ($0.009 \pm 0.001\%$) and winter ($0.007 \pm 0.000\%$). While extracting the oil from flowers collected during autumn provides the lowest yield ($0.006 \pm 0.001\%$), thus not recommended.

Our results are similar to other studies found in the literature, which report that seasonal variation has a significant influence on the quantitative EO yield [5–7].

3.2.2. Qualitative Analysis (by Antioxidant Activity)

DPPH Assay

The DPPH molecule (2,2-diphenyl-1-picryl hydrazyl) is one of the few stable organic nitrogen radicals. It is characterized by deep purple color, with maximum absorbance at 517 nm wavelength. The antioxidants cause the reduction reactions against the free radical DPPH by pairing; it has an odd electron with the free radical scavenging antioxidant. This reaction causes color loss of the purple DPPH due to the formation of the reduced DPPH-H. The higher the color loss, the higher the antioxidant concentration would be, thus higher DPPH scavenging activity [8].

In this work, the antioxidant concentration was calculated as Trolox equivalent from a calibration curve and the antioxidant activity was expressed as mg Trolox eq/g of extract.

The analysis of variance, by one-way ANOVA, showed a significant difference (at <0.0005 significance level) for the influence of seasonal variation on the antioxidant activity of *A. javanica* flowers' EO using DPPH assay. As shown in Figure 5, the highest antioxidant activities are obtained in the spring (12.20 ± 1.44 mg Aq/g), followed by winter (11.5 ± 0.15 mg Aq/g), autumn (7.77 ± 0.21 mg Aq/g), and summer (7.08 ± 0.21 mg Aq/g), respectively.

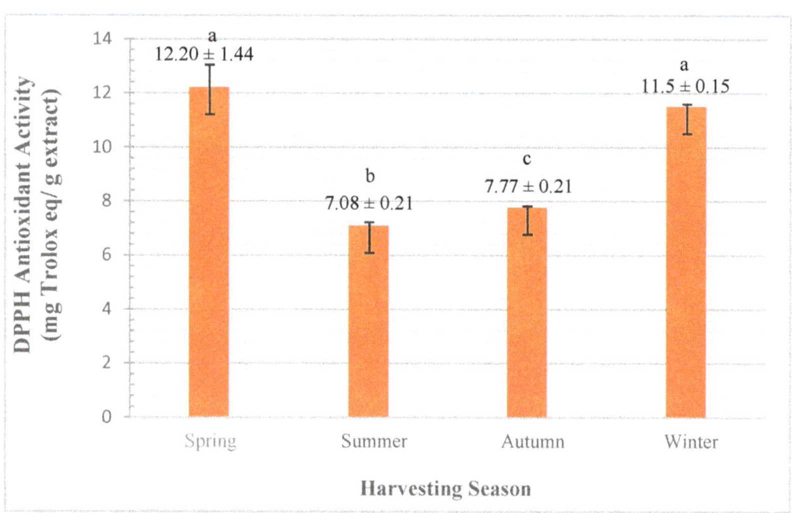

Figure 5. Effect of seasonal variation on *A. javanica* EO by DPPH assay. The different letters are indicators of the significant variation between groups.

The analysis of multiple comparisons between groups by Tukey HSD test showed significant variations of antioxidant activity recorded in spring with summer and autumn both at <0.0005 significance level, while no significant variation between spring and winter was observed. The antioxidant activity recorded in winter was significant with summer and autumn also at <0.0005 and 0.001 significance levels, respectively.

Although, there are many studies that link the influence of exposing the plant to biotic and abiotic stress factors with the high productivity of EO [9–11]. However, in our work there was a negative correlation between the severity of seasonal impacts and the EO yield, which is in agreement with the results obtained by Hussain et al. [5], who reported that during the year the highest antioxidant activity of basil EO was recorded in spring, while the lowest activity was recorded in summer.

Linking the quantitative and qualitative results of *A. javanica* EO obtained from flowers to our current antioxidant DPPH results shows that the high antioxidant activity recorded in the spring, which was then reduced significantly by the arrival of summer, could be due to the variation in amounts of major compounds, which causes the reduction in the availability of the electron-donation group.

It is worth mentioning that the *A. javanica* flowering stage starts approximately in November (end of autumn) and continues growing into the winter reaching maturity in spring, and completing the life cycle by the beginning of summer. This supported the trend of our obtained antioxidant results by DPPH assay.

Our results are similar to other studies found in the literature, which report that the seasonal variation has a significant influence on the qualitative EO yield [5–7]. According to Hussain et al. [5], basil EO extracted in spring was rich in oxygenated monoterpenes, while the oil obtained in summer was rich in sesquiterpene hydrocarbons. Meaning that, seasonal variation has a significant influence on the EO qualitative yield, which is similar to our obtained antioxidant results.

FRAP Assay

The analysis of variance, by one-way ANOVA, showed a significant difference (at <0.0005 significance level) for the influence of the seasonal variation on the antioxidant activity of *A. javanica* flowers' EO using FRAP assay. As shown in Figure 6, the highest antioxidant activities were obtained in the spring (7.76 ± 0.55 mg Aq/g), followed by summer (5.47 ± 0.57 mg Aq/g), winter (3.28 ± 0.148 mg Aq/g), and autumn (3.14 ± 0.38 mg Aq/g), respectively.

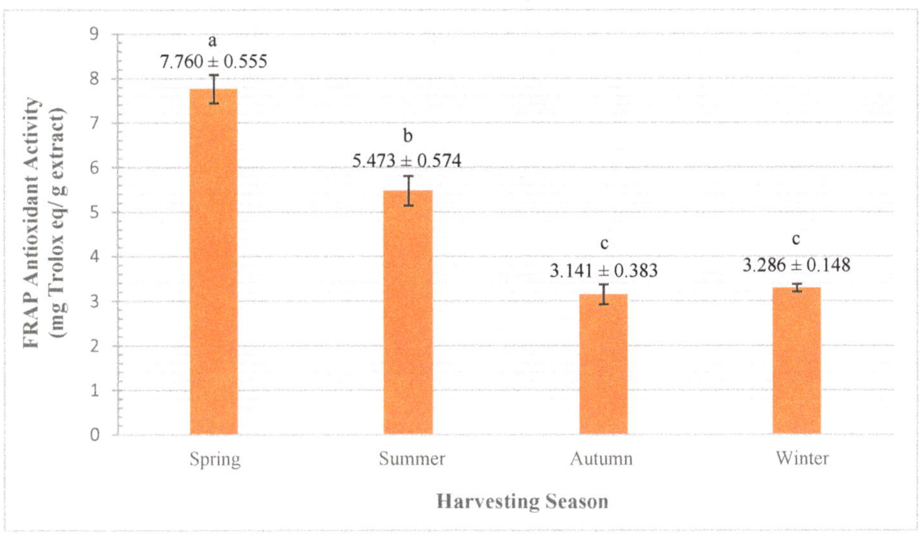

Figure 6. Effect of seasonal variation on *A. javanica* EO by FRAP assay. The different letters are indicators of the significant variation between groups.

The analysis of multiple comparisons between groups by Tukey HSD test showed significant variations of antioxidant activity recorded in spring with the antioxidant results recorded in summer at 0.001 significance level. Furthermore, the variation between the results of spring with autumn and winter were significant both at <0.0005 significance level. While no significant variation between autumn and winter was observed.

Although, there are many studies that link the influence of exposing the plant to biotic and abiotic stress factors with the high productivity of EO [9–11]. However, in our work there was a negative correlation between the severity of seasonal impacts and the EO yield, which is in agreement with the results obtained by Hussain et al. [5], who reported that during the year the highest antioxidant activity of basil EO was recorded in spring, which reduced afterward.

Our results are in agreement with the results obtained by Hussain et al. [5], who reported that during the year the highest antioxidant activity of basil EO was recorded in spring, which reduced afterward.

In general, our results are similar to other studies found in the literature, which report that the seasonal variation has a significant influence on the qualitative EO yield [5–7].

It is worth mentioning that the *A. javanica* flowering stage starts approximately in November (end of autumn) and continues growing in the winter reaching maturity in the spring, and completing the life cycle by the beginning of summer. This supports the trend of our obtained antioxidant results by FRAP assay.

In general, the antioxidant activity obtained from FRAP assay is due to the availability of hydrogen donating ability, which can be linked to the existence of adjacent substituted groups (e.g., hydroxyl).

ABTS Assay

The analysis of variance, by one-way ANOVA, showed a significant difference at <0.0005 significance level for the influence of seasonal variation on the antioxidant activity of *A. javanica* flowers' EO using ABTS assay. As shown in Figure 7, the highest antioxidant activities obtained in the spring (3.075 ± 0.263 mg Aq/g) are followed by autumn (1.719 ± 0.025 mg Aq/g), winter (1.685 ± 0.255 mg Aq/g), and summer (0.706 ± 0.248 mg Aq/g), respectively.

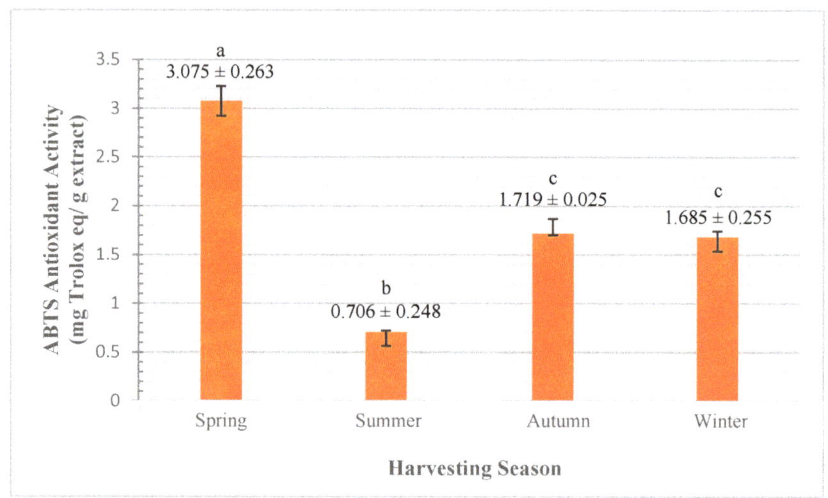

Figure 7. Effect of seasonal variation on *A. javanica* EO by ABTS assay. The different letters are indicators of the significant variation between groups.

The analysis of multiple comparisons between groups by Tukey HSD test showed significant variations of antioxidant activity recorded in spring with antioxidant results of all seasons at <0.0005 significance level. Furthermore, the variation between the results of summer with spring, autumn, and winter were significant at <0.0005, 0.002, and 0.003 significance levels, respectively. While no significant variation between autumn and winter was observed.

Still, there are many studies that link the influence of exposing the plant to biotic and abiotic stress factors with high production of EO [9,10]. However, in our work there was a negative correlation between the severity of seasonal impacts and the EO yield, which is in agreement with the results obtained by Hussain et al. [5], who reported that during the year the highest antioxidant activity of basil EO was recorded in spring, while the lowest activity was recorded in summer.

Linking the quantitative results of *A. javanica* EO obtained from flowers to our current antioxidant ABTS results shows that the high antioxidant activity recorded in the spring, which then reduced significantly by the arrival of the summer, could be due to the variation in the amounts of the major compounds, which causes the reduction in the availability of the electron-donation group.

It is worth mentioning that the *A. javanica* flowering stage started approximately in November (end of autumn) and continued growing into winter, reaching maturity in the spring, and completing the life cycle by the beginning of summer. This supports the trend of our obtained antioxidant results by ABTS assay.

Our results are similar to other studies found in the literature, which report that the seasonal variation has a significant influence on the qualitative EO yield [5–7].

Correlation between Antioxidant Assays

The correlation analysis using Pearson's correlation test was done to evaluate the strength of the relationship between the results of antioxidant activity obtained by applying the three antioxidant assays: DPPH, FRAP, and ABTS. The Pearson's correlation coefficients were calculated using SPSS statistical software (version 21) at ≤ 0.01 significance level, in which the significant effect will be illustrated in this section's tables using the sign (*).

Our correlation analysis of the flowers' EO antioxidant activity measured for the four seasons is shown in Tables 2–5. The illustrated correlation coefficients of spring flowers show strong positive relationship between all tested antioxidant assays. The correlation coefficient of spring flowers was found to be strong (at ≤ 0.01 significance level) between DPPH and FRAP, DPPH with ABTS, and FRAP with ABTS. While, the correlation coefficient measured for the remaining seasons shows a very strong positive relationship (significant at ≤ 0.01 significance level) between all the tested antioxidant assays.

Table 2. Pearson correlation coefficients of antioxidant activity of spring flowers.

	DPPH	**FRAP**	**ABTS**
DPPH	NA	+0.933	+0.994
FRAP	+0.933	NA	+0.968
ABTS	+0.994	+0.968	NA

(NA) means Not Applicable.

Table 3. Pearson correlation coefficients of antioxidant activity of summer flowers.

Antioxidant Assays	**DPPH**	**FRAP**	**ABTS**
DPPH	NA	+1 *	+1 *
FRAP	+1 *	NA	+1 *
ABTS	+1 *	+1 *	NA

(*) means significant at ≤ 0.01 significance level. (NA) means not applicable.

Table 4. Pearson correlation coefficients of antioxidant activity of autumn flowers.

Antioxidant Assays	**DPPH**	**FRAP**	**ABTS**
DPPH	NA	+1 *	+1 *
FRAP	+1 *	NA	+1 *
ABTS	+1 *	+1 *	NA

(*) means significant at ≤ 0.01 significance level. (NA) means not applicable.

Table 5. Pearson correlation coefficients of antioxidant activity of winter flowers.

Antioxidant Assays	**DPPH**	**FRAP**	**ABTS**
DPPH	NA	+1 *	+1 *
FRAP	+1 *	NA	+1 *
ABTS	+1 *	+1 *	NA

(*) means significant at ≤ 0.01 significance level. (NA) means not applicable.

3.3. Chemical Composition (by GC-MS)

Referring to our previous flowers' EO results (yield and antioxidant analysis), it was shown that the highest yield with the best antioxidant activity of the oil was recorded in spring, in which the chemical composition of the oil was investigated in this section.

The chromatogram result of *A. javanica* air-dried flowers' EO extracted during the spring is illustrated in Figure 8. The complete chemical composition of EO, retention time (RT) in minutes, retention indices (RI), and percentages (%) of identified compounds of the flowers' oil are all represented in Table 6.

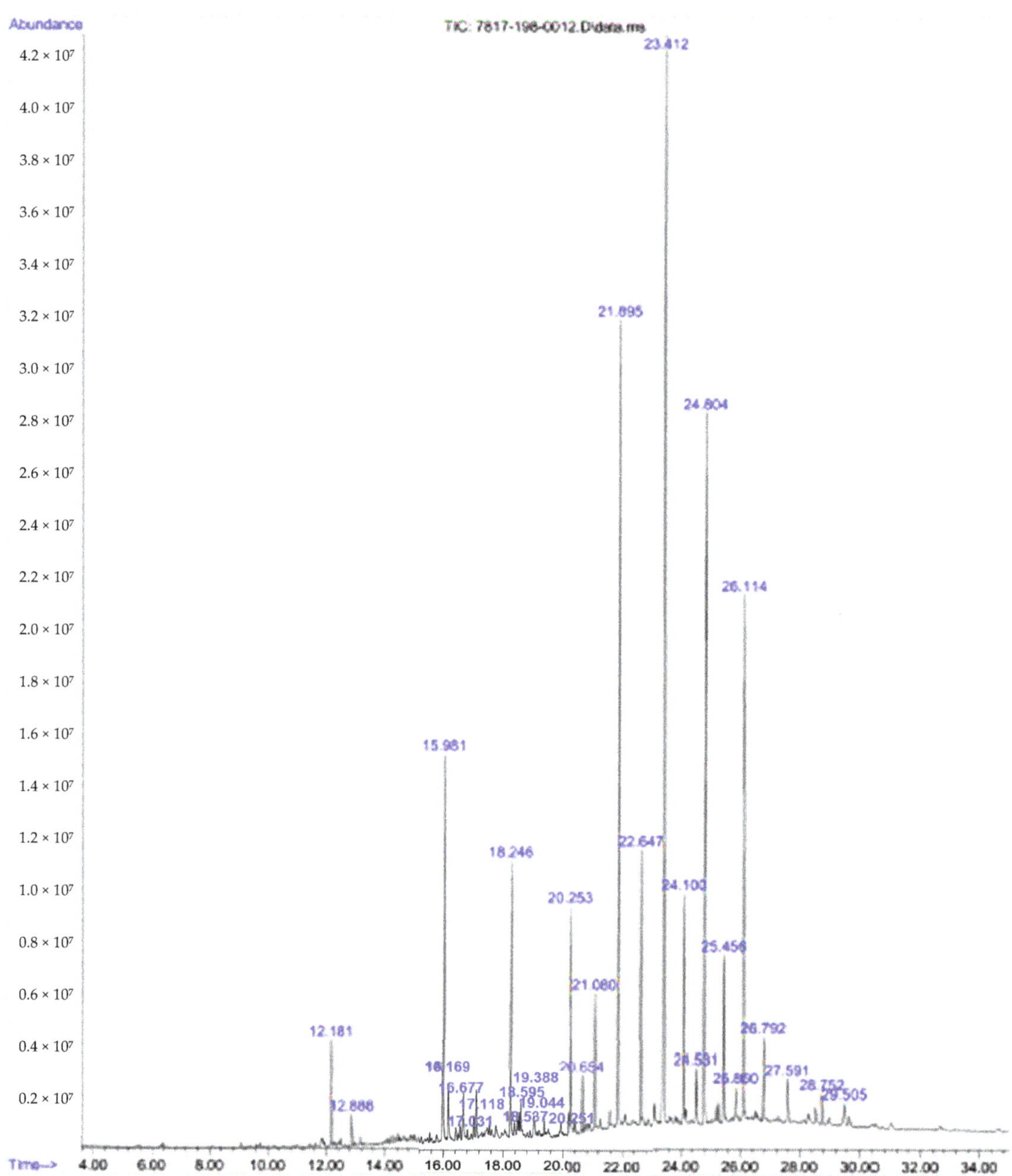

Figure 8. The chromatogram of *A. javanica* air-dried flowers' EO for spring.

Table 6. Composition of *A. javanica* EO obtained from air-dried flowers during spring.

No.	Compound	RT	RI	Percent Composition (%) Spring
1	2E,4E-Octadienol	12.184	1113	1.35
2	*cis*-Limonene oxide	12.883	1132	0.42
3	Verbenone	15.973	1204	6.06
4	(2E)-Octenol acetate	16.171	1208	1.14
5	endo-Fenchyl acetate	16.573	1218	—
6	*Methyl*-(2E)-nonenoate	16.678	1221	1.20
7	*nor*-Davanone	17.033	1228	0.39
8	Thymol methyl ether	17.115	1132	0.84
9	*tetrahydro*-Linalool acetate	17.121	1231	—
10	Pulegone	17.261	1233	—
11	Carvotanacetone	17.768	1244	—
12	(4Z)-Decen-1-ol	18.246	1255	4.07
13	(2E)-Decenal	18.485	1260	0.44
14	*cis*-Chrysanthenyl acetate	18.537	1261	0.43
15	Geranial	18.595	1264	0.97
16	*trans*-Carvone oxide	19.044	1273	0.71
17	(2Z)-Hexenyl valerate	19.388	1282	0.39
18	*n*-Tridecane	20.245	1300	3.79
19	Isoamyl bemzyl ether	20.653	1310	0.95
20	2E,4E-Decadienol	21.078	1319	2.08
21	3-oxo-ρ-Menth-1-en-7-al	21.580	1330	—
22	Evodone	21.877	1137	14.77
23	4-Hydroxybenzaldehyde	22.641	1355	4.56
24	Angustione	23.392	1372	20.72
25	(E)-Methyl cinnamate	23.556	1376	—
26	1-Tetradecene	24.092	1388	3.98
27	α-Chamipinene	24.448	1396	—
28	Cyperene	24.529	1398	1.16
29	*methyl*-Cresol acetate	24.791	1403	12.49
30	(2E,4E)-Undecadienal	25.281	1415	—
31	(E)-Trimenal	25.450	1419	2.77
32	*cis*-Thujopsene	25.858	1429	0.80
33	γ-Elemene	26.103	1434	9.15
34	α-Humulene	26.791	1452	1.65
35	(2E)-Dodecen-1-ol	27.589	1469	1.05
36	α-Selinene	28.755	1498	0.88
37	Menthyl isovalerate	29.507	1516	0.80

The GC-MS analysis of the flowers' EO showed the chemical composition of the oil with 29 identified volatiles (representing 100% of the oil). The major compounds were angustione (20.72%), evodone (14.77%), *methyl*-cresol acetate (12.49%), γ-elemene (9.15%), and verbenone (6.06%). Furthermore, 4-hydroxybenzaldehyde (4.56%), (4Z)-decen-1-ol (4.07%), and 1-tetradecene (3.98%) were identified as minor components.

Our results show that the *A. javanica* flowers' EO is a rich source of potential hydrocarbons, and supporting the results published by Samejo et al. [12] and Samejo et al. [13]. Our results are justifying the rich ethnomedicinal applications of the flowers [14,15]. Angustione is a ketone reported to have antioxidant, anticancer, and antiviral (anti-HIV) biological activity [16]. Therefore, spring is more favorable to be considered as a collection season of *A. javanica* flowers that provides EO with a higher percentage of angustione, and consequently, better biological activity.

It is worth mentioning that verbenone is a monoterpene (bicyclic ketone terpene) that has a pleasant characteristic aroma, thus it is used in a wide range of industries (e.g., herbal remedies, herbal teas, spices, aromatherapy, and perfumery) [17]. Furthermore, verbenone is an insect pheromone, thus used to control insects [18]. Similarly, γ-elemene is a sesquiterpene that has a floral aroma and reported to cause antifungal and antioxidant activity [19].

Truly, the chemical composition of the EO would be significantly varied according to the harvesting season [5–7]. Therefore, the decision of deciding the favorable harvesting season depends upon the proposed application of the extracted oil.; meaning that different applications can consider different harvesting seasons accordingly.

3.4. Morphology by SEM

SEM photos for the flower's epidermis are illustrated (Figures 9 and 10), in which the photos show pubescent indumentum, consisting of stellate (star-shaped) base cells with uniseriate tapered ending. The trichomes are prominent and appear at high density. The SEM photos for the flowers are illustrated for the first time in the literature.

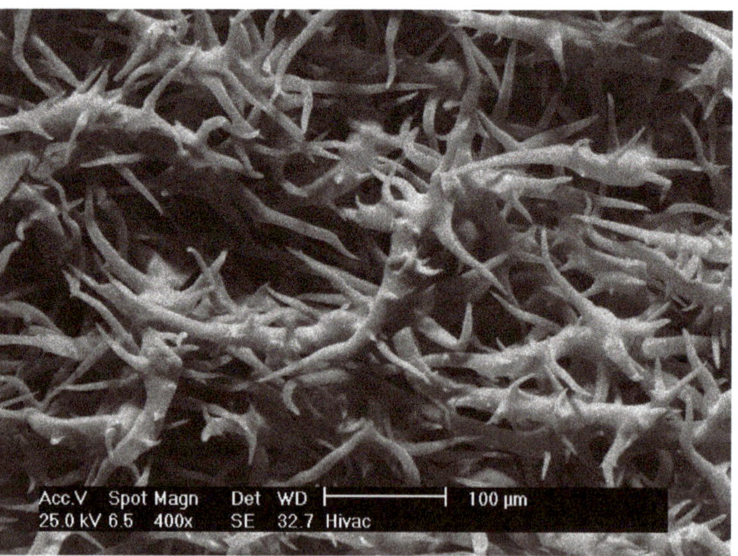

Figure 9. Flower epidermis trichomes of *A javanica* (magnification: 400×).

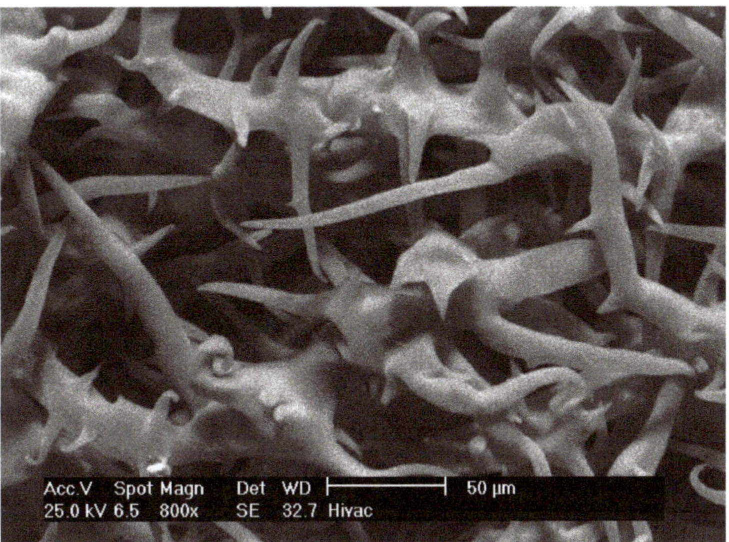

Figure 10. Flowers epidermis trichomes of *A javanica* (magnification: 800×).

The glandular trichomes indicating thero-tolerance by dissipating heat from the surface of the plant, which play a major role in enhancing the water use efficiency through adjusting the osmotic potential, which is a natural adaptation mechanism for desert plants in arid regions. Besides, the glandular trichomes are suspended to produce secondary metabolites, such as EOs, and acting as defense phytochemicals against biotic (e.g., herbivores) and abiotic (e.g., high temperature, low precipitation) stress factors [11,20].

The appearance of all the mentioned morphological features justifies the fact that *A. javanica* is a desert xerophytic plant that can survive high temperatures and high sun exposure rates with minimal water requirements. Furthermore, the existence of glandular trichomes on the epidermis of the flower supports the fact that *A. javanica* is an EO-bearing plant.

4. Conclusions

Aerva javanica is a xerophytic EO-bearing plant. The seasonal variation showed a significant effect on the flowers' EO yield, in which the best season to extract the highest EO yield is spring (0.011%). Similarly, the seasonal variation has a significant effect on the antioxidant activity of *A. javanica* flowers' EO, showing the best activity during the spring, tested in vitro using antioxidant assays (DPPH, FRAP, and ABTS).

The chemical composition of the flowers' EO extracted in the optimal season (during spring) is characterized with 29 identified volatiles, of which the major volatiles are angustione (20.72%) and evodone (14.77%).

A. javanica flowers' EO is a good resource of natural antioxidants, and has a high potential for the pharmaceuticals industry. Future clinical antioxidant studies for fractionate EO compounds are highly recommended.

Author Contributions: Conceptualization and methodology, S.M.S.; validation, A.J. and M.A.M.A.; writing—original draft preparation, S.M.S.; writing—review and editing, S.M.S., M.A.M.A. and A.J.; supervision, M.A.M.A. and A.J.; funding acquisition, M.A.M.A. All authors have read and agreed to the published version of the manuscript.

Funding: This research was funded by UAEU fund No. 31F029.

Acknowledgments: We express our appreciation to Ali El-Keblawy (University of Sharjah) and Shyam Kurup for the valuable knowledge related to the UAE native plants. Senthilkumar Annadurai (Research Associate, CAVM, UAEU) for his great support in the GC-MS analysis. Ali Al Marzouqi (Dean, College of Graduate Studies, UAEU) for sharing his valuable knowledge related to EO. Saeed Tariq (College of Medicine and Health Sciences, UAEU) for his assistance on SEM. Sajid Maqsood, Hina Kamal, and Babucarr Jobe (late) (Department of Nutrition, CAVM, UAEU), for the valuable assistance in conducting the antioxidant tests. Arshed el Daley (Agriculture Engineer, Al Foah Experimental Farm, CAVM), Abou-Messallam Azab (Lab Specialist, Horticulture Lab, CAVM), Saad Ismail (Lab Specialist, Soil and Water Lab, CAVM) and Rahaf Ajaj (Abu Dhabi University) for their cooperation and continuous support. This research was supported by the United Arab Emirates University (UAEU), College of Food and Agriculture, fund No. 31F029.

Conflicts of Interest: The authors declare no conflict of interest.

References

1. Jongbloed, M.; Feulner, G.; Böer, B.; Western, A.R. *The Comprehensive Guide to the Wild Flowers of the United Arab Emirates*; Environmental Research and Wildlife Development Agency: Abu Dhabi, United Arab Emirates, 2003.
2. Shahin, S. The United Arab Emirates Indigenous Essential Oil-bearing Plants: Essential oils of *Aerva javanica* and *Cleome ambolyocarpa* Grown in Sandy Soils under Arid Conditions. Ph.D. Thesis, United Arab Emirates University (UAEU), Al Ain, United Arab Emirates, 2018.
3. Saleem, M.; Parveen, S.; Riaz, N.; Tahir, M.N.; Ashraf, M.; Afzal, I.; Jabbar, A. New bioactive natural products from *Launaea nudicaulis*. *Phytochem. Lett.* **2012**, *5*, 793–799. [CrossRef]
4. Adams, R.P. *Identification of Essential Oil Compounds by Gas Chromatography and Mass Spectrometry*, 4th ed.; Allured Publishing Corporation: Carol Stream, IL, USA, 2009.
5. Hussain, A.I.; Anwar, F.; Sherazi, S.T.H.; Przybylski, R. Chemical composition, antioxidant and antimicrobial activities of basil (*Ocimum basilicum*) essential oils depends on seasonal variations. *Food Chem.* **2008**, *108*, 986–995. [CrossRef] [PubMed]
6. Omer, E.A.; Abou Hussein, E.A.; Hendawy, S.F.; Ezz, E.D.; Azza, A.; El Gendy, A.G. Effect of soil type and seasonal variation on growth, yield, essential oil and artemisinin content of *Artemisia annua* L. *Int. Res. J. Hortic.* **2013**, *1*, 15–27. [CrossRef]
7. Villa-Ruano, N.; Pacheco-Hernández, Y.; Cruz-Durán, R.; Lozoya-Gloria, E. Volatiles and seasonal variation of the essential oil composition from the leaves of *Clinopodium macrostemum* var. laevigatum and its biological activities. *Ind. Crops Prod.* **2015**, *77*, 741–747. [CrossRef]
8. Brand-Williams, W.; Cuvelier, M.E.; Berset, C. Use of free radical method to evaluate antioxidant activity. *Lebensm. Wiss. Technol.* **1995**, *28*, 25–30. [CrossRef]
9. Laribi, B.; Bettaieb, I.; Kouki, K.; Sahli, A.; Mougou, A.; Marzouk, B. Water deficit effects on caraway (*Carum carvi* L.) growth, essential oil and fatty acid composition. *Ind. Crops Prod.* **2009**, *30*, 372–379. [CrossRef]
10. Tátrai, Z.A.; Sanoubar, R.; Pluhár, Z.; Mancarella, S.; Orsini, F.; Gianquinto, G. Morphological and physiological plant responses to drought stress in *Thymus citriodorus*. *Int. J. Agron.* **2016**, *2016*, 4165750. [CrossRef]
11. Shahin, S.M.; Kurup, S.; Jaleel, A.; Alyafei, M.A.M. Chemical composition of *Cleome amblyocarpa* Barr. & Murb. essential oil under different irrigation levels in sandy soils with antioxidant activity. *J. Essen. Oil Bear. Plants* **2018**, *21*, 1235–1256.
12. Samejo, M.Q.; Memon, S.; Bhanger, M.I.; Khan, K.M. Chemical compositions of the essential oil of *Aerva javanica* leaves and stems. *Pak. J. Anal. Env. Chem.* **2012**, *13*, 48–52.
13. Samejo, M.Q.; Memon, S.; Bhanger, M.I.; Khan, K.M. Comparison of chemical composition of *Aerva javanica* seed essential oils obtained by different extraction methods. *Pak. J. Pharm. Sci.* **2013**, *26*, 757–760.
14. Hemmati, C.; Nikooei, M.; Bertaccini, A. Identification and transmission of phytoplasmas and their impact on essential oil composition in *Aerva javanica*. *3 Biotech* **2019**, *9*, 1–7. [CrossRef] [PubMed]
15. Al-Shehri, M.; Moustafa, M. Anticancer, antibacterial, and phytochemicals derived from extract of *Aerva javanica* (Burm. f.) Juss. ex Schult. grown naturally in Saudi Arabia. *Trop. Conserv. Sci.* **2019**, *12*, 1940082919864262. [CrossRef]
16. Robertshawe, P. Medicinal plants in Australia Volume 1: Bush Pharmacy. *J. Aust. Trad. Med. Soc.* **2011**, *17*, 173–174.
17. Selvaraj, M.; Kandaswamy, M.; Park, D.W.; Ha, C.S. Highly efficient and clean synthesis of verbenone over well ordered two-dimensional mesoporous chromium silicate catalysts. *Catal. Today* **2010**, *158*, 286–295. [CrossRef]
18. Huber, D.P.W.; Borden, J.H. Protection of lodgepole pines from mass attack by mountain pine beetle, *Dendroctonus ponderosae*, with nonhost angiosperm volatiles and verbenone. *Entomol. Exp. Appl.* **2001**, *99*, 131–141. [CrossRef]
19. Akpuaka, A.; Ekwenchi, M.M.; Dashak, D.A.; Dildar, A. Biological activities of characterized isolates of n-hexane extract of *Azadirachta indica* A. Juss (Neem) leaves. *N. Y. Sci. J.* **2013**, *6*, 119–124.
20. Gonzales, W.L.; Negritto, M.A.; Suarez, L.H.; Gianoli, E. Induction of glandular and non-glandular trichomes by damage in leaves of *Madia sativa* under contrasting water regimes. *Acta Oecol.* **2008**, *33*, 128–132. [CrossRef]

 *plants*

Article

Insight into Analysis of Essential Oil from *Anisosciadium lanatum* Boiss.—Chemical Composition, Molecular Docking, and Mitigation of Hepg2 Cancer Cells through Apoptotic Markers

Hany Ezzat Khalil [1,2,*], Hairul-Islam Mohamed Ibrahim [3,4], Hossam M. Darrag [5,6] and Katsuyoshi Matsunami [7]

[1] Department of Pharmaceutical Sciences, College of Clinical Pharmacy, King Faisal University, Al-Ahsa 31982, Saudi Arabia

[2] Department of Pharmacognosy, Faculty of Pharmacy, Minia University, Minia 61519, Egypt

[3] Biological Sciences Department, College of Science, King Faisal University, Al-Ahsa 31982, Saudi Arabia; himohamed@kfu.edu.sa

[4] Pondicherry Centre for Biological Sciences and Educational Trust, Kottakuppam 605104, India

[5] Research and Training Station, King Faisal University King Faisal University, Al-Ahsa 31982, Saudi Arabia; hdarag@kfu.edu.sa

[6] Pesticide Chemistry and Technology Department, Faculty of Agriculture, Alexandria University, Alexandria 21545, Egypt

[7] Department of Pharmacognosy, Graduate School of Biomedical & Health Sciences, Hiroshima University, 1-2-3 Kasumi, Minami-ku, Hiroshima 734-8553, Japan; matunami@hiroshima-u.ac.jp

* Correspondence: heahmed@kfu.edu.sa

Citation: Khalil, H.E.; Ibrahim, H.-I.M.; Darrag, H.M.; Matsunami, K. Insight into Analysis of Essential Oil from *Anisosciadium Lanatum* Boiss.—Chemical Composition, Molecular Docking, and Mitigation of Hepg2 Cancer Cells through Apoptotic Markers. *Plants* 2022, 11, 66. https://doi.org/10.3390/plants11010066

Academic Editors: Hazem Salaheldin Elshafie, Laura De Martino and Adriano Sofo

Received: 24 November 2021
Accepted: 23 December 2021
Published: 26 December 2021

Publisher's Note: MDPI stays neutral with regard to jurisdictional claims in published maps and institutional affil-iations.

Abstract: Essential oils have been used in various traditional healing systems since ancient times worldwide, due to their diverse biological activities. Several studies have demonstrated their plethora of biological activities—including anti-cancer activity—in a number of cell lines. *Anisosciadium lanatum* Boiss. is a perennial aromatic herb. Traditionally, it is an edible safe herb with few studies exploring its importance. The current study aims to investigate the chemical composition of essential oil isolated from *Anisosciadium lanatum* using GC-MS, as well as report its anti-cancer potential and its mechanistic effect on HepG2 liver cancer cell lines, and conduct molecular docking studies. To achieve this, the essential oil was isolated using a Clevenger apparatus and analyzed using GC-MS. The cell viability of HepG2 liver cancer and normal fibroblast NIH-3T3 cell lines was assessed by MTT cytotoxicity assay. The effects of the essential oil on cell migration and invasion were assessed using wound healing and matrigel assays, respectively. The effect of the essential oil on migration and apoptotic-regulating mRNA and proteins was quantified using quantitative real-time PCR and Western blot techniques, respectively. Finally, computational docking tools were used to analyze in silico binding of major constituents from the essential oil against apoptotic and migration markers. A total of 38 components were identified and quantified. The essential oil demonstrated regulation of cell proliferation and cell viability in HepG2 liver cancer cells at a sub-lethal dose of 10 to 25 μg/mL, and expressed reductions of migration and invasion. The treatment with essential oil indicated mitigation of cancer activity by aborting the mRNA of pro-apoptotic markers such as BCL-2, CASPASE-3, CYP-1A1, and NFκB. The algorithm-based binding studies demonstrated that eucalyptol, nerol, camphor, and linalool have potent binding towards the anti-apoptotic protein BCL-2. On the other hand, camphor and eucalyptol showed potent binding towards the pro-apoptotic protein CASPASE-3. These findings highlight the effectiveness of the essential oil isolated from *Anisosciadium lanatum* to drive alleviation of HepG2 cancer cell progression by modulating apoptotic markers. Our findings suggest that *Anisosciadium lanatum* could be used as a phytotherapeutic anti-cancer agent, acting through the regulation of apoptotic markers. More well-designed in vivo trials are needed in order to verify the obtained results.

Keywords: *Anisosciadium lanatum*; GC-MS; HepG2; BCL-2; CASPASE-3; apoptotic markers

1. Introduction

Traditionally, medicinal and aromatic plants have been considered to play essential roles in the field of therapeutics all over the world [1]. Essential oils (Eos), as secondary metabolites produced from such medicinal and aromatic plants, offer great value in terms of their various curative and biological properties. Several investigations have demonstrated the anti-inflammatory, anti-oxidant, anti-fungal, anti-microbial, and cytotoxic activities of such Eos [2–4]. Cancer is described as a fatal health condition, considered to be one of major factors leading to death. Cancer affects all human beings and does not differentiate between gender or age, leading to severe negative health and socio-economic impacts. More than 75% of anti-cancer drugs are directly or indirectly derived from medicinal plants [5]. In this context, the discovery of new natural product candidates with anti-cancer properties has unique interest for the purpose of medical care [6]. Many cytotoxic molecules that are of plant origin are widely used in chemotherapy [7]. The contribution of food components to cancer assessment through lifestyle patterns has become popular in diet–disease investigations [8]. Food enriched with vitamins and bioactive phytochemicals could act as tumor-controlling agents to reduce cancer progression, especially in the case of liver or colon cancer [9].

Moreover, a few reports have revealed that Mediterranean EOs and diets based on substances such as vegetables, nuts, whole grains, olive oil, and/or fish oils can reduce cancer-related and total mortality rates [10,11]. Several reports have also investigated the antioxidant activities of essential oils and have shown them to be potent natural sources of antioxidants to control cancer. The abnormal cellular stress causes non-nuclear DNA damage, which leads to inhibition of protein transport and reactive oxygen species formation. The Apiaceae family (formerly Umbelliferae) is one of the families of flowering plants, which consists of 3780 species in 434 genera. It is distributed worldwide [12,13]. Investigations have proven that this family is rich in its diversity of phytochemicals that have been considered as potential sources of new therapeutic drugs, including terpenoids, saponins, flavonoids, coumarins, and poly-acetylenes. In addition, numerous species of the Apiaceae family are reputed to be a significant source of EOs. EOs isolated and identified from this family have been shown to contain more than 760 components in various chemical classes. The identified and reported constituents have shown substantial pharmaceutical and nutritional values [13–16]. Many members of the Apiaceae family possess various biological activities, such as anti-bacterial, hepatoprotective, cyclo-oxygenase inhibitory, and anti-tumor activities [12]. Most of the members are safe and edible plants. The *Anisosciadium* genus (Apiaceae family) comprises three species—*Anisosciadium isosciadium* Bornm., *Anisosciadium orientale* DC., and *Anisosciadium lanatum* Boiss.—and is endemic to Southwest Asia [17]. Previous investigations have reported on cytotoxicity and antioxidant assessments pertaining to isolated Eos from *Anisosciadium orientale* [17,18]. In this aspect, *Anisosciadium lanatum* Boiss (*A. lanatum*) is a member of the *Anisosciadium* genus that is native and spread wildly throughout the Arabian Peninsula, including Saudi Arabia [19,20]. *A. lanatum* is a perennial herb. Anatomically, its leaves are characterized by a bipinnately parted incision and clasping base. The inflorescences are of a compound umbel type. Its flowers are tiny and demonstrate whitish-pink petals. The bracts are spiny-tipped. The fruitlets of each secondary umbel are aggregated before ripening, but are later separated into spiny units [21,22]. Traditionally, Bedouins have used *A. lanatum* as local medicinal herb: A water extract of the dried aerial parts, including the stem, leaves, flowers, and fruits, is used for skin sores and boils [23]. The young green leaves are a refreshing palatable herb for Bedouin children [24,25]. Additionally, *A. lanatum* has demonstrated veterinary importance, being used in livestock treatment. The extract from the leaves and shoots has been used to treat skin conditions in goats and sheep [19]. Recent studies have reported that *A. lanatum* contains guaiane sesquiterpene and shows anti-proliferative activity towards liver, colon, and lung cells, as well as anti-mutagenic activities [25,26]. A simple study reported an analysis of the EO from *A. lanatum* [20]; however, no previous reports have detailed advanced analyses of EO isolated from of *A. lanatum* and its modulatory cytotoxic effects, including those with respect to cell migration and invasion. In these contexts, the goal of the current exploration is to investigate *A. lanatum*, for the first time, with respect to its suggested potentials.

2. Results

2.1. Isolation and Identification of Chemical Components of EO

The EO obtained from *A. lanatum* was subjected to detailed gas chromatography–mass spectrometry (GC-MS) analysis (Figure 1). The oil yield was 0.46% volume per dried plant weight. Altogether, 38 components were identified and quantified, corresponding to 94.68% of the total components. The corresponding names of these components are listed in Table 1, according to their elution sequences and retention index (RI). The major identified monoterpene hydrocarbons (10 constituents representing 47.86%) were distributed as follows: α-pinene (10.14%), β-pinene (6.2%), camphene (4.4%), β-myrecene (8.45%), car-4-ene (5.8%), limonene (6.7%), and p-cymene (6.58%). On the other hand, the major identified oxygenated monoterpene components (11 constituents representing 16.44%) were eucalyptol (6.35%), linalool (4.34%), camphor (3.1%), and nerol (4.3%). Regarding the sesquiterpene hydrocarbons (11 constituents representing 22.13%), the results yielded β-farnesene (8.25%), β-caryophyllene (7.25%), and α-humulene (6.45%) whereas the major oxygenated sesquiterpenes (two constituents representing 6.48%) were caryophyllene oxide (7.25%) and α-eudesmol (0.8%). Phenylpropanoids and non-terpene derivatives showed very low percentages.

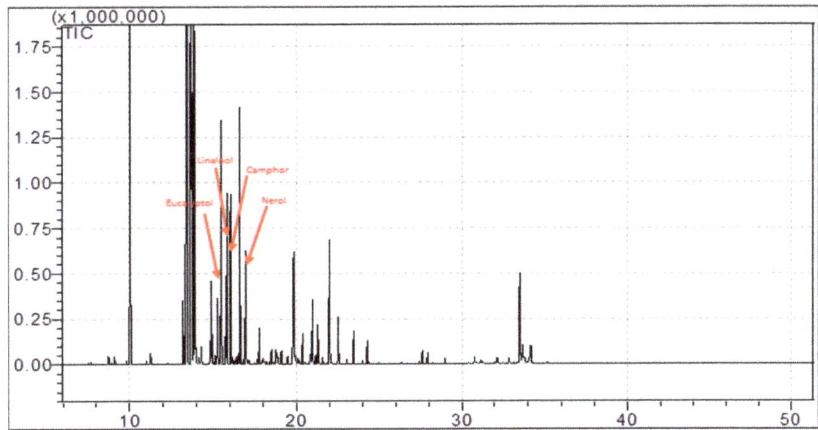

Figure 1. GC-MS chromatogram of EO from *A. lanatum*.

2.2. Proliferation and Cell Viability Assay

The suppressive effect of EO was assessed in HepG2 liver cancer cell lines and normal fibroblast NIH-3T3 cell lines. Cell viability and proliferation were determined colorimetrically by 3-(4,5-dimethylthiazol-2-yl)-2,5-diphenyltetrazolium bromide (MTT) assay. HepG2 and NIH-3T3 cell lines were treated with various concentrations (0, 5, 10, 25, 50, and 100 µg/mL) of EO for 24 h. The results revealed that proliferation was potentially inhibited in HepG2 cancer cells in a dose-dependent manner. The IC_{50} of EO in the cell lines was 11.3 µg/mL and 52.1 µg/mL for HepG2 and NIH-3T3, respectively. The sub-lethal dose was considered to be 10–25 µg/mL, showing survival rates of 52.2% and 36.2% in HepG2 and NIH-3T3, respectively. These concentrations were used for further experiments (see Figure 2). On the other hand, NIH-3T3 cell lines showed higher cell proliferation and viability, and it was insignificantly inhibited up to 25 µg/mL of EO.

Table 1. Volatile constituents of EO from *A. lanatum*.

No.	Constituents	Rt(min)	RI(Exp)	RI(Lit)	RA%	MS
1	α-pinene	9.03	937	932	10.14 ± 0.6	136.2340
2	camphene	9.29	952	946	4.4 ± 0.1	136.2340
3	β-pinene	9.76	977	977	6.2 ± 0.09	136.2340
4	β-myrcene	9.99	991	988	8.45 ± 0.2	136.2340
5	α-phellandrene	10.28	1005	1002	0.2 ± 0.02	136.2340
6	car-4-ene	13.41	1009	1004	5.8 ± 0.2	136.2340
7	α-terpinene	13.53	1017	1014	3.62 ± 0.07	136.2340
8	limonene	13.74	1030	1224	6.7 ± 0.3	136.2340
9	p-cymene	13.79	1026	1023	6.58 ± 0.1	134.2182
10	terpinolene	15.09	1088	1086	2.07 ± 0.07	136.2340
11	eucalyptol	15.36	1031	1031	6.35 ± 0.2	154.2493
12	linalool	16.37	1099	1095	4.34 ± 0.1	154.2493
13	camphor	16.57	1145	1141	4.3 ± 0.03	152.2334
14	isoborneol	17.03	1167	1165	0.1 ± 0.04	154.2493
15	terpinen-4-ol	17.29	1177	1174	2.4 ± 0.2	154.2493
16	α-terpineol	17.76	1189	1186	1.5 ± 0.05	154.2493
17	fenchyl acetate	17.81	1214	1214	0.1 ± 0.01	196.2860
18	nerol	17.92	1228	1227	3.1 ± 0.1	154.2493
19	bornyl acetate	18.57	1285	1284	0.6 ± 0.04	196.2860
20	methyl geranate	18.61	1321	1319	0.2 ± 0.02	182.2594
21	neryl acetate	18.74	1364	1359	0.1 ± 0.02	196.2860
22	α-copaene	19.56	1376	1374	0.9 ± 0.1	204.3511
23	β-cubebene	19.89	1389	1387	1.3 ± 0.2	204.3511
24	β-caryophyllene	20.45	1424	1424	7.25 ± 0.4	204.3511
25	trans-α-bergamotene	21.63	1435	1432	0.1 ± 0.04	204.3511
26	α-guaiene	21.21	1439	1437	2.02 ± 0.2	204.3511
27	α-humulene	22.69	1455	1452	6.45 ± 0.5	204.3511
28	β-farnesene	22.78	1457	1454	8.25 ± 0.4	204.3511
29	germacrene D	30.98	1481	1484	2.31 ± 0.07	204.3511
30	β-selinene	31.04	1486	1489	0.6 ± 0.03	204.3511
31	β-bisabolene	32.12	1509	1512	0.2 ± 0.03	204.3511
32	γ-cadinene	32.78	1513	1513	0.2 ± 0.04	204.3511
33	caryophyllene oxide	33.86	1640	1638	5.68 ± 0.02	220.3505
34	α-eudesmol	33.94	1653	1652	0.8 ± 0.06	222.3663
35	chavicol	17.73	1256	1247	0.07 ± 0.01	134.1751
36	eugenol	17.92	1357	1356	1.5 ± 0.02	164.2011
37	methyl eugenol	19.94	1406	1402	0.2 ± 0.02	178.2277
38	ethyl isovalerate	8.01	853	856	0.2 ± 0.02	130.1849

Classes of Constituents	RA% (No of Constituents)
Total monoterpene hydrocarbons	47.86 (10)
Total oxygenated monoterpenes	16.44 (11)
Total sesquiterpene hydrocarbons	22.13 (11)
Total oxygenated sesquiterpenes	6.48 (2)
Total phenylpropanoids	1.77 (3)
Total non-terpene derivatives	0.20 (1)
Total identified constituents	94.68 (38)

Values were obtained from three replicates. Mean ± standard deviation (SD) is shown. Rt, retention time; RI (exp), experimental relative retention index; RI (lit), literature relative retention index from MS libraries (Wiley) National Institute of Standards and Technology (NIST); RA, relative abundance; MS, mass spectra values.

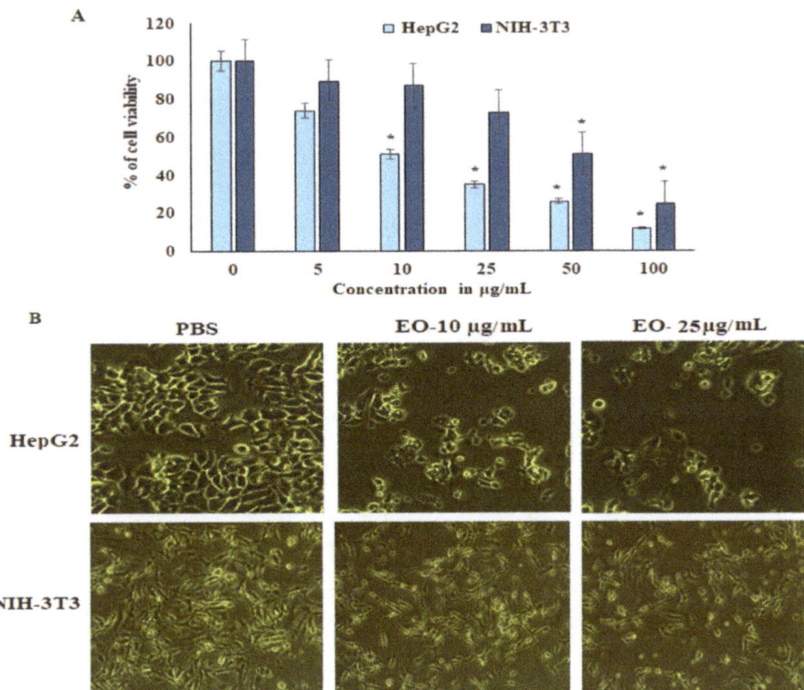

Figure 2. Effect of EO from *A. lanatum* on cell viability. EO inhibited cell growth and cell proliferation in HepG2 liver cancer cells and NIH-3T3 normal murine fibroblast cells: (**A**) HepG2 and NIH-3T3 cell lines were supplemented with EO (0–100 µg/mL) or a vehicle control (0.1% DMSO) for 48 h periods, and cell viability was assessed by MTT assay, and (**B**) HepG2 and NIH-3T3 cells were supplemented with EO (0–100 µg/mL) for 24 h, and cell morphology was examined beneath a phase-contrast microscope. The data are shown as the mean ± SD of triplicate measurements; * $p < 0.05$ when evaluated with respect to control cells.

2.3. Migration (Scratch Wound Assay)

The suppressive effect of the EO on migration was evaluated by scratch wound assay. In the scratch mobility assay, HepG2 liver cancer cells were treated with 25 µg/mL of EO supplemented with 1% FBS. The scratch wound was observed to determine the mobility of the cells in the junction, and scratch distance was measured to determine the closure disturbance, compared with that of the untreated control cells. The vehicle-treated HepG2 liver cancer cells significantly migrated after 24 h, whereas a distinct gap remained in the EO-treated groups after 24 h. The gaps were significantly suppressed (0.79 ± 0.05) by EO treatment (Figure 3). To confirm the inhibition effects, a gelatin-coated transwell insert assay was performed for invasion analysis. The relative migration fold of the EO-treated HepG2 liver cancer cells revealed significant differences compared with the untreated cells.

2.4. Invasion (Transwell Assay)

The invasive capacity of cells was evaluated by gelatin-coated transwell assay. As presented in Figure 4, the invasive ability of the EO-treated cells was significantly decreased after 24 h of treatment. Compared with the untreated cells, the relative invasion ratio of EO-treated HepG2 cells was 0.58 ± 0.046. There were significant differences between the treated and untreated cells, and EO significantly decreased the invasive capacity of HepG2 liver cancer cell lines.

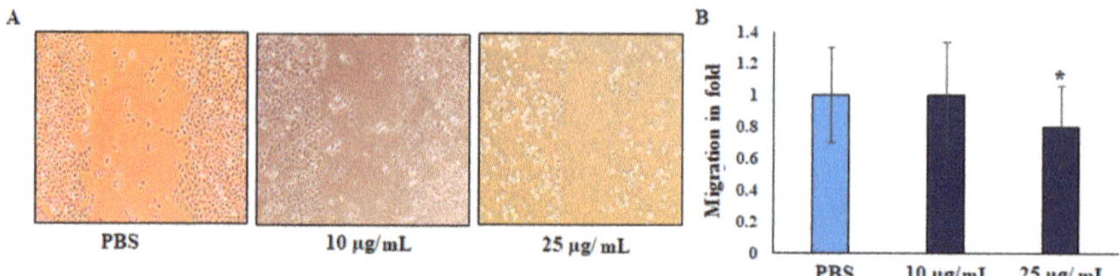

Figure 3. Effect of EO from *A. lanatum* on the migration of HepG2 liver cancer cell lines (scratch wound assay). (**A**) EO inhibits HepG2 human liver cancer cell proliferation and migration. HepG2 cell monolayers were scratched, and cells were supplemented with EO (10 and 25 μg/mL) for 24 h. Migration was determined using an optical microscope (200× magnification) by wound-healing assay. (**B**) Scratch distance was used to calculate the area of the wound and assess wound closure. The data are shown as the mean ± SD of triplicate values; * $p < 0.05$ when evaluated with respect to the control (PBS, phosphate-buffered saline).

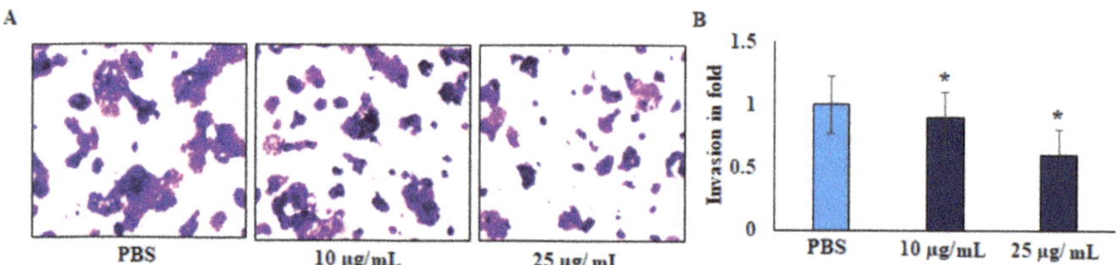

Figure 4. Effect of EO from *A. lanatum* on the inhibition of invasion and angiogenic capacity of HepG2 liver cancer cells. (**A**) EO inhibits human liver cancer cell migration and invasion. HepG2 liver cancer cell lines were supplemented with EO (10 and 25 μg/mL) and loaded to the upper chambers of matrigel-coated transwells. Invasion was determined by total cell counting. Cells invading the lower chamber after 24 h were counted. (**B**) The inhibition percentage of invasion was quantified and expressed relative to the control (untreated cells), whose level of invasion was set at 100%. Invading cells quantified using manual counting and values are noted as fold changes, compared to control. The data are shown as the mean ± SD of triplicate values; * $p < 0.05$ when evaluated against control (PBS, phosphate-buffered saline).

2.5. Immunoblotting and Localization of Cytochrome-c

Liver cancer markers are important tools for the evaluation of the migration and invasion of tumor cells. The apoptotic regulatory markers BCL-2 and CASPASE-3 were reciprocally regulated by the EO. Angiogenesis was lost through the regulation of CYP-1A1 and NFκB. EO negatively regulated the NFκB markers and increased CYP-1A1 expression levels in both mRNA and protein markers. These results demonstrate that EO regulates protein expression levels in HepG2 liver cancer cell lines; see Figure 5. The apoptotic marker CASPASE-3 was significantly up-regulated (2.5 ± 0.6-fold; $p \leq 0.05$), whereas the angiogenic marker NFκB was down-regulated in EO-treated HepG2 cells (0.65 ± 0.1-fold; $p \leq 0.05$). Furthermore, the metabolic marker CYP-1A1 was significantly up-regulated in EO-treated cells (3.2 ± 0.5-fold; $p \leq 0.05$). EO-treated HepG2 cells showed a loss of mitochondrial membrane potential (MMP) integrity at both concentrations (10 and 25 μg/mL). Untreated cells showed high MMP integrity by an emission of orange-red fluorescence. On the other hand, apoptotic cells stained with JC-1 showed green fluorescence. Consequently, the EO-treated cells demonstrated an emission of green fluorescence, indicating a loss of mitochondrial membrane integrity and release of the mitochondrial contents, including cytochrome-c (cyt-c), in the cytoplasm (Figure 5D,E). These changes revealed the possible participation of EO in induction of the apoptotic pathway. Therefore, it was suggested that EO treatment (at 25 μg/mL) decreased the migration and invasion abilities of HepG2 liver cancer cells by reciprocal regulation of angiogenesis and apoptotic markers.

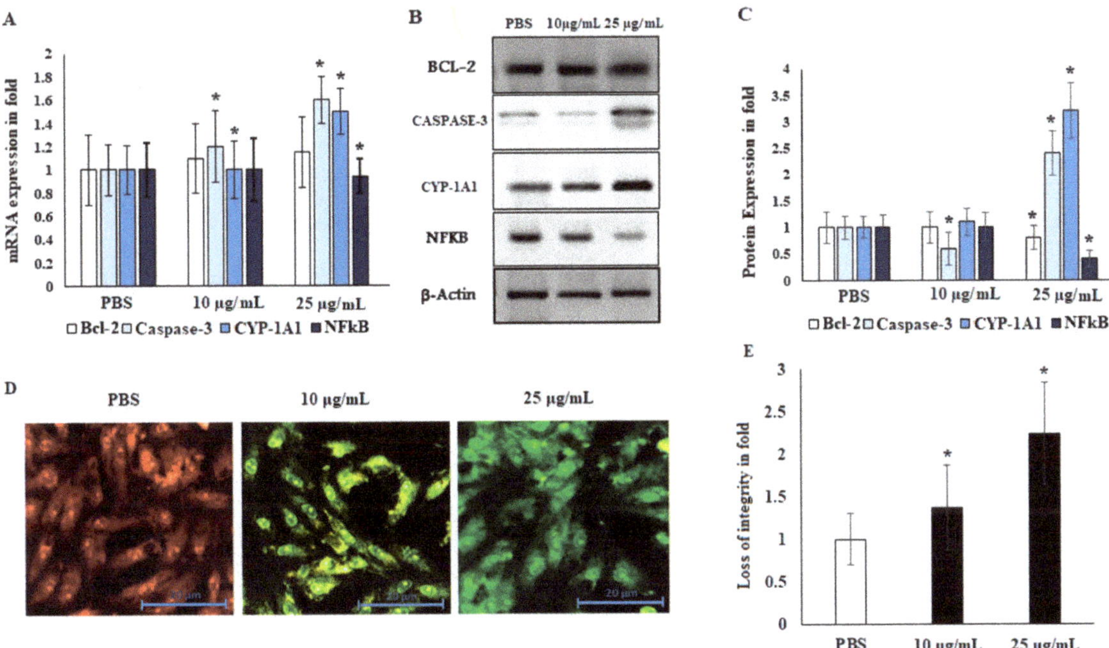

Figure 5. Effects of EO from *A. lanatum* on the mRNA and protein markers of HepG2 liver cancer cells. The effects of EO on the inhibition of apoptotic and angiogenic markers were evaluated by real-time PCR. (**A**) HepG2 liver cancer cells were supplemented with EO (10 and 25 µg/mL). The mRNA of apoptotic and angiogenic markers that was altered in EO-treated cells was quantified using quantitative real-time PCR. GAPDH was used as an internal mRNA control. (**B,C**) Alterations in the status of metastasis-associated proteins in response to EO supplementation were inspected using Western blot. HepG2 cells were supplemented with EO (10 and 25 µg/mL) for 24 h. β-actin was utilized as a control. (**D,E**) The mitochondrial membrane potential (MMP) was estimated for in EO-treated (10 and 25 µg/mL) HepG2 cell lines after an incubation period of 24 h. The mitochondrial membrane integrity was analyzed using the emission of green fluorescent by ImageJ software. The experimental data are shown as the mean ± SD of triplicate values; * $p < 0.05$ when evaluated against control (PBS; phosphate-buffered saline).

2.6. In Silico Docking of Major Constituents against BCL-2 and CASPASE-3

The computational interactions against ligands from the EO with BCL-2 and CASPASE-3 protein receptors were analyzed. In this study, the possible binding patterns and interaction mechanisms of major constituents—including α-pinene, camphene, β-pinene, β-myrcene, car-4-ene, α-terpinene, limonene, p-cymene, β-caryophyllene, α-humulene, β-farnesene, caryophyllene oxide, eucalyptol, linalool, nerol, and camphor—of the EO were analyzed using an auto-docking tool and evaluated using the binding energy and binding efficiency (Figure 6). In general, the binding intermol energy represents the best fit for a ligand in the active site of the target macromolecule. The in silico binding results revealed that, of the major constituents of EO, eucalyptol, linalool, nerol, and camphor showed the highest binding energies and high intermol energies (−3.76, −4.29, −3.37, and −5.5 kcal/mol, respectively). The docking simulation of BCL-2 to EO components resulted in the formation of eight hydrophobic interactions with potent binding energy, ligand efficiency, and intermol energy (Table 2). Several hydrophobic amino acid residues in chain-A and three predominant interactions with chain-B were observed in these EO interaction studies. Four major amino acid residues interacted with EO components in chain-A: Met-163, Arg-161, Lue-136, and Lys-137. Meanwhile, in chain-B, the predominant binding was against His-117, Val- 266, and Thr-266. Interactions against pro-apoptotic CASPASE-3 by the essential oil indicate that eucalyptol and camphor had potent binding compared to the other tested molecules (Figure 7). The binding energies of eucalyptol and camphor were found to be −4.29 and −3.81 kcal/mol, respectively (Table 3). We also observed interactions with the amino acid residues Leu-136 and Lys-137 in chain-A and Thr-195 in chain-B. From the above results, it can be suggested that molecules from EO of *A. lanatum* control liver cancer through apoptotic protein interactions, thus mitigating the migration and angiogenesis of HepG2 liver cancer cell lines.

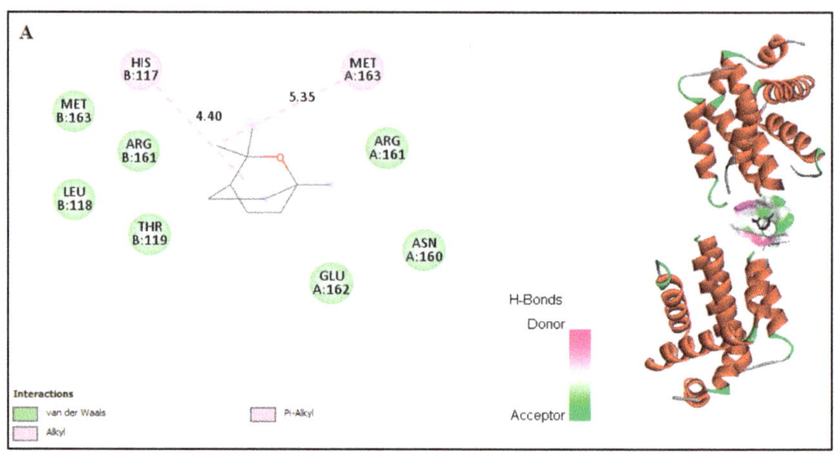

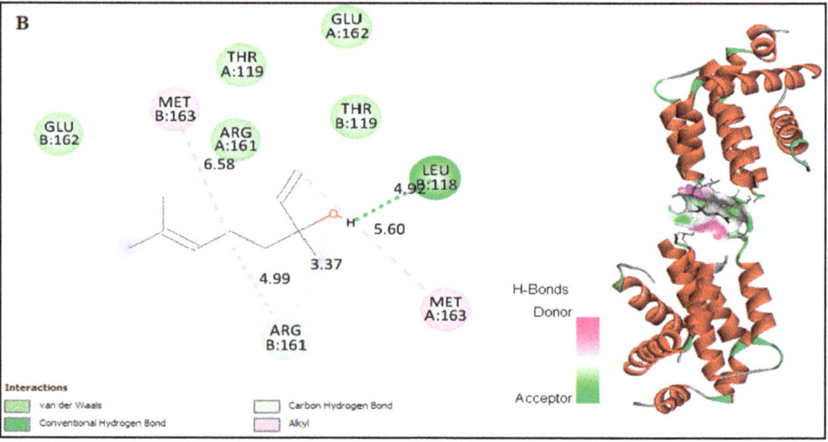

Figure 6. *Cont.*

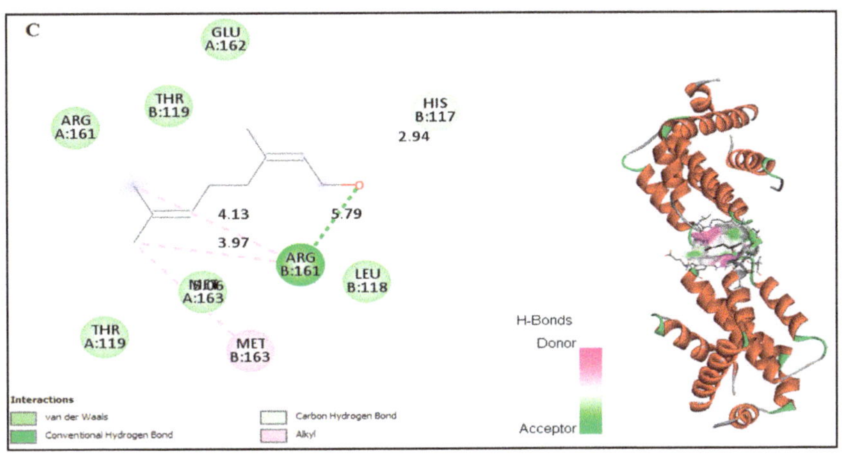

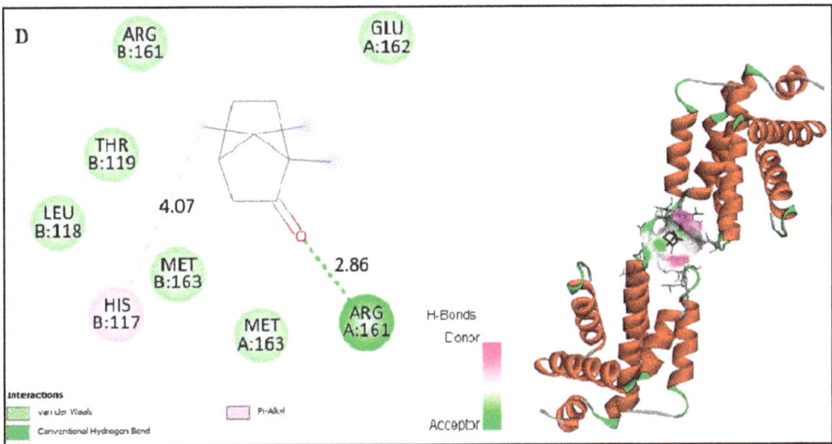

Figure 6. In silico docking of binding interactions of the constituents eucalyptol (**A**), linalool (**B**), nerol (**C**), and camphor (**D**) from EO of *A. lanatum* against anti-apoptotic protein BCL-2. To demonstrate the illustration of interactions in the hydrophobic bond and the other polar bond of BCL-2, we show the amino acid residue analysis of the interacted bond and its length, together with the binding pocket of ligand–receptor interactions.

Table 2. Hydrophobic interaction of potent binding constituents and amino acid residues of target proteins (BCL-2).

Ligand	Eucalyptol	Camphor	Linalool	Nerol
PubChem ID	CID:2758	CID:2537	CID:6549	CID:643820
Binding energy	−3.76	−4.29	−3.37	−5.5
Ligand efficiency	−0.34	−0.39	−0.31	−0.32
Intermol energy	−3.76	−4.29	−4.58	−6.1
Ligand atoms (ring)	Alkyl hydrophobic bond:C9 Pi-alkyl hydrophobic bond:C7	Hydrogen bonds:C2-O Pi-alkyl hydrophobic bond:C8	Alkyl hydrophobic bond:C8, C8 , C1 ,C1 ,C3′-O ,C3′-O Pi-alkyl hydrophobic bond:C5, C3′-O Carbon–hydrogen bond interaction: C3-OH	C-1 C-1-OH C-1 C-1-OH C-1

Table 2. *Cont.*

Ligand	Eucalyptol	Camphor	Linalool	Nerol
Docked amino acid residue (bond length)	Chain A: MET'163 (4.75 Å) Chain B: HIS'117 (5.21Å)	Chain A: ARG'161/ HE (2.86 Å) Chain B: HIS'117 (4.07Å)	Chain A: LEU'136 (4.62Å) Chain A: LYS'137 (4.16Å) Chain A: LEU'136 (4.48Å) Chain A: LYS'137 (4.23Å) Chain A: LYS'137/CE Chain B: VAL'266 (3.50 Å) Chain B: TYR'266 (3.50 Å) Chain B: TYR'195 (3.65Å) Chain B: TYR'195 (4.93Å) (3.18Å)	Chain A : MET Chain B : ARG Chain B : HIS

ARG, arginine; HIS, histidine; LEU, leucine; LYS, lysine; MET, methionine; TYR, tyrosine; VAL, valine.

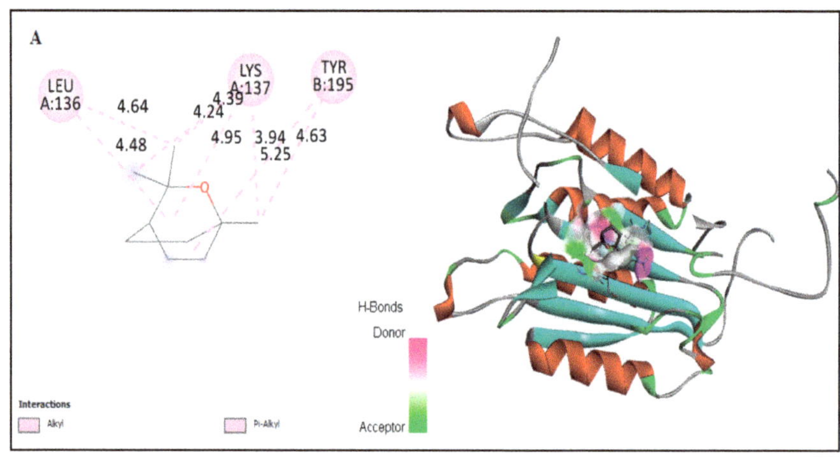

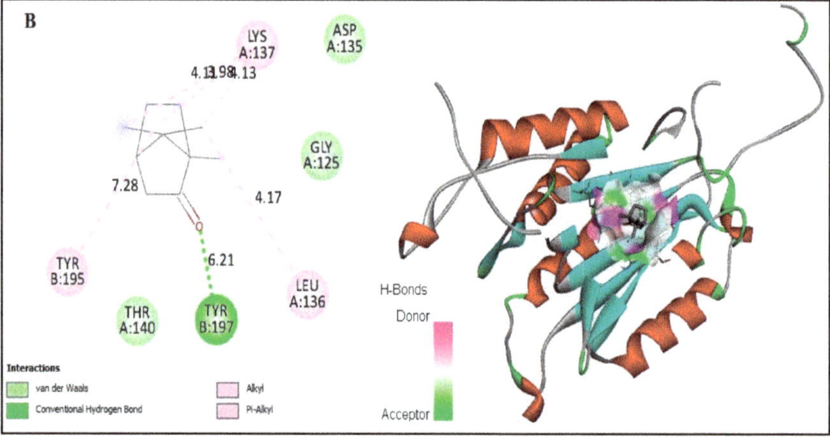

Figure 7. In silico docking of binding interactions of the constituents eucalyptol (**A**) and camphor (**B**) from EO of *A. lanatum* against pro-apoptotic protein CASPASE-3. To demonstrate the illustration of interactions in the hydrophobic bond and the other polar bond of CASPASE-3, we show the amino acid residue analysis of interacted bond and its length, together with the binding pocket of ligand–receptor interactions.

Table 3. Hydrophobic interaction of potent binding constituents and amino acid residues of target proteins (CASPASE-3).

Ligand	Camphor	Eucalyptol
PubChem ID	CID_2537	CID_2758
Binding energy	−4.29	−3.81
Ligand efficiency	−0.39	−0.35
Intermol energy	−4.29	−3.81
Ligand atoms (ring)	Hydrogen bonds:C2-O Alkyl hydrophobic bond:C9, C9, O, C8 Pi-alkyl hydrophobic bond: O	Alkyl hydrophobic bond:C7, O, C9, C10, C10, O Pi-alkyl hydrophobic bond:O
Docked amino acid residue (bond length)	Chain A: LEU′136/CG (4.56Å) Chain A: LYS′137/CG (3.88Å) Chain A: LYS′137 (4.38Å) Chain A: LYS′137 (3.95Å) Chain B: TYR′197/ HH (2.04 Å) Chain B: TYR′195 (5.01Å)	Chain A: LYS′137 (4.27Å) Chain A: LYS′137 (4.35Å) Chain A: LEU′136 (4.64Å) Chain A: LEU′136 (4.48Å) Chain B: LYS′137 (3.94 Å) Chain B: LYS′137 (4.95 Å) Chain B: TYR′195 (5.01Å)

LEU, leucine; LYS, lysine; TYR, tyrosine.

3. Discussion

Plant metabolites exhibit remarkable effects, including anti-cancer and other important biological activities. EOs, among the important secondary plant metabolites, through a long chain of evidence, have been shown to possess several different biological activities [1,27–29]. EOs are worth consideration in research in order to highlight their mechanisms of action and pharmacological targets. Chemically, EOs are a complex blend of hydrocarbons and oxygenated hydrocarbons, biosynthesized and arising from the isoprenoid pathways, and mainly consisting of monoterpenes and sesquiterpenes [30]. The EO obtained from *A. lanatum* was analyzed using GC-MS, and interpretation of the analysis revealed and quantified 38 components representing 94.68% of the total components, including monoterpene hydrocarbons (47.86%), oxygenated monoterpenes (16.44%), sesquiterpene hydrocarbons (22.13%), and oxygenated sesquiterpenes (6.48%); see Figure 1 and Table 1. Previous reports have discussed the molecular cytotoxicity effects of EOs, and some examples of isolated compounds from EOs towards various cancer cell lines have indicated the mediation of apoptosis, loss of mitochondrial membrane integrity, and several other mechanisms involving the anti-apoptotic factor BCL-2 and pro-apoptotic protein CASPASE-3 [30–33]. Camphor and eucalyptol have demonstrated the induction of apoptosis through the down-regulation of anti-apoptotic factor BCL-2 in a human oral epidermoid carcinoma cell line and activation of the CASPASE cascade in oral KB and colorectal cancer cell lines [30,34,35]. Linalool has been shown to lead to a reduction in BCL-2 protein expression in human dermal fibroblast cancer cell lines [36].

Few studies have investigated the anti-cancer activities of EO from *Anisosciadium*. The EO from *A. orientale* has shown antioxidant and anti-cancer activities against various human cancer cell lines [17,18]. In this study, the results from an MTT assay demonstrate that the proliferation of HepG2 was significantly inhibited by the EO of *A. lanatum*. Conversely, NIH-3T3 cell lines showed higher cell proliferation and viability, and were insignificantly inhibited up to 25 μg/mL EO. Considering these results, further studies were carried out on HepG2 liver cancer cell lines (see Figure 2). Cell migratory properties affect tumorigenesis and metastasis. Various studies have targeted cancer cell migration and invasion as potent tools for controlling the progression of cancers [37,38]. In the current study, the cellular functional results revealed that EO (25 μg/mL) from *A. lanatum* exhibited a regulation of cell migration and invasion in HepG2 liver cancer cell lines (Figures 3 and 4). Reliably, the activation of CASPASE-3 and an antagonist of BCL-2 inhibits the potential intrusion of HepG2 liver cancer cells [39]. BCL-2 is an anti-apoptotic protein that controls the release of cytochrome from mitochondria [40]. Mechanistically, the results showed that EO suppressed BCL-2 expression, which could have been reflected in the cell proliferation and survival of HepG2 liver cancer cells. As BCL-2 is a marker for tumorigenesis and neoplastic progression, it may be a potential marker for anti-cancer therapies. Several studies have confirmed that BCL-2 offers potential as a prospective drug target, as it activates the proto-oncogenic effect of the cancer environment [41–43]. In this study, EO potentially inhibited the BCL-2 mRNA and protein expression in HepG2 liver cancer cell lines (Figure 5). Many small molecular inhibitors of BCL-2, including ABT-737 and ABT-199, have been investigated extensively [44]. The suppressive effects of BCL-2 and CYP-1A1 are interlinked in cancer prognosis through migratory action and tumorigenesis [45]. Angiogenesis involves regulatory functions played by BCL-2 and CASPASE-3 in reverse roles [38,46]. In this study, the suppression effect of EO on the cell migration process may have been mediated by the suppression of BCL-2 and NFκB markers (Figure 5). The invasiveness activity was also inhibited in EO-treated HepG2 liver cancer cell lines. Blocking of the regulatory role of NFκB suppressed the invasiveness capacity and tumorigenesis in in vitro models. The liver cancer cells were inhibited by EO, which contributed to controlling cancer metastasis [47]. Such suppression regulates the apoptosis (also known as programmed cell death) of tumor cells, morphologically characterized by nuclear damage, chromatin condensation, cell shrinkage, and DNA endonuclease activation with apoptotic bodies [48]. Importantly, in this study, EO induced cell apoptosis in HepG2 liver cancer cells through the activation of mRNA of CASPASE-3 and CYP-1A1 (Figure 5). These genes are apoptotic markers involved in the spreading and invasiveness of cancer cells [44,48,49]. Subsequently high expression of CASPASE-3 and CYP-1A1 could be initiated due to mitochondrial membrane damage. Upon permeabilization of the mitochondrial membrane, caspase activators such as cyt-c were released from mitochondria into the cytosol [50]. The results verify the loss of mitochondrial membrane integrity. Consequently, the mitochondrial contents, including cyt-c, were released into cytoplasm and triggered an intrinsic apoptotic cascade. These findings express the possible activation of apoptosis via cyt-c, which in turn initiates other events in the programmed cell death in EO-treated HepG2 cell lines (Figure-5D, E). Further molecular modulations showed that EO increased the mRNA of CASPASE-3 expression levels, as well as increased and reciprocally regulated the protein expression levels of BCL-2 and NFκB (Figure 5). These results suggest that EO promoted apoptosis in and inhibited the invasiveness of HepG2 liver cancer cells. The previously discussed findings were further confirmed by in silico studies through the potential simulation of anti-proliferative properties of major molecules of EO (Figures 6 and 7). In some studies, β-sitosterol was shown to have potent binding against apoptotic regulating proteins, whereas isovitexin had the lowest binding affinity detected against BCL-2 and CASPASE-3 proteins [51]. In this study, of the major molecules, only camphor, eucalyptol, nerol, and linalool showed potential binding against BCL-2. Camphor and eucalyptol alone showed significant binding against pro-apoptotic CASPASE-3 regulatory molecules. These findings suggest that the EO from *A. lanatum* can control hepatoma HepG2 liver cancer cells through the reciprocal regulation of apoptotic markers. Taken together, the EO from *A. lanatum* was found to be able to control cancer progression and tumorigenesis, as mediated by modulatory effects on apoptotic markers.

4. Materials and Methods

4.1. Plant Material

Aerial parts of *A. lanatum* were collected from Riyadh, Saudi Arabia (April 2014) and were identified by Dr. Engineer Mamdouh Shokry, director of El-Zohria Botanical Garden, Giza, Egypt. A voucher specimen (14-Apr-AL) was deposited at the Herbarium Museum of the College of Clinical Pharmacy, King Faisal University, Al-Ahsa, Saudi Arabia.

4.2. Extraction of Essential Oil

One hundred grams of the carefully collected and air-dried whole *A. lanatum* plant were subjected to hydro-distillation using a Clevenger-type apparatus for 3 h. EO (yield

0.46% volume per dried plant weight) was recovered and dried over anhydrous sodium sulphate. The EO sample was kept in amber-colored vials in a refrigerator at 4 °C until further analyses [4].

4.3. Essential Oil Analysis

The EO was diluted with n-hexane (GC grade, 5 µL:1 mL) and 5 µL were injected into the GC (GC, Model CP-3800, Varian, Walnut Creek, CA, USA) and linked with a mass spectrometer (MS, Model Saturn 2200, Varian, Walnut Creek, CA, USA) equipped with a VF-5ms-fused silica capillary column (5% phenyl-dimethylpolysiloxane, with dimensions of 30 m × 0.25, film thickness was 0.25 µm, Varian). An electron impact (EI) ionization detector was used, with an ionization energy of 70 eV. Carrier gas (helium) was adjusted to have a steady flow rate (1 mL/min). The temperature of the oven was programed as follows: 1, 50, and 5 min at 50, 230, and 290 °C, respectively. The split ratio of injection samples was 1/500, with total time equal to 60 minutes. Identification of the constituents was conducted by comparison of Kovat's retention indices (RI) relative to a set of co-injected standard hydrocarbons (C10–C28, Sigma-Aldrich, Darmstadt, Germany) [52]. Components were identified by comparing their MS data and their corresponding retention indices with the Wiley Registry of Mass Spectral Data 10th edition (April 2013), the NIST 11 Mass Spectral Library (NIST11/2011/EPA/NIH), and literature data [53]. All of the identified constituents and their relative abundance percentages are listed in Table 1.

4.4. Cell Culture and MTT Assay

Human liver cancer cell lines (HepG2) and a normal fibroblast NIH-3T3 cell line (both cell lines procured from NCCS, Pune, India) were seeded in 96-well plates with a cell population of 1×10^4 cells/well in DMEM/F12 with antibiotic solution and 10% FCS (Invitrogen, CA, USA). Cells were incubated in a 5% CO_2 chamber at 37 °C. The monolayer cultured cells were washed with PBS, and then were treated with EO (10–100 µg/mL) with various test concentrations of tested samples in serum-free media and incubated for 24 h. The medium was aspirated, 0.5 mg/mL of MTT reagent was added, and the solution was incubated at 37 °C for 4 h. After the incubation period, measurements were carried out according to the method protocol described by Khalil et al. (2021) [5].

4.5. Migration (Scratch-Wound Assay)

The HepG2 cells were seeded in 6-well plates with a cell population of 1×10^5 cells/well in complete DMEM/F12 medium and allowed to attach overnight in a CO_2 incubator. The media was aspirated with DMEM/F12 with 25% charcoal-stripped FCS for 24 h. After the cells attained confluence, a scratch line was made in the centers of the wells using a sterile tip. The wells were then gently washed with serum-free DMEM/F12 medium and treated again with either DMSO or EO containing DMEM/F12 medium for another 24 h [37]. The scratch-line recovery was recorded using an Optika inverted microscope (200× magnification).

4.6. Invasion (Transwell Assay)

The transwell invasion assay was evaluated using a matrigel-coated 12-well Boyden chamber (8 µm PET; Corning, NY, USA), following the methodology described previously by Hanieh et al. (2016) [46], where 4×10^4 cells were cultured in the upper chamber in 600 µL of serum-free DMEM medium with EO. In the lower chamber, 800 µL of DMEM medium with 10% FBS were added.

4.7. Immunoblotting

After treatment of HepG2 liver cancer cell lines with different concentrations of EO for 24 h, cells were harvested using trypsin and washed twice with ice-cold PBS. For the immunoblotting analysis, treated cells were lysed in RIPA buffer (Santa Cruz Biotech, CA, USA) for 14 min on ice, then centrifuged at 8000× g for 15 min at 4 °C. Supernatants were collected and estimated using Bradford reagent. An equal amount of protein was denatured at 92 °C for 7 min. Denatured proteins were then separated on SDS-PAGE gels and transferred to PVDF membranes. Membranes were blocked with 5% non-fat milk at room temperature for 30 min, incubated with primary antibodies (BCL-2, 1:1000; CASPASE-3, 1:2000; CYP-1A1, 1:1000; and NFκB, 1:1500) from cell-signaling technology (CST; Beverly, MA, USA) overnight at 4 °C, and then washed and incubated with horseradish peroxidase-conjugated secondary antibody at room temperature for 1 h. Protein bands were visualized by enhanced chemiluminescence (Hisense, Thermo Scientific, Waltham, MA, USA) and analyzed using a Licor analyzer. β-actin (1:2000) was used as an internal control [54].

4.8. Mitochondrial Membrane Potential (MMP) Assessment for Localization of Cytochrome-c

The mitochondria-specific fluorescence dye, namely, 5,5′,6,6′-tetrachloro-1,1′3,3′- tetraethyl benzamidazol-carbocyanine iodide (JC-1) (JC-1 MMP assay kit, Abcam), was used to determine the MMP according to the method previously described [55]. HepG2 cells were seeded in 12-well plates and treated with EO (10 and 25 µg/mL) for 24 h. JC-1 working solution was added and incubated at 37 °C for 20 min. Treated cells were washed twice with PBS, replaced with fresh DMEM medium, and captured on a phase contrast inverted fluorescence microscope 200X (Leica 3000 fluorescence microscope). Mitochondrial membrane potentials were monitored by the ratio of red and green fluorescence intensity.

4.9. mRNA Expression

mRNA from treated cells was isolated using Trizol Reagent (Sigma, Darmstadt, Germany) and purified using chloroform and isopropanol. Synthesis of cDNA was carried out using an RT multiscreen kit and amplification was carried out using a thermocycler PCR machine (Eppendorf, Hamburg, Germany). Quantitative analysis of target mRNA such as BCL-2, CASPASE-3, CYP-1A1, and NFkB was performed using a SYBR green master mix (Bio-Rad, California, CA, USA). Details of the primers used for real-time PCR are given in Table 4. Quantification of mRNA expression was achieved using the quant studio software (Bio-Rad, California, CA, USA). Data were normalized to GAPDH levels. All qPCRs were performed at least in triplicate for each experiment [56].

Table 4. Real-time PCR primer details.

Primer Name	Forward	Reverse	Product Size
BCL-2	TGTGGATGACTGACTACCTGAACC	CAGCCAGGAGAAATCAAACAGAGG	186
CASPASE-3	GTGGAACTGACGATGATATGGC	CGCAAAGTGACTGGATGAACC	212
CYP-1A1	GGCCACTTTGACCCTTACAA	CAGGTAACGGAGGACAGGAA	236
NFκB	TGAAGAGAAGACACTGACCATGGAAA	TGGATAGAGGCTAAGTGTAGACACG	254
β-Actin	AAGATCCTGACCGAGCGTGG	CAGCACTGTGTTGGCATAGAGG	225

4.10. Computational Docking Analysis

4.10.1. Protein Preparation

For docking studies, three-dimensional structures of anti-apoptotic proteins, such as BCL-2 and CASPASE-3, were retrieved from the RCSB PDB protein data bank [57,58], with their respective PDB ID (4LVT and 3DEK). For docking simulations, water molecules and free hydrogen atoms were removed, and polar molecules were added to all protein structures using the pymol tool and automated AutoDock tool. Further active sites were analyzed in protein receptors using the web-based online tool Q-SiteFinder. All ions, except for the binding site, and non-relevant crystallographic materials were removed. These active sites were chosen as the most favorable binding residues for our docking simulation.

4.10.2. Ligand Preparation

The structures of the major constituents from EO of *A. lanatum*—namely, α-pinene, camphene, β-pinene, β-myrcene, car-4-ene, α-terpinene, limonene, p-cymene, β-caryophyllene, α-humulene, β-farnesene, caryophyllene oxide, camphor, eucalyptol, nerol, and linalool—were retrieved from the Pubchem compound database of the National Center for Biotechnology Information, with the respective IDs, as follows: CID—6654, CID—6616, CID—31253 14896, CID—16211587, CID—7462, CID—22311, CID—7463, CID—5281515, CID—10704181, CID—5281516, CID—1742210, CID—2537, CID—2758, CID—6549, and CID—643820, respectively. The downloaded structures were converted into PDB format using a freely available open-source tool, using pymol and Autodock docking tools.

4.10.3. Statistical Analysis

Representative data are shown as mean ± standard deviation ($n = 3$) from one experiment. For cell proliferation, migration, invasion, Western blot, and gene expression analyses, samples from the chosen replicate were used. Statistical analysis was conducted by one-way ANOVA followed by Dunnett's post-hoc test as a reference for comparison to the control group (undifferentiated). All analyses were performed using the GraphPad Prism 7.0 software (GraphPad Software Inc., San Diego, CA, USA), and the statistical results are shown as the corrected p value (* $p < 0.05$).

5. Conclusions

In the present study, EO was isolated from *Anisosciadium lanatum* using a hydro-distillation method and its components were identified using GC-MS. The yield of EO was 0.46% *v/w*. In total, 38 components were identified and quantified. The findings suggest that the EO from *A. lanatum* has a controlling effect on hepatoma cells (i.e., the HepG2 liver cancer cell line) through regulation of the apoptotic markers BCL-2 and CASPASE-3. Molecular docking analysis supported the experimental results, which revealed the oxygenated molecules camphor, eucalyptol, nerol, and linalool as having potential virtual binding against the anti-apoptotic marker BCL-2, whereas camphor and eucalyptol alone showed significant binding against the pro-apoptotic regulator CASPASE-3. The obtained results indicate that the EO extracted from *A. lanatum* should be further explored in vivo as a tool for the management of hepatoma cancer diseases.

Author Contributions: H.E.K., conceptualization, funding acquisition, methodology, writing—review and editing, writing—original draft preparation, and supervision; H.-I.M.I., formal analysis, methodology, validation, software, writing—review and editing, and writing—original draft preparation; H.M.D., methodology, software, and writing—review and editing; and K.M., conceptualization, supervision, and writing—original draft preparation. All authors have read and agreed to the published version of the manuscript.

Funding: This research was funded by the Deanship of Scientific Research (DSR), King Faisal University, Saudi Arabia, research group project (grant number 1811002).

Acknowledgments: The authors thank the Deanship of Scientific Research (DSR), King Faisal University, Saudi Arabia for funding this research group project (grant number 1811002) and the College of Clinical Pharmacy, King Faisal University.

Conflicts of Interest: The authors declare that there are no conflicts of interest.

References

1. Abd-ElGawad, A.M.; Elgamal, A.M.; Ei-Amier, Y.A.; Mohamed, T.A.; El Gendy, A.E.-N.G.; Elshamy, A.I. Chemical Composition, Allelopathic, Antioxidant, and Anti-Inflammatory Activities of Sesquiterpenes Rich Essential Oil of *Cleome amblyocarpa* Barratte & Murb. *Plants* **2021**, *10*, 1294. [PubMed]
2. Camele, I.; Grul'ová, D.; Elshafie, H.S. Chemical Composition and Antimicrobial Properties of *Mentha × piperita* cv.'Kristinka'Essential Oil. *Plants* **2021**, *10*, 1567. [CrossRef]
3. Abd-ElGawad, A.M.; Bonanomi, G.; Al-Rashed, S.A.; Elshamy, A.I. *Persicaria lapathifolia* Essential Oil: Chemical Constituents, Antioxidant Activity, and Allelopathic Effect on the Weed *Echinochloa colona*. *Plants* **2021**, *10*, 1798. [CrossRef]
4. Mohamed, M.E.; Mohafez, O.M.; Khalil, H.E.; Alhaider, I.A. Essential Oil from Myrtle Leaves Growing in the Eastern Part of Saudi Arabia: Components, Anti-inflammatory and Cytotoxic Activities. *J. Essent. Oil-Bear. Plants* **2019**, *22*, 877–892. [CrossRef]

5. Khalil, H.E.; Alqahtani, N.K.; Darrag, H.M.; Ibrahim, H.-I.M.; Emeka, P.M.; Badger-Emeka, L.I.; Matsunami, K.; Shehata, T.M.; Elsewedy, H.S. Date Palm Extract (*Phoenix dactylifera*) PEGylated Nanoemulsion: Development, Optimization and Cytotoxicity Evaluation. *Plants* **2021**, *10*, 735. [CrossRef]

6. Khalil, H.E.; Mohamed, M.E.; Morsy, M.A.; Kandeel, M. Flavonoid and phenolic compounds from *Carissa macrocarpa*: Molecular docking and cytotoxicity studies. *Pharmacogn. Mag.* **2018**, *14*, 304. [CrossRef]

7. Harvey, A.L. Natural products in drug discovery. *Drug Discov. Today* **2008**, *13*, 894–901. [CrossRef]

8. Bamia, C. Dietary patterns in association to cancer incidence and survival: Concept, current evidence, and suggestions for future research. *Eur. J. Clin. Nutr.* **2018**, *72*, 818–825. [CrossRef] [PubMed]

9. Clinton, S.K.; Giovannucci, E.L.; Hursting, S.D. The World Cancer Research Fund/American Institute for Cancer Research third expert report on diet, nutrition, physical activity, and cancer: Impact and future directions. *J. Nutr.* **2020**, *150*, 663–671. [CrossRef] [PubMed]

10. Karageorgou, D.; Magriplis, E.; Mitsopoulou, A.; Dimakopoulos, I.; Bakogianni, I.; Micha, R.; Michas, G.; Chourdakis, M.; Ntouroupi, T.; Tsaniklidou, S. Dietary patterns and lifestyle characteristics in adults: Results from the Hellenic National Nutrition and Health Survey (HNNHS). *Public Health* **2019**, *171*, 76–88. [CrossRef]

11. Buriani, A.; Fortinguerra, S.; Sorrenti, V.; Dall'Acqua, S.; Innocenti, G.; Montopoli, M.; Gabbia, D.; Carrara, M. Human adenocarcinoma cell line sensitivity to essential oil phytocomplexes from *Pistacia* species: A multivariate approach. *Molecules* **2017**, *22*, 1336. [CrossRef]

12. Amiri, M.S.; Joharchi, M.R. Ethnobotanical knowledge of Apiaceae family in Iran: A review. *Avicenna J. Phytomed.* **2016**, *6*, 621.

13. Sayed-Ahmad, B.; Talou, T.; Saad, Z.; Hijazi, A.; Merah, O. The Apiaceae: Ethnomedicinal family as source for industrial uses. *Ind. Crops. Prod.* **2017**, *109*, 661–671. [CrossRef]

14. Christensen, L.P.; Brandt, K. Bioactive polyacetylenes in food plants of the Apiaceae family: Occurrence, bioactivity and analysis. *J. Pharm. Biomed. Anal.* **2006**, *41*, 683–693. [CrossRef]

15. Vieira, J.N.; Gonçalves, C.; Villarreal, J.; Gonçalves, V.; Lund, R.; Freitag, R.; Silva, A.; Nascente, P. Chemical composition of essential oils from the apiaceae family, cytotoxicity, and their antifungal activity in vitro against *candida* species from oral cavity. *Braz. J. Biol.* **2018**, *79*, 432–443. [CrossRef] [PubMed]

16. Chizzola, R. Essential oil composition of wild growing Apiaceae from Europe and the Mediterranean. *Nat. Prod. Commun.* **2010**, *5*, 1934578X1000500925. [CrossRef]

17. Masroorbabanari, M.; Miri, R.; Firuzi, O.; Jassbi, A.R. Antioxidant and Cytotoxic Activities of the Essential Oil and Extracts of *Anisosciadium orientale*. *J. Chem. Soc. Pak.* **2014**, *36*, 457–461.

18. Rowshan, V.; Hatami, A.; Bahmanzadegan, A.; Yazdani, M. Chemical Constituents of the Essential Oil from Aerial Parts and Fruit of *Anisosciadium Orientale*. *Nat. Prod. Commun.* **2012**, *7*, 1934578X1200700127. [CrossRef]

19. Sher, H.; Aldosari, A. Ethnobotanical survey on plants of veterinary importance around Al-Riyadh (Saudi Arabia). *Afr. J. Pharm. Pharmacol.* **2013**, *7*, 1404–1410. [CrossRef]

20. Al-Mazroa, S. Chemical constituents of A. Lanatum, H. Tuberculatum, A. Garcini, S. Spinosa, H. Bacciferum, and A. Ludwigh grown in Saudi Arabia. *J. Saudi Chem.* **2003**, *7*, 255–258.

21. Al-Yemeny, M.N.; Al-Yemeny, A. A check list of weeds in Al-kharj area of Saudi Arabia. *Pak. J. Biol. Sci.* **1999**, *2*, 7–13. [CrossRef]

22. Joachim, W.; Kadereit, J.W.; Bittrich, V. *The Families and Genera of Vascular Plants*; Springer International Publishing AG: Steinhausen, Switzerland, 2018; p. 101.

23. Mandaville, J.P. *Bedouin Ethnobotany: Plant Concepts and Uses in a Desert Pastoral World*; University of Arizona Press: Tucson, AZ, USA, 2019; p. 355.

24. Middleditch, B.S. *Kuwaiti Plants: Distribution, Traditional Medicine, Pytochemistry, Pharmacology and Economic Value*; Elsevier: New York, NY, USA, 1991; p. 10.

25. El-Sayed, W.M.; Hussin, W.A.; Mahmoud, A.A.; AlFredan, M.A. Antimutagenic Activities of *Anisosciadium lanatum* Extracts Could Predict the Anticancer Potential in Different Cell Lines. *Int. J. Pharm. Pharm. Res.* **2017**, *9*, 197–206. [CrossRef]

26. Mahmoud, A.A.; El-Sayed, W.M. The anti-proliferative activity of anisosciadone: A new guaiane sesquiterpene from *Anisosciadium lanatum*. *Anti-Cancer Agents Med. Chem.* **2019**, *19*, 1114–1119. [CrossRef] [PubMed]

27. Mancianti, F.; Ebani, V.V. Biological activity of essential oils. *Molecules* **2020**, *25*, 678. [CrossRef]

28. De Souza, E.L. The effects of sublethal doses of essential oils and their constituents on antimicrobial susceptibility and antibiotic resistance among food-related bacteria: A review. *Trends Food Sci. Technol.* **2016**, *56*, 1–12. [CrossRef]

29. Spyridopoulou, K.; Fitsiou, E.; Bouloukosta, E.; Tiptiri-Kourpeti, A.; Vamvakias, M.; Oreopoulou, A.; Papavassilopoulou, E.; Pappa, A.; Chlichlia, K. Extraction, chemical composition, and anticancer potential of *Origanum onites* L. essential oil. *Molecules* **2019**, *24*, 2612. [CrossRef]

30. Sharifi-Rad, J.; Sureda, A.; Tenore, G.C.; Daglia, M.; Sharifi-Rad, M.; Valussi, M.; Tundis, R.; Sharifi-Rad, M.; Loizzo, M.R.; Ademiluyi, A.O. Biological activities of essential oils: From plant chemoecology to traditional healing systems. *Molecules* **2017**, *22*, 70. [CrossRef]

31. Pan, W.; Zhang, G. Linalool monoterpene exerts potent antitumor effects in OECM 1 human oral cancer cells by inducing sub-G1 cell cycle arrest, loss of mitochondrial membrane potential and inhibition of PI3K/AKT biochemical pathway. *J. Buon* **2019**, *24*, 323–328.

32. Bayala, B.; Bassole, I.H.; Scifo, R.; Gnoula, C.; Morel, L.; Lobaccaro, J.-M.A.; Simpore, J. Anticancer activity of essential oils and their chemical components-a review. *Am. J. Cancer Res.* **2014**, *4*, 591. [PubMed]

33. Bakkali, F.; Averbeck, S.; Averbeck, D.; Idaomar, M. Biological effects of essential oils—A review. *Food Chem. Toxicol.* **2008**, *46*, 446–475. [CrossRef]

34. Candrasari, D.S.; Mubarika, S.; Wahyuningsih, M.S.H. The effect of a-terpineol on cell cycle, apoptosis and Bcl-2 family protein expression of breast cancer cell line MCF-7. *J. Med. Sci.* **2015**, *47*, 59–67.

35. Kim, D.Y.; Kang, M.-K.; Lee, E.-J.; Kim, Y.-H.; Oh, H.; Kim, S.-I.; Oh, S.Y.; Na, W.; Kang, Y.-H. Eucalyptol Inhibits Amyloid-β-Induced Barrier Dysfunction in Glucose-Exposed Retinal Pigment Epithelial Cells and Diabetic Eyes. *Antioxidants* **2020**, *9*, 1000. [CrossRef]

36. Gunaseelan, S.; Balupillai, A.; Govindasamy, K.; Ramasamy, K.; Muthusamy, G.; Shanmugam, M.; Thangaiyan, R.; Robert, B.M.; Prasad Nagarajan, R.; Ponniresan, V.K. Linalool prevents oxidative stress activated protein kinases in single UVB-exposed human skin cells. *PLoS ONE* **2017**, *12*, e0176699. [CrossRef]

37. Fares, J.; Fares, M.Y.; Khachfe, H.H.; Salhab, H.A.; Fares, Y. Molecular principles of metastasis: A hallmark of cancer revisited. *Signal Transduct. Target. Ther.* **2020**, *5*, 28. [CrossRef] [PubMed]

38. Ibrahim, H.-I.M.; Ismail, M.B.; Ammar, R.B.; Ahmed, E.A. Thidiazuron suppresses breast cancer via targeting miR-132 and dysregulation of the PI3K–Akt signaling pathway mediated by the miR-202-5p–PTEN axis. *Biochem. Cell Biol.* **2021**, *99*, 1–11. [CrossRef]

39. Zhou, M.; Zhang, Q.; Zhao, J.; Liao, M.; Wen, S.; Yang, M. Phosphorylation of Bcl-2 plays an important role in glycochenodeoxycholate-induced survival and chemoresistance in HCC. *Oncol. Rep.* **2017**, *38*, 1742–1750. [CrossRef]

40. Tsuchiya, S.; Tsuji, M.; Morio, Y.; Oguchi, K. Involvement of endoplasmic reticulum in glycochenodeoxycholic acid-induced apoptosis in rat hepatocytes. *Toxicol. Lett.* **2006**, *166*, 140–149. [CrossRef]

41. Hata, A.N.; Engelman, J.A.; Faber, A.C. The BCL2 family: Key mediators of the apoptotic response to targeted anticancer therapeutics. *Cancer Discov.* **2015**, *5*, 475–487. [CrossRef] [PubMed]

42. Maji, S.; Panda, S.; Samal, S.K.; Shriwas, O.; Rath, R.; Pellecchia, M.; Emdad, L.; Das, S.K.; Fisher, P.B.; Dash, R. Bcl-2 antiapoptotic family proteins and chemoresistance in cancer. *Adv. Cancer Res.* **2018**, *137*, 37–75.

43. Kumar, V.S.; Kumaresan, S.; Tamizh, M.M.; Islam, M.I.H.; Thirugnanasambantham, K. Anticancer potential of NF-κB targeting apoptotic molecule "flavipin" isolated from endophytic *Chaetomium globosum*. *Phytomedicine* **2019**, *61*, 152830. [CrossRef] [PubMed]

44. Pan, R.; Hogdal, L.J.; Benito, J.M.; Bucci, D.; Han, L.; Borthakur, G.; Cortes, J.; DeAngelo, D.J.; Debose, L.; Mu, H. Selective BCL-2 inhibition by ABT-199 causes on-target cell death in acute myeloid leukemia. *Cancer Discov.* **2014**, *4*, 362–375. [CrossRef]

45. Zhang, S.; Kim, K.; Jin, U.H.; Pfent, C.; Cao, H.; Amendt, B.; Liu, X.; Wilson-Robles, H.; Safe, S. Aryl hydrocarbon receptor agonists induce microRNA-335 expression and inhibit lung metastasis of estrogen receptor negative breast cancer cells. *Mol. Cancer Ther.* **2012**, *11*, 108–118. [CrossRef] [PubMed]

46. Hanieh, H.; Mohafez, O.; Hairul-Islam, V.I.; Alzahrani, A.; Bani Ismail, M.; Thirugnanasambantham, K. Novel aryl hydrocarbon receptor agonist suppresses migration and invasion of breast cancer cells. *PLoS ONE* **2016**, *11*, e0167650. [CrossRef] [PubMed]

47. Gautam, N.; Mantha, A.K.; Mittal, S. Essential oils and their constituents as anticancer agents: A mechanistic view. *Biomed. Res. Int.* **2014**, *2014*, 154106. [CrossRef] [PubMed]

48. Wang, W.; Li, J.; Tan, J.; Wang, M.; Yang, J.; Zhang, Z.-M.; Li, C.; Basnakian, A.G.; Tang, H.-W.; Perrimon, N. Endonuclease G promotes autophagy by suppressing mTOR signaling and activating the DNA damage response. *Nat. Commun.* **2021**, *12*, 476. [CrossRef]

49. Zhang, X.; Fan, C.; Zhang, H.; Zhao, Q.; Liu, Y.; Xu, C.; Xie, Q.; Wu, X.; Yu, X.; Zhang, J. MLKL and FADD are critical for suppressing progressive lymphoproliferative disease and activating the NLRP3 inflammasome. *Cell Rep.* **2016**, *16*, 3247–3259. [CrossRef] [PubMed]

50. Kroemer, G.; Galluzzi, L.; Brenner, C. Mitochondrial membrane permeabilization in cell death. *Physiol. Rev.* **2007**, *87*, 99–163. [CrossRef]

51. Noor, Z.I.; Nithyanan, A.; Nur, N.M.Z.; Hasni, A. Molecular docking of selected compounds from *Clinacanthus Nutans* with Bcl-2, P53, Caspase-3 and Caspase-8 proteins in the apoptosis pathway. *J. Biol. Sci. Opin.* **2020**, *8*, 4–11.

52. Darrag, H.M.; Alhajhoj, M.R.; Khalil, H.E. Bio-Insecticide of *Thymus vulgaris* and *Ocimum basilicum* Extract from Cell Suspensions and Their Inhibitory Effect against Serine, Cysteine, and Metalloproteinases of the Red Palm Weevil (Rhynchophorus ferrugineus). *Insects* **2021**, *12*, 405. [CrossRef]

53. Adams, R.P. *Identification of Essential Oil Components by Gas Chromatography/Mass Spectrometry*; Allured Publishing Corporation: Carol Stream, IL, USA, 2007; Volume 456.

54. Ibrahim, H.-I.M.; Darrag, H.M.; Alhajhoj, M.R.; Khalil, H.E. Biomolecule from *Trigonella stellata* from Saudi Flora to Suppress Osteoporosis via Osteostromal Regulations. *Plants* **2020**, *9*, 1610. [CrossRef]

55. Marvibaigi, M.; Amini, N.; Supriyanto, E.; Abdul Majid, F.A.; Kumar Jaganathan, S.; Jamil, S.; Hamzehalipour Almaki, J.; Nasiri, R. Antioxidant activity and ROS-dependent apoptotic effect of Scurrula ferruginea (Jack) danser methanol extract in human breast cancer cell MDA-MB-231. *PLoS ONE* **2016**, *11*, e0158942. [CrossRef] [PubMed]

56. Ahmed, E.A.; Ibrahim, H.-I.M.; Khalil, H.E. Pinocembrin Reduces Arthritic Symptoms in Mouse Model via Targeting Sox4 Signaling Molecules. *J. Med. Food* **2021**, *24*, 282–291. [CrossRef] [PubMed]

57. Sivamani, P.; Singaravelu, G.; Thiagarajan, V.; Jayalakshmi, T.; Kumar, G.R. Comparative molecular docking analysis of essential oil constituents as elastase inhibitors. *Bioinformation* **2012**, *8*, 457–460. [CrossRef] [PubMed]
58. Kandeel, M.; Al-Taher, A. Metabolic drug targets of the cytosine metabolism pathways in the dromedary camel (*Camelus dromedarius*) and blood parasite Trypanosoma evansi. *Trop. Anim. Health Prod.* **2020**, *52*, 3337–3358. [CrossRef]

Article

Enantiomeric Composition, Antioxidant Capacity and Anticholinesterase Activity of Essential Oil from Leaves of Chirimoya (*Annona cherimola* Mill.)

Eduardo Valarezo *, Jeannette Ludeña, Estefanía Echeverria-Coronel, Luis Cartuche, Miguel Angel Meneses, James Calva and Vladimir Morocho

Departamento de Química, Universidad Técnica Particular de Loja, Loja 110150, Ecuador; isah4a@hotmail.com (J.L.); beecheverria@utpl.edu.ec (E.E.-C.); lecartuche@utpl.edu.ec (L.C.); mameneses@utpl.edu.ec (M.A.M.); jwcalva@utpl.edu.ec (J.C.); svmorocho@utpl.edu.ec (V.M.)
* Correspondence: bevalarezo@utpl.edu.ec; Tel.: +593-7-3701444

Abstract: *Annona cherimola* Mill. is a native species of Ecuador cultivated worldwide for the flavor and properties of its fruit. In this study, hydrodistillation was used to isolate essential oil (EO) of fresh *Annona cherimola* leaves collected in Ecuadorian Sierra. The EO chemical composition was determined using a non-polar and a polar chromatographic column and enantiomeric distribution with an enantioselective column. The qualitative analysis was carried out by gas chromatography coupled to a mass spectrometer and quantitative analysis using gas chromatography equipped with a flame ionization detector. The antibacterial potency was assessed against seven Gram-negative bacteria and one Gram-positive bacterium. ABTS and DPPH assays were used to evaluate the radical scavenging properties of the EO. Spectrophotometric method was used to measure acetylcholinesterase inhibitory activity. GC-MS analysis allowed us to identify more than 99% of the EO chemical composition. Out of the fifty-three compounds identified, the main were germacrene D (28.77 ± 3.80%), sabinene (3, 9.05 ± 1.69%), β-pinene (4, 7.93 ± 0.685), (E)-caryophyllene (10.52 ± 1.64%) and bicyclogermacrene (11.12 ± 1.39%). Enantioselective analysis showed the existence of four pairs of enantiomers, the (−)-β-Pinene (1S, 5S) was found pure (100%). Chirimoya essential oil exhibited a strong antioxidant activity and a very strong anticholinesterase potential with an IC_{50} value of 41.51 ± 1.02 µg/mL. Additionally, EO presented a moderate activity against *Campylobacter jejuni* and *Klebsiella pneumoniae* with a MIC value of 500 µg/mL.

Keywords: essential oil; *Annona cherimola*; chemical composition; enantioselective analysis; antibacterial activity; antioxidant activity; anticholinesterase activity; germacrene D; *Campylobacter jejuni*

Citation: Valarezo, E.; Ludeña, J.; Echeverria-Coronel, E.; Cartuche, L.; Meneses, M.A.; Calva, J.; Morocho, V. Enantiomeric Composition, Antioxidant Capacity and Anticholinesterase Activity of Essential Oil from Leaves of Chirimoya (*Annona cherimola* Mill.). *Plants* **2022**, *11*, 367. https://doi.org/10.3390/plants11030367

Academic Editors: Hazem Salaheldin Elshafie, Laura De Martino and Adriano Sofo

Received: 20 December 2021
Accepted: 26 January 2022
Published: 28 January 2022

Publisher's Note: MDPI stays neutral with regard to jurisdictional claims in published maps and institutional affiliations.

1. Introduction

Worldwide, the Annonaceae family comprises more than 128 genera and approximately 2106 species and they are mainly distributed in tropical and subtropical regions [1]. For Ecuador, 25 genera, 106 species and 20 endemic species are reported [2]. In a review of the antimalarial properties of the Annonaceae family, 11 species from *Annona* and *Xilopia* genus were recognized for their antiparasitic potential. Annonaceae species used in traditional medicine, over the tropical regions, are well documented for having potential for the treatment of parasitic diseases such as Malaria, Chagas and Leishmaniasis as well as other illnesses [3]. Indeed, *Annona muricata* was one of the most cited species with a variety of medicinal properties including the treatment for the symptoms of malarial infection, fever, liver ailments and headaches [4].

Annona cherimola, *A. crassiflora*, *A. muricata*, *A. squamosa* and *A. reticulata* are the commercial species, highly valued by their exotic edible fruit. Furthermore, different parts of the tree from these species have been used in folk medicine to treat several conditions including gastrointestinal diseases, diabetes and hypertension [5]. Many secondary metabolites have been reported such as phenols and other bioactive compounds but, the main chemical marker of the genus is a diverse group of polyketides called acetogenins, compounds closely associated to their antiproliferative effect on cancer cell lines [5,6].

In a related study, nine species from *Annona* genus, including *A. cherimola*, where reviewed in relation to their phytochemical composition and biological activity and found that mainly polar extracts obtained from these plants, induce a reduction in blood sugar levels in chemically induced type-2 diabetic rats which demonstrate the antidiabetic potential of species from this genus. Likewise, seven out of nine of the studied species reported a good antioxidant capacity profile in different in vitro assays. *A. cherimola* was only tested trough the oxygen radical absorbance capacity (ORAC) assay [7].

Annona cherimola Mill. is a native shrub, widely distributed in the Andean, Coastal, Amazon and Insular regions of Ecuador, between 0–3000 m a.s.l [2]. Currently, this species is cultivated in the subtropical and tropical regions worldwide, especially for its fruit, which is considered exotic. The *A. cherimola* species is commonly known as "chirimoya" or "chirimoyo" (Spanish language) and "custard-apple" (English language) [8]. The vernacular name "chirimoya" is derived from the Quechua (indigenous language) word "chirimuya", "chiri" that means cold and "muya" seeds. The fruit of this species is considered one the most appreciated inside the genus. The plant has been used in traditional medicine for the treatment of boils and others skin diseases [9].

Anthropological evidence suggests that the species *A. cherimola* was cultivated since the times of the Incan Empire and its fruit was considered as an active ingredient in the their diet [10]. Some important compounds have been isolated from this species such as alkaloids as cherimoline, annocherine A, annocherine B, cherianoine and rumocosine H [11], and some amides as cherinonaine, cheritamide, (N-trans-feruloyltyramine, N-trans-caffeoyltyramine, N-*cis*-caeoyltyramine, dihydro feruloyltyramine, N-trans-feruloylmethoxytyramine, N-*cis*-feruloylmethoxytyramine, and N-*p*-coumaroyltyramine [12]. Recent studies have shown that custard apple leaves contain flavonoids and other phenolic compounds with biological properties [13,14] and that alcoholic extracts from the leaves have proapoptotic and antidepressant activities [15].

Several studies on the species of the *Annona* genus have reported the occurrence of compounds with potential application for pharmaceuticals, food, agrochemicals products and cosmetics [3,5,6,16]. Extracts from these species are used for a wide range of beneficial purposes, however, the research have been focused mainly on the non-volatile fraction of the fruits, meanwhile, the enantiomeric distribution and biological properties of the EOof *Annona cherimola* have not been reported previously. This fact stimulated our interest in studying the chemical composition, enantiomeric distribution and antimicrobial, antioxidant and anticholinesterase activities of the essential oil of custard apple leaves.

2. Results

A total of 15,150 g of fresh custard apple leaves, with a moisture of 53 ± 5% (*w/w*), were used as raw material for the extraction. The isolation of the essential oil was carried out by hydrodistillation using a Clevenger-type apparatus. The amount of EO obtained was 38.5 mL, which represents a yield of 0.25 ± 0.02 (*v/w*).

2.1. Physical Properties of Essential Oil

The essential oil isolated from leaves of *A. cherimola* was presented as a viscous liquid with a characteristic texture. The physical properties of EO are shown in Table 1. In addition, this table shows the subjective color and its RGB and CMYK values.

2.2. Chemical Constituents of the Essential Oil

The compounds ocurring in *A. cherimola* EO were identified and quantified by GC-MS and GC-FID using nonpolar and polar columns. The Table 2 shows the quantitative and qualitative data of chemical constituents of custard apple obtained using nonpolar column DB-5ms. In the essential oil from chirimoya leaves fifty-three compounds were identified, which represent 99.80% of the total composition. According to their chemical nature, all the compounds were grouped in aliphatic monoterpene hydrocarbons (ALM), oxygenated monoterpenes (OXM), aliphatic sesquiterpene hydrocarbons (ALS) and oxygenated sesquiterpene (OXS). The ALS were the most representative compounds with twenty-seven compounds, which represents 69.40%, followed by ALM with 25.68%. Compounds belonging to the aromatic monoterpene hydrocarbons, aromatic sesquiterpene hydrocarbons and oxygenated sesquiterpene groups were not identified. The ALS germacrene D (compound 32, CF: $C_{15}H_{24}$, MM: 204.19 Da) was the main constituent with $28.77 \pm 3.80\%$. Other main compounds (>5%) were sabinene (3, $9.05 \pm 1.69\%$), β-pinene (4, $7.93 \pm 0.68\%$), (E)-caryophyllene (24, 10.52 ± 1.645) and bicyclogermacrene (36, $11.12 \pm 1.39\%$). Compounds 8 and 9 (limonene and β-phellandrene) co-eluted, both representing $0.66 \pm 0.13\%$.

Table 1. Physical properties of the essential oil.

	Annona cherimola EO	
	Mean [a]	**SD** [b]
Density, ρ (g/cm^3)	0.9472	0.0044
Refractive index, n^{20}	1.4713	0.0023
Specific rotation, $[\alpha]$ (°)	−61.8	0.8
Subjective color	Light-yellow	
RGB color values	R:255, G:255, B:224	
CMYK color values	C:0, M:0, Y:0.12, K:0	

[a] Mean of nine determinations: three distillations × three collections, [b] Standard deviation.

Table 2. Chemical composition of essential oil from *Annona cherimola* leaves.

CN	RT	Compounds	RI	RI [ref]	A. chirimola EO %	SD	Type	CF	MM (Da)
1	5.73	α-Thujene	924	924	0.89	0.15	ALM	$C_{10}H_{16}$	136.1
2	5.94	α-Pinene	932	932	4.02	0.36	ALM	$C_{10}H_{16}$	136.1
3	7.37	Sabinene	969	969	9.05	1.69	ALM	$C_{10}H_{16}$	136.1
4	7.50	β-Pinene	973	974	7.93	0.68	ALM	$C_{10}H_{16}$	136.1
5	8.13	Myrcene	987	988	0.78	0.19	ALM	$C_{10}H_{16}$	136.1
6	8.72	α-Phellandrene	1001	1002	0.11	0.03	ALM	$C_{10}H_{16}$	136.1
7	9.15	α-Terpinene	1013	1014	0.14	0.04	ALM	$C_{10}H_{16}$	136.1
8, 9	9.64	Limonene + β-Phellandrene [a]	1023	1024	0.66	0.13	ALM	$C_{10}H_{16}$	136.1
10	10.08	(Z)-β-Ocimene	1032	1032	0.84	0.23	ALM	$C_{10}H_{16}$	136.1
11	10.54	(E)-β-Ocimene	1043	1044	0.73	0.12	ALM	$C_{10}H_{16}$	136.1
12	10.93	γ-Terpinene	1053	1054	0.46	0.13	ALM	$C_{10}H_{16}$	136.1
13	12.15	Terpinolene	1083	1086	0.07	0.01	ALM	$C_{10}H_{16}$	136.1
14	13.10	Linalool	1099	1095	0.17	0.01	OXM	$C_{10}H_{18}O$	154.1
15	16.55	Terpinen-4-ol	1176	1174	0.07	0.01	OXM	$C_{10}H_{18}O$	154.1
16	23.00	Bicycloelemene	1332	1330	0.33	0.07	ALS	$C_{15}H_{24}$	204.2
17	23.17	δ-Elemene	1337	1335	1.26	0.41	ALS	$C_{15}H_{24}$	204.2
18	23.66	α-Cubebene	1348	1345	0.46	0.11	ALS	$C_{15}H_{24}$	204.2
19	24.81	α-Copaene	1377	1374	3.06	0.77	ALS	$C_{15}H_{24}$	204.2
20	25.17	β-Panasinsene	1384	1381	0.14	0.04	ALS	$C_{15}H_{24}$	204.2
21	25.38	β-Cubebene	1389	1387	1.38	0.27	ALS	$C_{15}H_{24}$	204.2
22	25.49	β-Elemene	1391	1389	2.41	0.62	ALS	$C_{15}H_{24}$	204.2
23	25.99	(Z)-Caryophyllene	1406	1408	0.04	0.01	ALS	$C_{15}H_{24}$	204.2
24	26.59	(E)-Caryophyllene	1419	1417	10.52	1.64	ALS	$C_{15}H_{24}$	204.2
25	27.05	β-Gurjunene	1429	1431	0.10	0.02	ALS	$C_{15}H_{24}$	204.2
26	27.16	γ-Elemene	1432	1434	0.20	0.04	ALS	$C_{15}H_{24}$	204.2
27	27.59	α-Guaiene	1439	1437	0.12	0.03	ALS	$C_{15}H_{24}$	204.2
28	27.86	Aromadendrene	1443	1439	0.33	0.08	ALS	$C_{15}H_{24}$	204.2
29	28.06	α-Humulene	1453	1452	2.05	0.51	ALS	$C_{15}H_{24}$	204.2
30	28.22	allo-Aromadendrene	1457	1458	0.20	0.03	ALS	$C_{15}H_{24}$	204.2

Table 2. *Cont.*

CN	RT	Compounds	RI	RI [ref]	*A. chirimola* EO %	*A. chirimola* EO SD	Type	CF	MM (Da)
31	28.78	4,5-di-epi-Aristolochene	1470	1471	tr	-	ALS	$C_{15}H_{24}$	204.2
32	29.01	γ-Gurjunene	1475	1475	0.45	0.07	ALS	$C_{15}H_{24}$	204.2
33	29.20	Germacrene D	1478	1480	28.77	3.80	ALS	$C_{15}H_{24}$	204.2
34	29.46	β-Selinene	1486	1489	0.38	0.12	ALS	$C_{15}H_{24}$	204.2
35	29.56	γ-Amorphene	1493	1495	0.20	0.08	ALS	$C_{15}H_{24}$	204.2
36	29.74	Bicyclogermacrene	1497	1500	11.12	1.39	ALS	$C_{15}H_{24}$	204.2
37	29.96	α-Muurolene	1501	1500	0.39	0.07	ALS	$C_{15}H_{24}$	204.2
38	30.17	Germacrene A	1506	1508	1.37	0.35	ALS	$C_{15}H_{24}$	204.2
39	30.49	δ-Amorphene	1510	1511	0.41	0.07	ALS	$C_{15}H_{24}$	204.2
40	30.63	γ-Cadinene	1512	1513	0.37	0.11	ALS	$C_{15}H_{24}$	204.2
41	30.73	δ-Cadinene	1517	1522	1.79	0.39	ALS	$C_{15}H_{24}$	204.2
42	31.26	trans-Cadina-1,4-diene	1530	1533	0.09	0.02	ALS	$C_{15}H_{24}$	204.2
43	32.03	Elemol	1546	1548	0.13	0.07	OXS	$C_{15}H_{26}O$	222.2
44	32.14	Germacrene B	1554	1559	1.45	0.24	ALS	$C_{15}H_{24}$	204.2
45	32.75	(E)-Nerolidol	1563	1561	0.54	0.08	OXS	$C_{15}H_{26}O$	222.2
46	32.97	Germacrene D-4-ol	1571	1574	1.41	0.42	OXS	$C_{15}H_{26}O$	222.2
47	33.45	Carotol	1590	1594	0.08	0.02	OXS	$C_{15}H_{26}O$	222.2
48	33.70	Ledol	1597	1602	1.12	0.17	OXS	$C_{15}H_{26}O$	222.2
49	34.75	1-epi-Cubenol	1621	1627	0.18	0.02	OXS	$C_{15}H_{26}O$	222.2
50	35.52	cis-Cadin-4-en-7-ol	1629	1635	0.12	0.01	OXS	$C_{15}H_{26}O$	222.2
51	35.59	epi-α-Cadinol	1632	1638	0.33	0.08	OXS	$C_{15}H_{26}O$	222.2
52	36.00	α-Muurolol (=Torreyol)	1640	1644	0.56	0.14	OXS	$C_{15}H_{26}O$	222.2
53	36.61	α-Cadinol	1653	1652	tr	-	OXS	$C_{15}H_{26}O$	222.2
		ALM			25.68				
		OXM			0.24				
		ALS			69.40				
		OXS			4.47				
		Total identified			99.80				

RT: Retention Time; RI: Calculated Retention Indices; RI [ref]: References Retention Indices; SD: Standard Deviation; CF: Chemical Formula; MM: Monoisotopic Mass; tr: traces; -: not calculated. [1] Co-eluted compounds.

Coeluting compounds (limonene and β-phellandrene) were separated using an HP-INNOWax polar column. The retention index in this column for limonene was 1192 and for β-phellandrene was 1201. Limonene (mixture of (+)-limonene and (−)-limonene) presented a percentage of 0.55 ± 0.09% and β-phellandrene a value of 0.12 ± 0.01%.

2.3. Enantioselective Analysis

The enantioselective analysis from *Annona cherimola* EO was achieved for the first time. The Table 3 shows the enantiomeric distribution, linear retention indices and enantiomeric excess (e.e.) of each enantiomer. Using a chiral column could be quantified four pairs of enantiomers, whose peaks were well separated at the base. The β-pinene (−) was found practically pure, while (−)-α-pinene and (−)-sabinene and (−)-germacrene D exhibited a high enantiomeric excess, whereas (−)-limonene and (+)-limonene were found in a racemic mixture.

2.4. Antibacterial Activity

Microdilution broth method was used to assess the antibacterial activity of essential oil of *A. cherimola* leaves. Tetracycline was used as a positive control and the maximum evaluated concentration was 1000 μg/mL. The minimum inhibitory concentration (MIC) values and the microorganisms used (seven Gram-negative bacteria and one Gram-positive bacterium) are shown in Table 4. The *A. cherimola* essential oil reported MIC values of 500 μg/mL against *Campylobacter jejuni* (ATCC 33560). EO dissolved in aqueous media caused the formation of an emulsion that difficulted the visual observation of bacterial growth particularly with *C. jejuni*. The reduction of 2,3,5-Triphenyl tetrazolium chloride

(TTZ) yield a red color product only in the wells were bacteria developed a well growth. A blank of EO with the same range of concentrations and media was included to discard interferences due to contamination which was also confirmed by reading at 405 nm (data not shown) For the other bacteria, the essential oil did not show activity at the maximum dose tested.

Table 3. Enantiomeric distribution of chiral constituents occurring in the EO of *A. cherimola*.

Enantiomers	RT min	RI	Enantiomeric Distribution %	e.e. %
(+)-α-Pinene (1R,5R)	4.60	916	18.01	63.99
(−)-α-Pinene (1S,5S)	5.10	921	81.99	
(−)-β-Pinene (1S,5S)	10.01	970	100.00	100.00
(+)-Sabinene (1R,5R)	10.51	975	2.05	95.91
(−)-Sabinene (1S,5S)	10.91	979	97.95	
(−)-Limonene (4S)	16.72	1037	63.75	27.50
(+)-Limonene (4R)	17.22	1042	36.25	
(+)-Germacrene D (8R)	61.61	1485	2.05	95.91
(−)-Germacrene D (8S)	62.01	1489	97.95	

Table 4. Antibacterial activity of essential oil from *Annona cherimola* leaves.

Bacteria	*Annona cherimola*	Positive Control [a]
	MIC (μg/mL)	
Gram-negative		
Campylobacter jejuni (ATCC 33560)	500	15.65
Escherichia coli (ATCC 25922)	>1000	1.95
Klebsiella pneumoniae (ATCC 9997)	500	1.95
Proteus vulgaris (ATCC 8427)	>1000	7.81
Pseudomonas aeruginosa (ATCC 27853)	>1000	15.62
Salmonella typhimurium (LT2)	>1000	3.90
Salmonella enterica (ATCC 29212)	>1000	1.95
Gram-positive		
Staphylococcus aureus (ATCC 25923)	>1000	1.95

[a] Erythromycin for *Campylobacter jejuni* and tetracycline for other bacteria.

2.5. Antioxidant Capacity

The results obtained for DPPH and ABTS radical scavenging of the EO are shown in Table 5. The results are expressed as the concentration of the EO that scavenge or decrease the concentration of the radical at 50% (SC_{50}). Trolox was used as a positive control.

Table 5. Antioxidant activity of essential oils of *Annona cherimola*.

Sample	DPPH	ABTS
	SC_{50} (μg/mL)	
Essential oil	470 ± 30	>1000
Trolox	232 ± 20	446 ± 30

Through the DPPH method, the essential oils of *A. cherimola* showed strong antioxidant activity with a SC_{50} value of 470 ± 30 μg/mL. Employing the ABTS technique the SC_{50} could not be calculated at the concentration ranges tested (Figure 1).

2.6. Anticholinesterase Activity

Three different concentrations of the essential oil from *A. cherimola* leaves were used to determine its anticholinesterase potential. The data obtained by measuring the rate of reaction of AChE against EO are shown in Figure 2. The results plotted as Log (concentration essential oil) vs. normalized response rate of reaction allowed us to calculate the IC_{50} value. The IC_{50} value obtained for chirimoya essential oil was 41.51 ± 1.02 µg/mL. The positive control (donepezil) exhibited an IC_{50} value of 13.80 ± 1.01 nM.

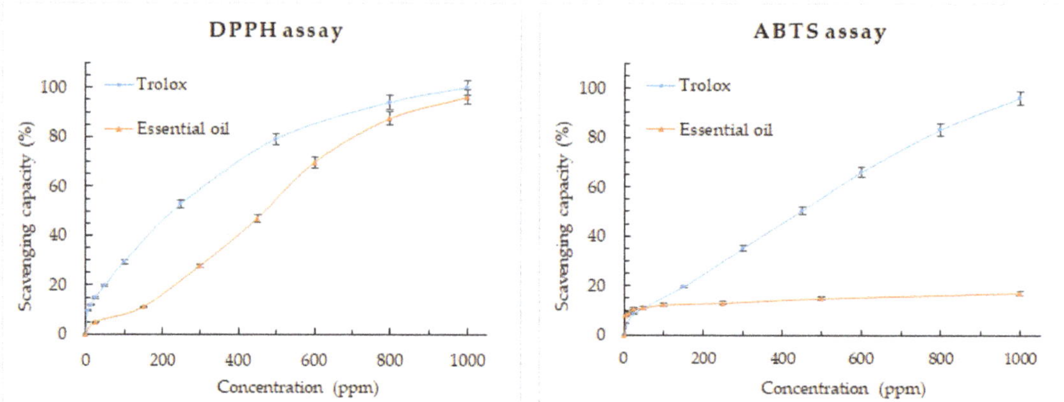

Figure 1. Scavenging capacity vs. concentration of *Annona cherimola* essential oil obtained by DPPH and ABTS assays.

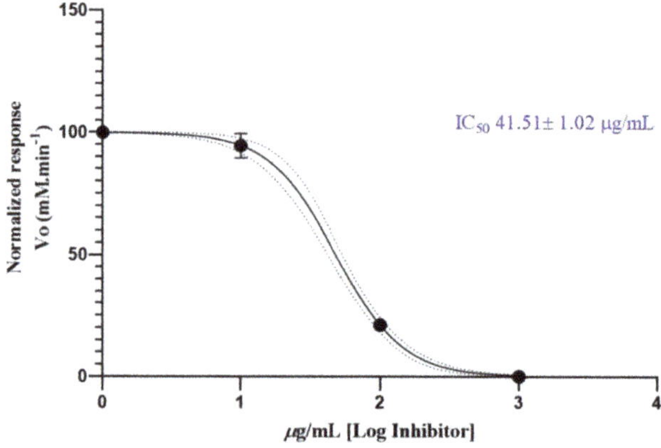

Figure 2. Half-maximum inhibitory concentration of *Annona cherimola* essential oil against acethylcholinesterase.

3. Discussion

The essential oil from *Annona cherimola* exhibited a low yield of 2.5 ± 0.2 mL/Kg [17]. The extraction yield of essential oils is very variable between plant species and depends on different aspects related to the plant such as the part, the age and the time after plant collection and other aspects related to the isolation process such as the pretreatment of the material (drying, grinding, etc.) and the extraction time [18].

The aroma of the *Annona* species is well recognized and has been studied in some species, however, little has been reported about the essential oil composition of *Annona cherimola*. In the present study, the main chemical components identified were aliphatic monoterpenes (25.68%) and aliphatic sesquiterpenes (69.40%), which was similar to the information reported by Rabelo et al. [19]. Furthermore, Rios et al. in 2003 [20] reported monoterpenes (6.09%) and sesquiterpenes (76.56%) as the main type of compounds in the

A. cherimola EO. On the other hand, the same type of volatile compounds were meaningful in fruits of *Annona cherimola* (monoterpene 40.3% and sesquiterpene 24.3%) [16].

The major components (>5%) identified in the *A. cherimola* EO were germacrene D (28.77%), bicyclogermacrene (11.12%), (E)-caryophyllene (10.52%), sabinene (9.05%) and β-pinene (7.93%). The results are different to those reported by Elhawary et al. fβ-caryophyllene with 9.50%, germacrene-D with 17.71% an β-elemene with 25.02% [21], and those reported by Rios et al. reported bicyclogermacrene (18.20%), trans-caryophyllene (11.50%), α-amorphene (7.57%), α-copaene (5.63%) and germacrene D (3.75%) [20]. In addition, Pino observed that the major compounds were α-thujene (18.7 ppm), α-pinene (23 ppm), terpinen-4-ol (19.8 ppm) and germacrene D (17.6 ppm) [16]. Despite the differences in their concentrations, the main component that is common in all the studies is germacrene D. It is well known the influence of different cultivation and climatic factors over the chemical composition of the essential oils.

Due to the relevance of aromatic compounds of the *Annona* species Ferreira et al. in 2009 compared the essential oil and the volatile compounds of the leaves and fruits of *Annona cherimola*. The chemical composition for the EO was different to the volatile compounds in fruits, the main compounds in the leaves essential oil were identified in lower quantities, germacrene-D (0.11% to 0.22%), sabinene (not identified), β-pinene (0.79% to 3.60%), (E)-caryophyllene (0.23% to 0.32%) and bicyclogermacrene (not identified) while the main compounds analyzed by headspace solid phase microextraction were methyl butanoate, butyl butanoate, 3-methylbutyl butanoate, 3-methylbutyl 3-methylbutanoate and 5-hydroxymethyl-2-furfural [22].

This is the first report of enantioselective GC-MS analysis of *A. cherimola* EO, this analysis showed the ocurrence of five pairs of enantiomers and one enantiomerically pure chiral monoterpenoid, β-pinene. The enantiomeric ratio of an essential oil is an important information which could be related with the biological activity, metabolism and organoleptic quality of the enantiomeric pairs [23]. The enantiomeric excess (e.e %) were (−)-α-pinene (1S,5S) (63.99%), (−)-sabinene (1S,5S) (95.91%), (−)-limonene (4R) (27.50%) and (−)-germacrene D (8S) (95.91%).

Regarding their biological activity, the essential oil of *Annona cherimola* showed moderate antibacterial activity against *Campylobacter jejuni* (ATCC 33560) and *Klebsiella pneumonia* (ATCC 9997), both with MIC at 500 µg/mL and no activity for the other bacteria tested (MIC was higher than 1000 µg/mL). Compared to data reported in the literature, Rios et al. in 2003 reported a significant activity against two Gram-positive bacteria *Staphylococcus aureus* (MIC 250 µg/mL) and *Enterococcus faecalis* (MIC 500 µg/mL), however, the MIC values for Gram-negative bacteria were higher than 5000 µg/mL [20]. Elhawary et al. in 2013 reported the MIC of EO *A. cherimola* for *Bacillus subtilis* (130 µg/mL), *Staphylococcus aureus* (285 µg/mL), *Escherichia coli* (110 µg/mL), *Pseudomonas aeruginosa* (140 µg/mL), and *Candida albicans* (152 µg/mL) [21]. When the antibacterial activity of pure compounds was analyzed [20] the MIC of trans-caryophyllene, β-pinene, linalool, and other compounds was higher than the value for the essential oil, therefore suggesting that the antibacterial potency could be exerted by a synergistic effect among the constituents above mentioned. The essential oil of *Annona* species showed a wide range of biological activity, for *A. vepretorum* Costa et al. in 2012 reported a moderate activity (MIC 500 µg/mL) against *Staphylococcus aureus* and *Staphylococcus epidermis* and a significant activity against *Candida tropicalis* (MIC 100 µg/mL) [24]. Another study, in 2013, Costa et al. observed the antibacterial activity of essential oil of *A. salzmannii* and *A. pickelii* against *Staphylococcus aureus*, *Staphylococcus epidermis* and *Candida tropicalis* with MIC of 500 µg/mL [25].

Some studies have shown that the enantiomers of a compound have different biological activities. Lis-Balcnin et al. reported that 18 out of 25 different bacteria were more affected by the (−)-α-pinene in comparison with the (+) enantiomer, 19 out of 20 different *Listeria monocytogenes* strains were affected more by (+)-α-pinene isomer and two of three filamentous fungi were affected more by the (+) enantiomer [26]. The MIC and minimal microbicidal concentration (MMC) showed that the positive enantiomers of pinene exerted a microbicidal effect against all the fungi and bacteria tested with MIC values ranging from 117 to 4150 µg/mL. However, with concentrations up to 20 mg/mL of the negative enantiomers, no antimicrobial activity was observed [27]. The MIC values against three Gram-positive (*B. cereus*, *E. faecalis* and *S. aureus*) and four Gram-negative (*E. coli*, *K. pneumoniae*, *M. catarrhalis* and *P. aeruginosa*) bacteria were in the ranges of 3 to 27 mg/mL for (+)-limonene and 2 to 27 mg/mL for (−)-limonene. The greatest difference was obtained against *Staphylococcus aureus* ATCC 12600 where the (+)-limonene showed a MIC of 14 mg/mL and the (−)-limonene a MIC of 4 mg/mL [28]. Omran et al. found that (−)-limonene had better antifungal activity than (+)-limonene against *Aspergillus niger*, *Aspergillus* sp., *Candida albicans* and *Penicillium* sp. [29]. It was not possible to find previous studies about the antifungal or antibacterial activity of the (+) and (−) enantiomers of the main compound germacrene D, however, Stranden et al. determined that the two enantiomers of this compound mediate the same kind of information to the receptor neurons of the moth *Helicoverpa armigera*, but (−)-germacrene D had approximately 10 times stronger effect than (+)-germacrene D [30]. The difference in biological activity of the enantiomers is maintained even when they are mixed with other compounds [28]. The enantiomers of a compound have different biological activities, then, the enantiomeric distribution of the compounds could influence the biological activity for an essential oil.

Regarding their antioxidant effect, the *Annona cherimola* essential oil showed an SC_{50} of 470 µg/mL for the DPPH assay while the SC_{50} was >1000 µg/mL in the ABTS assay. Costa et al. reported as strong the antioxidant activity of EO *A. salzamannii* and *A. pickelii* measured by a TLC-based DPPH assay, however, the individual components β-pinene and α-pinene did not show antioxidant activity [25]. Araújo, et al. [31] and Costa, et al. [24] reported a weak antioxidant activity for the EO of *A. vepretorum*. Another study, Gyesi, et al. in 2019 [32] reported an SC_{50} of 244.8 µg/mL from the DPPH assay for the EO of *A. muricata*. The differences between the antioxidant activity of EO and pure compounds could correspond to synergistic effects among the components in the essential oil.

The acetylcholinesterase inhibitory activity of *A. cherimola* EO has not been previously reported. Chirimoya EO showed an AChE IC_{50} value of 41.51 µg/mL, this inhibitory activity could be considered very strong compared to the related EO of *Piper carpunya* (IC_{50} of 36.42 µg/mL) [33]. The inhibition of AChE due to EO is of relevant interest in the treatment of Alzheimer disease since different studies report in vitro and clinical AChE inhibitory activity. Benny and Tomas summarize the neuroprotective effects of EO and its relevance on Alzheimer disease stating that EO could rebuild the antioxidant status of brain which confer neuroprotective effect as in the case of EO of *Coriandrum sativum* L., *Syzygium aromaticum* (L.), *Juniperus communis*, *Rosmarinis officinalis* (L.), and other species. The same activity has been observed for pure compounds such as thymol, linalool, α-terpinene, α-terpineol, carvacrol, (E)-β-caryophyllene, α-pinene, and eugenol [34].

4. Materials and Methods

4.1. Materials

Dichloromethane (DMC), methanol (MeOH), sodium sulfate anhydrous, DPPH (2,2-diphenyl-1-picrylhydryl), ABTS (2,2′-azinobis-3-ethylbenzothiazoline-6-sulfonic acid), acetylcholinesterase (AChE), acetylthiocholine (ATC), phosphate buffered saline, Ellman's reagent (donepezil, 5,5′-dithiobis(2-nitrobenzoic acid)), tris hydrochloride (Tris-HCl), magnesium chloride hexahydrate and 2,3,5-Triphenyl tetrazolium chloride (TTZ) were purchased from Sigma-Aldrich (San Luis, MO, USA). The microbiological media such as Mueller Himton broth, Mueller Hinton II broth and fluid thioglycollate medium were purchased from DIPCO (Quito, Ecuador). Horse serum and Oxoid CampyGen were purchased from Thermo Fisher Scientific (Waltham, MA, USA). The standard aliphatic hydrocarbons were purchased from ChemService (West Chester, PA, USA). Helium was purchased from INDURA (Quito, Ecuador). All chemicals were of analytical grade and used without further purifications.

4.2. Plant Material

Annona cherimola leaves were collected under permission granted for Ministerio del Ambiente de Ecuador (Ecuadorian Environmental Ministry) by means of the authorization No. 001-IC-FLO-DBAP-VS-DRLZCH-MA. The leaves of chirimoya were collected between the months of October and January in the surroundings of the locality of Cango Bajo, canton Calvas, province of Loja, Ecuador, at 1950 m a.s.l. at a latitude of $4°20'34''$ S and a longitude of $79°34'24''$ W. Collection, store and transfer of chirimoya leaves were performed according to what is described by Valarezo et al. [35]. Botanist Nixon Cumbicus made the identification of the plant material. A voucher specimen was deposited at the Herbarium of Universidad Técnica Particular de Loja (HUTPL).

4.3. Postharvest Treatments

Between 2 and 3 h after being collected, the plant material was subjected to the postharvest treatments, which consist of the separation of degraded leaves and foreign material.

4.4. Moisture Determination

Method Loss on drying (Moisture) in plants, AOAC 930.04-1930, was used to determine the moisture of plant material, for this an analytical balance (Mettler AC 100, Mettler Toledo, Columbus, OH, USA) was used. Moisture was calculated according to Equation (1).

$$\text{Moisture } (\%) = \frac{\text{wi} - \text{wo}}{\text{wi}} * 100 \tag{1}$$

where wi is the initial weight of sample and wo is weight of sample after drying.

4.5. Essential Oil Extraction

The isolation of the essential oil from leaves of *A. cherimola* was carried out by hydrodistillation using a Clevenger-type apparatus according to the procedure described by Valarezo et al. [36]. After being collected, the essential oil was dried using anhydrous sodium sulphate. Finally, the EO was stored at 4 °C in amber sealed vials until being used in the subsequent analysis.

4.6. Determination of the Physical Properties of the Essential Oil

Density of the essential oil was determined using the ISO 279:1998 standard (equivalent to the AFNOR NF T 75-111 standard). Density measurement was performed using an analytical balance (Mettler AC 100, Mettler Toledo, Columbus, OH, USA) and a pycnometer of 1 mL. Refractive index was determined using the standard ISO 280:1998 [37] (similary to AFNOR NF T 75-112), for which a refractometer (model ABBE, BOECO, Hamburg, Germany) was used. An automatic polarimeter (Mrc-P810, MRC, Holon, Israel) was used to measure the optical rotation of the EO according to the standard ISO 592:1998. All measurements were taken at 20 °C.

4.7. Identification of the Chemical Constituents of the Essential Oil

4.7.1. Quantitative and Qualitative Analysis

The dentification of the chemical constituents of the essential oil was carried out using an Agilent gas chromatograph (GC) (6890N series, Agilent Technologies, Santa Clara, CA, USA). For the quantitative analysis gas chromatograph was equipped with a flame ionization detector (FID) and for qualitative analysis gas chromatograph was coupled to a mass spectrometer (quadrupole) detector (MS) (model Agilent 5973 inert series, Agilent Technologies, Santa Clara, CA, USA). The GC-FID and GC-MS analyses were performed according to the procedure described by Valarezo et al. [35]. The injection of the samples was carried out by an automatic injector (Agilent 7683 automatic liquid sampler, Agilent Technologies, Santa Clara, CA, USA) in split mode. Chromatographic runs were performed using a nonpolar and a polar column. The nonpolar was an Agilent J&W DB-5ms Ultra Inert GC column with stationary phase 5%-phenyl-methylpolyxilosane and the polar was an Agilent J&W HP-INNOWax GC column with stationary phase polyethylene glycol. Both columns with a length of 30 m, an outer diameter of 0.25 mm and a stationary phase thickness of 0.25 μm. Identification of the EO compounds was based on a comparison of relative retention indices (RIs) and mass spectra data with those of the published literature [38,39] according as described by Valarezo et al. [36].

4.7.2. Enantioselective Analysis

The enantiomeric distribution was performed using an enantioselective column with stationary phase 2,3-diethyl-6-tert-butyldimethylsilyl-β-cyclodextrin. The chromatographic run was performed with a temperature ramp of 2 °C/min from 50 °C (maintained for 2 min) to 220 °C (maintained for 2 min) in an Agilent gas chromatograph (model 6890N series, Agilent Technologies, Santa Clara, CA, USA) coupled to a mass spectrometer (quadrupole) detector (model Agilent 5973 inert series, Agilent Technologies, Santa Clara, CA, USA). The injection of enantiomerically pure standards was used to determine the order of elution of the enantiomers.

4.8. Evaluation of Antibacterial Activity

The antibacterial activity of the essential oil from chirimoya leaves was assessed against Gram-negative and Gram-positive bacteria (Table 4) by the microdilution broth method according to the procedure described by Valarezo et al. [40]. The bacterial strains were incubated in Müeller-Hinton (MH) broth. Tetracycline was used as a positive control and DMSO was used as a negative control. Results are reported as minimum inhibitory concentration (MIC).

For *Campylobacter jejuni* (ATCC 33560) the broth microdilution method was carried out according to Valarezo et al. [40] with some specific requirements as described briefly. Fluid thioglycollate medium was used for reactivation of the strain, supplemented with 5% of Horse serum. An aliquot of a cryogenic reserve was resuspended in thioglycollate and incubated for 48 h at 37 °C in a microaerophilic atmosphere provided by an Oxoid CampyGen (2.5 L sachet). Sample solutions were made by dissolving 80 mg of EO in 1 mL of DMSO. Two-fold serial dilutions were employed to obtain decreasing concentrations of EO from 4000 to 31.25 μg/mL and cation-adjusted Muller Hinton II broth (pH 7.3) with 5% lyssed horse blood [41] as media for the antibacterial assay. The inoculum was prepared from thioglycollate culture and adjusted to 0.5 McFarland. Final concentration of bacteria was 5×10^5 CFU/mL. The microplate was incubated for 48 h at 37 °C in microaerophilic atmosphere (5% CO_2, Oxoid CampyGen). Erythromycin was used as positive control with a MIC value of 15.65 μg/mL and DMSO as negative control. MIC was determined by visual examination of growth and through addition of a 1% solution of TTZ as bacterial viability indicator after incubation time to confirm the visual results. A blank with the same range of concentrations of EO was prepared simultaneously and measurements at 405 nm were made to discard reduction of TTZ by contamination. Optical density (OD) for TTZ reduction was measured in a microplate reader (EPOCH 2, BioTek, Winooski, VT, USA).

4.9. Antioxidant Capacity

4.9.1. DPPH Radical Scavenging Capacity

The DPPH assay was performed using 2,2-diphenyl-1-picrylhydryl free radical (DPPH•) based on the technique described by Brand Williams et al. [42] and Thaipong et al. [43] according to what was described by Valarezo et al. [33]. The concentrations of the EO from *A. cherimola leaves* used were 1000, 800, 600, 450, 300, 150 and 25 ppm. Trolox was used as a positive control and methanol as a blank control. The samples were evaluated in a UV spectrophotometer (Genesys 10S UV.Vis Spectrophotometer, Thermo Scientific, Waltham, MA, USA) at a wavelength of 515 nm. The percentage of scavenging capacity was calculated according to Equation (2). SC_{50} is the EO concentration that provided 50% $DPPH^\bullet$ scavenging effect.

$$SC(\%) = \frac{(AEO - AMeOH)}{AEO} * 100 \tag{2}$$

where AEO is the absorbance of $DPPH^\bullet$ mixed with EO and As is absorbance of DPPH mixed with methanol.

4.9.2. ABTS Radical Cation Scavenging Capacity

The ABTS assay was performed using 2,2′-azinobis-3-ethylbenzothiazoline-6-sulfonic acid radical cation ($ABTS^{\bullet+}$) according to the procedure report by Arnao et al. [44] and Thaipong et al. [43], as described by Valarezo et al. [33]. The concentrations of the essential oil from *A. cherimola* used were 1000, 500, 250, 100, 50 and 25 ppm. The samples were evaluated in a UV spectrophotometer (Genesys 10S UV.Vis Spectrophotometer, Thermo Scientific, Waltham, MA, USA) at a wavelength of 734 nm. Deionized water was used as a blank control and trolox was used as a positive control. The percentage of scavenging capacity was calculated according to equation.

$$Sc(\%) = \frac{(ASO - ASA)}{ASA} * 100 \tag{3}$$

where ASO is the absorbance of $ABTS^{\bullet+}$ with solvent mixture and Ai is the absorbance after reaction of $ABTS^{\bullet+}$ with the sample.

4.10. Anticholinesterase Activity

The AChE inhibitory effect was measured based on the methodology designed by Ellman, et al. [45], with slight modifications as suggested by Rhee, et al. [46], as previously described by Valarezo et al. [33]. The inhibition of AChE was detected after the addition of Acetylthiocholine as the enzyme substrate and several concentrations of EO dissolved in MeOH. The enzyme reaction was monitored in a microplate reader (EPOCH 2, BioTek, Winooski, VT, USA) at 405 nm for 60 min. Final concentrations of 1000, 100, and 10 μg/mL of the EO in MeOH were prepared to assess the enzyme inhibition. The assay was carried out by triplicate in 96-well microplates. Donepezil was used as positive control. The IC_{50} value was calculated from the progression curve with the Graph Pad Prism software (v8.0.1.5., Graph Pad, San Diego, CA, USA)

4.11. Statistical Analysis

All procedures were repeated three times, except for the biological activity, which was repeated nine times. Microsoft Excel was used to collect the data and Minitab 17 (Version 17.1.0., Minitab LLC., State College, PA, USA) was used to calculate the measures of central tendency. The results are expressed as mean values.

5. Conclusions

Enantiomeric distribution, antibacterial activity against *Campylobacter jejuni*, antioxidant and anticholinesterase activities of the essential oil from *Annona cherimola* leaves were determined for the first time, in addition, the chemical composition and physical properties of this essential oil was studied. This research contributes to our knowledge about native species of Ecuador. The results obtained in this study motivate us to carry out new investigations in endemic and native species of this megadiverse country.

Author Contributions: Conceptualization, E.V. and V.M.; methodology, J.L., L.C., J.C. and M.A.M.; validation, E.V. and V.M.; formal analysis, J.L., E.E.-C., J.C., L.C. and M.A.M.; investigation, J.L., E.E.-C., M.A.M., L.C. and J.C.; resources, E.V.; data curation, V.M. and E.V.; writing—original draft preparation, M.A.M.; writing—review and editing, L.C. and E.V.; project management, V.M. and E.V. All authors have read and agreed to the published version of the manuscript.

Funding: This research received no external funding.

Institutional Review Board Statement: Not applicable.

Informed Consent Statement: Not applicable.

Data Availability Statement: Data are available from the authors upon reasonable request.

Acknowledgments: The authors are very grateful with the Ecuadorian Environmental Ministry (Ministerio del Ambiente de Ecuador, MAE) for the permission granted.

Conflicts of Interest: There is no conflict of interest to declare.

References

1. The Plant List. Annonaceae. Available online: http://www.theplantlist.org/ (accessed on 12 September 2021).
2. Jørgesen, P.M.; León-Yáñez, S. Catalogue of the Vascular Plants of Ecuador. Available online: http://legacy.tropicos.org/ProjectAdvSearch.aspx?projectid=2 (accessed on 11 July 2020).
3. Frausin, G.; Lima, R.B.S.; Hidalgo, A.F.; Maas, P.; Pohlit, A.M. Plants of the annonaceae traditionally used as antimalarials: A review. *Rev. Bras. Frutic.* **2014**, *36*, 315–337. [CrossRef]
4. Tajbakhsh, E.; Kwenti, T.E.; Kheyri, P.; Nezaratizade, S.; Lindsay, D.S.; Khamesipour, F. Antiplasmodial, antimalarial activities and toxicity of African medicinal plants: A systematic review of literature. *Malar. J.* **2021**, *20*, 349. [CrossRef] [PubMed]
5. Santos-Sánchez, N.F.; Salas-Coronado, R.; Hernández-Carlos, B.; Pérez-Herrera, A.; Rodríguez-Fernández, D.J. *Biological Activities of Plants from Genus Annona*; IntechOpen: London, UK, 2018; p. 168.
6. Moghadamtousi, S.Z.; Fadaeinasab, M.; Nikzad, S.; Mohan, G.; Ali, H.M.; Kadir, H.A. *Annona muricata* (Annonaceae): A Review of Its Traditional Uses, Isolated Acetogenins and Biological Activities. *Int. J. Mol. Sci.* **2015**, *16*, 15625. [CrossRef] [PubMed]
7. Chowdhury, S.S.; Tareq, A.M.; Tareq, S.M.; Farhad, S.; Sayeed, M.A. Screening of antidiabetic and antioxidant potential along with phytochemicals of *Annona* genus: A review. *Future J. Pharm. Sci.* **2021**, *7*, 144. [CrossRef]
8. Arunjyothi, B.; Venkatesh, K.; Chakrapani, P.; Anupalli, R.R. Phytochemical and Pharmacological potential of *Annona cherimola*—A Review. *Int. J. Phytomed.* **2012**, *3*, 9.
9. Gayoso Bazán, G.; Chang Chávez, L. *Annona cherimola* Mill. "chirimoya" (Annonaceae), una fruta utilizada como alimento en el Perú prehispánico. *Arnaldoa* **2017**, *24*, 619–634. [CrossRef]
10. Gupta-Elera, G.; Garrett, A.R.; Martinez, A.; Robison, R.A.; O'Neill, K.L. The antioxidant properties of the cherimoya (*Annona cherimola*) fruit. *Food Res. Int.* **2011**, *44*, 2205–2209. [CrossRef]
11. Chen, C.-Y.; Chang, F.-R.; Pan, W.-B.; Wu, Y.-C. Four alkaloids from Annona cherimola. *Phytochemistry* **2001**, *56*, 753–757. [CrossRef]
12. Chen, C.-Y.; Chang, F.-R.; Wu, Y.-C. Cherinonaine, a novel dimeric amide from the stems of Annona cherimola. *Tetrahedron Lett.* **1998**, *39*, 407–410. [CrossRef]
13. Díaz-de-Cerio, E.; Aguilera-Saez, L.M.; Gómez-Caravaca, A.M.; Verardo, V.; Fernández-Gutiérrez, A.; Fernández, I.; Arráez-Román, D. Characterization of bioactive compounds of Annona cherimola L. leaves using a combined approach based on HPLC-ESI-TOF-MS and NMR. *Anal. Bioanal. Chem.* **2018**, *410*, 3607–3619. [CrossRef]
14. Albuquerque, T.G.; Santos, F.; Sanches-Silva, A.; Beatriz Oliveira, M.; Bento, A.C.; Costa, H.S. Nutritional and phytochemical composition of Annona cherimola Mill. fruits and by-products: Potential health benefits. *Food Chem.* **2016**, *193*, 187–195. [CrossRef] [PubMed]
15. Ammoury, C.; Younes, M.; El Khoury, M.; Hodroj, M.H.; Haykal, T.; Nasr, P.; Sily, M.; Taleb, R.I.; Sarkis, R.; Khalife, R.; et al. The pro-apoptotic effect of a Terpene-rich Annona cherimola leaf extract on leukemic cell lines. *BMC Complement. Altern. Med.* **2019**, *19*, 365. [CrossRef] [PubMed]
16. Pino, J.A. Volatile Components of Cuban *Annona* Fruits. *J. Essent. Oil Res.* **2000**, *12*, 613–616. [CrossRef]
17. Molares, S.; González, S.B.; Ladio, A.; Agueda Castro, M. Etnobotánica, anatomía y caracterización físico-química del aceite esencial de Baccharis obovata Hook. et Arn. (Asteraceae: Astereae). *Acta Bot. Bras.* **2009**, *23*, 578–589. [CrossRef]
18. Valarezo, E.; Ojeda-Riascos, S.; Cartuche, L.; Andrade-González, N.; González-Sánchez, I.; Meneses, M.A. Extraction and Study of the Essential Oil of Copal (*Dacryodes peruviana*), an Amazonian Fruit with the Highest Yield Worldwide. *Plants* **2020**, *9*, 1658. [CrossRef]
19. Rabelo, S.V.; Quintans, J.d.S.S.; Costa, E.V.; Guedes da Silva Almeida, J.R.; Quintans Júnior, L.J. Chapter 24—*Annona* Species (Annonaceae) Oils. In *Essential Oils in Food Preservation, Flavor and Safety*; Preedy, V.R., Ed.; Academic Press: San Diego, CA, USA, 2016; pp. 221–229. [CrossRef]
20. Ríos, M.Y.; Castrejón, F.; Robledo, N.; León, I.; Rojas, G.; Navarro, V. Chemical Composition and Antimicrobial Activity of the Essential Oils from *Annona cherimola* (Annonaceae). *Rev. Soc. Quim. Mex.* **2003**, *47*, 139–142.
21. Elhaware, S.S.; El Tantawy, M.E.; Rabeh, M.A.; Fawaz, N.E. DNA fingerprinting, chemical composition, antitumor and antimicrobial activities of the essential oils and extractives of four Annona species from Egypt. *J. Nat. Sci. Res.* **2013**, *13*, 59–68.

22. Ferreira, L.; Perestrelo, R.; Câmara, J.S. Comparative analysis of the volatile fraction from *Annona cherimola* Mill. cultivars by solid-phase microextraction and gas chromatography—Quadrupole mass spectrometry detection. *Talanta* **2009**, *77*, 1087–1096. [CrossRef]

23. Barba, C.; Santa-María, G.; Herraiz, M.; Martínez, R.M. Direct enantiomeric analysis of Mentha essential oils. *Food Chem.* **2013**, *141*, 542–547. [CrossRef]

24. Costa, E.V.; Dutra, L.M.; Nogueira, P.C.d.L.; Moraes, V.R.D.S.; Salvador, M.J.; Ribeiro, L.H.G.; Gadelha, F.R. Essential Oil from the Leaves of *Annona vepretorum*: Chemical Composition and Bioactivity. *Nat. Prod. Commun.* **2012**, *7*, 1934578X1200700240. [CrossRef]

25. Costa, E.V.; Dutra, L.M.; Salvador, M.J.; Ribeiro, L.H.G.; Gadelha, F.R.; de Carvalho, J.E. Chemical composition of the essential oils of *Annona pickelii* and *Annona salzmannii* (Annonaceae), and their antitumour and trypanocidal activities. *Nat. Prod. Res.* **2013**, *27*, 997–1001. [CrossRef] [PubMed]

26. Lis-Balcnin, M.; Ochocka, R.J.; Deans, S.G.; Asztemborska, M.; Hart, S. Differences in Bioactivity between the Enantiomers of α-Pinene. *J. Essent. Oil Res.* **1999**, *11*, 393–397. [CrossRef]

27. Silva, A.C.R.d.; Lopes, P.M.; Azevedo, M.M.B.d.; Costa, D.C.M.; Alviano, C.S.; Alviano, D.S. Biological Activities of a-Pinene and β-Pinene Enantiomers. *Molecules* **2012**, *17*, 6305–6316. [CrossRef] [PubMed]

28. Vuuren, S.F.v.; Viljoen, A.M. Antimicrobial activity of limonene enantiomers and 1,8-cineole alone and in combination. *Flavour Fragr. J.* **2007**, *22*, 540–544. [CrossRef]

29. Mahdavi Omran, S.; Moodi, M.A.; Norozian Amiri, S.M.B.; Mosavi, S.J.; Ghazi Mir Saeed, S.A.M.; Jabbari Shiade, S.M.; Kheradi, E.; Salehi, M. The Effects of Limonene and Orange Peel Extracts on Some Spoilage Fungi. *Int. J. Mol. Clin. Microbiol.* **2011**, *1*, 82–86.

30. Stranden, M.; Borg-Karlson, A.-K.; Mustaparta, H. Receptor Neuron Discrimination of the Germacrene D Enantiomers in the Moth *Helicoverpa armigera*. *Chem. Senses* **2002**, *27*, 143–152. [CrossRef]

31. Araújo, C.d.S.; de Oliveira, A.P.; Lima, R.N.; Alves, P.B.; Diniz, T.C.; da Silva, A.J.R.G. Chemical constituents and antioxidant activity of the essential oil from leaves of *Annona vepretorum* Mart. (Annonaceae). *Pharmacogn. Mag.* **2015**, *11*, 615–618. [CrossRef]

32. Gyesi, J.N.; Opoku, R.; Borquaye, L.S. Chemical composition, total phenolic content, and antioxidant activities of the essential oils of the leaves and fruit pulp of *Annona muricata* L. (Soursop) from Ghana. *Biochem. Res. Int.* **2019**, *2019*, 4164576. [CrossRef]

33. Valarezo, E.; Rivera, J.X.; Coronel, E.; Barzallo, M.A.; Calva, J.; Cartuche, L.; Meneses, M.A. Study of Volatile Secondary Metabolites Present in *Piper carpunya* Leaves and in the Traditional Ecuadorian Beverage Guaviduca. *Plants* **2021**, *10*, 338. [CrossRef]

34. Benny, A.; Thomas, J. Essential oils as treatment strategy for Alzheimer's disease: Current and future perspectives. *Planta Med.* **2019**, *85*, 239–248. [CrossRef]

35. Valarezo, E.; Morocho, V.; Cartuche, L.; Chamba-Granda, F.; Correa-Conza, M.; Jaramillo-Fierro, X.; Meneses, M.A. Variability of the Chemical Composition and Bioactivity between the Essential Oils Isolated from Male and Female Specimens of Hedyosmum racemosum (Ruiz & Pav.) G. Don. *Molecules* **2021**, *26*, 4613. [CrossRef]

36. Valarezo, E.; Gaona-Granda, G.; Morocho, V.; Cartuche, L.; Calva, J.; Meneses, M.A. Chemical Constituents of the Essential Oil from Ecuadorian Endemic Species Croton ferrugineus and Its Antimicrobial, Antioxidant and α-Glucosidase Inhibitory Activity. *Molecules* **2021**, *26*, 4608. [CrossRef] [PubMed]

37. Association Française de Normalisation (AFNOR). *Huiles Essentielles. Tome 1, Échantillonnage et Méthodes D'analyse*; Paris-La Défense, AFNOR: Paris, France, 2000; p. 471.

38. Adams, R.P. *Identification of Essential Oil Components by Gas Chromatography/Mass Spectrometry*, 4th ed.; Allured Publishing Corporation: Carol Stream, IL, USA, 2007; p. 804.

39. NIST. Libro del Web de Química del NIST, SRD 69. in Base de Datos de Referencia Estándar del NIST Número 69. Available online: http://webbook.nist.gov (accessed on 19 March 2021).

40. Valarezo, E.; Merino, G.; Cruz-Erazo, C.; Cartuche, L. Bioactivity evaluation of the native Amazonian species of Ecuador: *Piper lineatum* Ruiz & Pav. essential oil. *Nat. Volatiles Essent. Oils* **2020**, *7*, 14–25. [CrossRef]

41. Ge, B.; Wang, F.; Sjölund-Karlsson, M.; McDermott, P.F. Antimicrobial resistance in Campylobacter: Susceptibility testing methods and resistance trends. *J. Microbiol. Methods* **2013**, *95*, 57–67. [CrossRef] [PubMed]

42. Brand-Williams, W.; Cuvelier, M.E.; Berset, C. Use of a free radical method to evaluate antioxidant activity. *LWT Food Sci. Technol.* **1995**, *28*, 25–30. [CrossRef]

43. Thaipong, K.; Boonprakob, U.; Crosby, K.; Cisneros-Zevallos, L.; Hawkins Byrne, D. Comparison of ABTS, DPPH, FRAP, and ORAC assays for estimating antioxidant activity from guava fruit extracts. *J. Food Compos. Anal.* **2006**, *19*, 669–675. [CrossRef]

44. Arnao, M.B.; Cano, A.; Acosta, M. The hydrophilic and lipophilic contribution to total antioxidant activity. *Food Chem.* **2001**, *73*, 239–244. [CrossRef]

45. Ellman, G.L.; Courtney, K.D.; Andres, V.; Featherstone, R.M. A new and rapid colorimetric determination of acetylcholinesterase activity. *Biochem. Pharmacol.* **1961**, *7*, 88–95. [CrossRef]

46. Rhee, I.K.; van de Meent, M.; Ingkaninan, K.; Verpoorte, R. Screening for acetylcholinesterase inhibitors from Amaryllidaceae using silica gel thin-layer chromatography in combination with bioactivity staining. *J. Chromatogr. A* **2001**, *915*, 217–223. [CrossRef]

Article

Chemical Composition, Allelopathic, Antioxidant, and Anti-Inflammatory Activities of Sesquiterpenes Rich Essential Oil of *Cleome amblyocarpa* Barratte & Murb.

Ahmed M. Abd-ElGawad [1,2,*], **Abdelbaset M. Elgamal** [3], **Yasser A. EI-Amier** [1], **Tarik A. Mohamed** [4], **Abd El-Nasser G. El Gendy** [5] and **Abdelsamed I. Elshamy** [6,*]

[1] Department of Botany, Faculty of Science, Mansoura University, Mansoura 35516, Egypt; yasran@mans.edu.eg
[2] Plant Production Department, College of Food & Agriculture Sciences, King Saud University, P.O. Box 2460, Riyadh 11451, Saudi Arabia
[3] Department of Chemistry of Microbial and Natural Products, National Research Centre, 33 El-Bohouth St., Dokki, Giza 12622, Egypt; algamalnrc_2010@nrc.sci.eg
[4] Chemistry of Medicinal Plants Department, National Research Centre, 33 El-Bohouth St., Dokki, Giza 12622, Egypt; ta.mourad@nrc.sci.eg
[5] Medicinal and Aromatic Plants Research Department, National Research Centre, 33 El Bohouth St., Dokki, Giza 12622, Egypt; ag.el-gendy@nrc.sci.eg
[6] Chemistry of Natural Compounds Department, National Research Centre, 33 El Bohouth St., Dokki, Giza 12622, Egypt
[*] Correspondence: aibrahim2@ksu.edu.sa (A.M.A.-E.); ai.el-shamy@nrc.sci.eg (A.I.E.); Tel.: +00966562680864 (A.M.A.-E.); +201005525108 (A.I.E.)

Citation: Abd-ElGawad, A.M.; Elgamal, A.M.; EI-Amier, Y.A.; Mohamed, T.A.; El Gendy, A.E.-N.G.; I. Elshamy, A. Chemical Composition, Allelopathic, Antioxidant, and Anti-Inflammatory Activities of Sesquiterpenes Rich Essential Oil of *Cleome amblyocarpa* Barratte & Murb. *Plants* **2021**, *10*, 1294. https://doi.org/10.3390/plants10071294

Academic Editors: Hazem Salaheldin Elshafie, Laura De Martino and Adriano Sofo

Received: 25 May 2021
Accepted: 23 June 2021
Published: 25 June 2021

Publisher's Note: MDPI stays neutral with regard to jurisdictional claims in published maps and institutional affiliations.

Abstract: The integration of green natural chemical resources in agricultural, industrial, and pharmaceutical applications allures researchers and scientistic worldwide. *Cleome amblyocarpa* has been reported as an important medicinal plant. However, its essential oil (EO) has not been well studied; therefore, the present study aimed to characterize the chemical composition of the *C. amblyocarpa*, collected from Egypt, and assess the allelopathic, antioxidant, and anti-inflammatory activities of its EO. The EO of *C. amblyocarpa* was extracted by hydrodistillation and characterized via gas chromatography–mass spectrometry (GC-MS). The chemometric analysis of the EO composition of the present studied ecospecies and the other reported ecospecies was studied. The allelopathic activity of the EO was evaluated against the weed *Dactyloctenium aegyptium*. Additionally, antioxidant and anti-inflammatory activities were determined. Forty-eight compounds, with a prespondence of sesquiterpenes, were recorded. The major compounds were caryophyllene oxide (36.01%), hexahydrofarnesyl acetone (7.92%), alloaromadendrene epoxide (6.17%), myrtenyl acetate (5.73%), isoshyobunone (4.52%), shyobunol (4.19%), and *trans*-caryophyllene (3.45%). The chemometric analysis revealed inconsistency in the EO composition among various studied ecospecies, where it could be ascribed to the environmental and climatic conditions. The EO showed substantial allelopathic inhibitory activity against the germination, seedling root, and shoot growth of *D. aegyptium*, with IC_{50} values of 54.78, 57.10, and 74.07 mg L^{-1}. Additionally, the EO showed strong antioxidant potentiality based on the IC_{50} values of 4.52 mg mL^{-1} compared to 2.11 mg mL^{-1} of the ascorbic acid as standard. Moreover, this oil showed significant anti-inflammation via the suppression of lipoxygenase (LOX) and cyclooxygenases (COX1, and COX2), along with membrane stabilization. Further study is recommended for analysis of the activity of pure authentic materials of the major compounds either singularly or in combination, as well as for evaluation of their mechanism(s) and modes of action as antioxidants or allelochemicals.

Keywords: phytotoxicity; environmental factors; volatile oils; *Cleome* genus; anti-inflammation

---

191

1. Introduction

Wild plants are considered green factories for the synthesis of thousands of bioactive compounds that have various biological activities and are integrated into the treatment of various diseases, controlling weeds as biocides, as well as being used in agricultural, industrial, and pharmaceutical applications [1]. The use of natural bioactive compounds instead of synthetic chemicals fascinates scientists, researchers, and policymakers because they are renewable, degradable, safe, and low toxic [2,3].

Essential oils (EOs) are the main constituents of the members of the plant kingdom [4]. Historically, EOs represented one of the main resources of significant pharmaceutical and biological agents because of their complicated chemical composition, basically as isoprenoids [5]. Many biological potentialities for EOs have been described as hepatoprotective, anticancer [4], antioxidant [6,7], anti-inflammatory [8], anti-aging [4], antipyretic [6], and antimicrobial [9], in addition to allelopathy [10–12].

Cleome amblyocarpa Barr. & Murb. (Syns: *Cleome arabica* var. *amblyocarpa* (Barratte & Murb.) Ozenda, *Cleome africana* Botsch., or *Cleome daryoushiana* Parsa) is a herbaceous plant of the Cleomaceae family [13]. It is widely growing in sandy or stony habitats of desert along with North Africa [14]. This plant has been used in folk medicine for the treatment of various diseases such as diabetes and colic, as a stomachic therapy, rheumatic fever, scabies, and inflammation [15]. Various bioactive compounds have been isolated from *C. amblyocarpa*, such as flavonoids and glucosinolates [16], saponins [17], triterpenoids [18]. Therefore, this herb has been reported to possess various biological activities, including anti-inflammatory [16], anti-COVID-19 [17], genotoxicity [9], antidiabetic and antioxidant [19], and antimicrobial effects [20].

The EOs of several species of genus *Cleome* have been studied, such as *C. droserifolia* [21], *C. trinervia* [22], *C. monophylla* [23], and *C. serrata* [24], *C. coluteoides* [25]. However, few studies have dealt with the EO of *C. amblyocarpa* [26]. Additionally, and to the best of our knowledge, the allelopathic activity of the EO of *C. amblyocarpa* has not been studied before. Therefore, the present study aimed to (1) characterize the chemical composition of the EO isolated from the Egyptian ecospecies of *C. amblyocarpa*, and (2) evaluate the allelopathic activity, antioxidant, and anti-inflammatory activities of its EO.

2. Results and Discussion

2.1. EO Composition of C. amblyocarpa

The hydrodistillation of the air-dried powder of the above-ground parts of *C. amblyocarpa* yielded 0.38% (*v/w*) of golden-yellow oil. Depending upon the GC-MS analysis (Figure 1), 48 compounds were characterized, representing 97.17% of the total mass (Table 1). The Egyptian ecospecies of the present study yielded higher EO yield compared to Iranian [27] and Saudi [28] ecospecies, which had 0.20 and 0.21%, respectively. This variation in oil production could be ascribed to environmental, climatic, or genetic factors [5,6,29–31]. The term "ecospecies" means that species of the plant can be divided into several ecotypes (a genetically distinct population of plants that is growing in a particular habitat).

Table 1. Chemical components of the essential oil of above-ground parts of *Cleome amblyocarpa*.

No.	RT [a]	RC [b] %	Compound Name	MF [c]	KI_{Lit} [d]	KI_{Exp} [e]	Identification [f]
			Oxygenated monoterpenes				
1	6.60	0.50	6-Camphenol	$C_{10}H_{16}O$	1113	1113	KI, MS
2	8.94	0.13	Eucalyptol	$C_{10}H_{18}O$	1031	1032	KI, MS
3	10.24	0.54	Camphor	$C_{10}H_{16}O$	1146	1165	KI, MS
4	11.00	0.45	Isopulegol	$C_{10}H_{18}O$	1149	1150	KI, MS
5	12.76	1.12	Borneol	$C_{10}H_{18}O$	1169	1169	KI, MS
6	11.99	0.25	2-ethyl-exo-Fenchol	$C_{10}H_{18}O$	1297	1299	KI, MS
7	26.43	5.73	Myrtenyl acetate	$C_{12}H_{18}O_2$	1326	1324	KI, MS
			Sesquiterpenes hydrocarbons				
8	16.34	0.13	Silphiperfol-5,7(14)-diene	$C_{15}H_{22}$	1360	1361	KI, MS
9	17.28	0.98	α-Copaene	$C_{15}H_{24}$	1376	1378	KI, MS
10	17.48	1.01	β-Maaliene	$C_{15}H_{24}$	1380	1381	KI, MS
11	17.70	0.47	Berkheyaradulene	$C_{15}H_{24}$	1388	1389	KI, MS
12	18.68	3.45	*trans*-Caryophyllene	$C_{15}H_{24}$	1408	1410	KI, MS
13	18.79	0.29	Widdrene	$C_{15}H_{24}$	1431	1430	KI, MS
14	19.22	0.46	β-Humulene	$C_{15}H_{24}$	1438	1436	KI, MS
15	19.85	0.49	β-Farnesene	$C_{15}H_{24}$	1442	1443	KI, MS
16	20.47	0.53	γ-Muurolene	$C_{15}H_{24}$	1479	1477	KI, MS
17	20.78	0.23	*ar*-Curcumene	$C_{15}H_{22}$	1480	1481	KI, MS
18	20.86	0.54	Valencene	$C_{15}H_{24}$	1496	1496	KI, MS
19	20.94	0.65	α-Selinene	$C_{15}H_{24}$	1498	1497	KI, MS
20	21.20	2.30	α-Muurolene	$C_{15}H_{24}$	1500	1500	KI, MS
21	21.78	1.17	α-Bulnesene	$C_{15}H_{24}$	1509	1510	KI, MS
22	22.00	1.02	γ-Cadinene	$C_{15}H_{24}$	1513	1514	KI, MS
23	23.55	0.42	*trans*-Calamenene	$C_{15}H_{22}$	1522	1520	KI, MS
24	32.80	1.01	α-Guaiene	$C_{15}H_{24}$	1600	1601	KI, MS
			Oxygenated sesquiterpenes				
25	20.21	0.13	β-Cubebene	$C_{15}H_{24}O$	1388	1389	KI, MS
26	21.53	1.53	*trans*-α-Bisabolene epoxide	$C_{15}H_{24}O$	1506	1508	KI, MS
27	21.67	1.54	*trans*-Nerolidol	$C_{15}H_{24}O$	1531	1533	KI, MS
28	23.17	4.52	Isoshyobunone	$C_{15}H_{24}O$	1571	1570	KI, MS
29	23.48	0.66	Spathulenol	$C_{15}H_{24}O$	1577	1579	KI, MS
30	23.78	36.01	Caryophyllene oxide	$C_{15}H_{24}O$	1582	1584	KI, MS
31	24.22	0.86	Humulene oxide	$C_{15}H_{24}O$	1608	1609	KI, MS
32	24.45	0.83	Junenol	$C_{15}H_{26}O$	1619	1621	KI, MS
33	24.53	0.56	Citronellyl valerate	$C_{15}H_{28}O_2$	1625	1626	KI, MS
34	24.63	1.18	Nerolidol-epoxyacetate	$C_{17}H_{28}O_4$	1638	1637	KI, MS
35	24.88	0.38	tau-Cadinol	$C_{15}H_{26}O$	1640	1639	KI, MS
36	25.34	6.17	Alloaromadendrene epoxide	$C_{15}H_{24}O$	1641	1643	KI, MS
37	25.57	0.71	α-Eudesmol	$C_{15}H_{26}O$	1653	1655	KI, MS
38	25.80	0.32	Aromadendrene oxide-(2)	$C_{15}H_{24}O$	1678	1680	KI, MS
39	25.99	4.19	Shyobunol	$C_{15}H_{24}O$	1688	1691	KI, MS
40	26.79	0.49	β-Santalol	$C_{15}H_{24}O$	1738	1735	KI, MS
41	29.32	0.54	Xanthorrhizol	$C_{15}H_{22}O$	1753	1752	KI, MS
			Diterpenes hydrocarbons				
42	33.47	0.22	Geranyl-α-terpinene	$C_{20}H_{32}$	1874	1872	KI, MS
			Oxygenated diterpenes				
43	37.49	0.17	Phytol	$C_{20}H_{40}O$	1942	1944	KI, MS
			Carotenoid derived compounds				
44	14.56	1.39	Theaspirane A	$C_{13}H_{22}O$	1305	1307	KI, MS
45	14.87	1.64	Dihydroedulan II	$C_{13}H_{22}O$	1318	1316	KI, MS
			Apocarotenoid derived compounds				
46	30.87	7.92	Hexahydrofarnesyl acetone	$C_{18}H_{36}O$	1835	1837	KI, MS
			Other compounds				
47	13.87	1.16	*p*-Isopropyl-benzaldehyde	$C_{10}H_{12}O$	1239	1240	KI, MS
48	38.11	0.18	9,12-Octadecadienoic acid	$C_{18}H_{32}O_2$	2085	2085	KI, MS
	Total	97.17					

[a] RT: Retention time. [b] RC: Relative concentration. [c] MF: Molecular formula. [d] KI_{Lit}: Kovats retention index according to Adams (2017) on a DB−5 column in reference to *n*-alkanes. [e] KI_{Exp}: Experimental calculated Kovats retention index. [f] EO constituent identification was constructed via compound mass spectra (MS) and Kovats retention indices (KI) with those of Wiley spectral library collection and NIST library databases.

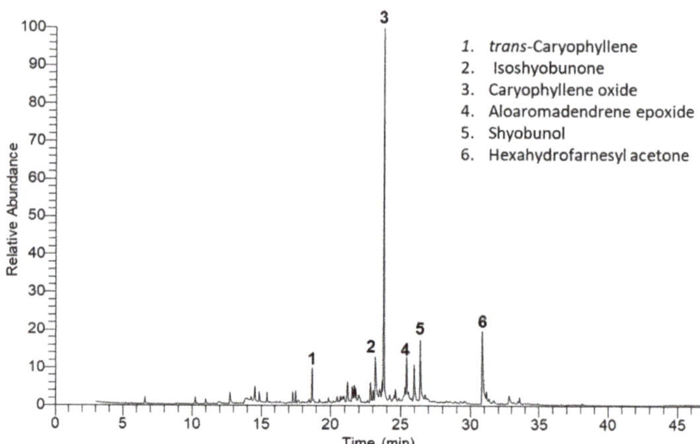

Figure 1. Chromatogram of the chemical compounds identified via GC-MS in the EO of *Cleome amblyocarpa* above-ground parts. The major compound peaks are numbered (1–6).

In the EO of *C. amblyocarpa*, six classes of components were determined, comprising oxygenated sesquiterpenes, sesquiterpenes hydrocarbons, oxygenated monoterpenes, diterpenes hydrocarbons, oxygenated diterpenes, apocarotenoid-derived compounds, carotenoid-derived compounds, and other compounds (Figure 2). These compounds pooled as 81.80% oxygenated compounds and 15.37% as non-oxygenated compounds. From overall mass, terpenoids represented the main constituents with a relative 84.88% with a preponderance of sesquiterpenes (75.77%), a remarkable concentration of monoterpenes (8.72%), and traces of diterpenes (0.39%). The abundance of terpenoids in the EO of *C. amblyocarpa* was in agreement with the data reported for samples collected from Iran [27] and the United Arab Emirates [26]. In contrast, the plurality of sesquiterpenes was inconsistent with the Iranian *C. amblyocarpa*, in which diterpenoids were determined as the major class [27], and *C. amblyocarpa* collected from the United Arab Emirates, in which monoterpenes were the main constituents [26]. These significant variations of chemical composition might be ascribed to environmental circumstances (such as temperature, rainfall, soil factors, altitude, etc.) and genetic characteristics [11,30,32,33].

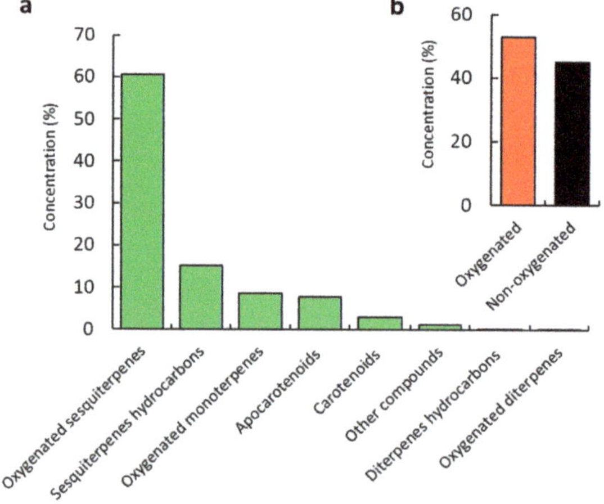

Figure 2. Concentrations of various identified classes of the chemical compounds of the *Cleome amblyocarpa* EO (**a**) and the percentage of oxygenated and non-oxygenated compounds (**b**).

Sesquiterpenes were assigned as the main components involving the oxygenated compounds as majors (60.62%) in addition to a relative concentration of 15.15% of sesquiterpene hydrocarbons. Out of the 16 identified oxygenated sesquiterpenes, caryophyllene oxide (36.01%), alloaromadendrene epoxide (6.17%), isoshyobunone (4.52%), and shyobunol (4.19%) represented the major compounds, whereas β-cubebene (0.13%) was the minor compound. Caryophyllene oxide is a common major compound in several EOs derived from plants such as *Cullen plicata* [34], *Schinus polygamus* [35], *Curcuma sahuynhensis* [36]. Caryophyllene oxide was documented as a minor compound in the EO of Iranian *C. amblyocarpa* [27], and totally absent from the EO of *C. amblyocarpa* collected from the United Arab Emirates [26], Tunisia [37], and Saudi Arabia [28]. On the other hand, *trans*-caryophyllene (3.45%), and α-muurolene (2.30%) were found to be the main sesquiterpene hydrocarbons, whereas silphiperfol-5,7(14)-diene (0.13%) was determined as a minor component. *trans*-caryophyllene has been reported in trace amounts of the EO of the Iranian ecospecies of *C. amblyocarpa* [27], whereas it is completely absent from the EO of Saudi [28], Emirati [26], and Tunisian [37] ecospecies.

The oxygenated monoterpenes were represented by 8.72%, which contained seven compounds, with myrtenyl acetate (5.73%) and borneol (1.12%) as major compounds. These two compounds are totally absent from the other ecospecies of *C. amblyocarpa* [26–28,37]. On the other hand, low diterpene contents were determined and represented by two compounds, geranyl-α-terpinene (0.22%) and phytol (0.17%). However, diterpenes were absent from other ecospecies of *C. amblyocarpa*. In other species of *Cleome* genus, phytol was reported in a high concentration such as *C. monophylla* [23], *C. serrata* [24], and *C. serrata* [38].

Carotenoid-derived compounds were determined in a concentration of 3.03%, that represented only two compounds, dihydroedulan II (1.64%) and theaspirane A (1.39%). Only one apocarotenoid-derived compound, hexahydrofarnesyl acetone, was identified with a high relative concentration (7.92%), whereas it was completely absent from the other reported ecospecies of *C. amblyocarpa* [26–28,37]. Hexahydrofarnesyl acetone is a widely distributed major compound in the EOs of several plants such as *Bassia muricata* [10], *Heliotropium curassavicum* [33], *Hildegardia barteri* [39], *Trianthema portulacastrum* [40].

Finally, traces of other non-terpenoid components were characterized including only two compounds, *p*-isopropy-l-benzaldehyde (1.16%) and 9,12-octadecadienoic acid (0.18%).

2.2. Chemometric Analysis of the EOs from Different C. amblyocarpa Ecospecies

The chemometric analysis of the EO composition of the present studied *C. amblyocarpa* and other reported ecospecies (Saudi, Iranian, Tunisian, and Emirati) was performed using cluster analysis and PCA (Figure 3). The cluster analysis revealed substantial variations among the studied ecospecies, and we can categorize them into three groups: group I comprising the present Egyptian and Tunisian ecospecies, group II containing Emirati and Iranian ecospecies, and finally the Saudi ecospecies separated alone as group III (Figure 3a). Interestingly, the chemometric analysis revealed that the EO compositions of ecospecies from the nearest countries were similar. This observation reflects the effect of environmental and climatic factors [30,32,33].

However, the present studied *C. amblyocarpa* ecospecies showed a correlation with the caryophyllene oxide, hexahydrofarnesyl acetone, and shyobunol, whereas the Tunisian ecospecies showed a correlation with ethyl 3-methylpentanoate, 7-epi-silphiperfol-5-ene, α-copaene, and 1,8-cineole (Figure 3b). The Saudi ecospecies was characterized by cis-dihydro carvone, 2-methoxy-4-vinyl phenol, and cubebene heptanal.

2.3. Allelopathic Effect of the C. amblyocarpa EO

The EO of *C. amblyocarpa* showed significant allelopathic activity of the seed germination ($p < 0.05$) as well as the shoot and root development of *D. aegyptium* in a dose-dependent manner (Figure 4a). At the highest concentration (100 μg mL^{-1}), the germination was inhibited by 70.18%, whereas the seedling root and shoot were reduced by 75.88% and 61.87%, respectively. Based on the IC$_{50}$ values, the root was more affected than the shoot, where the roots had an IC$_{50}$ value of 57.10 µg mL^{-1}, and the root attained 74.07 µg mL^{-1} (Figure 4b). Root has been reported to be more affected by allelochemicals than the shoot. This observation was reported for many plant species such as *Deverra tortuosa* [41], *Teucrium polium* [42], *Calotropis procera* [5], *Ficus carica* [43], and *C. plicata* [34]. This could be ascribed to the direct contact of the root with the medium and the high permeability of root cells [34,44].

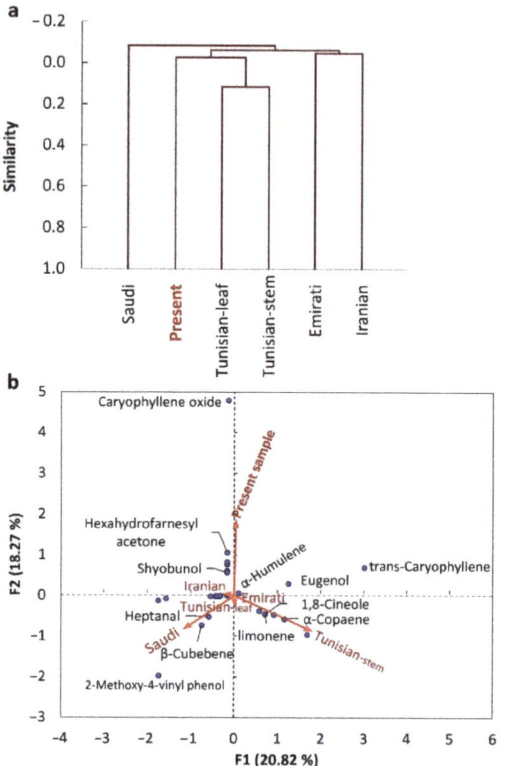

Figure 3. Chemometric analysis of various *Cleome amblyocarpa* ecospecies. (**a**) agglomerative hierarchical clustering (AHC), (**b**) and principal components analysis (PCA). (**F1**) and (**F2**) are factor 1 and 2.

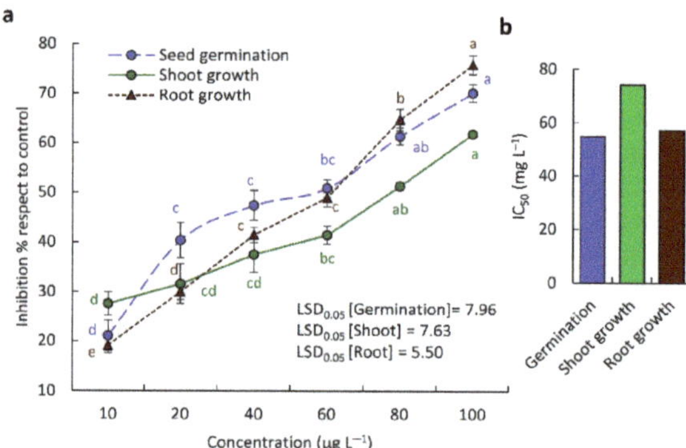

Figure 4. Allelopathic activity of the EO from the above-ground parts of *Cleome amblyocarpa* on the germination, root and shoot growth of *Dactyloctenium aegyptium*. (**a**) Various concentration and (**b**) IC$_{50}$. Different letters within each line mean values significantly different at $p < 0.05$ (Tukey's HSD test).

To the best of our knowledge, the allelopathic activity of the EO from *C. amblyocarpa* has not been studied yet. However, the aqueous, hexane, chloroform, and methanol extracts from *C. amblyocarpa* have been reported to inhibit lettuce germination and growth [45]. At a concentration of 6 g L^{-1}, ethyl acetate showed a complete inhibition of lettuce growth, whereas *Peganum harmala*, *Raphanus sativus*, and *Silybum marianum* were more resistant. Additionally, Ladhari et al. [46] identified some terpenoids and flavonoids from *C. amblyocarpa*, where dammarane-type triterpenes showed strong allelopathic activity, and flavonoid compounds exhibited <50% inhibition of the targeted species.

In our results, the allelopathic activity of *C. amblyocarpa* EO could be attributed to the activity of a single or combination of the major identified compounds. Caryophyllene oxide is reportedly a major compound of the EOs with substantial allelopathic activity such as *C. plicata* [34], *H. curassavicum* [33], *Acroptilon repens* [47], *Teucrium arduini*, *T. montbretii* [48], and *Nepeta curviflora* [49]. On the other hand, the EO from *Launaea mucronata* and *L. nudicaulis* showed the presence of hexahydrofarnesyl acetone as a major compound, where it revealed significant allelopathic activity against the weed: *Portulaca oleracea* [30]. Additionally, the EO of *H. curassavicum* had hexahydrofarnesyl acetone as the main compound, and showed marked allelopathic activity on *Chenopodium murale* [33]. The major compound, alloaromadendrene epoxide, in the EO of the present *C. amblyocarpa* has also been reported as the main compound (7.32%) of the EOs from *Lactuca serriola* that showed allelopathic activity against the weed *Bidens pilosa* [50]; however, in the EO of *Calamintha nepeta*, it did not show significant allelopathic activity against *Raphanus sativus*, *Lepidium sativum*, *Sinapis arvensis*, *Triticum durum*, and *Phalaris canariensis* [51]. This inconsistency could be attributed to the resistance of the weeds and shows that allelochemicals are species-specific [52].

Generally, the oxygenated terpene compounds have been reported to possess allelopathic activity compared to the non-oxygenated compounds [6]. In the present study, the EO of *C. amblyocarpa* was very rich in oxygenated compounds (81.80%), which could explain the notable allelopathic activity.

2.4. Antioxidant Activity of C. amblyocarpa EO

The activity of *C. amblyocarpa* EO in the reduction in the DPPH revealed significant antioxidant activity in a dose-dependent manner (Figure 5). At the lowest concentration of the EO (10 mg mL^{-1}), the EO showed a 15.28% scavenging activity of the DPPH, whereas at the highest concentration (100 mg mL^{-1}), the antioxidant activity was reduced by 3.7-fold of the lowest concentration of the EO. Based on the IC$_{50}$, the *C. amblyocarpa* EO had a value of 4.52 mg mL^{-1} compared to 2.11 mg mL^{-1} of the ascorbic acid as a standard antioxidant. The antioxidant activity of EOs in the present study were lower than those reported from Tunisian ecospecies [37]. This could be ascribed to the variation in the EO chemical compositions (Table 1).

The free radical scavenging activities of plant extracts and/or EOs were directly correlated with the concentration of the oxygenation of their constituents due to the increase in the free electrons [53]. The present data revealed 81.08% of oxygenated compounds, which means that a wealth of free electrons can act to diminish free radicals in the evaluation reaction. More specifically, the activity of this EO may be attributed to the activity of major compounds, either singularly or in synergy. Caryophyllene oxide has been reported to possess antioxidant activity [34]. EOs rich in hexahydrofarnesyl, such as *B. muricata* [10] and *H. curassavicum* [33], showed substantial antioxidant activity. The EO extracted from *L. serriola* has been reported to be rich in alloaromadendrene epoxide and isoshyobunone, where it expressed strong antioxidant activity [50].

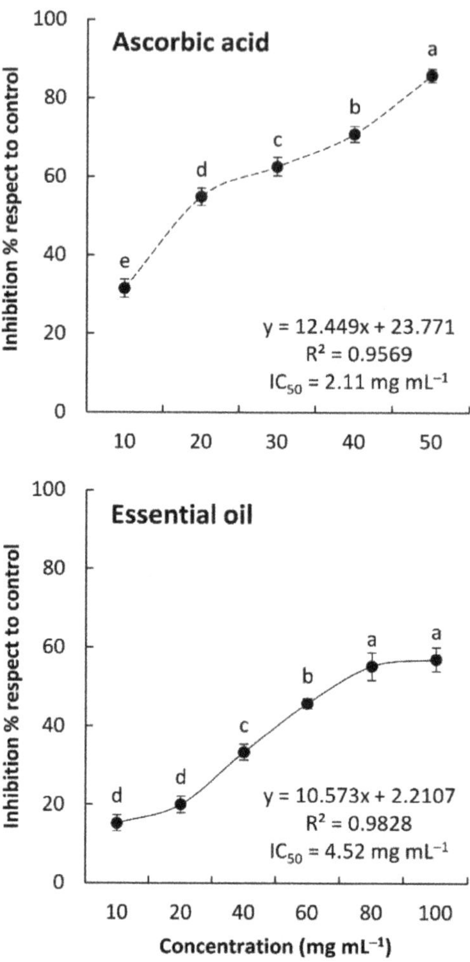

Figure 5. Antioxidant activity of the EO from the above-ground parts of *Cleome amblyocarpa*. Different letters of the line mean values significant at $p < 0.05$ (Tukey's HSD test).

2.5. Anti-Inflammatory effect of C. amblyocarpa EO

For the first time, the anti-inflammatory of EO of *C. amblyocarpa* has been evaluated via inhibition of the enzymes, lipoxygenase (LOX), cyclooxygenases (COX1, and COX2), as well as membrane stabilization. The results presented in Figure 6 revealed that the EO exhibited a significant anti-inflammatory action via the inhibition of LOX, COX1, and COX2 with respective IC_{50} values of 1.67, 12.77, 13.43 μg mL^{-1}, whereas ibuprofen showed inhibition with IC_{50} values of 1.53, 10.26, and 12.71 μg mL^{-1}, respectively. Additionally, the EO significantly inhibited the lysis of the hypotonic solution of the RBCs at an IC_{50} value of 15.25 μg mL^{-1}, compared with indomethacin which presented a result of 14.34 μg mL^{-1}.

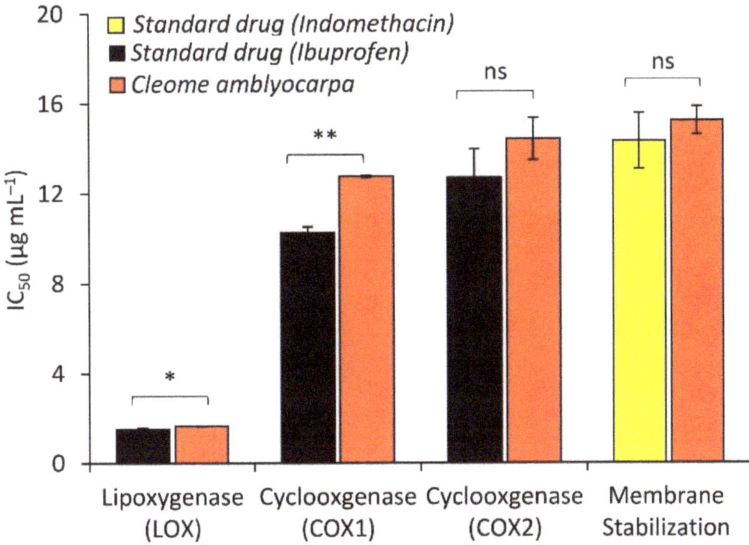

Figure 6. The anti-inflammatory activity of the EO of *Cleome amblyocarpa*, based on membrane stabilizing, and the inhibition of lipoxygenase (LOX) and cyclooxygenase (COX1 and COX 2). (ns) non-significant, * $p < 0.05$, ** $p < 0.01$ (two-tailed *t*-test). Data are mean values $\pm$ standard error ($n = 4$).

These data revealed that this EO has potent anti-inflammatory potentialities comparable with the two reference drugs, ibuprofen, and indomethacin, especially via the inhibition of lipoxygenase (LOX) and cyclooxygenase (COX1). This capability of EO for the inhibition of inflammations might be directly ascribed to the terpenoid contents as the main components (84.88%) [54]. Terpenoids represented major components of many of the documented plants with significant anti-inflammatory potentialities such as *Araucaria heterophylla* [8], *Ocimum basilicum* [55], and *Limnophila indica* [56].

Many studies have deduced that volatile sesquiterpene compounds have a potent anti-inflammation role in in vivo and in vitro models, especially caryophyllene and its oxide form [57]. Chavan et al. [58] described that the caryophyllene oxide, isolated from *Annona squamosa*, has significant in vivo anti-inflammatory activity. EO derived from *Cordia verbenacea* as well as its active constituent, caryophyllene, were demonstrated to exhibit strong anti-inflammatory activity was discussed with regard to their interfering with the production of TNF-α [59]. Moreover, Medeiros et al. [60] described that *trans*-caryophyllene showed potent anti-inflammatory in rats by significantly decreasing the migration of neutrophils as well as increasing the NF-κB-induced stimulation by lipopolysaccharides.

The EO of Sardinian *Santolina corsica* has been reported to contain a remarkable content of aromadendrene derivatives that exhibit a significant anti-inflammatory [61]. The enriched EO of Algerian *Myrtus communis* with myrtenyl acetate (38.7%) was shown to reduce the mice's inflammation and paw edema at a concentration of 100 mg/kg [62]. In addition to these significant roles of the main components, the other compounds were expected to have an important contribution via synergetic effects [8]. Based upon all these reported data, it is very clear that the present data agree with the previously documented. Additionally, the present results could be attributed to the prevalence of sesquiterpenes, especially *trans*-caryophyllene and caryophyllene oxide.

3. Materials and Methods

3.1. Plant Materials

Composite samples of the above-ground parts of the flowering *C. amblyocarpa* were collected in May 2018 from Wadi Hajoul, eastern desert, Egypt (29°57′51.2″ N 32°08′57.9″ E). The flowering plant was presented in Figure 7 The samples were collected from two populations in plastic bags, transferred to the laboratory, air-dried in a shaded place at room temperature for seven days, crushed into powder, and stored in paper bags until further analysis. The plant was identified according to [63,64]. A herbarium sheet (Mans.030301002) was prepared and deposited in the Herbarium of Botany Department, College of Science, Mansoura University, Egypt.

Figure 7. *Cleome amblyocarpa* Barratte & Murb. (**a**) Whole flowering plant, (**b**) enlarged flowering branch, and (**c**) enlarged fruiting branch. Photos were taken by Dr. Yasser El-Amier during spring of the year 2018 in Wadi Hajoul, eastern desert, Egypt.

3.2. EO Extraction Analysis and Characterization

The EO chemical compositions of the two extracted EO samples were analyzed and identified separately by gas chromatography–mass spectrometry (GC-MS), as described in our previous publication [65].

3.3. Allelopathic Bioassay

The allelopathic activity of the extracted EO from the above-ground parts of *C. amblyocarpa* was evaluated in vitro against the weed, *Dactyloctenium aegyptium*. The ripe seeds of the weed were collected from a cultivated field near Gamasa city, northern Mediterranean coast, Egypt (31°26′19.3″ N 31°34′12.9″ E). The uniform seeds were surface-sterilized with sodium hypochlorite (0.3%), rinsed with distilled sterilized water, and dried under sterile conditions [34]. To test the allelopathic activity, serial concentrations (10–100 µg mL^{-1}) of the EO were prepared using 1% polysorbate 80 (Sigma-Aldrich, Darmstadt, Germany) as an emulsifier. In a Petri plate, 20 sterilized seeds of *D. aegyptium* were arranged over wetted Whatman filter paper (Sigma-Aldrich, Darmstadt, Germany), either with each concentration or polysorbate 80 (as positive control). The plates were sealed with Parafilm® tape (Sigma, St. Louis, MO, USA) and kept in a growth chamber adjusted at 25 °C.

The plates were checked every day, and after 7 days, the number of germinated seeds was counted, and the lengths (mm) of seedling roots and shoots were measured. The inhibition of the germination, root growth, and shoot growth of seedlings was calculated according to the following equation:

$$Inhibition\ (\%) = 100 \times \left(\frac{Number/Length_{control} - (Number/Length_{treatment})}{Number/Length_{control}} \right) \quad (1)$$

Additionally, the IC_{50}, which is the concentration of the EO required for 50% inhibition of seed germination or seedling growth, was calculated by linear regression of the inhibition values versus various EO concentrations using MS-EXCEL 2016.

3.4. Antioxidant Activity

The antioxidant activity of the EO from *C. amblyocarpa* was estimated based on its ability to reduce the stable radical, 2,2-diphenyl-1-picrylhydrazyl (DPPH). According to the method of Miguel [66], a reaction mixture of an equal volume of 0.3 mM of freshly prepared DPPH and serial concentrations ($10–100\ mg\ mL^{-1}$) of the EO or ascorbic acid as standard was prepared and well shaken. The mixture was incubated in dark conditions at room temperature ($25 \pm 2\ °C$) for 30 min. The absorbance was measured at 517 nm using a spectrophotometer (Spectronic 21D model). The scavenging activity was estimated according to the following equation:

$$Scavenging\,activity\,(\%) = 100 \times \left(1 - \frac{Absorbance_{sample}}{Absorbance_{control}} \right) \quad (2)$$

The IC_{50}, which is the concentration of the EO required to reduce the color of the DPPH by 50%, was calculated graphically by linear regression using MS-EXCEL 2016.

3.5. Anti-Inflammatory Activity Estimation

The anti-inflammatory activity of the EO from the above-ground parts of *C. amblyocarpa* was evaluated by assessing the in vitro membrane stabilizing, and inhibition of lipoxygenase (LOX, EC: 1.13.11.12) and cyclooxygenase (COX1 and COX 2, EC: 1.14.99.1) enzymes.

3.5.1. Membrane Stabilization Inhibition Assay

The membrane-stabilizing activity of the samples was assessed using hypotonic solution-induced erythrocyte (RBCs) hemolysis [67]. For the preparation of erythrocyte suspension, whole blood was obtained with heparinized syringes from rats through cardiac puncture. The blood was washed three times with isotonic buffered solution (154 mM NaCl) in 10 mM sodium phosphate buffer (pH 7.4) and immediately centrifuged for 10 min at $3000\times\ g$. The test sample consisted of stock erythrocyte suspension (0.5 mL), 5 mL of hypotonic solution (50 mM NaCl), and the *C. amblyocarpa* EO ($7.81–1000\ \mu g\ mL^{-1}$ in ethanol) or indomethacin (as a standard drug). The control sample consisted of 0.5 mL of stock erythrocyte suspension and hypotonic-buffered saline solution alone. The mixtures were incubated for 10 min at room temperature ($25 \pm 2\ °C$) and centrifuged for 10 min at $3000\times\ g$. In 96-well plates, the absorbance of the supernatant was measured at 540 nm. The percentage inhibition of hemolysis or membrane stabilization were calculated according to the modified method described by Shinde et al. [67] as follows:

$$Inhibition\ \% = 100 \times [(A_{control} - A_{treatment}) \div A_{control}] \quad (3)$$

where $A_{control}$ is the absorbance control, and $A_{treatment}$ is the absorbance treatment.

The IC_{50} value was defined as the concentration of the EO required to inhibit 50% of the RBC hemolysis under the assay conditions. It was calculated graphically by linear regression of the inhibition values of different concentrations using MS-EXCEL 2016.

3.5.2. Lipoxygenase (LOX) Inhibition Assay

The activity of the *C. amblyocarpa* EO on the inhibition of the LOX enzyme (type I-B) was determined according to Granica et al. [68], with slight modifications. Briefly, in 96-well plates, 100 µL of soybean (*Glycine max*) LOX solution (1000 U/mL in borate buffer solution, pH 9) and 200 µL of borate buffer were mixed together with various concentrations of either EO (to a final concentration range of 0.98–125 µg mL^{-1}) or ibuprofen as a reference drug. Samples were pre-incubated with 100 µL of linoleic acid (substrate) to initiate the reaction and then were incubated at 25 °C for 15 min. The absorbance increase was measured at 234 nm using a microplate reader (BioTek Instruments Inc., Winooski, VT, USA). The inhibition percentage and IC$_{50}$ were calculated as previously mentioned in the membrane stabilization inhibition assay.

3.5.3. Cyclooxygenase (COX 1 and COX 2) Inhibition Assay

The COX activity was monitored as a result of the oxidation reaction of N,N,N,N-tetramethyl-p-phenylenediamine (TMPD) with arachidonic acid according to the protocol of Petrovic and Murray [69], with slight modifications. The activity of the EO or ibuprofen as a reference drug at a concentration range of 0.98–125 µg mL^{-1} was determined by monitoring the absorbance of TMPD oxidation reaction with arachidonic acid at 611 nm using a microplate reader (BioTek Instruments Inc., Winooski, VT, USA). The inhibition percentage and IC$_{50}$ were calculated as previously mentioned in the membrane stabilization inhibition assay.

3.6. Treatment of Data

The experiment of allelopathic bioassay was repeated three times with five replicas per each treatment. The data were subjected to one-way ANOVA followed by Tukey's HSD at the probability level of 0.05 using CoStat software program, version 6.311 (http://www.cohort.com, CoHort, Monterey, CA, USA, 1 April 2017). Additionally, the antioxidant scavenging data were subjected to one-way ANOVA followed by Tukey's HSD. The data of anti-inflammatory activity were compared using paired two-tailed *t*-tests using the XLSTAT 2018 program (https://www.xlstat.com/en/, Addinsoft Inc., New York, NY, USA, 15 January 2018). Chemometric analysis of the EO compositions of the studied Egyptian ecospecies in the present study and four other studied ecospecies collected from Saudi Arabia [28], Tunisia [37], Iran [27], and the United Arab Emirates [26] was performed via cluster analysis (agglomerative hierarchical clustering (AHC) and principal components analysis (PCA). We constructed a matrix of 65 compounds, from six samples of *C. amblyocarpa*, with a concentration >2%. The matrix was subjected to AHC and PCA using the XLSTAT 2018 program (https://www.xlstat.com/en/, Addinsoft Inc., New York, NY, USA, 15 January 2018).

4. Conclusions

For the first time, the analysis of the EO from the above-ground parts of the Egyptian ecospecies of *C. amblyocarpa* revealed the presence of 48 compounds, with a prevalence of sesquiterpenes. Caryophyllene oxide, hexahydrofarnesyl acetone, alloaromadendrene epoxide, myrtenyl acetate, isoshyobunone, shyobunol, and *trans*-caryophyllene have been identified as major compounds. The chemometric analysis of the presently studied ecospecies and other reported ecospecies revealed significant variation in the EO composition that could be ascribed to variation in the environmental and climatic conditions. EO showed substantial allelopathic inhibitory activity against the weed *D. aegyptium*, reflecting the potentiality of using this EO as an eco-friendly bioherbicide. Additionally, the EO showed significant antioxidant and anti-inflammatory activities. Further studies are recommended for evaluation (i) of anti-inflammatory effects of the *C. ambliocarpa* EO on an in vitro cell model (*ex.* RAW 264.7 cells); and (ii) the antioxidant, anti-inflammatory, and allelochemical activities along with the possible modes of action of the pure samples of the main EO compounds either singularly or in combination.

Author Contributions: Conceptualization, A.M.A.-E., A.I.E. and Y.A.E.-A.; formal analysis, A.M.A.-E., A.I.E., A.E.-N.G.E.G. and Y.A.E.-A.; investigation, A.M.A.-E., A.I.E., A.M.E., T.A.M. and Y.A.E.-A.; Validation, A.M.A.-E. and A.I.E.; writing—original draft preparation, A.M.A.-E. and A.I.E.; writing—

review and editing, A.M.A.-E., A.M.E., Y.A.E.-A., T.A.M., A.E.-N.G.E.G. and A.I.E. All authors have read and agreed to the published version of the manuscript.

Funding: This research received no external funding.

Institutional Review Board Statement: Not applicable.

Informed Consent Statement: Not applicable.

Data Availability Statement: Not applicable.

Conflicts of Interest: The authors declare no conflict of interest.

Sample Availability: Samples of the compounds are not available from the authors.

References

1. Poliakoff, M.; Licence, P. Green chemistry. *Nature* **2007**, *450*, 810–812. [CrossRef] [PubMed]
2. Poliakoff, M.; Fitzpatrick, J.M.; Farren, T.R.; Anastas, P.T. Green chemistry: Science and politics of change. *Science* **2002**, *297*, 807–810. [CrossRef]
3. Horváth, I.T.; Anastas, P.T. Innovations and green chemistry. *Chem. Rev.* **2007**, *107*, 2169–2173. [CrossRef]
4. Elgamal, A.M.; Ahmed, R.F.; Abd-ElGawad, A.M.; El Gendy, A.E.-N.G.; Elshamy, A.I.; Nassar, M.I. Chemical profiles, anticancer, and anti-aging activities of essential oils of *Pluchea dioscoridis* (L.) DC. and *Erigeron bonariensis* L. *Plants* **2021**, *10*, 667. [CrossRef]
5. Al-Rowaily, S.L.; Abd-ElGawad, A.M.; Assaeed, A.M.; Elgamal, A.M.; Gendy, A.E.-N.G.E.; Mohamed, T.A.; Dar, B.A.; Mohamed, T.K.; Elshamy, A.I. Essential oil of *Calotropis procera*: Comparative chemical profiles, antimicrobial activity, and allelopathic potential on weeds. *Molecules* **2020**, *25*, 5203. [CrossRef] [PubMed]
6. Assaeed, A.; Elshamy, A.; El Gendy, A.E.-N.; Dar, B.; Al-Rowaily, S.; Abd-ElGawad, A. Sesquiterpenes-rich essential oil from above ground parts of *Pulicaria somalensis* exhibited antioxidant activity and allelopathic effect on weeds. *Agronomy* **2020**, *10*, 399. [CrossRef]
7. Elshamy, A.; Essa, A.F.; El-Hawary, S.S.; El Gawad, A.M.A.; Kubacy, T.M.; El-Khrisy, E.E.-D.A.; Younis, I.Y. Prevalence of diterpenes in essential oil of *Euphorbia mauritanica* L.: Detailed chemical profile, antioxidant, cytotoxic and phytotoxic activities. *Chem. Biodivers.* **2021**, *18*, e2100238.
8. Elshamy, A.I.; Ammar, N.M.; Hassan, H.A.; Al-Rowaily, S.L.; Ragab, T.I.; El Gendy, A.E.-N.G.; Abd-ElGawad, A.M. Essential oil and its nanoemulsion of *Araucaria heterophylla* resin: Chemical characterization, anti-inflammatory, and antipyretic activities. *Ind. Crop. Prod.* **2020**, *148*, 112272. [CrossRef]
9. Edziri, H.; Mastouri, M.; Aouni, M.; Anthonissen, R.; Verschaeve, L. Investigation on the genotoxicity of extracts from *Cleome amblyocarpa* Barr. and Murb, an important Tunisian medicinal plant. *South Afr. J. Bot.* **2013**, *84*, 102–103. [CrossRef]
10. Abd El-Gawad, A.M.; El Gendy, A.E.-N.; El-Amier, Y.; Gaara, A.; Omer, E.; Al-Rowaily, S.; Assaeed, A.; Al-Rashed, S.; Elshamy, A. Essential oil of *Bassia muricata*: Chemical characterization, antioxidant activity, and allelopathic effect on the weed *Chenopodium murale*. *Saudi J. Biol. Sci.* **2020**, *27*, 1900–1906. [CrossRef]
11. Abd El-Gawad, A.M.; Elshamy, A.I.; El-Amier, Y.A.; El Gendy, A.E.-N.G.; Al-Barati, S.A.; Dar, B.A.; Al-Rowaily, S.L.; Assaeed, A.M. Chemical composition variations, allelopathic, and antioxidant activities of *Symphyotrichum squamatum* (Spreng.) Nesom essential oils growing in heterogeneous habitats. *Arab. J. Chem.* **2020**, *13*, 4237–4245. [CrossRef]
12. Abd El-Gawad, A.M.; El-Amier, Y.A.; Bonanomi, G. Allelopathic activity and chemical composition of *Rhynchosia minima* (L.) DC. essential oil from Egypt. *Chem. Biodivers.* **2018**, *15*, e1700438. [CrossRef]
13. IPNI. *The International Plant Names Index*; The Royal Botanic Gardens, Kew, Harvard University Herbaria & Libraries and Australian National Botanic Gardens: Richmond, UK, 2012.
14. IUCN. *A Guide to Medicinal Plants in North Africa*; IUCN: Barcelona, Spain, 2005.
15. Bouriche, H.; Arnhold, J. Effect of *Cleome arabica* leaf extract treated by naringinase on human neutrophil chemotaxis. *Nat. Prod. Commun.* **2010**, *5*, 415–418. [CrossRef] [PubMed]
16. Khlifi, A.; Pecio, Ł.; Lobo, J.C.; Melo, D.; Ayache, S.B.; Flamini, G.; Oliveira, M.B.P.; Oleszek, W.; Achour, L. Leaves of *Cleome amblyocarpa* Barr. and Murb. and *Cleome arabica* L.: Assessment of nutritional composition and chemical profile (LC-ESI-MS/MS), anti-inflammatory and analgesic effects of their extracts. *J. Ethnopharmacol.* **2021**, *269*, 113739. [CrossRef] [PubMed]
17. Zaki, A.A.; Al-Karmalawy, A.A.; El-Amier, Y.A.; Ashour, A. Molecular docking reveals the potential of *Cleome amblyocarpa* isolated compounds to inhibit COVID-19 virus main protease. *New J. Chem.* **2020**, *44*, 16752–16758. [CrossRef]
18. Ahmed, A.A.; Kattab, A.M.; Bodige, S.G.; Mao, Y.; Minter, D.E.; Reinecke, M.G.; Watson, W.H.; Mabry, T.J. 15α-Acetoxycleomblynol A from *Cleome amblyocarpa*. *J. Nat. Prod.* **2001**, *64*, 106–107. [CrossRef] [PubMed]
19. Seglab, F.; Hamia, C.; Khacheba, I.; Djeridane, A.; Yousfi, M. High in vitro antioxidant capacities of Algerian *Cleome arabica* leaves' extracts. *Phytothérapie* **2021**, *19*, 16–24. [CrossRef]
20. Takhi, D.; Ouinten, M.; Yousfi, M. Study of antimicrobial activity of secondary metabolites extracted from spontaneous plants from the area of Laghouat, Algeria. *Adv. Environ. Biol.* **2011**, *5*, 469–477.
21. Abd El-Gawad, A.M.; El-Amier, Y.A.; Bonanomi, G. Essential oil composition, antioxidant and allelopathic activities of *Cleome droserifolia* (Forssk.) Delile. *Chem. Biodivers.* **2018**, *15*, e1800392. [CrossRef] [PubMed]

22. Muhaidat, R.; Al-Qudah, M.A.; Samir, O.; Jacob, J.H.; Hussein, E.; Al-Tarawneh, I.N.; Bsoul, E.; Orabi, S.T.A. Phytochemical investigation and in vitro antibacterial activity of essential oils from *Cleome droserifolia* (Forssk.) Delile and *C. trinervia* Fresen. (Cleomaceae). *South Afr. J. Bot.* **2015**, *99*, 21–28. [CrossRef]

23. Ndungu, M.; Lwande, W.; Hassanali, A.; Moreka, L.; Chhabra, S.C. *Cleome monophylla* essential oil and its constituents as tick (*Rhipicephalus appendiculatus*) and maize weevil (*Sitophilus zeamais*) repellents. *Entomol. Exp. Appl.* **1995**, *76*, 217–222. [CrossRef]

24. McNeil, M.J.; Porter, R.B.; Williams, L.A. Chemical composition and biological activity of the essential oil from Jamaican *Cleome serrata*. *Nat. Prod. Commun.* **2012**, *7*, 1231–1232. [CrossRef]

25. Sodeifian, G.; Saadati Ardestani, N.; Sajadian, S.A.; Ghorbandoost, S. Application of supercritical carbon dioxide to extract essential oil from *Cleome coluteoides* Boiss: Experimental, response surface and grey wolf optimization methodology. *J. Supercrit. Fluids* **2016**, *114*, 55–63. [CrossRef]

26. Shahin, S.; Kurup, S.; Cheruth, A.-J.; Salem, M. Chemical composition of *Cleome amblyocarpa* Barr. & Murb. essential oils under different irrigation levels in sandy soils with antioxidant activity. *J. Essent. Oil Bear. Plants* **2018**, *21*, 1235–1256.

27. Rassouli, E.; Dadras, O.G.; Bina, E.; Asgarpanah, J. The essential oil composition of *Cleome brachycarpa* Vahl ex DC. *J. Essent. Oil Bear. Plants* **2014**, *17*, 158–163. [CrossRef]

28. Al-Humaidia, J.Y.; Al-Qudah, M.A.; Al-Saleem, M.S.; Alotaibi, S.M. Antioxidant activity and chemical composition of essential oils of selected *Cleome* species growing in Saudi Arabia. *Jordan J. Chem.* **2019**, *14*, 29–37.

29. Abd El-Gawad, A.M.; El Gendy, A.E.-N.G.; Assaeed, A.M.; Al-Rowaily, S.L.; Alharthi, A.S.; Mohamed, T.A.; Nassar, M.I.; Dewir, Y.H.; Elshamy, A.I. Phytotoxic effects of plant essential oils: A systematic review and structure-activity relationship based on chemometric analyses. *Plants* **2021**, *10*, 36. [CrossRef] [PubMed]

30. Elshamy, A.; Abd El-Gawad, A.M.; El-Amier, Y.A.; El Gendy, A.; Al-Rowaily, S. Interspecific variation, antioxidant and allelopathic activity of the essential oil from three *Launaea* species growing naturally in heterogeneous habitats in Egypt. *Flavour Fragr. J.* **2019**, *34*, 316–328. [CrossRef]

31. Hassan, E.M.; El Gendy, A.E.-N.G.; Abd El-Gawad, A.M.; Elshamy, A.I.; Farag, M.A.; Alamery, S.F.; Omer, E.A. Comparative chemical profiles of the essential oils from different varieties of *Psidium guajava* L. *Molecules* **2021**, *26*, 119. [CrossRef]

32. Fernández-Sestelo, M.; Carrillo, J.M. Environmental effects on yield and composition of essential oil in wild populations of spike lavender (*Lavandula latifolia* Medik.). *Agriculture* **2020**, *10*, 626. [CrossRef]

33. Abd-El-Gawad, A.M.; Elshamy, A.; Al-Rowaily, S.; El-Amier, Y.A. Habitat affects the chemical profile, allelopathy, and antioxidant properties of essential oils and phenolic enriched extracts of the invasive plant *Heliotropium Curassavicum*. *Plants* **2019**, *8*, 482. [CrossRef]

34. Abd El-Gawad, A.M. Chemical constituents, antioxidant and potential allelopathic effect of the essential oil from the aerial parts of *Cullen plicata*. *Ind. Crops Prod.* **2016**, *80*, 36–41. [CrossRef]

35. El-Nashar, H.A.S.; Mostafa, N.M.; El-Badry, M.A.; Eldahshan, O.A.; Singab, A.N.B. Chemical composition, antimicrobial and cytotoxic activities of essential oils from *Schinus polygamus* (Cav.) cabrera leaf and bark grown in Egypt. *Nat. Prod. Res.* **2020**, *34*, 1–4. [CrossRef]

36. Sam, L.N.; Huong, L.T.; Minh, P.N.; Vinh, B.T.; Dai, D.N.; Setzer, W.N.; Ogunwande, I.A. Chemical composition and antimicrobial activity of the rhizome essential oil of *Curcuma sahuynhensis* from Vietnam. *J. Essent. Oil Bear. Plants* **2020**, *23*, 803–809. [CrossRef]

37. Khlifi, A.; Chrifa, A.B.; Lamine, J.B.; Thouri, A.; Adouni, K.; Flamini, G.; Oleszek, W.; Achour, L. Gas chromatography-mass spectrometry (GM-MS) analysis and biological activities of the aerial part of *Cleome amblyocarpa* Barr. and Murb. *Environ. Sci. Pollut. Res.* **2020**, *27*, 22670–22679. [CrossRef]

38. McNeil, M.J.; Porter, R.B.; Williams, L.A.; Rainford, L. Chemical composition and antimicrobial activity of the essential oils from *Cleome spinosa*. *Nat. Prod. Commun.* **2010**, *5*, 1301–1306. [CrossRef]

39. Balogun, O.S.; Ajayi, O.S.; Adeleke, A.J. Hexahydrofarnesyl acetone-rich extractives from *Hildegardia barteri*. *J. Herbs Spices Med. Plants* **2017**, *23*, 393–400. [CrossRef]

40. Abd El-Gawad, A.; El Gendy, A.; Elshamy, A.; Omer, E. Chemical composition of the essential oil of *Trianthema portulacastrum* L. Aerial parts and potential antimicrobial and phytotoxic activities of its extract. *J. Essent. Oil Bear. Plants* **2016**, *19*, 1684–1692. [CrossRef]

41. Fayed, E.M.; Abd EI-Gawad, A.M.; Elshamy, A.I.; El-Halawany, E.S.F.; EI-Amier, Y.A. Essential oil of *Deverra tortuosa* aerial parts: Detailed chemical profile, allelopathic, antimicrobial, and antioxidant activities. *Chem. Biodivers.* **2021**, *18*, e2000914. [CrossRef]

42. Saleh, I.; Abd El-Gawad, A.; El Gendy, A.E.-N.; Abd El Aty, A.; Mohamed, T.; Kassem, H.; Aldosri, F.; Elshamy, A.; Hegazy, M.-E.F. Phytotoxic and antimicrobial activities of *Teucrium polium* and *Thymus decussatus* essential oils extracted using hydrodistillation and microwave-assisted techniques. *Plants* **2020**, *9*, 716. [CrossRef]

43. Ladhari, A.; Gaaliche, B.; Zarrelli, A.; Ghannem, M.; Ben Mimoun, M. Allelopathic potential and phenolic allelochemicals discrepancies in *Ficus carica* L. cultivars. *South Afr. J. Bot.* **2020**, *130*, 30–44. [CrossRef]

44. Chon, S.U.; Kim, Y.M.; Lee, J.C. Herbicidal potential and quantification of causative allelochemicals from several Compositae weeds. *Weed Res.* **2003**, *43*, 444–450. [CrossRef]

45. Ladhari, A.; Omezzine, F.; DellaGreca, M.; Zarrelli, A.; Zuppolini, S.; Haouala, R. Phytotoxic activity of *Cleome arabica* L. and its principal discovered active compounds. *South Afr. J. Bot.* **2013**, *88*, 341–351. [CrossRef]

46. Ladhari, A.; Tufano, I.; DellaGreca, M. Influence of new effective allelochemicals on the distribution of *Cleome arabica* L. community in nature. *Nat. Prod. Res.* **2020**, *34*, 773–781. [CrossRef] [PubMed]

47. Razavi, S.M.; Narouei, M.; Majrohi, A.A.; Chamanabad, H.R.M. Chemical constituents and phytotoxic activity of the essential oil of *Acroptilon repens* (L.) Dc from Iran. *J. Essent. Oil Bear. Plants* **2012**, *15*, 943–948. [CrossRef]

48. de Almeida, L.F.R.; Frei, F.; Mancini, E.; De Martino, L.; De Feo, V. Phytotoxic activities of Mediterranean essential oils. *Molecules* **2010**, *15*, 4309–4323. [CrossRef]

49. Mancini, E.; Arnold, N.A.; De Martino, L.; De Feo, V.; Formisano, C.; Rigano, D.; Senatore, F. Chemical composition and phytotoxic effects of essential oils of *Salvia hierosolymitana* Boiss. and *Salvia multicaulis* Vahl. var. simplicifolia Boiss. growing wild in Lebanon. *Molecules* **2009**, *14*, 4725–4736. [PubMed]

50. Abd El-Gawad, A.M.; Elshamy, A.; El Gendy, A.E.-N.; Al-Rowaily, S.L.; Assaeed, A.M. Preponderance of oxygenated sesquiterpenes and diterpenes in the volatile oil constituents of *Lactuca serriola* L. revealed antioxidant and allelopathic activity. *Chem. Biodivers.* **2019**, *16*, e1900278. [CrossRef]

51. Mancini, E.; De Martino, L.; Malova, H.; De Feo, V. Chemical composition and biological activities of the essential oil from *Calamintha nepeta* plants from the wild in southern Italy. *Nat. Prod. Commun.* **2013**, *8*, 129–142. [CrossRef]

52. Peng, S.L.; Wen, J.; Guo, Q.F. Mechanism and active variety of allelochemicals. *Acta Bot. Sin.* **2004**, *46*, 757–766.

53. El-Kashak, W.A.; Elshamy, A.I.; Mohamed, T.A.; El Gendy, A.E.-N.G.; Saleh, I.A.; Umeyama, A. Rumpictuside A: Unusual 9,10-anthraquinone glucoside from *Rumex pictus* Forssk. *Carbohydr. Res.* **2017**, *448*, 74–78. [CrossRef]

54. de Santana Souza, M.T.; Almeida, J.R.G.d.S.; de Souza Araujo, A.A.; Duarte, M.C.; Gelain, D.P.; Moreira, J.C.F.; dos Santos, M.R.V.; Quintans-Júnior, L.J. Structure–activity relationship of terpenes with anti-inflammatory profile—A systematic review. *Basic Clin. Pharmacol. Toxicol.* **2014**, *115*, 244–256. [CrossRef] [PubMed]

55. Rodrigues, L.B.; Martins, A.O.B.P.B.; Ribeiro-Filho, J.; Cesário, F.R.A.S.; e Castro, F.F.; de Albuquerque, T.R.; Fernandes, M.N.M.; da Silva, B.A.F.; Júnior, L.J.Q.; de Sousa Araújo, A.A. Anti-inflammatory activity of the essential oil obtained from *Ocimum basilicum* complexed with β-cyclodextrin (β-CD) in mice. *Food Chem. Toxicol.* **2017**, *109*, 836–846. [CrossRef]

56. Kumar, R.; Kumar, R.; Prakash, O.; Srivastava, R.; Pant, A. Chemical composition, in vitro antioxidant, anti-inflammatory and antifeedant properties in the essential oil of Asian marsh weed *Limnophila indica* L. (Druce). *J. Pharmacogn. Phytochem.* **2019**, *8*, 1689–1694.

57. De Cássia Da Silveira e Sá, R.; Andrade, L.N.; De Sousa, D.P. Sesquiterpenes from essential oils and anti-inflammatory activity. *Nat. Prod. Commun.* **2015**, *10*, 1767–1774. [CrossRef]

58. Chavan, M.; Wakte, P.; Shinde, D. Analgesic and anti-inflammatory activity of caryophyllene oxide from *Annona squamosa* L. bark. *Phytomedicine* **2010**, *17*, 149–151. [CrossRef]

59. Passos, G.F.; Fernandes, E.S.; da Cunha, F.M.; Ferreira, J.; Pianowski, L.F.; Campos, M.M.; Calixto, J.B. Anti-inflammatory and anti-allergic properties of the essential oil and active compounds from *Cordia verbenacea*. *J. Ethnopharmacol.* **2007**, *110*, 323–333. [CrossRef]

60. Medeiros, R.; Passos, G.; Vitor, C.; Koepp, J.; Mazzuco, T.; Pianowski, L.; Campos, M.; Calixto, J. Effect of two active compounds obtained from the essential oil of *Cordia verbenacea* on the acute inflammatory responses elicited by LPS in the rat paw. *Br. J. Pharmacol.* **2007**, *151*, 618–627. [CrossRef]

61. Foddai, M.; Marchetti, M.; Ruggero, A.; Juliano, C.; Usai, M. Evaluation of chemical composition and anti-inflammatory, antioxidant, antibacterial activity of essential oil of Sardinian *Santolina corsica* Jord. & Fourr. *Saudi J. Biol. Sci.* **2019**, *26*, 930–937.

62. Touaibia, M. Composition and anti-inflammatory effect of the common myrtle (*Myrtus communis* L.) essential oil growing wild in Algeria. *Phytotherapie* **2020**, *18*, 156–161. [CrossRef]

63. Tackholm, V. *Students' Flora of Egypt*, 2nd ed.; Cairo University Press: Cairo, Egypt, 1974.

64. Boulos, L. *Flora of Egypt*; All Hadara Publishing: Cairo, Egypt, 1995; Volume 1.

65. Abd El-Gawad, A.; Elshamy, A.; El Gendy, A.E.-N.; Gaara, A.; Assaeed, A. Volatiles profiling, allelopathic activity, and antioxidant potentiality of *Xanthium strumarium* leaves essential oil from Egypt: Evidence from chemometrics analysis. *Molecules* **2019**, *24*, 584. [CrossRef]

66. Miguel, M.G. Antioxidant activity of medicinal and aromatic plants. *Flavour Fragr. J.* **2010**, *25*, 219–312. [CrossRef]

67. Shinde, U.; Phadke, A.; Nair, A.; Mungantiwar, A.; Dikshit, V.; Saraf, M. Membrane stabilizing activity—A possible mechanism of action for the anti-inflammatory activity of *Cedrus deodara* wood oil. *Fitoterapia* **1999**, *70*, 251–257. [CrossRef]

68. Granica, S.; Czerwińska, M.E.; Piwowarski, J.P.; Ziaja, M.; Kiss, A.K. Chemical composition, antioxidative and anti-inflammatory activity of extracts prepared from aerial parts of *Oenothera biennis* L. and *Oenothera paradoxa* Hudziok obtained after seeds cultivation. *J. Agric. Food Chem.* **2013**, *61*, 801–810. [CrossRef] [PubMed]

69. Petrovic, N.; Murray, M. Using N,N,N′,N′-tetramethyl-p-phenylenediamine (TMPD) to assay cyclooxygenase activity In Vitro. In *Advanced Protocols in Oxidative Stress II*; Armstrong, D., Ed.; Humana Press: Totowa, NJ, USA, 2010; pp. 129–140. [CrossRef]

Review

Phytotoxic Effects of Plant Essential Oils: A Systematic Review and Structure-Activity Relationship Based on Chemometric Analyses

Ahmed M. Abd-ElGawad [1,2,*], Abd El-Nasser G. El Gendy [3], Abdulaziz M. Assaeed [1], Saud L. Al-Rowaily [1], Abdullah S. Alharthi [1], Tarik A. Mohamed [4], Mahmoud I. Nassar [5], Yaser H. Dewir [1,6] and Abdelsamed I. Elshamy [5,7]

[1] Plant Production Department, College of Food and Agriculture Sciences, King Saud University, P.O. Box 2460, Riyadh 11451, Saudi Arabia; assaeed@ksu.edu.sa (A.M.A.); srowaily@ksu.edu.sa (S.L.A.-R.); 437105762@ksu.edu.sa (A.S.A.); ydewir@ksu.edu.sa (Y.H.D.)
[2] Department of Botany, Faculty of Science, Mansoura University, Mansoura 35516, Egypt
[3] Medicinal and Aromatic Plants Research Department, National Research Centre, Cairo 11865, Egypt; aggundy_5@yahoo.com
[4] Chemistry of Medicinal Plants Department, National Research Centre, 33 El-Bohouth St., Dokki, Giza 12622, Egypt; ta.mourad@nrc.sci.eg
[5] Chemistry of Natural Compounds Department, National Research Centre, 33 El Bohouth St., Dokki, Giza 12622, Egypt; mnassar_eg@yahoo.com (M.I.N.); ai.el-shamy@nrc.sci.eg (A.I.E.)
[6] Department of Horticulture, Faculty of Agriculture, Kafrelsheikh University, Kafr El-Sheikh 33516, Egypt
[7] Faculty of Pharmaceutical Sciences, Tokushima Bunri University, Yamashiro-Cho, Tokushima 770-8514, Japan
[*] Correspondence: aibrahim2@ksu.edu.sa; Tel.: +20-1003438980

Citation: Abd-ElGawad, A.M.; El Gendy, A.E.-N.G.; Assaeed, A.M.; Al-Rowaily, S.L.; Alharthi, A.S.; Mohamed, T.A.; Nassar, M.I.; Dewir, Y.H.; Elshamy, A.I. Phytotoxic Effects of Plant Essential Oils: A Systematic Review and Structure-Activity Relationship Based on Chemometric Analyses. *Plants* **2021**, *10*, 36. https://dx.doi.org/10.3390/plants10010036

Received: 16 November 2020
Accepted: 23 December 2020
Published: 25 December 2020

Publisher's Note: MDPI stays neutral with regard to jurisdictional claims in published maps and institutional affiliations.

Abstract: Herbicides are natural or synthetic chemicals used to control unwanted plants (weeds). To avoid the harmful effects of synthetic herbicides, considerable effort has been devoted to finding alternative products derived from natural sources. Essential oils (EOs) from aromatic plants are auspicious source of bioherbicides. This review discusses phytotoxic EOs and their chemical compositions as reported from 1972 to 2020. Using chemometric analysis, we attempt to build a structure-activity relationship between phytotoxicity and EO chemical composition. Data analysis reveals that oxygenated terpenes, and mono- and sesquiterpenes, in particular, play principal roles in the phytotoxicity of EOs. Pinene, 1,8 cineole, linalool, and carvacrol are the most effective monoterpenes, with significant phytotoxicity evident in the EOs of many plants. Caryophyllene and its derivatives, including germacrene, spathulenol, and hexahydrofarnesyl acetone, are the most effective sesquiterpenes. EOs rich in iridoids (non-terpene compounds) also exhibit allelopathic activity. Further studies are recommended to evaluate the phytotoxic activity of these compounds in pure forms, determine their activity in the field, evaluate their safety, and assess their modes of action.

Keywords: allelopathy; bioherbicides; volatile oils; terpenes; aromatic plants

1. Introduction

Humans have been cultivating plants for nearly 10,000 years ago. Today, any plant growing where it is not wanted is defined as a weed. Weeds represent an important constraint to agricultural production [1]. Weeds represent approximately 0.1% of the world's flora and they evolve with agricultural practices. Weeds can cause declines in crop yields via competition for resources such as light, water, space, and nutrients, and by producing chemical weapons known as allelopathic compounds [2]. Weed management is achieved using several techniques to limit infestation and minimize competition. These techniques evolved to mitigate crop yield losses, but weed control is typically used only after a problem has been identified.

Plants **2021**, *10*, 36. https://dx.doi.org/10.3390/plants10010036   https://www.mdpi.com/journal/plants

Scientists and researchers address weed control through physical, chemical, and biological methods. Controlling weeds in an environmentally friendly way is often considered a challenge. Natural resources offer new approaches to producing eco-friendly, and safe bioherbicides that are effective against nuisance weeds. Plants produce the essential oils (EOs) in their various organs as a complex mixture of secondary metabolites such as mono-, sesquit-, and di-terpenoids in addition to hydrocarbons [3,4]. In plants, EOs were biosynthesized via different isoprenoid pathways such as methylerythritol phosphate (MEP) pathway and mevalonic acid (MVA) pathway [5]. The EOs have been described as potent biological agents such as phytotoxic [6–9], antimicrobial [10], anti-inflammatory, antipyretic [11], antiulcer [12], and hepatoprotective [13]. The bioactivities potential of EOs are directly correlated with the quality and quantity of their chemical constituents [6]. Many studies have been performed using the extracted EOs from various plants as phytotoxic chemicals (allelochemicals), where the phytotoxicity is usually correlated to the whole EO profile that contained a mixture of compounds. However, the activity of the EO could be ascribed to a specific compound(s) in the EO. Therefore, in the present review, we try to elucidate a framework of the most frequent and major allelochemicals that were identified in the EOs with a substantial phytotoxic activity using chemometric tools. Additionally, the activities of the authentic identified major compounds are discussed.

2. Materials and Methods

This review focuses on reports of EOs from plants that exhibit phytotoxic activity published between 1972 and early 2020, using Google, Sci-finder, Google Scholar, PubMed, Elsevier, and Springer databases. Based on the major compounds (those constituting > 5% of the total mass of the EO), the plants were categorized into three groups; mono-, sesqui-, and non-terpenoid–rich compounds. Firstly, the database of EOs rich in monoterpenes derived from plants comprised 46 species belonging to 12 botanical families, including Lamiaceae (18 species), Myrtaceae (nine species), Asteraceae (eight species), Anacardiaceae (three species), and Cannabaceae, Euphorbiaceae, Monimiaceae, Pinaceae, Poaceae, Verbenaceae, Winteraceae, and Apiaceae (a single species each). Additionally, the EOs of these plants were tested against 49 plant species.

Secondly, the plant EOs rich in sesquiterpenes from 25 plant species belonging to eight botanical families were studied. The most represented botanical families were Lamiaceae and Asteraceae (nine plant species each), while Anacardiaceae, Boraginaceae, Fabaceae, Myrtaceae, Simaroubaceae, Verbenaceae, and Chenopodiaceae were represented by a single species. All the EOs of these plants were investigated against 13 plant species. Thirdly, six plant EOs rich in non-terpenoid compounds were identified belonging to Lamiaceae (three species), Apiaceae (two), and Cucurbitaceae (one), were tested against 17 plants.

To assess the correlation of EOs phytotoxic activity and structural compounds, a data matrix of each group was performed as a spreadsheet in MS-EXCEL. A matrix of 42 major monoterpene compounds from 45 plant species was assembled, while a matrix of 26 sesquiterpene compounds, identified in the EOs of 22 plant species was prepared. These matrices were subjected to PCA using XLSTAT software version 14 (Addinsoft, New York, NY, USA).

3. Phytotoxic EOs Derived from Plants Rich in Monoterpenes

The EOs from different plant species with monoterpenes as the main compounds that exhibited significant phytotoxic activity against various target plant species are presented in Table 1. Zhang, et al. [14] concluded that monoterpene-rich EOs derived from *Eucalyptus salubris*, *E. dundasii*, *E. spathulata*, and *E. brockwayii* strongly inhibited germination and seedling growth in *Solanum elaeagnifolium* relative to commercial Eucalyptus oil and 1,8-cineole. Moreover, the EO of *E. salubris* was found to be the most powerful inhibitor of germination and roots and shoot growth, while *E. spathulata* exhibited the lowest effect [14].

Table 1. Monoterpene-rich EOs derived from various reported plants with significant allelopathic activity.

Plant Name	Main Monoterpenoid Compounds	Phytotoxic against	Reference
Euphorbia heterophylla	1,8-cineole, camphor,	*Cenchrus echinatus* *	[6]
Symphyotrichum squamatum	β-pinene	*Bidens pilosa* *	[15]
Salvia sclarea	l-linalool, linalyl acetate, α-terpineol, and geraniol	*Lactuca sativa, Lepidium sativum,* and *Portulaca oleracea* *	[16]
Schinus terebinthifolius	3-carene, α-pinene, limonene, and β-pinene	*Bidens pilosa* *	[17]
Cannabis sativa	myrcene, terpinolene, and (E)-β-ocimene	*Avena sativa, Zea mays, Brassica oleracea, Avena fatua* *, *Bromus secalinus* *, *Echinochloa crus-galli* *, *Amaranthus retroflexus* *, *Centaurea cyanus* *	[18]
Callistemon viminalis	1,8-cineole α-pinene, and d-limonene	*Bidens pilosa* *, *Cassia occidentalis* *, *Echinochloa crus-galli* *, and *Phalaris minor* *	[19]
Cymbopogon citratus	neral, geranial, and α-pinene	*Sinapis arvensis* *	[20]
Eucalyptus cladocalyx	1.8-cineole, and *p*-cymene		
Origanum vulgare	carvacrol, γ-terpinene, and *p*-cymene		
Artemisia absinthium	β-thujone, and linalool		
Cymbopogon citratus	neral, geranial, and β-myrcene	*Echinochloa crus-galli* *	[21]
Origanum acutidens	carvacrol, *p*-cymene, linalool acetate	*Amaranthus retroflexus* *, *Chenopodium album* *, and *Rumex crispus* *	[22]
Eriocephalus africanus	carvacrol, *p*-cymene, linalool acetate	*Amaranthus hybridus* * and *Portulaca oleracea* *	[23]
Vitex agnus-castus	1,8-cineole, sabinene, and α-pinene	*Lactuca sativa* and *Lepidium sativum*	[24]
Thymus daenensis	thymol, carvacrol, and *p*-cymene	*Amaranthus retroflexus* *, *Avena fatua* *, *Datura stramonium* *, and *Lepidium sativum*	[25]
Thymus eigii	thymol, carvacrol, *p*-cymene, γ-terpinene, and borneol	*Lactuca sativa, Lepidium sativum,* and *Portulaca oleracea* *	[26]
Thymbra spicata	carvacrol, γ-terpinene, *p*-cymene	*Triticum aestivum, Zea mays, Lactuca sativa, Lepidium sativum,* and *Portulaca oleracea* *	[27]
Nepeta flavida	linalool, 1,8-cineole, and sabinene	*Lepidium sativum, Raphanus sativus,* and *Eruca sativa*	[28]
Heterothalamus psiadioides	β-pinene, δ^3-carene, and limonene	*Lactuca sativa* and *Allium cepa*	[29,30]
Salvia hierosolymitana	α-pinene, myrtenol, and sabinyl acetate,	*Raphanus sativus* and *Lepidium sativum*	[31]
Artemisia scoparia	*p*-cymene, β-myrcene, and (+)-limonene	*Achyranthes aspera, Cassia occidentalis* *, *Parthenium hysterophorus* *, *Echinochloa crus-galli* *, and *Ageratum conyzoides* *	[32]
Eucalyptus grandis	α-pinene, γ-terpinene, and *p*-cymene	*Lactuca sativa*	[33]
Eucalyptus citriodora	β-citronellal, geraniol, and citronellol		
Plectranthus amboinicus	carvacrol	*Lactuca sativa* and *Sorghum bicolor*	[34]

Table 1. *Cont.*

Plant Name	Main Monoterpenoid Compounds	Phytotoxic against	Reference
Pinus brutia	α-pinene, and β-pinene	*Lactuca sativa, Lepidium sativum,* and *Portulaca oleracea* *	[35]
Pinus pinea	limonene, α-pinene, and β-pinene	*Sinapis arvensis* *, *Lolium rigidum* *, and *Raphanus raphanistrum* *	[36]
Cotinus coggyria Scop.	α-pinene, limonene, and β-myrcene	*Silybum marianum* *, and *Portulaca oleracea* *	[37]
Zataria multiflora	carvacrol, linalool and *p*-cymene	*Hordeum spontaneum* *, *Secale cereale* *, and *Amaranthus retroflexus* *, and *Cynodon dactylon* *	[38]
Mentha × *piperita*	menthol, mentone, and menthofuran	*Lycopersicon esculentum, Raphanus sativus, Convolvulus arvensis* *, *Portulaca oleracea* *, and *Echinochloa colonum* *	[39]
Hyssopus officinalis	β-pinene, iso-pinocamphone, and, *trans*-pinocamphone	*Raphanus sativus, Lactuca sativa,* and *Lepidium sativum*	[40]
Lavandula angustifolia	β-pinene, iso-pinocamphone, and *trans*-pinocamphone		
Majorana hortensis	1,8-cineole, β-phellandrene, and α-pinene		
Melissa officinalis	(-)-citronellal, carvacrol, and citronellol		
Ocimum basilicum	linalol, and borneol		
Origanum vulgare	*o*-cymene, carvacrol, and linalyl acetate		
Salvia officinalis	*trans*-thujone, camphor, and borneol		
Thymus vulgaris	*o*-cymene, and α-pinene		
Verbena officinalis	isobornyl formate , and (E)-citral		
Shinus molle	β-phellendrene, α-phellendrene, and myrcene	*Triticum aestivum*	[41]
Syzygium aromaticum	eugenol, and eugenol acetate	*Mimosa pudic* **a and *Senna obtusifolia* *	[42]
Peumus boldus	ascaridole, *p*-cymene and 1,8-cineole	*Amaranthus hybridus* * and *Portulaca oleracea* *	[43]
Drimys winterii	terpinen-4-ol, γ-terpinene, and sabinene		
Agastache rugosa	*d*-limonene, and linalool	*Majorana hortensis* *, *Trifolium repens* *, *Rudbeckia hirta, Chrysanthemum zawadskii, Melissa officinalis* *, *Taraxacum platycarpum* *, and *Tagetes patula*	[44]
Eucalyptus lehmanii	1,8-cineole, α-thujene, and α-pinene	*Sinapis arvensis* *, *Diplotaxis harra* *, *Trifolium campestre* *, *Desmazeria rigida* *, and *Phalaris canariensis* *	[45]
Tanacetum aucheranum	1,8-cineole, camphor, and terpinen-4-ol	*Amaranthus retroflexus* *, *Chenopodium album* *, and *Rumex crispus* *	[46]
Tanacetum chiliophyllum	camphor, 1,8-cineole and borneol		

Table 1. *Cont.*

Plant Name	Main Monoterpenoid Compounds	Phytotoxic against	Reference
Heterothalamus psiadioides	β-pinene, δ^3-carene, and limonene	*Lactuca sativa* and *Allium cepa*	[29]
Baccharis patens	linalool		
Senecio amplexicaulis	α-phellandrene, o-cymene and β-ocimene	*Phalaris minor* * and *Triticum aestivum*	[47]
Eucalyptus salubris	1,8-cineole, α-pinene and p-cymene	*Solanum elaeagnifolium* *	[14]
Eucalyptus dundasii	1,8-cineole, α-pinene and trans-pinocarveol		
Eucalyptus spathulata	1,8-cineole and α-pinene		
Eucalyptus brockwayii	α-pinene, 1,8-cineole and isopentyl isovalerate		
Carum carvi	estragole, limonene, and β-pinene	*Raphanus sativus, Lactuca sativa,* and *Lepidium sativum*	[40]

* Reported as a weed.

Hydro-distilled EO from *Senecio amplexicaulis* with a high content of monoterpenes, including α-phellandrene, O-cymene, and β-ocimene, reportedly exhibited strong allelopathic activity at higher concentrations, with a significant ability to inhibit germination of *Phalaris minor* and *Triticum aestivum* seeds [47]. The EOs of *Heterothalamus psiadioides*, composed mainly of the monoterpenes β-pinene, Δ3-carene, and limonene, showed cidal effects against *Lactuca sativa* and *Allium cepa* by inhibiting germination as well as growth of shoots and roots [29]. Moreover, a strong herbicidal activity against *Amaranthus retroflexus*, *Chenopodium album*, and *Rumex crispus* was reported in the EOs of the two *Tanacetum* species (*T. aucheranum* and *T. chiliophyllum*) by completely inhibiting seed germination and seedling growth, an ability that may be attributable to their monoterpene content, including 1,8-cineole, camphor, borneol, and terpinen-4-ol [46]. Significant reduction of seedling emergence and growth of *Sinapis arvensis*, *Diplotaxis harra*, *Trifolium campestre*, *Desmazeria rigida*, and *Phalaris canariensis* were reported via the EO derived from *E. lehmanii* in which monoterpenes represented the major constituents, including 1,8-cineole, α-thujene, and α-pinene [45].

The EOs from *Agastache rugosa* leaves collected over different seasons reportedly achieved partial or complete prevention of germination and growth of hypocotyl and radicles in *Majorana hortensis*, *Trifolium repens*, *Rudbeckia hirta*, *Chrysanthemum zawadskii*, *Melissa officinalis*, *Taraxacum platycarpum*, and *Tagetes patula*. These extracted EOs were described to be rich in the monoterpenes methylchavicol, *d*-limonene, and linalool as the main compounds [44]. In the same manner, the chemical profiles of the EOs of Chilean *Peumus boldus* and *Drimys winterii* were reported to be composed primarily of the monoterpenes ascaridole, *p*-cymene, and 1,8-cineole and γ-eudesmol, elemol, and terpinen-4-ol. These two EOs were found to exhibit inhibitory effects against *Amaranthus hybridus* and *Portulaca oleracea* [43]. The EO of *Peumus boldus* was found to inhibit germination and seedling growth in two weeds at all used concentrations, while *Drimys winterii* EO exhibited inhibitory activity against germination activity in *Portulaca oleracea* only at the highest dose (1 μL mL^{-1}) [43]. In addition, de Oliveira, et al. [42] reported that different samples of EOs extracted by supercritical CO_2 from *Syzygium aromaticum* at varying temperatures and pressures displayed allelopathic activities by inhibiting germination and radicle elongation in *Mimosa pudica* and *Senna obtusifolia*, with extraction of the EO at 50 °C and 300 bars associated with the most effective activity. The monoterpenes eugenol, eugenol acetate, and (E)-caryophyllene were reported as the main constituents of these samples. The EOs of the leaves and fruits of *Shinus molle* were reported to cause a concentration-dependent decline in the germination and radicle elongation of *Triticum aestivum* with more activity seen in leafy samples. Both oils were found to be composed of monoterpenes as the main components, with an abundance of β-phellendrene, α-phellendrene, and myrcene [41].

The chemical components as well as the phytotoxic activities of EOs derived from 12 Mediterranean plants, including *Hyssopus officinalis, Lavandula angustifolia, Ocimum basilicum, Majorana hortensis, Origanum vulgare, Salvia officinalis, Foeniculum vulgare, Thymus vulgaris, Melissa officinalis, Verbena officinalis, Pimpinella anisum,* and *Carum carvi,* on germination and radicle growth in *Raphanus sativus, Lactuca sativa,* and *Lepidium sativum* seeds have been documented [40]. The EOs reportedly have an inhibitory effect against germination and initial radicle elongation at different doses through different mechanisms, with samples of *Melissa officinalis, Thymus vulgaris, Verbena officinalis,* and *Carum carvi* demonstrating the strongest effect. Monoterpenes were described as the main components of *Hyssopus officinalis, Lavandula angustifolia, Majorana hortensis, Melissa officinalis, Ocimum basilicum, Origanum vulgare, Salvia officinalis, Thymusvulgaris, Verbena officinalis,* and *Carum carvi,* while non-terpenoid phenols were the main constituents in *Foeniculum vulgare* and *Pimpinella anisum* [40]. Similarly, peppermint EO is reportedly rich in menthone, menthol, and menthofuran, and has been described as having a potent allelopathic effect on seed germination and seedling growth in *Lycopersicon esculentum, Raphanus sativus, Convolvulus arvensis, Portulaca oleracea,* and *Echinochloa colonum* [39]. In 2010, the main components of *Zataria multiflora* EO were reported as the monoterpenes carvacrol, linalool, and *p*-cymene, all of which exhibited significant herbicidal activities against *Hordeum spontaneum, Secale cereal, Amaranthus retroflexus,* and *Cynodon dactylon* [38]. This activity was associated with significant inhibition of the rate of germination, seedling length, root and stem fresh and dry weights at all used concentrations, and 320 and 640 mL L^{-1} in particular [38].

The dose-dependent toxicity of the EOs of *Cotinus coggyria,* which consist mainly of monoterpenes such as limonene, α-pinene, and β-myrcene, against the weeds *of Silybum marianum* and *Portulaca oleracea* reportedly [37] decreased germination in radishes by 83% and 60%, seedling radicle length by 93% and 84%, and plumule length by 84% and 91% at 32 µL mL^{-1}. The EOs of *Pinus brutia* and *Pinus pinea* were documented to have monoterpenes as the major components, with a preponderance of α- and β-pinene and caryophyllene [35]. At higher dose, these EOs were found to have a potent inhibitory effect on germination by 53% and 22% of *Lactuca sativa,* 60% and 33% of *Lepidium sativum,* and 13% and 3% of *Portulaca oleracea,* respectively [35]. An evaluation of Tunisian *Pinus pinea* EO, rich in limonene, α- and β-pinene, revealed a dose-related gradual inhibition of *Lolium rigidum, Sinapis arvensis,* and *Raphanus raphanistrum,* and seed germination was completely inhibited at low concentrations [36]. The EO of *Plectranthus amboinicus,* which is composed primarily of monoterpenes, and carvacrol in particular (88.61%), was reported to significantly inhibit germination and reduce the growth of *Lactuca sativa* and *Sorghum bicolor* roots and shoots [34]. Another study [33] described the EO chemical composition of two Eucalyptus plants in which monoterpenes, including p-cymene, β-myrcene, and (+)-limonene, were the main components in *Eucalyptus grandis,* and α-pinene, γ-terpinene, and *p*-cymene in *E. citriodora.* These EOs were found to dose-dependently inhibit germination of *Lactuca sativa,* with a concentration of 0.1 µL mL^{-1} of both oils suppressing germination by 74% and 68%, respectively [33].

With a preponderance of the monoterpenoids *p*-cymene, β-myrcene, and (+)-limonene, along with acenaphthene, EO extracted from *Artemisia scoparia* was found to have significant phytotoxic activities, primarliy against the roots of *Achyranthes aspera, Cassia occidentalis, Echinochloa crus-galli, Ageratum conyzoides,* and *Parthenium hysterophorus* [32] with the latter species suffering the most effects [32]. In the same manner, an artemisia ketone–rich EO derived from *Eucalyptus africanus* reportedly exhibited a potent effect similar to that *of Eucalyptus camaldulensis* on *Amaranthus hybridus,* but without a noticeable effect on *Portulaca oleracea* [23].

The phytotoxic activity of the EO of the Turkish *Origanum acutidens* and its monoterpenoid components (carvacrol, *p*-cymene, and thymol) were studied by Kordali, et al. [22]. Their results revealed that carvacrol and thymol completely inhibited seed germination and seedling growth in *Chenopodium album, Amaranthus retroflexus,* and *Rumex crispus,* but no effect was observed with *p*-cymene [22]. The EO of *Cymbopogon citratus,* which is composed mostly of the monoterpenes neral, geranial, and *β*-myrcene, was reported to delay the germination of seeds and inhibit seedling growth in *Echinochloa crus-galli* [21]. In another study, the EOs of the aerial parts of four plants—*Cymbopogon citratus, Origanum vulgare, Eucalyptus cladocalyx,* and *Artemisia absinthium*—were determined to be potential bioherbicides against the seeds of *Sinapis arvensis,* with the EOs of *Cymbopogon citratus* and *Eucalyptus cladocalyx* [20] exhibiting the most activity. Neral, geranial, and *α*-pinene, along with other monoterpenes, were found to be the main components of *Cymbopogon citratus.* However, the sesquiterpene spathulenol, as well as the monoterpenes 1,8-cineole and *p*-cymene, were found to be the main components of the EO of *Eucalyptus cladocalyx.* The three monoterpenoids carvacrol, *γ*-terpinene, and *p*-cymene were described as major constituents of the EO of *Origanum vulgare,* while *Artemisia absinthium* EO was reported to be composed largely of monoterpenes, including *β*-thujone, chamazulene, and linalool [20].

Potential herbicidal activity of the EO from *Nepeta flavida* was reported against *Raphanus sativus, Lepidium sativum,* and *Eruca sativa,* in which it completely inhibited germination at a concentration of 4.0 μL mL^{-1} [28], an effect that may be attributed to the presence of the monoterpenes linalool, 1,8-cineole, and sabinene [28]. Monoterpenes including carvacrol, *γ*-terpinene, *p*-cymene were found to be the predominant constituents of the EO of *Thymbra spicata* and may be responsible for the strong phytotoxic activity reported against *Zea mays, Triticum aestivum, Lactuca sativa, Lepidium sativum,* and *Portulaca oleracea* [27]. Ulukanli, et al. [26] reported that the EO of *Thymus eigii* exhibited significant toxic effects against *Lepidium sativum, Lactuca sativa,* and *P. oleracea.* This oil was found to be rich with monoterpenes, with thymol, carvacrol, and *p*-cymene the major constituents [26]. Moreover, EOs extracted from *Thymus daenensis* collected from four different habitats inhibited germination in *Avena fatua, Amaranthus retroflexus, Datura stramonium,* and *Lepidium sativum.* These four ecospecies of *Thymus daenensis* were found to be rich with monoterpenoids, including thymol, carvacrol, and *p*-cymene in particular [25].

Foliar volatiles of *Callistemon viminalis* and EOs reportedly reduced seed germination, seedling growth, and accumulation of dry matter in *Bidens pilosa, Cassia occidentalis, Echinochloa crusgalli,* and *Phalaris minor,* with the greatest sensitivity observed in *B. pilosa* [19]. Monoterpenoids, and 1,8-cineole, *α*-pinene, and *d*-limonene in particular, have been described as major components in the EO of this plant. Pinheiro, et al. [17] reported an allelopathic effect of the EO extracted from *Cannabis sativa* on germination and seedling growth in *Amaranthus retroflexus, Bromus secalinus, Avena sativa,* and *Brassica oleracea.* Based on gas chromatography-mass spectroscopy analysis, this oil is rich in monoterpenes, including myrcene, terpinolene, and (E)-*β*-Ocimene [18].

Monoterpene-rich EOs derived from *Schinus terebinthifolius* collected from two different areas of Brazil reportedly produced an inhibitory effect on germination, root and hypocotyl growth, and production of biomass in *Bidens pilosa.* In the Cerrado biome, the EO of *Schinus terebinthifolius* was found to be rich with *trans*-caryophyllene, 3-carene, and germacrene B, while *Schinus terebinthifolius* from the country's Atlantic forest biome is rich with *α*-pinene, limonene, and *β*-pinene [17]. Additionally, the EO of *Vitex agnus-castus* was reported to be rich with monoterpenes, particularly 1,8-cineole, sabinene, *trans-β*-farnesene, and *α*-pinene. This EO was found to exhibit significant inhibitory activity on *Lactuca sativa* and *Lepidium sativum* [24].

The EO of *Salvia sclarea* was described to have significant phytotoxic effects against *Lepidium sativum, Lactuca sativa,* and *Portulaca oleracea* at a concentration of 0.16 mg mL^{-1}, reducing seed germination by 94%, 100%, and 50%, respectively [16]. The main constituent of this EO was reported to be monoterpenes, including L-linalool, linalyl acetate, *α*-terpineol, and geraniol [16].

The EO from the aerial parts of *Euphorbia heterophylla* reportedly inhibited germination (93.9%), root (84.6%), and shoot growth (57.8%) in *Cenchrus echinatus* weeds at 100 μL L^{-1}.

The authors described monoterpenes as the major components (69.48 %), and 1,8-cineole was the primary monoterpene, representing 32.03% of the total mass [6].

4. Monoterpene-Rich EO-Allelopathy Correlation

Application of a PCA to a dataset of the 46 different plant species with EOs comprised mainly of monoterpenes found allelopathic activity is presented in Figure 1. The results show that that α- and β-pinene, 1,8 cineole, linalool, and carvacrol were the most effective allelopathic monoterpene compounds. They also showed that *Eucalyptus africanus*, *Origanum acutidens*, *Zataria multiflora*, and *Plectranthus amboinicus* were correlated to each other linalool and carvacrol predominating (Figure 1). Meanwhile, *Pinus brutia*, *Schinus terebinthifolius*, *Thymus vulgaris*, *Eucalyptus lehmanii*, *Eucalyptus lehmanii*, *Eucalyptus lehmanii*, *Callistemon viminalis*, *Majorana hortensis*, *Vitex agnus-castus*, and *Eucalyptus brockwayii* showed a close correlation with each other with respect to the composition of their EOs; 1,8-cineole and α, and β-pinene were the major monoterpenoid compounds. The analysis found α-, and β-pinene and 1,8 cineole in most of the allelopathic plants in which monoterpenes are major EO compounds.

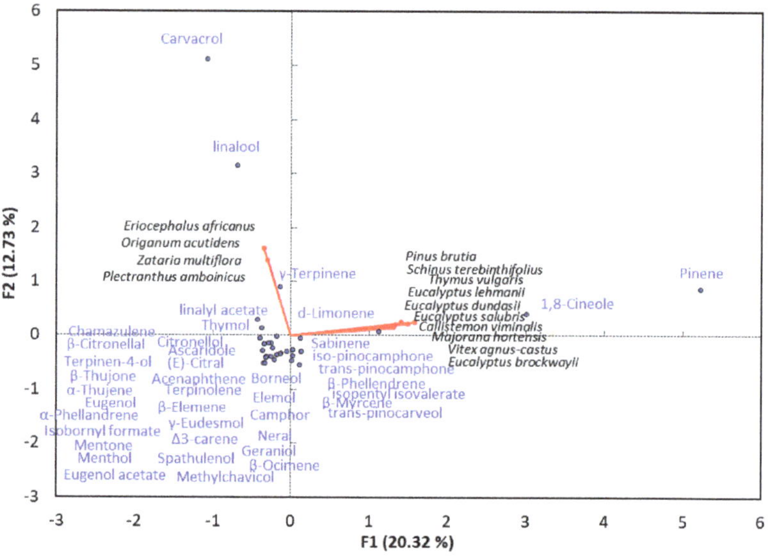

Figure 1. Principal component analysis of reported plants with essential oils containing monoterpenes as major compounds and showing allelopathic activity.

5. Phytotoxic EOs Derived from Plants Rich in Sesquiterpenes

Sesquiterpene-rich EOs from different plants associated with notable phytotoxic activities are listed in Table 2. The EO of *Eupatorium adenophorum* was described as being composed primarily of sesquiterpenes, with γ-cadinene, γ-muurolene, and 3-acetoxyamorpha-4,7(11)-diene-8-one as the main compounds. This oil reportedly exhibited strong phytotoxic activity against *Phalaris minor* and *Triticum aestivum*, with a stronger effect observed against *Phalaris minor* [48]. Elshamy and his co-workers reported the EO composition and allelopathic activities of three *Launaea* plants (*Launaea mucronata*, *Launaea nudicaulis*, and *Launaea spinosa*) collected from different habitats. Results showed that these EOs had significant and concentration-dependent effects on *Portulaca oleracea* weeds. The EOs of two samples of *Launaea mucronata* collected from the desert and coastal regions were found to have the highest activity, inhibiting germination by 96.1% and 87.9% and radicle growth by 92.6% and 89.7%, respectively, at 250 µL L^{-1} [6]. The authors found that sesquiterpenes were the main components, and hexahydrofarnesyl acetone the main compound, in *Launaea mucronata* [6]. The EO of *Schinus lentiscifolius* was reported to be associated with a 19.35% reduction in the mitotic index in onions and 25.14% in lettuce, compared with negative control. This EO was found to comprise sesquiterpenoid compounds as the main components, and δ-cadinene in particular [49].

Table 2. Sesquiterpene-rich EOs derived from various plants and exhibiting phytotoxic activity.

Plant Name	Major Sesquiterpenes Compounds	Phytotoxic Against	Reference
Lactuca serriola	isoshyobunone, and alloaromadendrene oxide-1	*Bidens pilosa* *	[7]
Launaea mucronata	hexahydrofarnesyl acetone and (-)-spathulenol	*Portulaca oleracea* *	
Launaea nudicaulis	hexahydrofarnesyl acetone and γ-gurjunen epoxide (2)		[6]
Launaea spinosa	α-acorenol, trans-longipinocarveol, and γ-eudesmol		
Heliotropium curassavicum	Hexahydrofarnesyl acetone, (-)-caryophyllene oxide, farnesyl acetone	*Chenopodium murale* *	[6]
Xanthium strumarium	α-eudesmol, (-)-spathulenol, and ledene alcohol	*Bidens pilosa* *	[3]
Cullen plicata	(−)-caryophyllene oxide, z-nerolidol, tau.cadinol and α-cadinol	*Bidens pilosa* * and *Urospermum picroides* *	[50]
Scutellaria strigillosa	germacrene D, bicyclogermacrene, and β-caryophyllene	*Amaranthus retroflexus* * and *Poa annua* *	[51]
Acroptilon repens	caryophyllene oxide, β-cubebene, β-caeyophyllen, and α-copaen	*Amaranthus retroflexus* * and *Cardaria draba* *	[52]
Lantana camara	α-curcumene, β-caryophyllene, and γ-curcumene	*Amaranthus hybridus* * and *Portulaca oleracea* *	[23]
Eucalyptus camaldulensis	spathulenol, and isobicyclogermacrenal		
Eupatorium adenophorum	γ-cadinene, γ-muurolene, and 3-acetoxyamorpha-4,7(11)-diene-8-one	*Phalaris minor* * and *Triticum aestivum* *	
Baccharis patens	β-caryophyllene, and spathulenol	*Lactuca sativa* and *Allium cepa*	[29]
Salvia multicaulis	α-Copaene, β-caryophyllene, and aromadendrene	*Raphanus sativus* and *Lepidium sativum*	[40]
Teucrium arduini	caryophyllene, caryophyllene oxide, germacrene D, and spathulenol		
Teucrium maghrebinum	germacrene d, δ-cadinene, γ-cadinene, and caryophyllene		
Teucrium polium	caryophyllene, torreyol, and α-cadinol		
Teucrium montbretii	carvacrol, caryophyllene, and caryophyllene oxide		
Nepeta curviflora	β-caryophyllene, caryophyllene oxide, and (E)-β-farnesene		[31]
Nepeta nuda	β-bisabolene		
Ailanthus altissima	β-caryophyllene, (Z)-caryophyllene, and germacrene D,	*Lactuca sativa*	[53]
Schinus lentiscifolius	δ-cadinene, α-cadinol, and β-caryophyllene		[49]
Pulicaria somalensis	Juniper camphor (24.7%), α-sinensal (7.7%), 6-epi-shyobunol (6.6%), and α-zingiberene (5.8%)	*Dactyloctenium aegyptium* * and *Bidens pilosa* *	[4]
Bassia muricata	hexahydrofarnesyl acetone, and α-gurjunene	*Chenopodium murale* *	[54]

* Reported as a weed.

A study of the chemical profiles of EOs of Tunisian *Ailanthus altissima* [53] deduced the presence of a high concentration of sesquiterpenes such as *β*-caryophyllene, (Z)-caryophyllene, germacrene D, and hexahydrofarnesyl acetone. The phytotoxic activities of the EOs (at a concentration of 1 mg mL^{-1}) of the roots, stems, leaves, flowers, and fruits completely inhibited seed germination in *Lactuca sativa* [53]. *Raphanus sativus* and *Lepidium sativum* root growth was reduced under the effects of the EOs of *Nepeta curviflora* and *Nepeta nuda*. These EOs were found to have sesquiterpenes, and *β*-caryophyllene, caryophyllene oxide, and *β*-bisabolene in particular, as the main components. [31]. In addition, the EOs of *Teucrium maghrebinum*, *Teucrium polium*, and *Teucrium montbretii* were reported to be rich sources of sesquiterpenes, including caryophyllene, caryophyllene oxide, and carvacrol in particular. These EOs were found to significantly reduce the radicle growth of *Raphanus sativus* and *Lepidium sativum* with mild effect on germination [55].

The herbicidal effects of the EOs derived from *Lantana camara*, *Eucalyptus camaldulensis*, and *Eriocephalus africanus* were determined by Verdeguer, et al. [23]. The EO of *Eucalyptus camaldulensis*, which is reportedly composed primarily of spathulenol, had the greatest impact among the three plants, completely inhibiting seedling growth and germination in *Amaranthus hybridus* and *Portulaca oleracea*. With a high concentration of sesquiterpenes, and sesquiterpene hydrocarbons in particular. The EO of *Lantana camara* reportedly exhibited significant allelopathic activity against *Amaranthus hybridus* [23]. Recently, Elshamy, et al. [7] reported significant allelopathic effects of *Lactuca serriola* EO against *Bidens pilosa*, with half-maximal inhibitory concentrations (IC$_{50}$) of 104.3, 92.3, and 140.3 µL L^{-1} for germination, growth of roots, and growth of shoots, respectively. The EO of *Lactuca serriola* was described to be rich with sesquiterpenes, with isoshyobunone and alloaromadendrene oxide-1 as major components. The EO of the invasive noxious plant *Heliotropium curassavicum*, collected from an inland area, demonstrated remarkable phytotoxic activities against *Chenopodium murale*, with IC$_{50}$ values of 2.66, 0.59, and 0.70 mg mL^{-1} for germination, growth of roots, and growth of shoots, respectively. A coastal sample of the same species exhibited more allelopathic activity, with IC$_{50}$ values of 1.58, 0.45, and 0.66 mg mL^{-1} [7]. Sesquiterpenes were determined to be the main class of EOs of *Heliotropium curassavicum*, and hexahydrofarnesyl acetone, (-)-caryophyllene oxide, and farnesyl acetone were the major compounds. In 2019, Abd El-Gawad and his co-authors reported that EOs from the leaves of the Egyptian *Xanthium strumarium* exhibited allelopathic effects against *Bidens pilosa*, and a concentration of 1000 µL L^{-1} inhibited seed, root, and shoot germination growth by 97.34%, 98.45%, and 93.56%, respectively [3]. In the EO of *Xanthium strumarium*, the sesquiterpenoids 1,5-dimethyltetralin, eudesmol, and l-borneol were the major identified compounds.

The EO of *Symphyotrichum squamatum* collected from Egypt was analyzed and found to be enriched in sesquiterpenes such as humulene, epoxide, (-)-spathulenol, and (-)-caryophyllene oxide [15]. The EO of this plant was reported to have a strong and concentration -dependent allelopathic effect against *Bidens pilosa* weeds. The EO of *Cullen plicata*, rich in sesquiterpenes such as (−)-caryophylleneoxide, Z-nerolidol, tau cadinol, and *α*-cadinol, was reported to completely inhibit germination in *Bidens pilosa* and *Urospermum picroides* at 200 µL L$_{-1}$ with respective IC$_{50}$ values of 49.39 and 17.86 µL L^{-1} [50]. The EO derived from *Scutellaria strigillosa* was found to have significant phytotoxic potential against *Amaranthus retroflexus* and *Poa annua* [51]. These weeds were inhibited by 86.6 % and 20.0%, respectively, when treated with 1 µL ml^{-1} of *Scutellaria strigillosa* EOs. This active EO was found to be rich in sesquiterpenes, and germacrene D, 1-octen-3-ol, bicyclogermacrene, and *β*-caryophyllene in particular. In another study, the extracted EO of *Acroptilon repes* was examined by Razavi, et al. [52] for its phytotoxic activity against *Amaranthus retroflexus* and *Cardaria draba*. They reported that EOs from *Acroptilon repes* had a significant inhibitory effect on seed germination in *Amaranthus retroflexus*. Sesquiterpenes including caryophyllene oxide, *β*-cubebene, *β*-caeyophyllen, and *α*-copaen were reported as the main constituents of this EO [52].

Recently, Assaeed et al. reported that sesquiterpene-rich EOs of the aerial parts of *Pulicaria somalensis* had significant phytotoxic effects on the weeds of *Dactyloctenium aegyptium* and *Bidens poilosa*, with an IC_{50} of 0.6 mg mL^{-1} for root growth in both weeds, and 0.7 and 1.0 mg mL^{-1} for shoot growth, respectively. Juniper camphor (24.7%), α-sinensal (7.7%), 6-epi-shyobunol (6.6%), and α-zingiberene (5.8%) were reported to be the main chemical constituents of the EO of this plant [54].

Lastly, the EO of aboveground parts of *Bassia muricata* (Chenopodiaceae) was found to have a significant reduction effect on root growth, shoot growth, and germination in *Chenopodium murale* weed, with IC_{50} values of 175.60 μL L^{-1}, 246.65 μL L^{-1}, and 308.33 μL L^{-1}, respectively. Sesquiterpenes were found to be the main constituents of the EO, with an abundance of hexahydrofarnesyl acetone, and α-gurjunene [54].

6. Sesquiterpene-Rich EO-Allelopathy Correlation

The application of PCA to a dataset of 15 different plant species with EOs composed mainly of sesquiterpenoid compounds showed allelopathic activity is presented in Figure 2. Caryophyllene, caryophyllene oxide, germacrene D, spathulenol, and hexahydrofarnesyl acetone were the sesquiterpenoids most associated with allelopathic activity. Most tested plant species were correlated to each other regarding these major EO compounds.

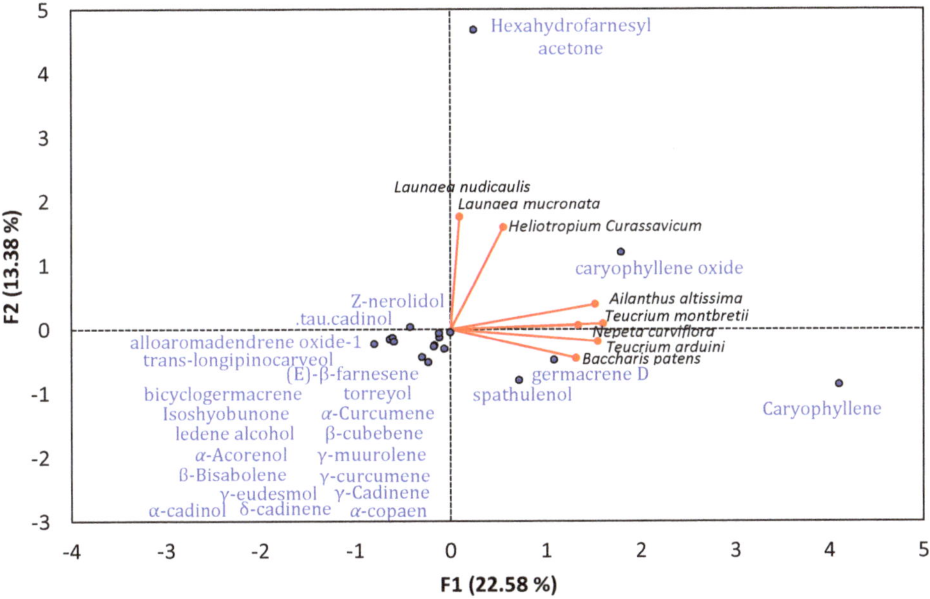

Figure 2. Principal component analysis of reported plants with essential oils containing sesquiterpenes as major compounds and showing allelopathic activity.

7. Phytotoxic EOs Derived from Plants Rich in Non-Terpenoids

Phytotoxic EOs with non-terpenoid major compounds are listed in Table 3. The EOs of leaves and fruits of *Ecballium elaterium* reportedly contain phenolics and hydrocarbons, including E-anethol, octyl octanoate, 3-(6,6-dimethyl-5-oxohept-2-enyl)-cyclohexanone, and tetracosane as major components [56]. The EO of the leaves was found to have an allelopathic effect on *Lactuca sativa* that was stronger than that of fruits, with a significant (12%) decrease in seed germination. In another study, Mutlu, et al. [57] found that EO rich in iridoids from *Nepeta meyeri* had a strong inhibitory effect (>50%) on seed germination of

Bromus danthoniae, *Bromus intermedius*, and *Lactuca serriola* at a concentration of 0.01% and 0.02%. Kordali, et al. [58] reported that the EO of the Turkish plant *Nepeta meyeri* contained 4a-α,7-α,7a-β nepetalactone and 4a-α,7-α,7a-α nepetalactone as major compounds. This EO completely inhibited germination of *Amaranthus retroflexus*, *Chenopodium album*, *Cirsium arvense*, and *Sinapsis arvensis* at a concentration of 0.5 mg mL^{-1}. Iridoids, and 4a-α,7-α,7a-β-nepetalactone and 4a-α,7-β,7a-α-nepetalactone in particular, were determined to be the major compounds of the EO of *Nepeta cataria* [59]. This EO can reportedly act as an allelochemical agent against *Hordeum spontaneum*, *Taraxacum officinale*, *Avena fatua*, and *Lipidium sativum*, with dose-dependent suppression of germination [59]. Similarly, Bozok, et al. [60] reported a strong herbicidal activity for the EO of *Nepeta nuda* on germination and seedling growth in *Raphanus sativus*, *Triticum aestivum*, *Lactuca sativa*, *Portulaca oleracea*, and *Lepidium sativum*. This EO was rich in iridoids 4a-α,7-α,7a-β-nepetalactone, 2(1H)-naphthalenone, and trans-octahydro-8a-methyl. Finally, Mancini, et al. [31] reported that the EOs of the two Salvia species (*Salvia hierosolymitana* and *Salvia multicaulis*) displayed phytotoxic effects on *Raphanus sativus* by reducing radicle elongation and seed germination. These EOs are characterized by an abundance of carbonylic compounds.

Table 3. Nonterpenoidial-rich EOs derived from plants and with significant phytotoxic activity.

Plant Name	Main Components	Major Compounds	Phytotoxic Against	Reference
Nepeta nuda	Iridoids	4a-α,7-α,7a-β-nepetalactone, 2(1h)-naphthalenone, and octahydro-8a-methyl-trans-	*Triticum aestivum*, *Raphanus sativus*, *Lactuca sativa*, *Lepidium sativum*, and *Portulaca oleracea* *	[60]
Nepeta cataria	Iridoids	4a-α,7-α,7a-β-nepetalactone and 4a-α,7-β,7a-α-nepetalactone	*Hordeum spontaneum* *, *Taraxacum officinale* *, *Avena fatua* *, and *Lipidium sativum*	[59]
Nepeta meyeri	Iridoids	4a-α,7-α,7a-β-nepetalactone and 4a-α,7-β,7a-α-nepetalactone	*Amaranthus retroflexus* *, *Bromus danthoniae* *, *Bromus intermedius* *, *Chenopodium album* *, *Cynodon dactylon* *, *Lactuca serriola* *, *Portulaca oleracea* *, *Cirsium arvense* *, and *Sinapsis arvensis* *	[57,58]
Ecballium elaterium	Phenolics and hydrocarbons	*e*-anethol, octyl octanoate, 3-(6,6-dimethyl-5-oxohept-2-enyl)-cyclohexanone, and tetracosane	*Lactuca sativa*	[56]
Pimpinella anisum	Non-terpenoidial phenols	*cis*-anethole	*Raphanus sativus*, *Lactuca sativa*, and *Lepidium sativum*	[40]
Foeniculum vulgare	Non-terpenoidial phenols	*cis*-anethole		

* Reported as a weed.

8. Structure-Activity Relationship Summary

Based on the data presented in Tables 1–3 and correlation analysis between phytotoxic EOs derived from different plants and their major chemical constituents (Figures 1 and 2), we concluded that the phytotoxic activities of EOs increase with terpenoid content, particularly oxygenated terpenoid content. Almost all previous studies found that increasing the oxygenation of terpenoids led to an increase in allopathic activities via inhibition of germination and growth of noxious weeds [6,16,23].

As can be seen in Table 1, oxygenated monoterpenoids are the main components of phytotoxic EOs, and their phytotoxicity was observed to increase with the degree of oxygenation. For example, the mono-oxygenated monoterpenoid 1,8-cineole (eucalyptol, $C_{10}H_{18}O$), was to be the main compound in several allopathic EOs derived from plants from different botanical families such as *Euphorbia heterophylla* [6], *Callistemon viminalis* [19], *Eucalyptus cladocalyx* [20], *Nepeta flavida* [28], *Majorana hortensis* [40], *Peumus boldus* [43], *Eucalyptus lehmanii* [45], *Tanacetum aucheranum, Tanacetum chiliophyllum* [46], *Eucalyptus salubris, Eucalyptus brockwayii,* and *Eucalyptus dundasii* [14].

In addition, linalool and borneol were found to be the major compounds in numerous phytotoxic oils, such as *Salvia sclarea* [16], *Artemisia absinthium* [20], *Origanum acutidens* [22], *Eriocephalus africanus* [23], *Nepeta flavida* [28], *Zataria multiflora* [38], *Agastache rugosa* [44], *Salvia officinalis,* and *Ocimum basilicum* [40], and *Tanacetum chiliophyllum* [46]. It is therefore clear that the oxygenated monoterpenoids 1,8-cineole, linalool, and borneol play significant and effective roles as allopathic agents and more research into their phytotoxic activity and phytotoxic mechanism(s) is recommended.

Similarly, careful analysis of sesquiterpene-rich phytotoxic EOs revealed that an increase in oxygenated sesquiterpene levels can enhance the phytotoxic activity of an EO. The data supplied in Table 2 and PCA analysis suggest the major oxygenated sesquiterpenes caryophyllene and its derivatives, as well as hexahydrofarnesyl acetone, can be potent phytotoxic agents. The phytotoxic EOs derived from *Baccharis patens* [29,30], *Heliotropium curassavicum* [7], *Cullen plicata* [50], *Scutellaria strigillosa* [51], *Acroptilon repens* [52], *Lantana camara* [23], *Teucrium arduini, Teucrium maghrebinum, Teucrium polium, Teucrium montbretii* [31], and *Ailanthus altissima* [53] were reported to have all or one of β-caryophyllene, (-)-caryophyllene, and caryophyllene oxide as primary compounds. These reports indicate a strong correlation between the phytotoxic activities of EOs and the presence of these compounds as main components. Hexahydrofarnesyl acetone has been described as an essential compound in the phytotoxic EOs of *Launaea mucronata, Launaea nudicaulis* [6], *Heliotropium curassavicum* [7], and *Ailanthus altissima* [53]. The authors of these studies also concluded that compounds with hexahydrofarnesyl acetone as a main constituent can play a major role as phytotoxic mediators.

EOs derived from aromatic plants typically consist of low-molecular-weight terpenoids, including mono, sesqui-, and diterpenoids as well as non-terpenoid components [14]. Two plants belonging to the *Nepeta* genus were reported that containing iridoid-rich EOs such as *Nepeta meyeri* [57,58] and *Nepeta cataria* [59]. The two iridoids 4a-α,7-α,7a-β-nepetalactone and 4a-α,7-β,7a-α-nepetalactone were reported to be the main phytotoxic mediators in the EOs of these two species. The two compounds should, therefore, be subjected to further study to evaluate their allopathic abilities against several noxious weeds.

Author Contributions: A.I.E. and A.M.A.-E. contributed to the conceptualization, data collection, analysis of data, visualization, and writing of the first draft of the manuscript. A.I.E., A.M.A.-E., A.E.-N.G.E.G., A.M.A., S.L.A.-R., A.S.A., T.A.M., M.I.N., and Y.H.D. contributed to writing—review and editing of the manuscript. All authors discussed the results, commented on the paper, and approved the final manuscript. All authors have read and agreed to the published version of the manuscript.

Funding: This research was funded by Deputyship for Research & Innovation, "Ministry of Education" in Saudi Arabia for funding this research work through the project number IFKSURP-113 and the APC was funded also by the same project.

Acknowledgments: The authors extend their appreciation to the Deputyship for Research & Innovation, "Ministry of Education" in Saudi Arabia for funding this research work through the project number IFKSURP-113. The authors thank the Deanship of Scientific Research and RSSU at King Saud University for their technical support.

Conflicts of Interest: The authors declare no conflict of interest.

References

1. Jugulam, M. *Biology, Physiology and Molecular Biology of Weeds*; CRC Press: Boca Raton, FL, USA, 2017.
2. Rice, E. *Allelopathy*, 2nd ed.; Academic Press: New York, NY, USA, 1984.
3. Abd El-Gawad, A.M.; Elshamy, A.I.; El Gendy, A.E.-N.; Gaara, A.; Assaeed, A.M. Volatiles profiling, allelopathic activity, and antioxidant potentiality of *Xanthium strumarium* leaves essential oil from Egypt: Evidence from chemometrics analysis. *Molecules* **2019**, *24*, 584. [CrossRef] [PubMed]
4. Assaeed, A.; Elshamy, A.; El Gendy, A.E.-N.; Dar, B.; Al-Rowaily, S.; Abd El-Gawad, A. Sesquiterpenes-rich essential oil from above ground parts of *Pulicaria somalensis* exhibited antioxidant activity and allelopathic effect on weeds. *Agronomy* **2020**, *10*, 399. [CrossRef]
5. Sharifi-Rad, J.; Sureda, A.; Tenore, G.C.; Daglia, M.; Sharifi-Rad, M.; Valussi, M.; Tundis, R.; Sharifi-Rad, M.; Loizzo, M.R.; Ademiluyi, A.O. Biological activities of essential oils: From plant chemoecology to traditional healing systems. *Molecules* **2017**, *22*, 70. [CrossRef]
6. Abd El-Gawad, A.M.; Elshamy, A.; El Gendy, A.E.-N.; Al-Rowaily, S.L.; Assaeed, A.M. Preponderance of oxygenated sesquiterpenes and diterpenes in the volatile oil constituents of *Lactuca serriola* L. revealed antioxidant and allelopathic activity. *Chem. Biodivers.* **2019**, *16*, e1900278. [CrossRef] [PubMed]
7. Elshamy, A.; Abd El-Gawad, A.M.; El-Amier, Y.A.; El Gendy, A.; Al-Rowaily, S. Interspecific variation, antioxidant and allelopathic activity of the essential oil from three *Launaea* species growing naturally in heterogeneous habitats in Egypt. *Flavour Fragr. J.* **2019**, *34*, 316–328. [CrossRef]
8. Abd El-Gawad, A.M.; El-Amier, Y.A.; Bonanomi, G. Essential oil composition, antioxidant and allelopathic activities of *Cleome droserifolia* (Forssk.) Delile. *Chem. Biodivers.* **2018**, *15*, e1800392. [CrossRef]
9. Abd El-Gawad, A.M.; El-Amier, Y.A.; Bonanomi, G. Allelopathic activity and chemical composition of *Rhynchosia minima* (L.) DC. essential oil from Egypt. *Chem. Biodivers.* **2018**, *15*, e1700438. [CrossRef]
10. Deng, W.; Liu, K.; Cao, S.; Sun, J.; Zhong, B.; Chun, J. Chemical composition, antimicrobial, antioxidant, and antiproliferative properties of grapefruit essential oil prepared by molecular distillation. *Molecules* **2020**, *25*, 217. [CrossRef]
11. Elshamy, A.I.; Ammar, N.M.; Hassan, H.A.; Al-Rowaily, S.L.; Raga, T.R.; El Gendy, A.; Abd El-Gawad, A.M. Essential oil and its nanoemulsion of *Araucaria heterophylla* resin: Chemical characterization, anti-inflammatory, and antipyretic activities. *Ind. Crop. Prod.* **2020**, *148*, 112272. [CrossRef]
12. Arunachalam, K.; Balogun, S.O.; Pavan, E.; de Almeida, G.V.B.; de Oliveira, R.G.; Wagner, T.; Cechinel Filho, V.; de Oliveira Martins, D.T. Chemical characterization, toxicology and mechanism of gastric antiulcer action of essential oil from *Gallesia integrifolia* (Spreng.) Harms in the in vitro and in vivo experimental models. *Biomed. Pharmacother.* **2017**, *94*, 292–306. [CrossRef]
13. Damtie, D.; Braunberger, C.; Conrad, J.; Mekonnen, Y.; Beifuss, U. Composition and hepatoprotective activity of essential oils from Ethiopian thyme species (*Thymus serrulatus* and *Thymus schimperi*). *J. Essent. Oil Res.* **2019**, *31*, 120–128. [CrossRef]
14. Zhang, J.; An, M.; Wu, H.; Li Liu, D.; Stanton, R. Chemical composition of essential oils of four *Eucalyptus* species and their phytotoxicity on silverleaf nightshade (*Solanum elaeagnifolium* Cav.) in Australia. *Plant. Growth Regul.* **2012**, *68*, 231–237. [CrossRef]
15. Abd El-Gawad, A.M.; Elshamy, A.; El-Amier, Y.A.; El Gendy, A.; Al-Barati, S.; Dar, B.; Al-Rowaily, S.; Assaeed, A. Chemical composition variations, allelopathic, and antioxidant activities of *Symphyotrichum squamatum* (Spreng.) Nesom essential oils growing in heterogeneous habitats. *Arab. J. Chem.* **2020**, *13*, 237–4245. [CrossRef]
16. Bozok, F.; Ulukanli, Z. Volatiles from the aerial parts of east Mediterranean clary sage: Phytotoxic activity. *J. Essent. Oil Bear. Plants* **2016**, *19*, 1192–1198. [CrossRef]
17. Pinheiro, P.F.; Costa, A.V.; Tomaz, M.A.; Rodrigues, W.N.; Fialho Silva, W.P.; Moreira Valente, V.M. Characterization of the Essential Oil of Mastic Tree from Different Biomes and its Phytotoxic Potential on Cobbler's Pegs. *J. Essent. Oil Bear. Plants* **2016**, *19*, 972–979. [CrossRef]
18. Agnieszka, S.; Magdalena, R.; Jan, B.; Katarzyna, W.; Malgorzata, B.; Krzysztof, H.; Danuta, K. Phytotoxic effect of fiber hemp essential oil on germination of some weeds and crops. *J. Essent. Oil Bear. Plants* **2016**, *19*, 262–276. [CrossRef]
19. Bali, A.S.; Batish, D.R.; Singh, H.P.; Kaur, S.; Kohli, R.K. Chemical characterization and phytotoxicity of foliar volatiles and essential oil of *Callistemon viminalis*. *J. Essent. Oil Bear. Plants* **2017**, *20*, 535–545. [CrossRef]
20. Fouad, R.; Bousta, D.; Lalami, A.E.O.; Chahdi, F.O.; Amri, I.; Jamoussi, B.; Greche, H. Chemical composition and herbicidal effects of essential oils of *Cymbopogon citratus* (DC) Stapf, *Eucalyptus cladocalyx*, *Origanum vulgare* L and *Artemisia absinthium* L. cultivated in Morocco. *J. Essent. Oil Bear. Plants* **2015**, *18*, 112–123. [CrossRef]
21. Poonpaiboonpipat, T.; Pangnakorn, U.; Suvunnamek, U.; Teerarak, M.; Charoenying, P.; Laosinwattana, C. Phytotoxic effects of essential oil from *Cymbopogon citratus* and its physiological mechanisms on barnyardgrass (*Echinochloa crus-galli*). *Ind. Crop. Prod.* **2013**, *41*, 403–407. [CrossRef]
22. Kordali, S.; Cakir, A.; Ozer, H.; Cakmakci, R.; Kesdek, M.; Mete, E. Antifungal, phytotoxic and insecticidal properties of essential oil isolated from Turkish *Origanum acutidens* and its three components, carvacrol, thymol and *p*-cymene. *Bioresour. Technol.* **2008**, *99*, 8788–8795. [CrossRef]
23. Verdeguer, M.; Blázquez, M.A.; Boira, H. Phytotoxic effects of *Lantana camara*, *Eucalyptus camaldulensis* and *Eriocephalus africanus* essential oils in weeds of Mediterranean summer crops. *Biochem. Syst. Ecol.* **2009**, *37*, 362–369. [CrossRef]

24. Ulukanli, Z.; Çenet, M.; Öztürk, B.; Bozok, F.; Karabörklü, S.; Demirci, S.C. Chemical characterization, phytotoxic, antimicrobial and insecticidal activities of *Vitex agnus-castus'* essential oil from East Mediterranean Region. *J. Essent. Oil Bear. Plants* **2015**, *18*, 1500–1507. [CrossRef]
25. Kashkooli, A.B.; Saharkhiz, M.J. Essential oil compositions and natural herbicide activity of four Denaei Thyme (*Thymus daenensis* Celak.) ecotypes. *J. Essent. Oil Bear. Plants* **2014**, *17*, 859–874. [CrossRef]
26. Ulukanli, Z.; Cenet, M.; Ince, H.; Yilmaztekin, M. Antimicrobial and herbicidal activities of the essential oil from the Mediterranean *Thymus eigii*. *J. Essent. Oil Bear. Plants* **2018**, *21*, 214–222. [CrossRef]
27. Ulukanli, Z.; Bozok, F.; Cenet, M.; Ince, H.; Demirci, S.C.; Sezer, G. Secondary metabolites and bioactivities of *Thymbra spicata* var. *spicata* in Amanos mountains. *J. Essent. Oil Bear. Plants* **2016**, *19*, 1754–1761. [CrossRef]
28. Bozok, F. Herbicidal Activity of *Nepeta flavida* Essential Oil. *J. Essent. Oil Bear. Plants* **2018**, *21*, 1687–1693. [CrossRef]
29. Silva, E.R.; Overbeck, G.E.; Soares, G.L.G. Phytotoxicity of volatiles from fresh and dry leaves of two Asteraceae shrubs: Evaluation of seasonal effects. *S. Afr. J. Bot.* **2014**, *93*, 14–18. [CrossRef]
30. Da Silva, E.R.; Lazarotto, D.C.; Pawlowski, A.; Schwambach, J.; Soares, G.L.G. Antioxidant Evaluation of *Baccharis patens* and *Baccharis psiadioides* Essential Oils. *J. Essent. Oil Bear. Plants* **2018**, *21*, 485–492. [CrossRef]
31. Mancini, E.; Arnold, N.A.; De Martino, L.; De Feo, V.; Formisano, C.; Rigano, D.; Senatore, F. Chemical composition and phytotoxic effects of essential oils of *Salvia hierosolymitana* Boiss. and *Salvia multicaulis* Vahl. var. *simplicifolia* Boiss. growing wild in Lebanon. *Molecules* **2009**, *14*, 4725–4736. [CrossRef]
32. Kaur, S.; Singh, H.P.; Mittal, S.; Batish, D.R.; Kohli, R.K. Phytotoxic effects of volatile oil from *Artemisia scoparia* against weeds and its possible use as a bioherbicide. *Ind. Crop. Prod.* **2010**, *32*, 54–61. [CrossRef]
33. Aragão, F.; Palmieri, M.; Ferreira, A.; Costa, A.; Queiroz, V.; Pinheiro, P.; Andrade-Vieira, L. Phytotoxic and cytotoxic effects of Eucalyptus essential oil on lettuce (*Lactuca sativa* L.). *Allelopath. J.* **2015**, *35*, 259–272.
34. Pinheiro, P.F.; Costa, A.V.; Alves, T.d.A.; Galter, I.N.; Pinheiro, C.A.; Pereira, A.F.; Oliveira, C.M.R.; Fontes, M.M.P. Phytotoxicity and cytotoxicity of essential oil from leaves of *Plectranthus amboinicus*, carvacrol, and thymol in plant bioassays. *J. Agric. Food Chem.* **2015**, *63*, 8981–8990. [CrossRef]
35. Ulukanli, Z.; Karabörklü, S.; Bozok, F.; Burhan, A.; Erdogan, S.; Cenet, M.; KARAASLAN, M.G. Chemical composition, antimicrobial, insecticidal, phytotoxic and antioxidant activities of Mediterranean *Pinus brutia* and *Pinus pinea* resin essential oils. *Chin. J. Nat. Med.* **2014**, *12*, 901–910. [CrossRef]
36. Amri, I.; Gargouri, S.; Hamrouni, L.; Hanana, M.; Fezzani, T.; Jamoussi, B. Chemical composition, phytotoxic and antifungal activities of *Pinus pinea* essential oil. *J. Pest. Sci.* **2012**, *85*, 199–207. [CrossRef]
37. Ulukanli, Z.; Karabörklü, S.; Bozok, F.; Çenet, M.; Öztürk, B.; Balcilar, M. Antimicrobial, insecticidal and phytotoxic activities of *Cotinus coggyria* Scop. essential oil (Anacardiaceae). *Nat. Prod. Res.* **2014**, *28*, 2150–2157. [CrossRef]
38. Saharkhiz, M.J.; Smaeili, S.; Merikhi, M. Essential oil analysis and phytotoxic activity of two ecotypes of *Zataria multiflora* Boiss. growing in Iran. *Nat. Prod. Res.* **2010**, *24*, 1598–1609. [CrossRef]
39. Mahdavikia, F.; Saharkhiz, M.J. Phytotoxic activity of essential oil and water extract of peppermint (*Mentha× piperita* L. CV. Mitcham). *J. Appl. Res. Med. Aromat. Plants* **2015**, *2*, 146–153. [CrossRef]
40. De Almeida, L.F.R.; Frei, F.; Mancini, E.; De Martino, L.; De Feo, V. Phytotoxic activities of Mediterranean essential oils. *Molecules* **2010**, *15*, 4309–4323. [CrossRef]
41. Zahed, N.; Hosni, K.; Ben Brahim, N.; Kallel, M.; Sebei, H. Allelopathic effect of *Schinus molle* essential oils on wheat germination. *Acta Physiol. Plant.* **2010**, *32*, 1221–1227. [CrossRef]
42. De Oliveira, M.S.; da Costa, W.A.; Pereira, D.S.; Botelho, J.R.S.; de Alencar Menezes, T.O.; de Aguiar Andrade, E.H.; da Silva, S.H.M.; da Silva Sousa Filho, A.P.; de Carvalho Junior, R.N. Chemical composition and phytotoxic activity of clove (*Syzygium aromaticum*) essential oil obtained with supercritical CO_2. *J. Supercrit. Fluid.* **2016**, *118*, 185–193. [CrossRef]
43. Verdeguer, M.; García-Rellán, D.; Boira, H.; Pérez, E.; Gandolfo, S.; Blázquez, M.A. Herbicidal activity of *Peumus boldus* and *Drimys winterii* essential oils from Chile. *Molecules* **2011**, *16*, 403–411. [CrossRef] [PubMed]
44. Kim, J. Phytotoxic and antimicrobial activities and chemical analysis of leaf essential oil from *Agastache rugosa*. *J. Plant. Biol.* **2008**, *51*, 276–283. [CrossRef]
45. Grichi, A.; Nasr, Z.; Khouja, M.L. Phytotoxic effects of essential oil from *Eucalyptus lehmanii* against weeds and its possible use as a bioherbicide. *Bull. Environ. Pharmacol. Life Sci.* **2016**, *5*, 17–23.
46. Salamci, E.; Kordali, S.; Kotan, R.; Cakir, A.; Kaya, Y. Chemical compositions, antimicrobial and herbicidal effects of essential oils isolated from Turkish *Tanacetum aucheranum* and *Tanacetum chiliophyllum* var. *chiliophyllum*. *Biochem. Syst. Ecol.* **2007**, *35*, 569–581. [CrossRef]
47. Singh, R.; Ahluwalia, V.; Singh, P.; Kumar, N.; Prakash Sati, O.; Sati, N. Antifungal and phytotoxic activity of essential oil from root of *Senecio amplexicaulis* Kunth.(Asteraceae) growing wild in high altitude-Himalayan region. *Nat. Prod. Res.* **2016**, *30*, 1875–1879. [CrossRef] [PubMed]
48. Ahluwalia, V.; Sisodia, R.; Walia, S.; Sati, O.P.; Kumar, J.; Kundu, A. Chemical analysis of essential oils of *Eupatorium adenophorum* and their antimicrobial, antioxidant and phytotoxic properties. *J. Pest. Sci.* **2014**, *87*, 341–349. [CrossRef]
49. Pawlowski, Â.; Kaltchuk-Santos, E.; Brasil, M.; Caramão, E.; Zini, C.; Soares, G. Chemical composition of *Schinus lentiscifolius* March. essential oil and its phytotoxic and cytotoxic effects on lettuce and onion. *S. Afr. J. Bot.* **2013**, *88*, 198–203. [CrossRef]

50. Abd El-Gawad, A.M. Chemical constituents, antioxidant and potential allelopathic effect of the essential oil from the aerial parts of *Cullen plicata*. *Ind. Crops Prod.* **2016**, *80*, 36–41. [CrossRef]
51. Zhu, X.; Han, C.; Gao, T.; Shao, H. Chemical composition, phytotoxic and antimicrobial activities of the essential oil of *Scutellaria strigillosa* Hemsley. *J. Essent. Oil Bear. Plants* **2016**, *19*, 664–670. [CrossRef]
52. Razavi, S.M.; Narouei, M.; Majrohi, A.A.; Chamanabad, H.R.M. Chemical constituents and phytotoxic activity of the essential oil of *Acroptilon repens* (L.) Dc from Iran. *J. Essent. Oil Bear. Plants* **2012**, *15*, 943–948. [CrossRef]
53. El Ayeb-Zakhama, A.; Ben Salem, S.; Sakka-Rouis, L.; Flamini, G.; Ben Jannet, H.; Harzallah-Skhiri, F. Chemical composition and phytotoxic effects of essential oils obtained from *Ailanthus altissima* (Mill.) swingle cultivated in Tunisia. *Chem. Biodivers.* **2014**, *11*, 1216–1227. [CrossRef] [PubMed]
54. Abd El-Gawad, A.; El Gendy, A.; El-Amier, Y.; Gaara, A.; Omer, S.; Al-Rowaily, S.; Assaeed, A.; Al-Rashed, S.; Elshamy, A. Essential oil of *Bassia muricata*: Chemical characterization, antioxidant activity, and allelopathic effect on the weed *Chenopodium murale*. *Saudi J. Biol. Sci.* **2020**, *27*, 1900–1906. [CrossRef] [PubMed]
55. Martino, L.D.; Formisano, C.; Mancini, E.; Feo, V.D.; Piozzi, F.; Rigano, D.; Senatore, F. Chemical composition and phytotoxic effects of essential oils from four *Teucrium* species. *Nat. Prod. Commun.* **2010**, *5*, 1969–1976. [CrossRef] [PubMed]
56. Razavi, S.M.; Nejad-Ebrahimi, S. Phytochemical analysis and allelopathic activity of essential oils of *Ecballium elaterium* A. Richard growing in Iran. *Nat. Prod. Res.* **2010**, *24*, 1704–1709. [CrossRef] [PubMed]
57. Mutlu, S.; Atici, Ö.; Esim, N.; Mete, E. Essential oils of catmint (*Nepeta meyeri* Benth.) induce oxidative stress in early seedlings of various weed species. *Acta Physiol. Plant.* **2011**, *33*, 943–951. [CrossRef]
58. Kordali, S.; Tazegul, A.; Cakir, A. Phytotoxic effects of *Nepeta meyeri* Benth. extracts and essential oil on seed germinations and seedling growths of four weed species. *Rec. Nat. Prod.* **2015**, *9*, 404–418.
59. Saharkhiz, M.J.; Zadnour, P.; Kakouei, F. Essential oil analysis and phytotoxic activity of catnip (*Nepeta cataria* L.). *Am. J. Essent. Oils Nat. Prod.* **2016**, *4*, 40–45.
60. Bozok, F.; Cenet, M.; Sezer, G.; Ulukanli, Z. Essential oil and bioherbicidal potential of the aerial parts of *Nepeta nuda* subsp. *albiflora* (Lamiaceae). *J. Essent. Oil Bear. Plants* **2017**, *20*, 148–154. [CrossRef]

Review

Curcuma longa L. Rhizome Essential Oil from Extraction to Its Agri-Food Applications. A Review

María Dolores Ibáñez and María Amparo Blázquez *

Departament de Farmacologia, Facultat de Farmàcia, Universitat de València, Avd. Vicent Andrés Estellés s/n, 46100 Burjassot, València, Spain; mijai@alumni.uv.es
* Correspondence: amparo.blazquez@uv.es; Tel.: +34-963-544-945

Abstract: *Curcuma longa* L. rhizome essential oil is a valuable product in pharmaceutical industry due to its wide beneficial health effects. Novel applications in the agri-food industry where more sustainable extraction processes are required currently and safer substances are claimed for the consumer are being investigated. This review provides information regarding the conventional and recent extraction methods of *C. longa* rhizome oil, their characteristics and suitability to be applied at the industrial scale. In addition, variations in the chemical composition of *C. longa* rhizome and leaf essential oils regarding intrinsic and extrinsic factors and extraction methods are also analysed in order to select the most proper to obtain the most efficient activity. Finally, the potential applications of *C. longa* rhizome oil in the agri-food industry, such as antimicrobial, weedicide and a food preservative agent, are included. Regarding the data, *C. longa* rhizome essential oil may play a special role in the agri-food industry; however, further research to determine the application threshold so as not to damage crops or affect the organoleptic properties of food products, as well as efficient encapsulation techniques, are necessary for its implementation in global agriculture.

Keywords: *Curcuma longa*; essential oil; extraction methods; chemical composition; agri-food industry; antimicrobial; herbicidal; antioxidant

Citation: Ibáñez, M.D.; Blázquez, M.A. *Curcuma longa* L. Rhizome Essential Oil from Extraction to Its Agri-Food Applications. A Review. *Plants* **2021**, *10*, 44. https://doi.org/10.3390/plants10010044

Received: 29 November 2020
Accepted: 24 December 2020
Published: 28 December 2020

Publisher's Note: MDPI stays neutral with regard to jurisdictional claims in published maps and institutional affiliations.

1. Introduction

Medicinal and aromatic plant species (MAPs) have been broadly exploited as food flavourings, medicinal agents, preservatives and ornaments, as well as beauty and personal delight products, becoming natural alternatives that offer reliability, safety and sustainability [1,2]. Amongst them, turmeric (*Curcuma longa* L., *Zingiberaceae*) is especially popular worldwide because of its attractive culinary, cosmetic and medicinal uses [3]. Specifically, the interest of this tuberous species resides in its exploitation as a colouring and flavouring agent, as well as in its numerous pharmacological activities, such as antioxidant, anticancer, anti-inflammatory, neuro- and dermoprotective, antiasthmatic or hypoglycaemic [4–10], being recently reported that turmeric can even potentially contribute against the life-threatening viral disease COVID-19 by inhibiting the main protease enzyme [11]. Most of these interesting features and properties principally come from the rhizome [3,12], a horizontal underground stem from which the shoots and roots arise [13]. It has distinctive organoleptic properties: a yellow/brown colour externally, with a deep orange inner part, a special aromatic smell and a bitter, hot taste. These characteristics make *C. longa* rhizome ideal for gastronomy. Especially, it is the principal ingredient of curry, for which it is probably popularly known [14–16].

Furthermore, rhizomes are a rich source of two major products with remarkable attributes: curcuminoids and essential oils [17]. On the one hand, curcuminoids are the responsible for the previously described orange-yellow colour [15]. They particularly refer to a group of three phenolic compounds, curcumin, desmethoxycurcumin and bis-desmethoxycurcumin, belonging to the diarylheptanoid family. They consist of a diketonic hydroxycarbon skeleton with different functional groups, depending on the curcuminoid [18,19]. Their content in the *C. longa* rhizome may vary according to many factors, such as the variety and geographic location, as well as cultivation and postharvest processing conditions [17,20]. For these secondary metabolites, turmeric is commonly employed as a spice and additives that provide colour and flavour in the food industry [21]. Additionally, they have demonstrated promising antioxidant and anti-inflammatory activities, being considered a valuable complementary therapy to pharmaceuticals in Crohn's, diabetes and cancer between other disorders [15,21,22]. Unfortunately, their poor solubility, low absorption and bioavailability, as well as high metabolic rate, limit their use for therapeutic purposes [23–26]. In fact, the major component curcumin has not been approved as a therapeutic agent yet due to its pharmacokinetics and physicochemical properties, despite it is generally considered a safe substance [24,27]. In response, curcuminoids have been associated with lipids, micelles, nanoparticles and other molecules to enhance their effects. An example is the binding of curcumin with phosphocaseins. This combination represents a suitable vector to deliver efficiently the compound, as well as other drugs and nutrients in general. New analogues with improved activity are being tried to develop from the original ones [21,28–32].

On the other hand, the essential oil is the one that provides the *C. longa* rhizome a particular spicy and aromatic flavour [3,15] with its distinctive chemical composition. In general, sesquiterpenes are the predominant phytochemical group in *C. longa* rhizome oil [33]. More concretely, ar-, α- and β–turmerones are usually the major and most representative components [34,35], although numerous intrinsic and extrinsic elements may influence in their quality and quantity [36–41]. Nevertheless, this chemical composition is different from the essential oil extracted from the aerial parts in which monoterpenes (α-phellandrene, terpinolene, 1,8-cineole, etc.) stand out [42–46]. Countless beneficial health effects have been attributed to *C. longa* rhizome oil as a consequence of this particular chemical composition: cardiovascular protection, antihyperlipidemic, antiglycaemic, antioxidant, antiplatelet, anti-inflammatory, antioxidant, antiarthritic, etc. [47]. Especially, abundant research has been focused on ar-turmerone, demonstrating its promising interesting medicinal properties, like the protection against the development of certain tumours [48,49], antifungal activity against dermatophytes [50], antiangiogenic effects [51], anticonvulsant properties [52] and treatment of neurodegenerative and other inflammatory diseases, such as psoriasis [53,54].

Nowadays, there is a growing demand of essential oils in the perfume and cosmetics, agriculture, pharmacy, food and beverage, as well as in many other, industries. One of the principal aims is to replace synthetic products with detrimental health and environmental effects [55]. In particular, numerous essential oils such as winter savoury, peppermint, oregano, wintergreen and eucalypt, as well as many of their principal components (carvacrol, limonene, etc.) have already exhibited attractive and useful antimicrobial, herbicidal and antioxidant activities for the agri-food industry [56–62]. These data favour their potential use as natural preservatives to prevent the crop and food spoilage and extend the shelf-life, as well as weed control without significantly affecting the harvests.

The medicinal and culinary properties of *C. longa* rhizome oil are well-known. However, its potential applications in the agri-food industry are still under investigation. Therefore, the attempt of the present review is to present detailed literature dealing with the extraction, chemical composition and biological activity of *C. longa* rhizome essential oil in order to highlight the potential application in the agri-food industry as natural, safer and more sustainable antimicrobial, herbicidal and antioxidant agents. Specifically, the different possible extraction methods of the essential oil from the rhizomes *C. longa* and their characteristics will be discussed first. Then, the qualitative and quantitative chemical compositions of *C. longa* rhizome oil and the factors that influence it, as well as the difference with other parts of the plant and other *Curcuma* spp. will be described. Finally, the antimicrobial, herbicidal and food preservative properties of *C. longa* rhizome oil will also be discussed to assess a prospective application in the emergent "bio" agri-food industry.

2. Extraction Methods to Obtain Essential Oil from *C. longa* Rhizomes

The characteristic aroma of turmeric's rhizomes is provided mainly by its essential oil, representing an excellent marker of quality of this spice and its derived products. Several extraction processes have been carried out with the subterranean plant stems to obtain this mixture of flavouring compounds, steam distillation being the most commonly chosen one [63–65]. In this process, a blast of steam goes through the plant material placed on a perforated plate above, dragging the organic compounds [66,67]. It presents certain disadvantages on an industrial scale (Table 1), including the huge amounts of raw material and time required and, consequently, the high price [68]. In addition, this process can present difficulties sometimes, either the evaporation of the steam-volatile compounds by the remaining latent heat or the collapse by their excessive elevation in the flask [69,70]. With the aim of avoiding these drawbacks and consequently increasing the quality and quantity of the essential oil, this technique has usually been modified and/or combined with others. For instance, Chandra et al. incorporated a continuous water circulation process to the regular steam distillation of the essential oil of turmeric rhizomes and leaves, achieving 13% and 29% more yield, respectively, compared to the conventional process [71]. Moreover, a subsequent distillation with vacuum allows a more efficient extraction of turmeric monoterpenes and sesquiterpenes [72]. The addition of a packed bed of turmeric rhizomes above the steam source has been also key to maximize the essential oil yield [73]. In general, a steam jacket is formed, helping reach a constant elevated temperature of the distillation and avoiding the degradation of the oil and, therefore, the unwanted odours that emerge from it [73,74] (Table 1).

On the other hand, hydrodistillation is also widely employed in the extraction of the essential oils from turmeric rhizomes on an industrial scale, due to its low-cost efficiency and easy implementation [75]. Unfortunately, it may sometimes mean longer extraction times and the production of wastewater, as well as loss and alteration in the composition of essential oils because the raw material is in contact with the boiling water [74,75]. Despite this, distillation in the Clevenger apparatus gives better results in the deodorisation process of turmeric relative to other distillation methods, such as distillation using the Kjeldahl apparatus or under high vacuum [76].

The most recent extraction methods appear to overcome the limitations of the conventional ones, such as heat transfer, time and quality of the resulting essential oil [77,78]. These advantages have also been observed in the extraction of oleoresins and, more particularly, curcuminoids, active components of the dried rhizome of *C. longa* extracts [79–81].

Amongst these methods, supercritical fluid extraction (SFE) [65,70,82] has shown many advantages for the extraction of essential oils on an industrial scale, including the reduction of extraction times, higher quality extracts and, principally, the use of carbon dioxide (CO_2) as a nontoxic, non-flammable and free-of-residues solvent [74,83–85]. In relation to turmeric, superior yields but no significant differences in the relative composition or higher concentrations of most of the essential oil components [79] were obtained using SFE rather than the conventional systems of steam distillation and ultrasound extraction. However, the turmeric oil yield was higher with Soxhlet extraction than SFE [86,87]. Particularly, the combination of 320 K and 26 MPa gives an optimum production of turmeric oil with 71% turmerones' purity [88] or 67.7% with 313 K and 20.8 MPa [89]. Similar optimal conditions to obtain the highest-quality essential oils from turmeric rhizomes (75% of ar-, α- and β-turmerone) were reported by Carvalho et al. (333 K and 25 MPa) [90]. Nevertheless, this technique is still under study to achieve a higher optimization. The influence of the variation of different operating parameters (temperature, extraction time, pressure, solubility and particle size) together with the integration of other techniques, such as SFE assisted by pressing (SFEAP), are investigated to reach higher yields, the quality of turmeric essential oil and its main compounds [82,83,91,92] (Table 1).

Among SFE, subcritical water extraction (SWE) also demonstrated many advantages over traditional methods in the recovery of bioactive compounds from plants, excepting implementation on the industrial scale for the moment [93]. Specifically, it takes advantage of the special properties of supercritical water under high temperature and pressure conditions (100–374 °C, >50 bar) to extract nonpolar compounds [94]. After a deep study of the influence of operating conditions in the extraction of *C. longa* essential oil from rhizomes (temperature, flow rate, particle size, time, etc.), SWE has demonstrated its selectivity to enhance a target compound and its suitability as a green and effective method for the extraction of essential oil and curcumin from turmeric rhizome [86] (Table 1).

Table 1. Different extraction methods to obtain *Curcuma longa* essential oil: advantages and limitations. SFME: solvent-free microwave extraction, MAE: microwave-assisted extraction, HDAM: hydrodistillation assisted by microwave, SDAM: steam distillation assisted by microwave, VMHD: vacuum microwave hydrodistillation and MHG: microwave by hydrodiffusion and gravity. ↑: Increase, ↓: Decrease.

Extraction method				
	Steam Distillation	Advantages	• Can be modified and/or combined with other techniques to maximize the yield and efficiency, e.g., ↑13–29% yield	[63–65,71,74]
		Limitations	• Huge amounts of raw material needed • Time-consuming • High price • Evaporation of steam-volatile compounds and even collapse	[68–70]
	Hydrodistillation	Advantages	• Low-cost efficiency • Easy implementation • Clevenger gives better deodorization results than other processes	[75,76]
		Limitations	• Long extraction times • Production of wastewater • Loss and/or alteration in the composition of essential oils	[74,75]
	Supercritical Fluid Extraction	Advantages	• Reduction of extraction times • Higher quality extracts • CO_2 as nontoxic, non-flammable and free-of-residues solvent • Superior yields	[73,79,83–85]
		Limitations	• No significant differences in qualitative and quantitative composition of turmeric essential oil with respect to other methods: 67.7–75% turmerone purity at 313–320 K and 20.8–26 MPa • Under study to achieve higher optimisation	[79,82,83,88,89,91,92]
	Subcritical Water Extraction	Advantages	• Especially useful to extract non-polar compounds • Selective to enhance a target compound • Green and effective to extract the essential oil and curcumin	[86,94]
		Limitations	• Low implementation in industry currently	[93]

Table 1. *Cont.*

Ultrasonic Extraction	Advantages	• Improved mass transfer between plant cell and solvent • Combination with other techniques: ↑ efficiency, ↓ processing time, ↓costs	[68,82,95]
	Limitations		
Microwave Energy (SFME, MAE)	Advantages	• ↓ Costs • ↓ Extraction times • ↓ Energy consumption • ↓ CO_2 emissions • Combination with other techniques to improve the performance: HDAM, SDAM, VMHD, MHG • ↓ Extraction time from 4 h of hydrodistillation to 1 h • No degradation products • Maximum yield	[96–98]
	Limitations		
Solvent Extraction	Advantages	• Overcomes the problem of excessive heat; avoids the loss of compounds and properties of the essential oil • Suitable and safe extractants: chloroform and freons	[99–101]
	Limitations		

Ultrasonic extraction is another method of extraction of essential oils and other bioactive compounds [77,95]. It is based on ultrasonic cavitation: a bubble implosion produces micro-jets that destroy the lipid glands in the plant cell tissue, releasing the essential oil [68]. It has overcome low extraction kinetics and yields of SFE-resulting essential oils by enhancing the mass transfer between the plant cell and solvent [68,95]. Moreover, it is usually combined with other extraction techniques, enhancing the efficiency and reducing the processing time and costs [68,95] (Table 1).

The use of microwave energy (solvent-free microwave extraction (SFME) and microwave-assisted extraction (MAE)) shares similar advantages as the previous cases: a reduction of costs, extraction times, energy consumption and CO_2 emissions [96]. A microwave reactor is the source of heat that promotes the bursting and release of accumulations of essential oils [97]. It represents a more efficient method for extraction of essential oils from the *Zingibereaceace* family, because it is able to reduce the extraction time from four h in hydrodistillation to one h, avoiding the formation of degradation products and obtaining the maximum yield [97]. Furthermore, the use of microwave extraction gives rise to other categories of techniques to improve its performance, such as hydrodistillation assisted by microwave (HDAM), steam distillation assisted by microwave (SDAM), vacuum microwave hydrodistillation (VMHD) or microwave by hydrodiffusion and gravity (MHG) [96,98] (Table 1).

Finally, solvent extraction was also used for the extraction of *C. longa* essential oil. It overcomes the problem of excessive heat reached with certain conventional techniques and, consequently, avoids the loss of the compounds and properties of the essential oil [99]. Ethanol, hexane or chloroform are some of the solvents used to extract turmeric essential oil, being the last one with which a higher yield of turmeric essential oil was obtained [100]. Recently, a group of researchers proposed freons as suitable and safe extractants of the essential oil from the roots of *C. longa* and its main components [101] (Table 1).

In general, the result is a yellow to orange-coloured liquid having a fresh, peppery and aromatic odour with sweet orange and ginger notes and a sharp and burning bitter taste [63,64]. These physical characteristics, as well as the chemical composition and related properties of the essential oil, may vary depending on the extraction technique. For this reason, the selection of both the most adequate method and operating conditions is key to obtain the maximum amount and quality of the *C. longa* essential oil [17]. Together with the extraction method, other factors such as drying and storage processes also influence the chemical composition of turmeric essential oil, being necessary the subsequent identification of the chemical composition to identify the variations and the quality control of turmeric essential oil.

3. Chemical Analysis of the Essential Oil Obtained from *C. longa* Rhizomes

The chemical composition of the essential oil obtained from *C. longa* rhizomes has been widely determined through gas chromatography-mass spectrometry (GC-MS) (Table 2), which is normally used for a sesquiterpenoid analysis [102] alone or combined with gas chromatography-flame ionisation detector (GC-FID) [103–105] to achieve a quantitative analysis. The determination of the chemical composition is key, because the components of the essential oil and their concentration can be considered a fingerprint conferring specific characteristics and properties [106].

As a general rule, oxygenated sesquiterpenes have been identified as the predominant ones (Table 2) and the principal reason of the biological activity of turmeric essential oil [107]. Concretely, turmerones (α-, β- and ar-) represent the major and the most distinctive individual components [108,109] (Table 2 and Figure 1). They give interesting properties to *C. longa* essential oil, such as anticancer, anti-inflammatory, antioxidant and the prevention of dementia [72,110–113]. Even they enhance the bioavailability and activity of other important turmeric components like curcumin [114–116]. In particular, ar-turmerone (6S-2-methyl-6-(4-methylphenyl) hept-2-en-4-one) has been identified as the leading one, followed by α- and β-, in *C. longa* rhizome oil (Table 2). Many authors have reported about the therapeutic potential of ar-turmerone and its numerous benefits for human health [113]. Lee demonstrated its antibacterial activity against human pathogens like *Clostridium perfringens* and *Escherichia coli* [117]. In the same year, he also reported a higher inhibitory effect than aspirin in platelet aggregation induced by collagen and arachidonic acid [118]. Other researchers have proposed ar-turmerone as a natural anticancer and cancer-preventive agent, being considered the α,β-unsaturated ketone of the molecule, the principal pharmacophore, for this activity [51,119–121]. ar-Turmerone has also been observed as useful in the prevention and attenuation of inflammatory diseases like psoriasis and neuronal ones [122,123].

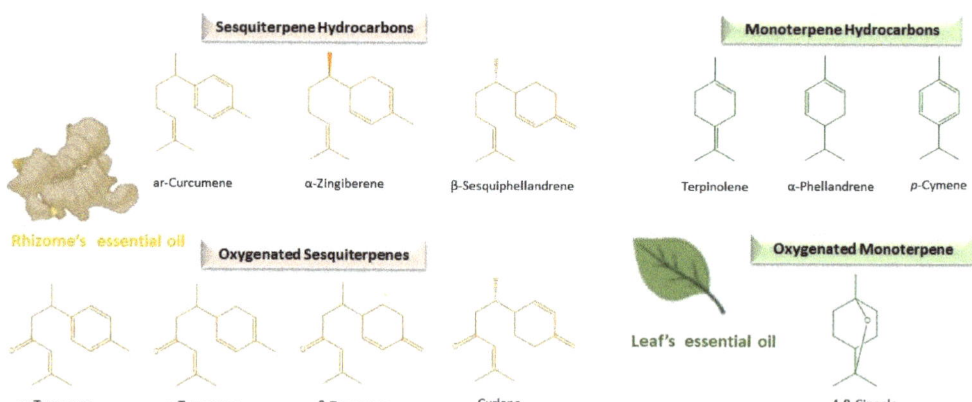

Figure 1. Main compounds found in the rhizomes and leaves of turmeric essential oils.

Oxygenated sesquiterpenes also constitute the predominant group in the essential oils obtained from the rhizome of other species included in the genus *Curcuma* [124]. For instance, curzerenone was the main compound in the rhizome oil of *C. angustifolia* and *C. zedoaria*; curdione was the major one in *C. nankunshanensis*, *C. wenyujin* and *C. kwangsiensis*; germacrone in *C. sichuanensis* and *C. leucorhiza*; β-elemenone in *C. nankunshanensis* var. *nanlingensis*; xanthorrhizol in *C. xanthorrhiza* and velleral in *C. attenuata* [124–128]. Turmerones are normally present, being considered the most representative components in general. Nevertheless, their amount may vary between species, probably due to the intrinsic differences between them [129]. The quantification of oxygenated sesquiterpenes, together with the identification of the secondary components, are key for the distinction and quality control of *Curcuma* spp. [17,130].

The sesquiterpenoids are generally followed by smaller quantities of sesquiterpene hydrocarbons in *C. longa* rhizome oil (Table 2 and Figure 1). This group is characterised by great structural diversity, providing a variety of fragrances and characteristic aromas to the essential oil [131]. Specifically, monocyclic bisabolane derivatives with a C_6-ring formed in analogy to the menthane skeleton highlighted in turmeric essential oil obtained from rhizomes. Some examples are bisabolene isomers (β-bisabolene), α-zingiberene and ar-curcumene, characteristic of *Curcuma* spp. and ginger. β-caryophyllene is also common, widely spread in food plants and derived from α-humulene, with a C_9-ring fused to a cyclobutane ring [132]. Sesquiterpene hydrocarbons predominate over oxygenated ones in the rhizome oil of other *Curcuma* spp., such as *C. aromatica* (Sesquiterpene Hydrocarbons (SH): 8.30% ± 1.90% and Oxygenated Sesquiterpenes (OS): 7.10% ± 2.14%) and *C. kwangsiensis* var *nanlingensis* (SH: 9.76% ± 1.89% and OS: 6.80% ± 1.27%) [124].

The amount of monoterpene hydrocarbons and oxygenated monoterpenes are usually lower in most samples of rhizome essential oil of *C. longa* (Table 2). Contrarily, they constitute the most abundant group in the rhizome oil of other different *Curcuma* spp., such as *C. amada* [133], as well as in the essential oils obtained from the aerial parts of *C. longa* [17,134–137]. Regarding this, the yield of *C. longa* essential oil varied between the leaves (23%), rhizomes (48%) and rhizoids (27%), and the chemical composition was different between the leaf petiole, lamina and rhizoid oils (myrcene, *p*-cymene, etc.) compared to the stem and rhizome ones in which turmerones predominated [138]. α-Phellandrene, terpinolene and 1,8-cineole (Figure 1) are usually the most abundant compounds detected in the essential oil extracted from the leaves of *C. longa* [36,39,43,44], whereas turmerones are found in minor concentrations (Table 2) [109], being also usually found in the essential oils of the aerial parts of *C. longa p*-cymene, α-terpinene, myrcene and pinenes (Table 2) [134,135,137,139,140]. However, in samples of *C. longa* grown in Nigeria, the leaf essential oil was dominated by turmerones, like in rhizomes (Table 2) [141,142]. In addition, important concentrations of C_8-aldehyde (20.58%) were found in the essential oil of *C. longa* leaves in a high-altitude research station in Odisha, India [140]. The concentration of these compounds can be increased by enhancing the leaf biomass production [143].

The aerial parts of *C. longa* normally end as waste products. An interest approach is their recycling to obtain biologically active compounds. In this sense, *C. longa* leaf essential oil and its principal component α-phellandrene have demonstrated remarkable insecticidal activity against *Cochliomya macellaria*, causative agents of myasis in humans and animals, as well as against *Lucilia cuprina* [144,145], being also a *C. longa* leaf essential oil highlight because of its medicinal and food-preservation properties, with a significant inhibition of microbial growth and toxin production [146,147].

On the other hand, several studies corroborate that the qualitative and quantitative chemical compositions of turmeric rhizomes essential oil may fluctuate according to many factors [124,148,149]. Sometimes, different chemical compositions come from the intrinsic characteristics of each genotype. In fact, certain traits of a specific variety of *C. longa* can influence the content of rhizome oil, representing good criteria for the selection of high-yield ones. Regarding this, an interesting study observed a direct relationship between plant height and rhizome oil content, as well as a negative correlation between the amount of essential oil in the dry leaf with the one contained in the fresh rhizome [150]. A clear example of genotype influence is the dissimilar chemical composition between yellow *C. longa* rhizome oil rich in oxygenated sesquiterpenes (ar-turmerone, turmerone, curlone, etc.) and red one with oxygenated monoterpenes (carvacrol, citral, methyl eugenol, geraniol, etc.) as principal compounds more similar to *Origanum* or *Thymus* spp. [151]. Indeed, the rhizome colour is closely related to the beneficial properties of *C. longa* [152]. The influence of the genotype or cultivars have also been reported by other authors who observed significant variations in the yield and chemical composition of rhizome oils of *C. longa* under similar climatic conditions [153–155].

Together with the genetic and environmental factors, the geographic location contributes to the different yields and quality of *C. longa* rhizome oils, even developing different chemotypes [39,109]. In India, the region of production determines the type of turmeric [156]. Samples from Nepal included α- and β-turmerones (8.19% and 17.74%, respectively) between other compounds like *epi*-α-patshutene (7.19%), β-sesquiphellandrene (4.99%), 1,4-dimethyl-2-isobutylbenzene (4.4%), (±)-dihydro-ar-turmerone (4.27%) and zingiberene (4.03%) [33]. The main components of the essential oil from Nigeria were ar-turmerone, α-turmerone and β-turmerone [141,157], while turmerones (approximately 37%), together with terpinolene (15.8%), zingiberene (11.8%) and β-sesquiphellandrene (8.8%), predominated in the rhizome oil from Reunion Island [134]. Turmerones still are also the predominant compounds in samples from Faisalabad (Pakistan) and Turkey [104,158]. In the South American continent, the essential oil isolated from rhizomes grown in Ecuador was rich in ar-turmerone (45.5%) and α-turmerone (13.4%), similar to Colombian samples, while that from Brazil was dominated by zingiberene (11%), sesquiphellandrene (10%), β-turmerone (10%) and α-curcumene (5%) [105,107,159].

The analysis of each *C. longa* habitat's conditions can help to predict the features of the resulting essential oil and enhance its yield and quality; what results especially important for its optimisation and commercialisation. Altitude, humidity, rainfall, temperature, soil pH, organic carbon, nitrogen, phosphorous and potassium are some of the factors that lead to wide variations in the yield and chemical composition of rhizome essential oil. From the development of predictive models and in vivo tests, the altitude, soil pH, nitrogen and organic carbon have been observed as enhancers of rhizome essential oil production. Amongst them, nitrogen and organic carbon raise the turmerone content concretely and phosphorous and potassium the oil yield [40,160–162]. Land configurations involving furrows and thatches surrounding *C. longa* reduce the loss of these soil nutrients, enhancing the rhizome yield [41].

The stage of maturity of *C. longa* rhizomes can also influence in the yield, chemical composition and properties of the essential oil. In relation to this, Garg et al. demonstrated that the percentage of the essential oil content widely varied between fresh and dried rhizomes of 27 accessions of *C. longa* in North India [163]. Similarly, Sharma et al. also observed certain variations in the qualitative and quantitative chemical compositions between the essential oils extracted from a mix of 5–10 month-old rhizomes and eight ones [139]. Furthermore, Singh et al. confirmed that fresh rhizome essential oil contained a major quantity of the active compound turmerone than dry ones, consequently having stronger activity [164]. A different trend was observed by Gounder et al., who reported the higher activity of cured (fresh rhizome boiled in water, dried in shade and polished) and dried rhizome oils over fresh ones [165], probably due to the lower percentage of ar-turmerone and β-turmerone. Anyway, the control of the drying conditions constitutes an important parameter in order to obtain the highest content of essential oil in the minimum time possible [166,167]. The sun and mechanical drying coexist as drying methods of *C. longa* rhizomes [156]. In particular, Monton et al. confirmed that one hour of microwave drying without conventional drying represented the optimum conditions to obtain the highest content of turmeric essential oil [167].

Table 2. Main components of *C. longa* essential oil according to the part of the plant used, origin, method of extraction and analysis. GC-MS: gas chromatography-mass spectrometry, CG-FID: flame ionisation detector, SFE: supercritical fluid extraction, SWE: supercritical water extraction and: CG-FTIR: gas chromatography-Fourier-transform infrared.

Part of Turmeric	Origin	Method of Extraction	Analysis	Yield	Main Components	Ref.
Powdered rhizomes	Nepal	Hydrodistillation Clevenger	GC-MS	3.0%	β–turmerone (17.74%), α-turmeron (8.19%), *epi*-α-patschutene (7.19%), β–sesquiphellandrene (4.99%), 1,4-dimethyl-2-isobutylbenzene (4.4%)	[33]
Pulverized rhizome	India	Steam distillation + vacuum distillation	GC-MS	1.6–46.6%	Turmerones, *l*-zingiberene, β–sesquiphellandrene, ar-curcumene	[72]
Rhizomes	Brazil	Hydrodistillation assisted by microwave (HDAM)	GC-MS	0.6%	ar-turmerone (50.37 ± 0.99%), β–turmerone (14.39 ± 0.33%), ar-curcumene (6.24 ± 0.21%)	[98]
Rhizomes	Brazil	HDAM + Cryogenic grinding (CG)	GC-MS	1.00%	ar-turmerone (47.97 ± 1.19%), β–turmerone (13.70 ± 0.55%), ar-curcumene (5.94 ± 0.27%)	[98]
Rhizomes	Brazil	Steam distillation assisted by microwave (SDAM)	GC-MS	0.9%	-	[98]
Rhizomes	Brazil	SDAM + CG	GC-MS	1.45%	-	[98]
Powdered dried rhizome	Serbia	Hydrodistillation Clevenger	GC-MS and GC-FID	0.3 cm^3/100 g	ar-turmerone (22.7%), turmerone (26%) and curlone (16.8%)	[104]
Rhizomes	Pakistan	Hydrodistillation	GC-MS	0.673%	ar-turmerone (25.3%), α-turmerone (18.3%) and curlone (12.5%)	[158]
Powdered rhizomes	Thailand	Hydrodistillation Clevenger	GC-MS	-	ar-turmerone (43–49%), turmerone (13–16%) and curlone (17–18%)	[166,167]
Dried rhizomes	Brazil	SFE	GC-MS	0.5–6.5 g/100 g	ar-turmerone (20%) and ar-, α- and β–turmerones (~75%)	[91]
Dried rhizomes	Brazil	Extraction with volatile solvents	GC-MS and CG-FID	5.49%	α-turmerone and β–turmerone (~8.7%), ar-turmerone (~3.6%)	[104]
Dried rhizomes	Brazil	Steam distillation	GC-MS and CG-FID	0.46%	ar-turmerone (~12.8%), α-turmerone and β–turmerone (~4.1%)	[104]
Dried rhizomes	China	Steam distillation	GC-MS	4.50% *w/w*	ar-turmerone (11.81%)	[124]
Dried rhizomes	Nigeria	Hydrodistillation Clevenger	GC-MS	1.33% *w/w*	ar-turmerone (44.4%), α-turmerone (20.8%), β–turmerone (26.5%)	[141]
Dry rhizomes	India	Hydrodistillation Clevenger	GC-MS	2.9%	ar-turmerone (21.4%), α-santalene (7.2%) and ar-curcumene (6.6%)	[164]

Table 2. *Cont.*

Part of Turmeric	Origin	Method of Extraction	Analysis	Yield	Main Components	Ref.
Dried rhizomes	India	Hydrodistillation Clevenger	GC-MS	3.05 ± 0.15%	ar-turmerone (30.3%), α-turmerone (26.5%), β–turmerone (19.1%)	[167]
Cured rhizomes	India	Hydrodistillation Clevenger	GC-MS	4.45 ± 0.37%	ar-turmerone (28.3%), α-turmerone (24.8%), β–turmerone (21.1%)	[167]
Dried root	-	SFE	GC-MS	2–5.3 wt%	ar-turmerone (31–67.1%), β–turmerone (2–37.9%), α-turmerone (0–21.3%)	[87]
Fresh rhizomes	Brazil	Hydrodistillation Clevenger	GC-MS	1000 μL	α-turmerone (42.6%), β –turmerone (16.0%) and ar-turmerone (12.9%)	[34]
Fresh rhizomes	India	Hydrodistillation Clevenger	GC-MS	0.6–2.1%	Turmerone (35.24–44.22%)	[39]
Fresh rhizomes	India	Hydrodistillation Clevenger	GC-MS	0.8%	α-turmerone (44.1%), β–turmerone (18.5%) and ar-turmerone (5.4%)	[43]
Fresh rhizomes	India	Hydrodistillation Clevenger	GC-MS	0.36%	ar-turmerone (31.7%), α-turmerone (12.9%), β–turmerone (12.0%) and (Z)-β–ocimene (5.5%)	[44]
Fresh rhizomes	India	Modified distillation process	GC-MS	2.09–2.50%	ar-turmerone (45.27%), curlone (5.6%), turmerone (4.4%), zingiberene (4.01%), ar-curcumene (4.01%), dehydrocurcumene (2.0%)	[73]
Fresh rhizome	Malaysia	SFE	GC-MS	-	ar-turmerone (10.84–21.50%), turmerone (36.14–45.68%) and curlone (21.27–22.30%)	[79]
Fresh rhizomes	Iran	SWE	GC-MS	0.98%	ar-turmerone (62.88%), curcumin (10.49%), β–sesquiphellandrene (9.62%), α-phellandrene (6.50%)	[86]
Fresh rhizomes	Ecuador	Steam distillation	GC-FID and GC-MS	0.8% *v/w*	ar-turmerone (45.5%) and α-turmerone (13.4%)	[105]
Fresh rhizomes	France	Steam distillation	GC-MS and GC-FTIR	1.1%	α-turmerone (21.4%), zingiberene (11.8%), terpinolene (15.8%), β–sesquiphellandrene (8.8%), ar-turmerone (7.7%) and β–turmerone (7.1%)	[134]
Fresh mature rhizomes	Bhutan	Hydrodistillation Clevenger	GC-MS	2–5.5%	α-turmerone (30–32%), ar-turmerone (17–26%) and β–turmerone (15–18%)	[139]
Fresh rhizome	India	Steam distillation	-	2.03–6.50%	-	[156]

Table 2. *Cont.*

Part of Turmeric	Origin	Method of Extraction	Analysis	Yield	Main Components	Ref.
Dried rhizomes	India	Hydrodistillation Clevenger	GC-MS	3.05 ± 0.15%	ar-turmerone (30.3%), α-turmerone (26.5%), β–turmerone (19.1%)	[167]
Cured rhizomes	India	Hydrodistillation Clevenger	GC-MS	4.45 ± 0.37%	ar-turmerone (28.3%), α-turmerone (24.8%), β–turmerone (21.1%)	[167]
Dried root	-	SFE	GC-MS	2–5.3 wt%	ar-turmerone (31–67.1%), β–turmerone (2–37.9%), α-turmerone (0–21.3%)	[87]
Fresh rhizomes	Brazil	Hydrodistillation Clevenger	GC-MS	1000 μL	α-turmerone (42.6%), β–turmerone (16.0%) and ar-turmerone (12.9%)	[34]
Fresh rhizomes	India	Hydrodistillation Clevenger	GC-MS	0.6–2.1%	Turmerone (35.24–44.22%)	[39]
Fresh rhizomes	India	Hydrodistillation Clevenger	GC-MS	0.8%	α-turmerone (44.1%), β–turmerone (18.5%) and ar-turmerone (5.4%)	[43]
Fresh rhizomes	India	Hydrodistillation Clevenger	GC-MS	0.36%	ar-turmerone (31.7%), α-turmerone (12.9%), β–turmerone (12.0%) and (Z)-β–ocimene (5.5%)	[44]
Fresh rhizomes	India	Modified distillation process	GC-MS	2.09–2.50%	ar-turmerone (45.27%), curlone (5.6%), turmerone (4.4%), zingiberene (4.01%), ar-curcumene (4.01%), dehydrocurcumene (2.0%)	[73]
Fresh rhizome	Malaysia	SFE	GC-MS	-	ar-turmerone (10.84–21.50%), turmerone (36.14–45.68%) and curlone (21.27–22.30%)	[79]
Fresh rhizomes	Iran	SWE	GC-MS	0.98%	ar-turmerone (62.88%), curcumin (10.49%), β–sesquiphellandrene (9.62%), α-phellandrene (6.50%)	[86]
Fresh rhizomes	Ecuador	Steam distillation	GC-FID and GC-MS	0.8% *v/w*	ar-turmerone (45.5%) and α-turmerone (13.4%)	[105]
Fresh rhizomes	France	Steam distillation	GC-MS and GC-FTIR	1.1%	α-turmerone (21.4%), zingiberene (11.8%), terpinolene (15.8%), β–sesquiphellandrene (8.8%), ar-turmerone (7.7%) and β–turmerone (7.1%)	[134]
Fresh mature rhizomes	Bhutan	Hydrodistillation Clevenger	GC-MS	2–5.5%	α-turmerone (30–32%), ar-turmerone (17–26%) and β–turmerone (15–18%)	[139]
Fresh rhizome	India	Steam distillation	-	2.03–6.50%	-	[156]

C. longa nutrition also has a significant impact in the yield and composition of rhizome oil. Especially, fertilizer use can enhance the productivity of volatile oil of *C. longa* rhizomes 6% [148]. Furthermore, a prior treatment with minerals during in vitro rhizome development followed by a fertilizer treatment in a greenhouse increases the percentage of volatiles in *C. longa* rhizomes. Particularly remarkable is the interaction of KNO_3 and Ca^{2+}, which favours the accumulation of sesquiterpenes in turmeric rhizome [168]. An interesting research proposed the use of arbuscular mycorrhizal fungi instead of chemical fertilizers in the cultivation of *C. longa* rhizomes. These optimise the absorption of nutrients and water, augment the metabolic activity of the plant, etc. In consequence, the root system becomes more robust, and the chemical composition of the essential oil is improved, increasing the production of certain compounds, including caryophyllene, α-curcumene, β-bisabolene and β-curcumene, using sustainable technologies [169,170]. Finally, the postharvest management of turmeric rhizomes also has a noteworthy influence on the quality of the derived products. Concretely, the boiling conditions, way of slicing, type of mill and speed of crushing and presence of heat and oxygen need to be controlled and standardised to obtain essential oils with certain characteristics [156].

In conclusion, the study of the chemical composition of the essential oil from the rhizome of *C. longa* gives us an idea of the characteristics and possible properties that it possesses. Sesquiterpenes are usually the main compounds in *C. longa* rhizome essential oil, highlighting the oxygenated turmerones followed by sesquiterpene hydrocarbons (Figure 1). However, the qualitative and quantitative chemical compositions of the essential oil can vary depending on the genetic and commented on factors. The knowledge of these can help to achieve a high-yield product with useful composition and properties for the agri-food industry.

4. Potential Applications of *C. longa* Essential Oil Obtained from Rhizomes in the Agri-Food Industry

Foodborne diseases, spoilage, insect and weed infestation are some common problems that cause significant economic losses to the agri-food industry. Chemical preservatives and pesticides have been widely exploited to maintain and enhance yields and productivity. However, the numerous handicaps derived from their overuse have been extensively described. As a result, sustainability has become an increasingly important subject in the agri-food industry. The characteristics of certain natural products, especially essential oils (zero waste), have become a matter of study as sustainable alternatives [171–176]. Amongst them, *C. longa* rhizome oil can take part in the safer and eco-friendly emergent agri-food industry due to its promising antimicrobial, herbicidal and antioxidant activities (Figure 2).

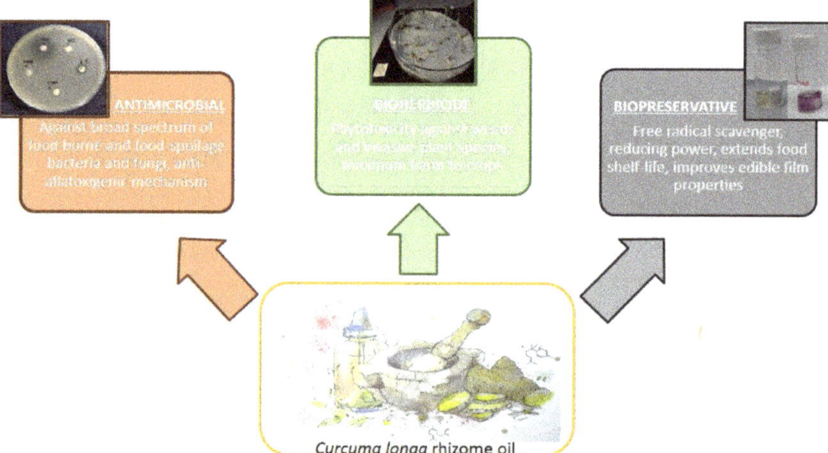

Figure 2. Representation of the roles that *Curcuma longa* rhizome oil can play in the safer and more sustainable emerging agri-food industry: antimicrobial, herbicidal and antioxidant activities.

4.1. Prevention and Inhibition of Microbial Attack in Crops and Food-Spoilage Microorganisms

Microbial contamination can affect any step of the food chain, from seed germination to food processing and storage [177–180]. Initially, the seed-borne pathogens endanger the correct development of grains, affecting both yield and quality [179]. Besides, bacteria and fungi are the principal causative agents of many postharvest diseases. *Erwinia*, *Pseudomonas*, *Corynebacterium*, *Aspergillus* and *Fusarium* are some of the most common food spoiler species, seriously compromising the quality of food products. Moreover, human health may be disturbed by the contact with these contaminated products [181,182], along with a deleterious impact in the reliability and economics of the agri-food industry [177,183,184]. Synthetic pesticides and preservatives, some of them with detrimental effects, have been the most commonly used formulations to prevent and stop the growth of these microorganisms. Therefore, natural, safer and eco-friendly antimicrobials are demanded by the consumers [178,180,185,186]. Essential oils constitute a potential alternative, because they possess antimicrobial activity, individually or in combinations between them and with antibiotics. They prevent food deterioration, maintain their appearance and quality and are able to be used in biopreservation and biocontrol in the agri-food industry [187–191].

In general, the essential oils proceeding from the rhizomes of the genus *Curcuma* have demonstrated noteworthy antimicrobial activity [192–194]. Amongst them, the essential oil of *C. longa* rhizome with 58% of ar-turmerone, together with limonene and borneol as the principal compounds, has presented a dose-dependent antimicrobial activity against a broad spectrum of food-borne and food-spoilage bacteria and fungi, including *Bacillus subtilis*, *Salmonella choleraesuis*, *Escherichia coli*, *A. niger* and *Saccharomyces cerevisiae* but at higher doses than the traditional chloramphenicol and amphotericin antibiotics [195] (Table 3). The addition of *C. longa* essential oil (33.42% ar-turmerone, 22.35% α-turmerone and 20.14% β-turmerone) to an edible film with sorbitol and egg white protein power improved both the properties of the film (thickness and lipophilicity) and its antibacterial activity against *E. coli* and *Staphylococcus aureus* [196] (Table 3).

Usually, bacterial contamination is more difficult to detect, because food generally appears normal until advanced infection. In contrast, fungal contamination can be easily perceived, as it normally alters the odour, appearance and texture of food [177,178,197]. *C. longa* essential oil has already demonstrated its strong fungicidal effect against the causal agents of important diseases in crops [192,198]. In particular, the radial growth of *Colletotrichum gloeosporioides*, *Sphaceloma cardamomi* and *Pestalotia palmarum* were completely inhibited after the treatment with essential oil from *C. longa* rhizomes at 1–5%. Other phytopathogenic fungi, such as *Rhizoctonia solani*, *Aspergillus* sp. and *Fusarium* sp., were also notably affected, especially at the highest concentration (5%) assayed [199] (Table 3).

It is interesting to note that essential oils represent a natural alternative to the usually employed weak-acid preservatives in the prevention of *A. niger*, a common contaminant of yogurt, ready-to-drink beverages and, especially, bakery products [200,201]. Particularly, packaging with a biopolymer film containing turmeric essential oil (35.46% turmerone, 20.61% cumene and 13.82% ar-turmerone) constitutes a sustainable and efficient technology to protect these food products against attacks of the filamentous fungus. The biopolymer film acts as a carrier, releasing in a sustained way the antimicrobial agent turmeric essential oil [202] (Table 3). In fact, the addition of turmeric essential oil in edible coating films could enhance food protection from microbial contamination in general. In relation to this, the fungal growth of common spoilers of pumpkin *Penicillium* and *Cladosporium* spp. were reduced 60.3% and 41.6%, respectively, for 15 days with an edible coating based on achira starch (*Canna indica* L.) containing 0.5% *w/w C. longa* rhizome oil [203] (Table 3).

The antifungal effect of *C. longa* essential oil has been tested in other *Aspergillus* spp., such as *A. flavus*, a common contaminant of cereals, legumes, juices, and fresh and dried fruits [182,204–208], as well as one of the major source of aflatoxins in agricultural crops, considered the most problematic mycotoxins worldwide [181,209,210]. The growth rate of *A. flavus* was significantly reduced with only 0.10% *v/v* of *C. longa* rhizome oil (33.2%

ar-turmerone, 23.5% α-turmerone and 22.7% β-turmerone). Furthermore, the germination and sporulation were completely inhibited at 0.5% *v/v* [211] (Table 3).

Regarding *Fusarium* spp., versatile spoilers of fruit, vegetables, cereals, etc. [212,213] generating important economic losses in the agri-food industry, *C. longa* rhizome essential oil has exhibited also promising results. The mycotoxin production, particularly of thrichothecenes and fumonisins, with serious health impacts in humans and livestock by their potentially carcinogenic and inhibition of the protein synthesis, respectively [214], is another problem to solve. The essential oil obtained from the fresh rhizomes of *C. longa* (42.6% α-turmerone, 16.0% β-turmerone and 12.9% ar-turmerone) significantly affected the development of *F. vericillioides* by decreasing the thickness and length of the microconidia, as well as the fungal biomass. The fumonisin production was also significantly inhibited [34] (Table 3). Likewise, *C. longa* rhizome oil (53.10% ar-turmerone) had a considerable effect in the morphology of the mycelia and spores, as well as in the zearalenone production of *F. graminearum* [215], being the mycelial growth of *F. moniliforme* and *F. oxysporum* inhibited at 1000 and 2000 ppm, respectively [192] (Table 3).

On the other hand, the essential oil obtained from other parts of *C. longa* with different chemical compositions has also shown antimicrobial activity. In this sense, *C. longa* essential oil dominated by oxygenated monoterpenes (82.0%) displayed promising in vivo antifungal activity against *P. expansum* and *Rhizopus stolonifer* when combined with *A. sativum* essential oil, representing a natural alternative to chemical fungicides in tomato protection [216]. Similarly, *C. longa* essential oil rich in monoterpenes (20.4% α-phellandrene, 10.3% 1,8-cineole and 6.19% terpinolene) and with considerable quantities of α- and β-turmerone (19.8% and 7.35%) presented one of the highest MICs (0.06–0.36 mg/mL) with respect to 11 different essential oils against five food-spoilage yeasts [217] (Table 3).

Therefore, the high antimicrobial activity of *C. longa* essential oil may be due to a synergism between the usual main compounds ar-turmerone, turmerone and curlone and the other phenolic group [218].

Regarding these data, the essential oil from the rhizome of *C. longa* can be considered a green alternative for biopreservation in the agri-food industry. It has demonstrated promising dose-dependent antimicrobial activity against a wide range of microorganisms. This efficacy is not always shared with the essential oils extracted from other parts of *C. longa*. Therefore, its efficacy may be due to its particular chemical composition, especially to the predominance of turmerones and combinations with other oxygenated compounds. This makes *C. longa* rhizome oil the subject of incorporation in edible coating films and other encapsulating technologies for future applications.

4.2. Herbicidal Activity

The resistance and tolerance development of weeds, crop damage or environmental pollution are the main problems due to the continuous use of synthetic herbicides in global agriculture [219–221]. Alternatives to synthetic herbicides for weed management and food security require the research of natural sources such as essential oils to develop safer and more sustainable herbicides without significantly affecting crops yields. Several essential oils have demonstrated promising herbicidal properties, inhibiting seed germination and/or seedling growth of a broad number of weeds [175,222–225]. In fact, some of them are already the main components of several commercial herbicidal compositions, taking part in the construction of a harmless and eco-friendlier emergent agri-food industry [226,227].

Regarding turmeric, the rhizome essential oil (38.7% ar-turmerone, 18.6% β-turmerone and 14.2% α-turmerone) has proven to be a potential post-emergent treatment in the control of weeds such as common purslane (*Portulaca oleracea* L.), especially aggressive in agriculture because of its versatility in affecting a wide variety of scenarios due to its tolerance to changes and rapid growing [228,229], Italian ryegrass (*Lolium multiflorum* Lam.), rapidly growing weed with the capacity of producing large quantities of seeds, being particularly competitive in small grain and vegetable harvests, where it represents a great problem due to the development of herbicide resistance [230,231] and barnyard grass (*Echinochloa crus-galli* (L.) Beauv.), considered one of the world's worst weeds infesting cropping systems [232], especially detrimental in rice paddies, where it interferes with canopy light transmission, triggering a series of metabolic alterations in rice that can lead to severe losses of even 55.2% [233]. Concretely, it reduced the hypocotyl development of the three weeds 56.55%, 40.45% and 39.33%, respectively, from 0.125 to 1 μL/mL, without affecting either the seed germination or the hypocotyl growth of the tomato, cucumber and rice crops [234] (Table 3).

The harmlessness of *C. longa* rhizome essential oil for food crops, a great challenger in the search for natural herbicides, has been corroborated by other authors. For instance, Prakash et al. confirmed that it did not affect the germination of chickpea seeds. The mean length of both hypocotyl and radicle were not significantly reduced after three days of exposure to the essential oil regarding control (3.65 and 0.82 cm *vs.* 3.75 and 0.93 cm, respectively). Only its combination with *Z. officinale* essential oils showed certain phytotoxicity against the seeds, probably due to the activity of *Z. officinale* [235]. However, the essential oil proceeding from other species included in the genus *Curcuma* have shown phytotoxic actions against food crops. For instance, *C. zedoaria* essential oil with a predominance of oxygenated compounds (18.20% *epi*-curzerenone and 15.75% 1,8-cineole) severely depressed the germination, germination rate and seedling development of lettuce and tomatoes. Particularly, the seed germination of both crops decreased from 80% to 0% and from 100% to 40%, correspondingly, at the highest dose of *C. zedoaria* essential oil (1.00%) assayed, and the hypocotyl and radicle growths were significantly reduced, with the essential oil at only 0.73–0.86% [236].

Furthermore, *C. longa* rhizome oil constitutes a potential candidate for biological control of the emerging invasive alien plant species. Specifically, it is outstanding in the inhibitory effect in the development of pampas grass (*Cortaderia selloana* (Schult. & Schult. f.) Asch. & Graebn.) and tree tobacco (*Nicotiana glauca* Graham.) from the lowest dose (0.125 μL/mL) assayed. Among them, *C. selloana* exhibited a special sensitivity to *C. longa* essential oil. The seed germination was drastically inhibited in a dose-dependent manner, achieving 81.71% of reduction at the highest dose (1 μL/mL) applied [234] (Table 3). It is interesting to note that the management of invasive species with sustainable alternatives is another important step in global agriculture, because these species are becoming naturalized in a wide number of areas with serious consequences: they influence the environment, change soil properties, affect diversity, etc. and, finally, are reverberating in socioeconomic factors, as well as human health [237–239].

On the other hand, other products derived from *C. longa* have demonstrated phytotoxic activity. In this way, the ethanolic extract completely inhibited the growth of the floating weed common duckweed (*Lemna minor* (L.) Griff.) at 100 and 1000 μg/mL [240], whereas the ethyl acetate extract (1000–10,000 ppm) showed the highest inhibitory effect vs. the seed germination and seedling growth of radishes in comparison to cyclohexane and *n*-hexane, which stimulated germination and elongation at 10,000 and 7500 ppm, respectively [241]; more recently, Akter et al. remarked on the potent inhibitory effect of the methanolic extract against the seed germination and seedling growth of both weed beggarticks (*Bidens pilosa* L.) and crops cress, radishes and lettuce. Especially, the major curcuminoids present in *C. longa*'s Ryudai gold variety strongly reduced the seed germination, as well as root and shoot growth of the weed (IC$_{50}$ 8.7–12.9 and 15.5–38.9 μmol/L, respectively) [242].

Therefore, *C. longa* can be considered an important source of bioproducts with interesting phytotoxic properties. Especially, the rhizome essential oil has demonstrated apt herbicidal activity against specific weed and invasive plant species, without significantly harming food crops. These observations make the essential oil of *C. longa* rhizome a reference of investigation for new weedicide compounds. Further research involving more weed and crop species, as well as different conditions, is needed to keep demonstrating its potential as a bioherbicide.

4.3. Food Decay Prevention: Antioxidant Activity

Stored food products are subject to oxidation, involving a loss of quality, alteration of the organoleptic properties and nutritional value, as well as of food safety problems. Synthetic antioxidant additives commonly used to avoid this process are under controversy currently, which has led to an increased interest in the agri-food industry to use the preservative properties of plant products.

Several essential oils and their components have already demonstrated their potential role in overcoming storage losses and enhancing food shelf-lives in the near future [243–245]. Some have even been approved as flavour or food additives, and others are under validation. Nowadays, the encapsulation of essential oils is also being studied to try to stabilise their antioxidant activity and even enhance it [246].

In general, *C. longa* and its products have shown their antioxidant potential as biopreservatives of physicochemical and organoleptic properties of food items, such as paneer, white hard clams, rainbow trout, cuttlefish and mashed potatoes, either alone or in combination with other plant products. This property can be improved even more with the help of nanotechnology that may control the aqueous solubility and stability of turmeric derivatives [245,247–255].

The antioxidant properties of the turmeric essential oils have been widely studied. The leaf essential oil with 22.8% β-sesquiphellandrene and 9.5% terpinolene as the main compounds has been proposed as a potential option to prevent the oxidative deterioration of fat-containing food products because of its hydrogen-donating properties and reducing power [256] (Table 3). Likewise, *C. longa* rhizome oil is able to decrease lipid peroxidation and other processes related to free-radical formation, achieving the extending shelf-lives of food products. In fact, it has exhibited the lowest peroxide value with respect to oleoresins and synthetic antioxidants, meaning a more efficient inhibitory effect of the formation of the secondary oxidation product malondialdehyde [164] (Table 3). This effect has been corroborated by means of diverse methods that evaluate both the scavenging capacity for different free radicals and the metal ion-chelating ability of the essential oil. Particularly, the essential oil obtained from the fresh rhizomes of *C. longa* (α-turmerone (42.6%), β-turmerone (16.0%) and ar-turmerone (12.9%)) has exhibited satisfactory dose-dependent DPPH (2,2-diphenyl-1-picrylhydrazyl) and ABTS (2,2′-azino-bis-3-ethylbenzothiazoline-6-sulphonic acid) radical-scavenging activities (IC_{50} 10.03 and 0.54 mg/mL, respectively), as well as reducing power [34] (Table 3). Both the DPPH and ABTS methods are between the most carried out antioxidant capacity assays [257]. These are good estimators of the antioxidant activity of any extract in general, using a simple redox reaction between the antioxidant and reactive oxygen species (ROS), being considered the DPPH method as the first line for evaluating the ability of a compound and extract or other biological source to act as a free-radical scavenger or hydrogen donor because of its accuracy, simplicity and low cost [258]. On the other hand, ABTS has been observed as especially useful to track changes in the antioxidant system itself during the storage and processing steps [257]. Reducing power is usually a complementary test to the previous ones to further evaluate the antioxidant activity [259].

In this way, the antioxidant activity of *C. longa* rhizome essential oil stood out over 10 other different essential oils. Its free radical-scavenging potential was twice higher than that of Trolox (~60% vs. 28.2%, respectively), and the antioxidant activity (72.4%) was near the values of the reference essential oil *Thymus vulgaris* (90.9%) and butylated hydroxyanisole (BHA) (86.74%) [217] (Table 3). Similarly, the reducing potential of *C. longa* rhizome oil was also highlighted over *Eucalyptus* spp., such as *E. sideroxylon*, *E. tereticornis* and *E. citriodora* [130.5 ± 1.2, 122.1 ± 1.4 and 95.8 ± 1.0 µM ferric reducing antioxidant power (FRAP) equivalents, respectively], with 138.4 ± 1.1 µM FRAP equivalents. This value was even higher than the one of other *Curcuma* spp.—for instance, *C. aromatica* (130.6 ± 1.5 µM FRAP equivalents) [260]. This antioxidant potential has also been demonstrated in vivo. A starch/carboxymethyl cellulose (CMC) edible coating including *C. longa* oil suppressed the oxidase enzyme activity of fresh-cut "Fuji" apples by 9% [261] (Table 3).

The luminol-photochemiluminiscence (PLC) assay corroborated afterwards the high antioxidant activity of *C. longa* essential oil [217]. It results in an easy, fast and sensitive method to know the scavenging activity of antioxidants against the radical anion superoxide, especially for hydrophobic-like essential oils [217]. This property may be due to the total phenolic content of *C. longa* essential oil that also highlights over more than 15 essential oils from different plant species [235]. However, the phenolic compounds of *C. longa* essential oil and, consequently, the antioxidant activity can vary depending on the cultivation conditions. Especially, the substrate type, together with the presence of fungi, have significantly influenced the composition and activity of *C. longa* leaf essential oil [169,170]. The antioxidant activity can also change according to many other factors, such as the degree of dryness of *C. longa* rhizome. Specifically, the essential oil from the fresh rhizomes (24.4% ar-turmerone, 20.5% α-turmerone and 11.1% β-turmerone) exhibited higher DPPH radical scavenging, as well as Fe^{2+}-chelating abilities, than the dry ones (21.4% ar-turmerone, 7.2% α-santalene and 6.6% ar-curcumene). The antioxidant activity of both essential oils was significantly higher than the commercial antioxidants BHA and butylated hydroxytoluene (BHT) [164] (Table 3). Nevertheless, other authors reported a different trend. Gounder et al. demonstrated throughout several tests that dried and cured rhizomes had higher antioxidant activity than the fresh ones (ar-turmerone (21.0–30.3%), α-turmerone (26.5–33.5%) and β-turmerone (18.9–21.1%)). Specifically, ABTS radical cation scavenging [Trolox equivalent antioxidant capacity (TEAC) 68.0, 66.9 and 38.9 μM at 1 mg/mL]; ferric-reducing antioxidant potential (TEAC 276.8, 264.1 and 178.4 μM at 1 mg/mL); total antioxidant capacity by phosphomolybdenum assay (686, 638 and 358 ascorbic acid equivalents per 1 mg of oil) and reducing power were stronger in dried and cured rhizome than in fresh ones, respectively [165]. These differences are mainly due to the different compositions reported by the authors [164,165] in the fresh and dry rhizome essential oils used in the test.

Several research carried out with turmeric rhizome essential oil without β-turmerone among the main compounds [105,262] reported lower DPPH bleaching potential and ferric-reducing antioxidant power of *C. longa* rhizome oil (45.5% ar-turmerone and 13.4% α-turmerone) principally, in comparison to those of Trolox (IC_{50} 14.5 ± 2.9 mg/mL *vs.* 0.012 ± 0.004 mg/mL and 389.0 ± 112.0 *vs.* 402.3 ± 20.1 μM ascorbic acid equivalents, respectively) [105], as well as negligible DPPH radical scavenging activity (38.7% ar-turmerone and 14.2% α-turmerone) with respect to other, different essential oils, among which were cinnamon, clove, green tea, lemon eucalyptus, rosemary, oregano and its main compound carvacrol [262] (Table 3).

Table 3. Antimicrobial, herbicidal and antioxidant activities of *C. longa* essential oil in the agri-food industry. DPPH: 2,2-diphenyl-1-picrylhydrazyl, ABTS: 2,2′-azino-bis-3-ethylbenzothiazoline-6-sulphonic acid, BHA: butylated hydroxyanisole and BHT: butylated hydroxytoluene.

Antimicrobial Activity			
Chemical Composition	**Concentration**	**Effect**	**Ref.**
42.6% α-Turmerone 16.0% β-Turmerone 12.9% ar-Turmerone	17.9 and 294.9 µg/mL	Decrease the development of *Fusarium verticillioides* by 56.0 and 79.3%, respectively, as well as the thickness and length of microconidia, fungal biomass and fumonisin production	[34]
51.8% ar-Turmerone 11.9% ar-Turmerol	1000 ppm	Complete mycelial growth inhibition of *Colletotrichum falcatum* and *F. moniliforme*	[192]
51.8% ar-Turmerone 11.9% ar-Turmerol	2000 ppm	Complete mycelial growth inhibition of *Curvularia pallescens*, *Aspergillus niger* and *F. oxysporium*	[192]
58% ar-Turmerone Limonene Borneol	>45–90 µg/disc	Significant inhibition of *Bacillus subtilis*, *Salmonella choleraesuis*, *Escherichia coli*, *A. niger* and *Saccharomyces cerivisiae* at higher doses than chloramphenicol and amphotericin	[195]
33.42% ar-Turmerone 22.35% α-Turmerone 20.14% β-Turmerone	1–2% *(v/v)*	Antibacterial activity against *E. coli* and *Staphylococcus aureus* when incorporated to an edible film with sorbitol and egg white protein	[196]
-	1–5%	Complete radial growth inhibition of *C. gloeosporoides*, *Sphaceloma cardamomi*, *Pestalotia palmarum*, *Rhizoctonia solani*, *Aspergillus* sp. and *Fusarium* sp.	[199]
35.46% Turmerone 20.61% Cumene 13.82% ar-Turmerone	>0.5 µL	Antifungal effect against *A. niger* when incorporated to a biopolymer film	[202]
-	0.5% *w/w*	Reduction of the growth of *Penicillium* and *Cladosporium* spp. in 60.3 and 41.6%, respectively, for 15 days when incorporated to an edible coating based on achira starch (*Canna indica* L.)	[203]
33.2% ar-Turmerone 23.5% α-Turmerone 22.7% β-Turmerone	0.10–0.5% *v/v*	Significant reduction of the growth rate of *A. flavus*, as well as complete inhibition of germination and sportulation	[211]
53.10% ar-Turmerone	2450 and 3300 µg/mL	Minimum inhibitory and minimum fungicidal concentration against *F. graminearum*	[215]
53.10% ar-Turmerone	3500 and 3000 µg/mL	Complete inhibition of fungal biomass and zearalenone production in *F. graminearum*, respectively	[215]
20.4% α-Phellandrene 19.8% α-Turmerone 10.3% 1.8-Cineole 7.35% β-Turmerone	0.06–0.36 µg/mL	One of the highest minimum inhibitory concentrations with respect to 11 different essential oils against five-food spoilage yeasts	[217]
Herbicidal Activity			
38.7% ar-Turmerone 18.6% β-Turmerone 14.2% α-Turmerone	0.125–1 µL/mL	Reduction of the hypocotyl growth of *Portulaca oleracea*, *Lolium multiflorum* and *Echinochloa crus-galli* in 56.55, 40.45 and 39.33%, respectively, without affecting neither seed germination nor hycopotyl growth of tomato, cucumber and rice crops	[234]
38.7% ar-Turmerone 18.6% β-Turmerone 14.2% α-Turmerone	1 µL/mL	Significant inhibition of *Cortaderia selloana* seed germination (81.71%)	[234]
38.7% ar-Turmerone 18.6% β-Turmerone 14.2% α-Turmerone	>0.125 µL/mL	Outstanding inhibitory effect in the development of *C. selloana* and *Nicotiana glauca*	[234]

Table 3. *Cont.*

Antimicrobial Activity			
Chemical Composition	**Concentration**	**Effect**	**Ref.**
Antioxidant Activity			
42.6% α-Turmerone 16.0% β-Turmerone 12.9% ar-Turmerone	IC$_{50}$ 10.03 mg/mL (DPPH) IC$_{50}$ 0.54 mg/mL (ABTS)	Dose-dependent DPPH and ABTS radical scavenging activities, as well as reducing power	[34]
45.5% ar-Turmerone 13.4% α-Turmerone	IC$_{50}$ 14.5 ± 2.9 mg/mL (DPPH) 389.0 ± 12.0 μM Ascorbic Acid (AA) eq.	Low DPPH bleaching potential and ferric-reducing antioxidant power in comparison to Trolox	[105]
24.4% ar-Turmerone 20.5% α-Turmerone 11.1% β-Turmerone	<100 Meq/kg (peroxide value) 0.04–0.08 TBA value 5–20 μL (DPPH) 10–100 μL (Fe^{2+} chelating effect)	The lowest peroxide value with respect to oleoresins, synthetic antioxidants and essential oil from dry rhizomes. More efficient inhibitory effect of malondialdehyde Higher DPPH radical scavenging, as well as Fe^{2+} chelating abilities than the dry ones (21.4% ar-turmerone, 7.2% α-santalene and 6.6% ar-curcumene) Higher DPPH radical scavenging activity than BHA and BHT	[164]
21.4% ar-Turmerone 7.2% α-Santalene 6.6% ar-Curcumene	100–200 Meq/kg (peroxide value) 0.04–0.08 TBA value 15–20 μL (DPPH) 10–100 μL (Fe^{2+} chelating effect)	Higher DPPH radical scavenging activity than BHA and BHT	[164]
20.4% α-Phellandrene 19.8% α-Turmerone 10.3% 1,8-Cineole 7.35% β-Turmerone	28.1 ± 1.45 mmol Trolox/L (PLC)	Free radical-scavenging potential twice higher than that of Trolox (~60 *vs.* 28.2%, respectively) Antioxidant activity (72.4%) near the values of the reference essential oil *Thymus vulgaris* (90.9%) and butylated hydroxyanisole (BHA) (86.74%)	[217]
22.8% β-Sesquiphellandrene 9.5% Terpinolene	IC$_{50}$ 3.227 mg/mL (DPPH) IC$_{50}$ 1.541 mg/mL (ABTS) 1 mg/mL (antiperoxidative)	Hydrogen donating properties and reducing power. Potential option to prevent oxidative deterioration of fat containing food products	[256]
35.46% Turmerone 20.61% Cumene 13.82% ar-Turmerone	30 μL/mL	Suppression of oxidase enzyme activity of the fresh-cut "Fuji" apples by 9% when incorporated in a starch/carboxymethyl cellulose edible coating	[261]
38.7% ar-Turmerone 14.2% α-Turmerone	10 μL	Negligible DPPH radical scavenging activity with respect to other different essential oils (cinnamon, clove, green tea, lemon eucalyptus, rosemary, oregano and its main compound carvacrol)	[262]

On the other hand, other *Curcuma* spp. essential oil with very different chemical compositions have also demonstrated strong and dose-dependent antioxidant abilities. In this sense, *C. zedoaria* (17.72% curzerenone, 15.85% γ-eudesmol acetate and 6.50% germacrone) and *C. angustifolia* (29.62% epicurzerenone, 10.79% curzerenone and 6.12% *trans*-β-terpineol) rhizome essential oils showed higher DPPH (IC$_{50}$ 2.58 ± 077 μg/mL and 12.53 ± 0.14 μg/mL) and ABTS (IC$_{50}$ 1.28 ± 0.05 μg/mL and 5.53 ± 0.29 μg/mL) radical scavenging ability, as well as reducing power (EC$_{50}$ 4.77 ± 0.14 μg/mL and 5.68 ± 0.11 μg/mL) than BHT and ascorbic acid (DPPH: 19.07 ± 0.17 and 5.31 ± 0.2 μg/mL, ABTS: 14.19 ± 0.21 and 1.51 ± 0.32 μg/mL and reducing power: 9.61 ± 0.18 and 5.21 ± 0.13 μg/mL, respectively) [194]. The leaf essential oil of *C. angustifolia* (33.2% curzerenone, 18.6% 14-hydroxy-δ-cadinene and 7.3% γ-eudesmol acetate) showed even higher DPPH and ABTS free-radical scavenging (4.06 ± 0.06 and 1.35 ± 0.14 µg/mL, respectively), as well as reducing (EC_{50} 2.62 ± 0.25 µg/mL) activities, than the rhizome oil and the standard references [128]; *C. amada* rhizome oil (40% β-myrcene, 11.78% β-pinene and 10% ar-curcumene) and the essential oil obtained from the pulverized rhizome of *C. petiolata* (83.99% 2-methyl-5-pentanol) presented moderate antioxidant activity in comparison to the extracts and standard references [133,263].

C. longa rhizome essential oil has also exhibited strong antioxidant potential when combined with other essential oils—for instance, *Z. officinale*. In this case, the combination of both showed higher DPPH radical scavenging activity (IC_{50} 3.75 µL/mL vs. 4.28 and 7.19 µL/mL), as well as stronger β-carotene–linoleic acid bleaching (65.24% vs. 59.88 and 55.82%) than *C. longa* and *Z. officinale* oils alone, respectively [235]. This last test has been commonly used to compare the lipid peroxidation inhibitory activity of either individual compounds or mixtures, despite possible scattered results due to different factors like the chemical composition and extracting solvent [264].

Overall, the genus *Curcuma* and its derived products have been popularly used as food additives to confer special beneficial properties, which include colouring, preservation and healthy effects. Particularly, the biopreservative properties of *C. longa* rhizome oil can meet the needs of the agri-food industry. Its suitability as a natural alternative to synthetic antioxidants has been broadly corroborated through many in vitro and in vivo tests, obtaining interesting results replacing the reference synthetic antioxidants. So much so that this essential oil is being included in food coatings to keep them much longer. Moreover, further investigation regarding the most appropriate application of *C. longa* rhizome oil, as well as combinations with other, different essential oils, is being carried out, with the aim of trying to enhance its antioxidant potential and being finally implemented in the sustainable agri-food industry.

5. Conclusions

Consumers demand natural, safer and greener products, as well as sustainable food technologies, from the agri-food industry. However, an equilibrium between meeting consumer expectations and achieving the maximum efficiency in industrial production according to Green Chemistry is required.

The potential applications of numerous plant products in the agri-food industry have been widely investigated. Amongst them, the essential oil extracted from the rhizome of *C. longa* (species popularly known for its medicinal and culinary benefits) has demonstrated a high antimicrobial potential against a broad spectrum of plagues in crops and food-spoilage microorganisms, as well as significant phytotoxic effects against diverse weeds that are considered truly a threat for agricultural production and ecology. Besides, it has exhibited interesting antioxidant activity that would avoid postharvest decay and extend food shelf-lives.

This versatility is mainly due to the characteristic chemical composition of *C. longa* rhizome essential oil. Usually, sesquiterpenes constitute the main phytochemical group identified, and turmerones are the most representative components. However, this pattern is subjected to changes depending on countless internal (genetics) and external (geographic location, cultivation conditions, post-harvest processing, etc.) factors. For this reason, predictive models need to be developed to previse the chemical composition of *C. longa* rhizome essential oil according to the conditions surrounding the plant. In this way, the control of these factors is useful to obtain a high-yield essential oil with the aimed chemical composition, convenient for carrying out a specific activity in the agri-food industry in an optimum way.

Given the nature of these products (complex mixtures of volatile compounds), one of the first processes to take into account is the extraction technique chosen. Despite that the conventional methods (steam distillation, hydrodistillation, etc.) are still the most commonly used, there is a current tendency to employ the novel ones (SFE, SWE, SFME,

MAE, etc.) that offer several advantages, such as the reduction of costs, of extraction times, energy consumption, etc., in an attempt to offer higher-quality *C. longa* rhizome essential oil in the lowest time possible and with the minimum residues produced. For its total implementation, further research is needed to achieve the most efficient extraction that allows obtaining a chemical composition enriched in the active component to elucidate its mechanisms of action, encapsulating techniques of *C. longa* essential oil for its preservation and/or release against external conditions (temperature, oxygen, etc.), as well as to determine the threshold application with which it would neither damage crops nor affect the organoleptic properties of food products, are necessary research prior to their employment in the agri-food industry.

The sustainable and efficient encapsulation of *C. longa* rhizome oil represents the ultimate step for its implementation in the agri-food industry. Current research is oriented to solve the limitations when applying turmeric essential oil (volatility, instability under certain conditions and hydrophobicity), with the aim of longer preserving its numerous benefits and improving its performance. Biodegradable and biocompatible products as edible alginate-based films with turmeric represent advantages over traditional plastic containers, increasing the antioxidant capacity and extending the shelf-lives of the final products. Many encapsulation methods, including β-cyclodextrines, chitosan–alginate, microemulsions, nanoparticles etc., have been described to enhance the curcumin bioavailability. They represent potential options to also enhance the beneficial properties of the essential oil of *C. longa* rhizome and its components, as well as controlling their release. A complex study regarding the cost-efficiency and sustainability, as well as threshold concentrations not to harm crops and food, have to be taken into account.

Author Contributions: Both authors contributed equally to this work. All authors have read and agreed to the published version of the manuscript.

Funding: This research received no external funding.

Institutional Review Board Statement: Not applicable.

Informed Consent Statement: Not applicable.

Data Availability Statement: No new data were created or analysed in this study. Data is not applicable to this article.

Conflicts of Interest: The authors declare no conflict of interest.

References

1. Singab, A.N.B.I. Medicinal & Aromatic Plants. *Med. Aromat. Plants* **2012**, *1*, 1000e109.
2. Inoue, M.; Hayashi, S.; Craker, L.E. Role of medicinal and aromatic plants: Past, present, and future. In *Pharmacognosy-Medicinal Plants*; Perveen, S., Al-Taweel, A., Eds.; IntechOpen: London, UK, 2019; pp. 13–26. [CrossRef]
3. Saiz de Cos, P. Cúrcuma I (*Curcuma longa* L.). *Reduca Ser. Botánica* **2014**, *7*, 84–99.
4. Araújo, C.A.C.; Leon, L.L. Biological activities of *Curcuma longa* L. *Memórias Inst. Oswaldo Cruz* **2001**, *96*, 723–728. [CrossRef] [PubMed]
5. Luthra, P.M.; Singh, R.; Chandra, R. Therapeutic uses of *Curcuma longa* (turmeric). *Indian J. Clin. Biochem.* **2001**, *16*, 153–160. [CrossRef] [PubMed]
6. Wickenberg, J.; Ingemansson, S.L.; Hlebowicz, J. Effects of *Curcuma longa* (turmeric) on postprandial plasma glucose and insulin in healthy subjects. *Nutr. J.* **2010**, *9*, 43. [CrossRef]
7. Vaughn, A.R.; Branum, A.; Sivamani, R.K. Effects of turmeric (*Curcuma longa*) on skin health: A systematic review of the clinical evidence. *Phytother. Res.* **2016**, *30*, 1243–1264. [CrossRef]
8. Dasgupta, A. Antiinflammatory herbal supplements. In *Translational Inflammation*; Actor, J.K., Smith, K.C., Eds.; Elsevier Inc.: London, UK, 2019; pp. 69–91. [CrossRef]
9. El-Kenawy, A.E.-M.; Hassan, S.M.A.; Mohamed, A.M.M.; Mohammed, H.M.A.M. Tumeric or *Curcuma longa* Linn. In *Nonvitamin and Nonmineral Nutritional Supplements*; Nabavi, S.M., Sanches Silva, A., Eds.; Elsevier Inc.: London, UK, 2019; pp. 447–453. [CrossRef]
10. Ortega, A.M.M.; Segura Campos, M.R. Medicinal plants and their bioactive metabolites in cancer prevention and treatment. In *Bioactive Compounds: Health Benefits and Potential Applications*; Campos, M.R.S., Ed.; Elsevier Inc.: Duxford, UK, 2019; pp. 83–109. [CrossRef]

11. Rajagopal, K.; Varakumar, P.; Baliwada, A.; Byran, G. Activity of phytochemical constituents of *Curcuma longa* (turmeric) and *Andrographis paniculata* against coronavirus (COVID-19): An in silico approach. *Future J. Pharm. Sci.* **2020**, *6*, 104. [CrossRef]

12. Alvis, A.; Arrazola, G.; Martínez, W. Evaluation of antioxidant activity and potential hydro-alcoholic extracts of curcuma (*Curcuma longa*). *Inf. Tecnol.* **2012**, *23*, 11–18. [CrossRef]

13. Sawant, R.S.; Godghate, A.G. Qualitative phytochemical screening of rhizomes of *Curcuma longa* Linn. *Int. J. Sci. Environ. Technol.* **2013**, *2*, 634–641.

14. Abraham, A.; Samuel, S.; Mathew, L. Pharmacognostic evaluation of *Curcuma longa* L. rhizome and standardization of its formulation by HPLC using curcumin as marker. *Int. J. Pharmacogn. Phytochem. Res.* **2018**, *10*, 38–42.

15. Meng, F.; Zhou, Y.; Ren, D.; Wang, R. Turmeric: A review of its chemical composition, quality control, bioactivity, and pharmaceutical application. In *Natural and Artificial Flavoring Agents and Food Dyes*; Grumezescu, A.M., Holban, A.M., Eds.; Elsevier Inc.: London, UK, 2018; pp. 299–350. [CrossRef]

16. Carballido, E. Características de la Planta de la Cúrcuma. Available online: https://www.botanical-online.com/plantas-medicinales/curcuma-caracteristicas (accessed on 18 November 2020).

17. Li, S.; Yuan, W.; Deng, G.; Wang, P.; Aggarwal, B.B. Chemical composition and product quality control of turmeric (*Curcuma longa* L.). *Pharm. Crops* **2011**, *2*, 28–54. [CrossRef]

18. González-Albadalejo, J.; Sanz, D.; Claramunt, R.M.; Lavandera, J.L.; Alkorta, I.; Elguero, J. Curcumin and curcuminoids: Chemistry, structural studies and biological properties. *An. Real Acad. Nac. Farm.* **2015**, *81*, 278–310.

19. Ramesh, T.N.; Paul, M.; Manikanta, K.; Girish, K.S. Structure and morphological studies of curcuminoids and curcuminoid mixture. *J. Cryst. Growth* **2020**, *547*, 125812. [CrossRef]

20. Bambirra, M.L.A.; Junqueira, R.G.; Glória, M.B. Influence of post harvest processing conditions on yield and quality of ground turmeric (*Curcuma longa* L.). *Braz. Arch. Biol. Technol.* **2002**, *45*, 423–429. [CrossRef]

21. Amalraj, A.; Pius, A.; Gopi, S. Biological activities of curcuminoids, other biomolecules from turmeric and their derivatives—A review. *J. Tradit. Chin. Med. Sci.* **2017**, *7*, 205–233. [CrossRef] [PubMed]

22. Bengmark, S.; Mesa, M.D.; Gil, A. Plant-derived health: The effects of turmeric and curcuminoids. *Nutr. Hosp.* **2009**, *24*, 273–281. [PubMed]

23. Tayyem, R.F.; Heath, D.D.; Al-delaimy, W.K.; Rock, C.L.; Tayyem, R.F.; Heath, D.D.; Al-delaimy, W.K.; Rock, C.L.; Tayyem, R.F.; Heath, D.D.; et al. Curcumin content of turmeric and curry powders. *Nutr. Cancer* **2006**, *55*, 126–131. [CrossRef]

24. Siviero, A.; Gallo, E.; Maggini, V.; Gori, L.; Mugelli, A.; Firenzuoli, F.; Vannacci, A. Curcumin, a golden spice with a low bioavailability. *J. Herb. Med.* **2015**, *5*, 57–70. [CrossRef]

25. Akbar, A.; Kuanar, A.; Joshi, R.K.; Sandeep, I.S.; Mohanty, S. Development of prediction model and experimental validation in predicting the curcumin content of turmeric (*Curcuma longa* L.). *Front. Plant Sci.* **2016**, *7*, 1507. [CrossRef]

26. Tung, B.T.; Nham, D.T.; Hai, N.T.; Thu, D.K. *Curcuma longa*, the polyphenolic curcumin compound and pharmacological effects on liver. In *Dietary Interventions in Liver Disease*; Watson, R.R., Preedy, V.R., Eds.; Elsevier Inc.: London, UK, 2019; pp. 125–134. [CrossRef]

27. Soleimani, V.; Sahebkar, A.; Hosseinzadeh, H. Turmeric (*Curcuma longa*) and its major constituent (curcumin) as nontoxic and safe substances: Review. *Phytother. Res.* **2018**, *32*, 985–995. [CrossRef]

28. Changtam, C.; De Koning, H.P.; Ibrahim, H.; Sajid, M.S.; Gould, M.K.; Suksamrarn, A. Curcuminoid analogs with potent activity against *Trypanosoma* and *Leishmania* species. *Eur. J. Med. Chem.* **2010**, *45*, 941–956. [CrossRef] [PubMed]

29. Priyadarsini, K.I. Chemical and structural features influencing the biological activity of curcumin. *Curr. Pharm. Des.* **2013**, *19*, 2093–2100. [PubMed]

30. Arshad, L.; Nasir, S.; Bukhari, A.; Jantan, I. An overview of structure-activity relationship studies of curcumin analogs as antioxidant and anti-inflammatory agents. *Future Med. Chem.* **2017**, *9*, 605–626. [CrossRef] [PubMed]

31. Noureddin, S.A.; El-shishtawy, R.M.; Al-footy, K.O. Curcumin analogues and their hybrid molecules as multifunctional drugs. *Eur. J. Med. Chem.* **2019**, *182*, 111631. [CrossRef]

32. Benzaria, A.; Maresca, M.; Taieb, N.; Dumay, E. Interaction of curcumin with phosphocasein micelles processed or not by dynamic high-pressure. *Food Chem.* **2013**, *138*, 2327–2337. [CrossRef]

33. Devkota, L.; Rajbhandari, M. Composition of essential oils in turmeric rhizome. *Nepal J. Sci. Technol.* **2015**, *16*, 87–94. [CrossRef]

34. Avanço, G.B.; Ferreira, F.D.; Bomfim, N.S.; Santos, P.A.D.S.R.D.; Peralta, R.M.; Brugnari, T.; Mallmann, C.A.; de Filho, B.A.A.; Mikcha, J.M.G.; Machinski, M., Jr. *Curcuma longa* L. essential oil composition, antioxidant effect, and effect on *Fusarium verticillioides* and fumonisin production. *Food Control* **2017**, *73*, 806–813. [CrossRef]

35. Ferreira Guimares, A.; Andrade Vinhas, A.C.; Ferraz Gomes, A.; Souza, L.H.; Baier Krepsky, P. Essential oil of *Curcuma longa* L. rhizomes chemical composition, yield variation and stability. *Quim. Nova* **2020**, *43*, 909–913. [CrossRef]

36. Babu, G.D.K.; Shanmugam, V.; Ravindranath, S.D.; Joshi, V.P. Comparison of chemical composition and antifungal activity of *Curcuma longa* L. leaf oils produced by different water distillation techniques. *Flavour Fragr. J.* **2007**, *22*, 191–196. [CrossRef]

37. Usman, L.A.; Hamid, A.A.; George, O.C.; Ameen, O.M.; Muhammad, N.O.; Zubair, M.F.; Lawal, A. Chemical composition of rhizome essential oil of *Curcuma longa* L. growing in North Central Nigeria. *World J. Chem.* **2009**, *4*, 178–181.

38. Niranjan, A.; Singh, S.; Dhiman, M.; Tewari, S.K. Biochemical composition of *Curcuma longa* L. accessions. *Anal. Lett.* **2013**, *46*, 1069–1083. [CrossRef]

39. Akbar, A.; Kuanar, A.; Sandeep, I.S.; Kar, B.; Singh, S.; Mohanty, S.; Patnaik, J.; Nayak, S. GC-MS analysis of essential oil of some high drug yielding genotypes of turmeric (*Curcuma longa* L.). *Int. J. Pharm. Pharm. Sci.* **2015**, *7*, 35–40.

40. Sandeep, I.S.; Das, S.; Nayak, S.; Mohanty, S. Chemometric profile of *Curcuma longa* L. towards standardization of factors for high essential oil yield and quality. *Proc. Natl. Acad. Sci. India Sect. B Biol. Sci.* **2018**, *88*, 949–957. [CrossRef]

41. Choudhary, V.K.; Kumar, P.S. Weed suppression, nutrient leaching, water use and yield of turmeric (*Curcuma longa* L.) under different land configurations and mulches. *J. Clean. Prod.* **2019**, *210*, 795–803. [CrossRef]

42. Leela, N.K.; Tava, A.; Shafi, P.M.; Sinu, P.J.; Chempakam, B. Chemical composition of essential oils of turmeric. *Acta Pharm.* **2002**, *52*, 137–141.

43. Raina, V.K.; Srivastava, S.K.; Syamsundar, K.V. Rhizome and leaf oil composition of *Curcuma longa* from the lower himalayan region of northern India. *J. Essent. Oil Res.* **2005**, *17*, 556–559. [CrossRef]

44. Awasthi, P.; Dixit, S. Chemical composition of *Curcuma longa* leaves and rhizome oil from the plains of Northern India. *J. Young Pharm.* **2009**, *1*, 312–319. [CrossRef]

45. Singh, S.; Panda, M.K.; Subudhi, E.; Nayak, S. Chemical composition of leaf and rhizome oil of an elite genotype *Curcuma longa* L. from South Eastern ghats of Orissa. *J. Pharm. Res.* **2010**, *3*, 1630–1633.

46. Mishra, R.; Gupta, A.K.; Kumar, A.; Lal, R.K.; Saikia, D.; Chanotiya, C.S. Genetic diversity, essential oil composition, and in vitro antioxidant and antimicrobial activity of *Curcuma longa* L. germplasm collections. *J. Appl. Res. Med. Aromat. Plants* **2018**, *10*, 75–84. [CrossRef]

47. Dosoky, N.S.; Setzer, W.N. Chemical composition and biological activities of essential oils of *Curcuma* species. *Nutrients* **2018**, *10*, 1196. [CrossRef]

48. Kim, D.; Suh, Y.; Lee, H.; Lee, Y. Immune activation and antitumor response of ar-turmerone on P388D1 lymphoblast cell implanted tumors. *Int. J. Mol. Med.* **2013**, *31*, 386–392. [CrossRef] [PubMed]

49. Nair, A.; Amalraj, A.; Jacob, J.; Kunnumakkara, A.B.; Gopi, S. Non-curcuminoids from turmeric and their potential in cancer therapy and anticancer drug delivery formulations. *Biomolecules* **2019**, *9*, 13. [CrossRef] [PubMed]

50. Jankasem, M.; Wuthi-udomlert, M.; Gritsanapan, W. Antidermatophytic properties of ar-turmerone, turmeric oil, and *Curcuma longa* preparations. *ISRN Dermatol.* **2013**, *2013*, 250597. [CrossRef] [PubMed]

51. Yue, G.G.L.; Kwok, H.F.; Lee, J.K.M.; Jiang, L.; Chan, K.M.; Cheng, L.; Wong, E.C.W.; Leung, P.C.; Fung, K.P.; Lau, C.B.S. Novel anti-angiogenic effects of aromatic-turmerone, essential oil isolated from spice turmeric. *J. Funct. Foods* **2015**, *15*, 243–253. [CrossRef]

52. Orellana-paucar, A.M.; Afrikanova, T.; Thomas, J.; Aibuldinov, Y.K.; Dehaen, W.; De Witte, P.A.M.; Esguerra, C.V. Insights from zebrafish and mouse models on the activity and safety of ar-turmerone as a potential drug candidate for the treatment of epilepsy. *PLoS ONE* **2013**, *8*, e81634. [CrossRef]

53. Hucklenbroich, J.; Klein, R.; Neumaier, B.; Graf, R.; Fink, G.R.; Schroeter, M.; Rueger, M.A. Aromatic-turmerone induces neural stem cell proliferation in vitro and in vivo. *Stem Cell Res. Ther.* **2014**, *5*, 100. [CrossRef]

54. Li, Y.; Du, Z.; Li, P.; Yan, L.; Zhou, W.; Tang, Y. Aromatic-turmerone ameliorates imiquimod-induced psoriasis-like inflammation of BALB/c mice. *Int. Immunopharmacol.* **2018**, *64*, 319–325. [CrossRef]

55. Khayreddine, B. *Essential Oils, An Alternative to Synthetic Food Additives and Thermal Treatments*; MedCrave Group LLC: Edmond, OK, USA, 2018; pp. 1–44.

56. Boskovic, M.; Baltic, Z.M.; Ivanovic, J.; Duric, J.; Loncina, J.; Dokmanovic, M.; Markovic, R. Use of essential oils in order to prevent foodborne illnesses caused by pathogens in meat. *Tehnol. Mesa* **2013**, *54*, 14–20. [CrossRef]

57. Mihai, A.L.; Popa, M.E. Essential oils utilization in food industry—A literature review. *Sci. Bull. Ser. F Biotechnol.* **2013**, *17*, 187–192.

58. Mihai, A.L.; Popa, M.E. Inhibitory effects of essential oils with potential to be used in food industry. *Sci. Pharm.* **2014**, *18*, 220–225.

59. Bhavaniramya, S.; Vishnupriya, S.; Al-aboody, M.S. Role of essential oils in food safety: Antimicrobial and antioxidant applications. *Grain Oil Sci. Technol.* **2019**, *2*, 49–55. [CrossRef]

60. Boukhatem, M.N.; Boumaiza, A.; Nada, H.G.; Rajabi, M. *Eucalyptus globulus* essential oil as a natural food preservative: Antioxidant, antibacterial and antifungal properties *in vitro* and in a real food matrix (orangina fruit juice). *Appl. Sci.* **2020**, *10*, 5581. [CrossRef]

61. Ibáñez, M.D. Commercial Essential Oils: Sustainable Alternatives in the Agri-Food Industry. Ph.D. Thesis, Universitat de València, Valencia, Spain, 2019.

62. Ibáñez, M.D.; Sanchez-Ballester, N.M.; Blázquez, M.A. Encapsulated limonene: A pleasant lemon-like aroma with promising application in the agri-food industry. A review. *Molecules* **2020**, *25*, 2598. [CrossRef]

63. Weiss, E.A. Turmeric. In *Spice Crops*; CABI Publishing: Wallingford, CT, USA, 2002; pp. 338–352.

64. Balakrishnan, K. Postharvest technology and processing of turmeric. In *Turmeric. The Genus Curcuma*; Ravindran, P., Nirman Babu, K., Sivaraman, K., Eds.; CRC Press: Boca Raton, FL, USA, 2007; pp. 193–256.

65. Pereira, C.G.; Meireles, M.A.A. Supercritical fluid extraction of bioactive compounds: Fundamentals, applications and economic perspectives. *Food Bioprocess Technol.* **2010**, *3*, 340–372. [CrossRef]

66. Sorensen, E. Principles of Binary Distillation. In *Distillation: Fundamentals and Principles*; Górak, A., Sorensen, E., Eds.; Elsevier: Oxford, UK, 2014; pp. 145–186.

67. Akdag, A.; Ozturk, E. Distillation methods of essential oils. *Nisan* **2019**, *45*, 22–31.

68. Hielscher Ultrasonic Hydrodistillation of Essential Oils. Available online: https://www.hielscher.com/ultrasonic-hydrodistillation-of-essential-oils.htm (accessed on 5 July 2020).

69. Gujarathi, D.B.; Ilay, N.T. Continuous water circulation distillation: A modification of steam distillation. *J. Chem. Educ.* **1993**, *70*, 86. [CrossRef]

70. Santos, D.T.; Angela, A.; Meireles, M. Extraction of volatile oils by supercritical fluid extraction: Patent survey. *Recent Patents Eng.* **2011**, *5*, 17–22. [CrossRef]

71. Chandra, A.; Prajapati, S.; Garg, S.K.; Rathore, A.K. Extraction of turmeric oil by continuous water circulation distillation. *Int. J. Sci. Eng. Appl. Sci.* **2016**, *2*, 2395–3470.

72. Matsumura, S.; Murata, K.; Zaima, N.; Yoshioka, Y.; Morimoto, M.; Kugo, H.; Yamamoto, A.; Moriyama, T.; Matsuda, H. Inhibitory activities of essential oil obtained from turmeric and its constituents against β-secretase. *Nat. Prod. Commun.* **2016**, *11*, 1785–1788. [CrossRef]

73. Sehgal, H.; Jain, T.; Malik, N.; Chandra, A.; Singh, S. Isolation and chemical analysis of turmeric oil from rhizomes. In Proceedings of the Chemical Engineering Towards Sustainable Development, Chennai, India, 27–30 December 2016; pp. 1–14.

74. Masango, P. Cleaner production of essential oils by steam distillation. *J. Clean. Prod.* **2005**, *13*, 833–839. [CrossRef]

75. De Oliveira, M.S.; Silva, S.G.; da Cruz, J.N.; Ortiz, E.; da Costa, W.A.; Bezerra, F.W.F.; Cunha, V.M.B.; Cordeiro, R.M.; Al, E. Supercritical CO_2 application in essential oil extraction. In *Industrial Applications of Green Solvents: Volume II*; Inamuddin, D., Mobin, R., Asiri, A.M., Eds.; Materials Research Forum LLC: Millersville, PA, USA, 2019; pp. 1–28.

76. Silva, L.V.; Nelson, D.L.; Drummond, M.F.B.; Dufossé, L.; Glória, M.B.A. Comparison of hydrodistillation methods for the deodorization of turmeric. *Food Res. Int.* **2005**, *38*, 1087–1096. [CrossRef]

77. Gupta, A.; Naraniwal, M.; Kothari, V. Modern extraction methods for preparation of bioactive plant extracts. *Int. J. Appl. Nat. Sci.* **2012**, *1*, 8–26.

78. Tran, T.H.; Nguyen, P.T.N.; Pham, T.N.; Nguyen, D.C.; Dao, T.P.; Nguyen, T.D.; Nguyen, D.H.; Vo, D.V.N.; Le, X.T.; Le, N.T.H.; et al. Green technology to optimize the extraction process of turmeric (*Curcuma longa* L.) oils. In *Proceedings of the IOP Conference Series: Materials Science and Engineering, Sanya, China, Sanya, China, 17–19 November, 2018*; IOP Publishing Ltd.: Bristol, UK, 2019; Volume 479, pp. 1–6. [CrossRef]

79. Haiyee, Z.A.; Shah, S.H.M.; Ismail, K.; Hashim, N.; Ismail, W.I.W. Quality parameters of *Curcuma longa* L. extracts by supercritical fluid extraction (SFE) and ultrasonic assisted extraction (UAE). *Malays. J. Anal. Sci.* **2016**, *20*, 626–632. [CrossRef]

80. Sahne, F.; Mohammadi, M.; Najafpour, G.D.; Moghadamnia, A.A. Extraction of bioactive compound curcumin from turmeric (*Curcuma longa* L.) via different routes: A comparative study. *Pak. J. Biotechnol.* **2016**, *13*, 173–180.

81. Lee, W.-J.; Suleiman, N.; Hadzir, N.H.N.; Chong, G.-H. Supercritical fluids for the extraction of oleoresins and plant phenolics. In *Green Sustainable Process for Chemical and Environmental Engineering and Science*; Inamuddin, D., Asiri, A.M., Isloor, A.M., Eds.; Elsevier Inc.: Cambridge, MA, USA, 2020; pp. 279–328. [CrossRef]

82. Khezeli, T.; Ghaedi, M.; Bahrani, S.; Daneshfar, A. Supercritical fluid extraction in separation and preconcentration of organic and inorganic species. In *New Generation Green Solvents for Separation and Preconcentration of Organic and Inorganic Species*; Soylak, M., Yilmaz, E., Eds.; Elsevier Inc.: Cambridge, MA, USA, 2020; pp. 425–451. [CrossRef]

83. Capuzzo, A.; Maffei, M.E.; Occhipinti, A. Supercritical fluid extraction of plant flavors and fragrances. *Molecules* **2013**, *18*, 7194–7238. [CrossRef] [PubMed]

84. Wrona, O.; Rafińska, K.; Możeński, C.; Buszewski, B. Supercritical fluid extraction of bioactive compounds from plant materials. *J. AOAC Int.* **2017**, *100*, 1624–1635. [CrossRef] [PubMed]

85. Yousefi, M.; Rahimi-Nasrabadi, M.; Pourmortazavi, S.M.; Wysokowski, M.; Jesionowski, T.; Ehrlich, H.; Mirsadeghi, S. Supercritical fluid extraction of essential oils. *TrAC Trends Anal. Chem.* **2019**, *118*, 182–193. [CrossRef]

86. Mottahedin, P.; Asl, A.H.; Khajenoori, M. Extraction of curcumin and essential oil from *Curcuma longa* L. by subcritical water via response surface methodology. *J. Food Process. Preserv.* **2017**, *41*. [CrossRef]

87. Khanam, S. Influence of operating parameters on supercritical fluid extraction of essential oil from turmeric root. *J. Clean. Prod.* **2018**, *188*, 816–824. [CrossRef]

88. Chang, L.H.; Jong, T.T.; Huang, H.S.; Nien, Y.F.; Chang, C.M.J. Supercritical carbon dioxide extraction of turmeric oil from *Curcuma longa* Linn and purification of turmerones. *Sep. Purif. Technol.* **2006**, *47*, 119–125. [CrossRef]

89. Kao, L.; Chen, C.R.; Chang, C.M.J. Supercritical CO_2 extraction of turmerones from turmeric and high-pressure phase equilibrium of CO_2+ turmerones. *J. Supercrit. Fluids* **2007**, *43*, 276–282. [CrossRef]

90. Carvalho, P.I.N.; Osorio-Tobón, F.J.; Rostagno, M.A.; Mauricio, A.; Petenate, A.J.; Meireles, A.A. Optimization of the ar-turmerone extraction from turmeric (*Curcuma longa* L.) using supercritical carbon dioxide. In Proceedings of the 14th European Meeting on Supercritical Fluids, Marselha, France, 18–21 May 2014.

91. Carvalho, P.I.N.; Osorio-Tobón, J.F.; Rostagno, M.A.; Petenate, A.J.; Meireles, M.A.A. Techno-economic evaluation of the extraction of turmeric (*Curcuma longa* L.) oil and ar-turmerone using supercritical carbon dioxide. *J. Supercrit. Fluids* **2015**, *105*, 44–54. [CrossRef]

92. Topiar, M.; Sajfrtova, M.; Karban, J.; Sovova, H. Fractionation of turmerones from turmeric SFE isolate using semi-preparative supercritical chromatography technique. *J. Ind. Eng. Chem.* **2019**, *77*, 223–229. [CrossRef]

93. Essien, S.O.; Young, B.; Baroutian, S. Recent advances in subcritical water and supercritical carbon dioxide extraction of bioactive compounds from plant materials. *Trends Food Sci. Technol.* **2020**, *97*, 156–169. [CrossRef]

94. Gbashi, S.; Adebo, O.A.; Piater, L.; Madala, N.E.; Njobeh, P.B. Subcritical water extraction of biological materials. *Sep. Purif. Rev.* **2017**, *46*, 21–34. [CrossRef]

95. Baysal, T.; Demirdoven, A. Ultrasound in food technology. In *Handbook on Applications of Ultrasound: Sonochemistry for Sustainability*; Chen, D., Sharma, S.K., Mudhoo, A., Eds.; CRC Press: Boca Raton, FL, USA, 2011; pp. 163–182.

96. Cardoso-Ugarte, G.A.; Juárez-Becerra, G.P.; Sosa-Morales, M.E.; López-Malo, A. Microwave-assisted extraction of essential oils from herbs. *J. Microw. Power Electromagn. Energy* **2013**, *47*, 63–72. [CrossRef]

97. Aziz, N.A.A.; Hassan, J.; Osman, N.H.; Abbas, Z. Extraction of essential oils from *Zingiberaceace* family by using solvent-free microwave extraction (SFME), microwave-assisted extraction (MAE) and hydrodistillation (HD). *Asian J. Appl. Sci.* **2017**, *5*, 97–100.

98. Akloul, R.; Benkaci-Ali, F.; Eppe, G. Kinetic study of volatile oil of *Curcuma longa* L. rhizome and *Carum carvi* L. fruits extracted by microwave-assisted techniques using the cryogrinding. *J. Essent. Oil Res.* **2014**, *26*, 473–485. [CrossRef]

99. Ching, W.-Y.; Bin-Yusoff, Y.; Wan-Amarina, W.-N.B. Extraction of essential oil from *Curcuma longa*. *J. Food Chem. Nutr.* **2014**, *2*, 1–10.

100. Hafizuddin Mokhtar, M. Extraction of Essential Oil from Turmeric (*Curcuma longa*). Bachelor's Thesis, Universiti Teknologi MARA, Selangor, Malaysia, 2017.

101. Zhilyakova, E.; Novikov, O.; Pisarev, D.; Boyko, N.; Nikitin, K. Prospects of using freons as extractants of *Curcuma longa* L. root essential oil. *Adv. Biol. Sci. Res.* **2019**, *7*, 373–376. [CrossRef]

102. Shang, Z.-P.; Xu, L.-L.; Lu, Y.-Y.; Guan, M.; Li, D.-Y.; Le, Z.-Y.; Bai, Z.-L.; Qiao, X.; Ye, M. Advances in chemical constituents and quality control of turmeric. *World J. Tradit. Chin. Med.* **2019**, *5*, 116–121. [CrossRef]

103. Manzan, A.C.C.M.; Toniolo, F.S.; Bredow, E.; Povh, N.P. Extraction of essential oil and pigments from *Curcuma longa* [L.] by steam distillation and extraction with volatile solvents. *J. Agric. Food Chem.* **2003**, *51*, 6802–6807. [CrossRef] [PubMed]

104. Stanojevic, J.; Stanojevic, L.; Cvetkovic, D.; Danilovic, B. Chemical composition, antioxidant and antimicrobial activity of the turmeric essential oil (*Curcuma longa* L.). *Adv. Technol.* **2015**, *4*, 19–25. [CrossRef]

105. Pino, J.A.; Fon-fay, F.M.; Falco, A.S.; Rodeiro, I. Chemical composition and biological activities of essential oil from turmeric (*Curcuma longa* L.) rhizomes grown in Amazonian Ecuador. *Rev. CENIC* **2018**, *49*, 1–8.

106. Afzal, A.; Oriqat, G.; Khan, M.A.; Jose, J.; Afzal, M. Chemistry and biochemistry of terpenoids from *Curcuma* and related species. *J. Biol. Act. Prod. Nat.* **2013**, *3*, 1–55. [CrossRef]

107. Coy Barrera, C.C.A.; Eunice Acosta, G. Antibacterial activity and chemical composition of essential oils of rosemary (*Rosemary officinalis*), thyme (*Thymus vulgaris*) and turmeric (*Curcuma longa*) from Colombia. *Rev. Cuba. Plantas Med.* **2013**, *18*, 237–246.

108. Jain, V.; Prasad, V.; Singh, S.; Pal, R. HPTLC method for the quantitative determination of ar-turmerone and turmerone in lipid soluble fraction from *Curcuma longa*. *Nat. Prod. Commun.* **2007**, *2*, 927–932. [CrossRef]

109. Bahl, J.R.; Bansal, R.P.; Garg, S.N.; Gupta, M.M.; Singh, V.; Goel, R.; Kumar, S. Variation in yield of curcumin and yield and quality of leaf and rhizome essential oils among Indian land races of turmeric *Curcuma longa* L. *Proc. Indian Natl. Sci. Acad.* **2014**, *80*, 143–156. [CrossRef]

110. Wegmann, M.; Lersch, P.; Wenk, H.H.; Klee, S.K.; Maczkiewitz, U.; Farwick, M.; Goldschmidt, E. Protective turmerones from Curcuma longa. *Pers. Care* **2009**, 37–40.

111. Bagad, A.S.; Joseph, J.A.; Bhaskaran, N.; Agarwal, A. Comparative evaluation of anti-inflammatory activity of curcuminoids, turmerones, and aqueous extract of *Curcuma longa*. *Adv. Pharmacol. Sci.* **2013**, *2013*. [CrossRef]

112. Del Prete, D.; Millán, E.; Pollastro, F.; Chianese, G.; Luciano, P.; Collado, J.A.; Munoz, E.; Appendino, G.; Taglialatela-Scafati, O. Turmeric sesquiterpenoids: Expeditious resolution, comparative bioactivity, and a new bicyclic turmeronoid. *J. Nat. Prod.* **2016**, *79*, 267–273. [CrossRef] [PubMed]

113. Prabhakaran Nair, K. *Turmeric (Curcuma longa L.) and ginger (Zingiber o cinale Rosc.)—World's Invaluable Medicinal Spices the Agronomy and Economy of Turmeric*; Springer: Cham, Switzerland, 2019; pp. 1–568.

114. Murakami, A.; Furukawa, I.; Miyamoto, S.; Tanaka, T.; Ohigashi, H. Curcumin combined with turmerones, essential oil components of turmeric, abolishes inflammation-associated mouse colon carcinogenesis. *BioFactors* **2013**, *39*, 221–232. [CrossRef] [PubMed]

115. Yue, G.G.L.; Cheng, S.W.; Yu, H.; Xu, Z.S.; Lee, J.K.M.; Hon, P.M.; Lee, M.Y.H.; Kennelly, E.J.; Deng, G.; Yeung, S.K.; et al. The role of turmerones on curcumin transportation and P-glycoprotein activities in intestinal caco-2 cells. *J. Med. Food* **2012**, *15*, 242–252. [CrossRef] [PubMed]

116. Yue, G.G.L.; Jiang, L.; Kwok, H.F.; Lee, J.K.M.; Chan, K.M.; Fung, K.P.; Leung, P.C.; Lau, C.B.S. Turmeric ethanolic extract possesses stronger inhibitory activities on colon tumour growth than curcumin—The importance of turmerones. *J. Funct. Foods* **2016**, *22*, 565–577. [CrossRef]

117. Lee, H.-S. Antimicrobial properties of turmeric (*Curcuma longa* L.) rhizome-derived ar-turmerone and curcumin. *Food Sci. Biotechnol.* **2006**, *15*, 559–563.

118. Lee, H.S. Antiplatelet property of *Curcuma longa* L. rhizome-derived ar-turmerone. *Bioresour. Technol.* **2006**, *97*, 1372–1376. [CrossRef] [PubMed]

119. Baik, K.U.; Jung, S.H.; Ahn, B.Z. Recognition of pharmacophore of ar-turmerone for its anticancer activity. *Arch. Pharm. Res.* **1993**, *16*, 254–256. [CrossRef]

120. Aratanechemuge, Y.; Komiya, T.; Moteki, H.; Katsuzaki, H.; Imai, K.; Hibasami, H. Selective induction of apoptosis by ar-turmerone isolated from turmeric (*Curcuma longa* L) in two human leukemia cell lines, but not in human stomach cancer cell line. *Int. J. Mol. Med.* **2002**, *9*, 481–484. [CrossRef]

121. Cheng, S.B.; Wu, L.C.; Hsieh, Y.C.; Wu, C.H.; Chan, Y.J.; Chang, L.H.; Chang, C.M.J.; Hsu, S.L.; Teng, C.L.; Wu, C.C. Supercritical carbon dioxide extraction of aromatic turmerone from *Curcuma longa* Linn. Induces apoptosis through reactive oxygen species-triggered intrinsic and extrinsic pathways in human hepatocellular carcinoma HepG2 cells. *J. Agric. Food Chem.* **2012**, *60*, 9620–9630. [CrossRef]

122. Chen, M.; Chang, Y.Y.; Huang, S.; Xiao, L.H.; Zhou, W.; Zhang, L.Y.; Li, C.; Zhou, R.P.; Tang, J.; Lin, L.; et al. Aromatic-turmerone attenuates LPS-induced neuroinflammation and consequent memory impairment by targeting TLR4-dependent signaling pathway. *Mol. Nutr. Food Res.* **2018**, *62*, 1700281. [CrossRef]

123. Yang, S.; Liu, J.; Jiao, J.; Jiao, L. Ar-turmerone exerts anti-proliferative and anti-inflammatory activities in HaCaT keratinocytes by inactivating hedgehog pathway. *Inflammation* **2020**, *43*, 478–486. [CrossRef] [PubMed]

124. Zhang, L.; Yang, Z.; Wei, J.; Su, P.; Chen, D.; Pan, W.; Zhou, W.; Zhang, K.; Zheng, X.; Lin, L.; et al. Contrastive analysis of chemical composition of essential oil from twelve *Curcuma* species distributed in China. *Ind. Crops Prod.* **2017**, *108*, 17–25. [CrossRef]

125. Devi, L.R.; Rana, V.S.; Devi, S.I.; Blázquez, M.A.; Devi, L.R.; Rana, V.S.; Devi, S.I. Chemical composition and antimicrobial activity of the essential oil of *Curcuma leucorhiza* Roxb. *J. Essent. Oil Res.* **2012**, *24*, 533–538. [CrossRef]

126. Jantan, I.; Saputri, F.C.; Qaisar, M.N.; Buang, F. Correlation between chemical composition of *Curcuma domestica* and *Curcuma xanthorrhiza* and their antioxidant effect on human low-density lipoprotein oxidation. *Evidence-Based Complement. Altern. Med.* **2012**, *2012*. [CrossRef] [PubMed]

127. Singh, P.; Singh, S.; Kapoor, I.P.S.; Singh, G.; Isidorov, V.; Szczepaniak, L. Chemical composition and antioxidant activities of essential oil and oleoresins from *Curcuma zedoaria* rhizomes, part-74. *Food Biosci.* **2013**, *3*, 42–48. [CrossRef]

128. Jena, S.; Ray, A.; Banerjee, A.; Sahoo, A.; Nasim, N.; Sahoo, S.; Kar, B.; Patnaik, J.; Panda, P.C.; Nayak, S. Chemical composition and antioxidant activity of essential oil from leaves and rhizomes of *Curcuma angustifolia* Roxb. *Nat. Prod. Res.* **2017**, *31*, 2188–2191. [CrossRef]

129. Zhu, J.; Lower-Nedza, A.D.; Hong, M.; Jiec, S.; Wang, Z.; Yingmao, D.; Tschiggerl, C.; Bucar, F.; Brantner, A.H. Chemical composition and antimicrobial activity of three essential oils from *Curcuma wenyujin*. *Nat. Prod. Commun.* **2013**, *8*, 523–526. [CrossRef]

130. Hong, S.L.; Lee, G.S.; Rahman, S.N.S.A.; Hamdi, O.A.A.; Awang, K.; Nugroho, N.A.; Malek, S.N.A. Essential oil content of the rhizome of *Curcuma purpurascens* Bl. (Temu Tis) and its antiproliferative effect on selected human carcinoma cell lines. *Sci. World J.* **2014**, *2014*, 397430. [CrossRef]

131. König, W.A.; Rieck, A.; Hardt, I.; Gehrcke, B.; Kubeczka, K.-H.; Muhle, H. Enantiomeric composition of the chiral constituents of essential oils. Part 2: Sesquiterpene hydrocarbons. *J. High Resolut. Chromatogr.* **1994**, *17*, 315–320. [CrossRef]

132. Chizzola, R. Regular monoterpenes and sesquiterpenes. In *Natural Products: Phytochemistry, Botany and Metabolism of Alkaloids, Phenolics and Terpenes*; Ramawat, K.G., Mérillon, J.M., Eds.; Springer: Heidelberg, Germany, 2013; pp. 2973–3008. [CrossRef]

133. Tamta, A.; Prakash, O.; Punetha, H.; Pant, A.K. Chemical composition and *in vitro* antioxidant potential of essential oil and rhizome extracts of *Curcuma amada* Roxb. *Cogent Chem.* **2016**, *2*, 1–11. [CrossRef]

134. Chane-Ming, J.; Vera, R.; Chalchat, J.C.; Cabassu, P. Chemical composition of essential oils from rhizomes, leaves and flowers of *Curcuma longa* L. from Reunion Island. *J. Essent. Oil Res.* **2002**, *14*, 249–251. [CrossRef]

135. Garg, S.N.; Mengi, N.; Patra, N.K.; Charles, R.; Kumar, S. Chemical examination of the leaf essential oil of *Curcuma longa* L. from the North Indian plains. *Flavour Fragr. J.* **2002**, *17*, 103–104. [CrossRef]

136. Pande, C.; Chanotiya, C.S. Constituents of the leaf oil of *Curcuma longa* L. from Uttaranchal. *J. Essent. Oil Res.* **2006**, *18*, 166–167. [CrossRef]

137. Sharma, S.K.; Singh, S.; Tewari, S.K. Study of leaf oil composition from various accessions of *Curcuma longa* L. grown on partially reclaimed sodic soil. *Int. J. Plant Environ.* **2019**, *5*, 293–296. [CrossRef]

138. Bansal, R.P.; Bahl, J.R.; Garg, S.N.; Naqvi, A.A.; Kumary, S. Differential chemical compositions of the essential oils of the shoot organs, rhizomes and rhizoids in the turmeric *Curcuma longa* grown in indo-gangetic plains. *Pharm. Biol.* **2002**, *40*, 384–389. [CrossRef]

139. Sharma, R.K.; Misra, B.P.; Sarma, T.C.; Bordoloi, A.K.; Pathak, M.G.; Leclercq, P.A. Essential oils of *Curcuma longa* L. from Bhutan. *J. Essent. Oil Res.* **1997**, *9*, 589–592. [CrossRef]

140. Behura, S.; Sahoo, S.; Srivastava, V.K. Major constituents in leaf essential oils of *Curcuma longa* L. and *Curcuma aromatica* Salisb. *Curr. Sci.* **2002**, *83*, 1312–1313.

141. Ajaiyeoba, E.O.; Sama, W.; Essien, E.E.; Olayemi, J.O.; Ekundayo, O.; Walker, T.M.; Setzer, W.N. Larvicidal activity of turmerone-rich essential oils of *Curcuma longa* leaf and rhizome from Nigeria on *Anopheles gambiae*. *Pharm. Biol.* **2008**, *46*, 279–282. [CrossRef]

142. Essien, E.; Newby, J.; Walker, T.; Setzer, W.; Ekundayo, O. Chemotaxonomic characterization and *in-vitro* antimicrobial and cytotoxic activities of the leaf essential oil of *Curcuma longa* grown in Southern Nigeria. *Medicines* **2015**, *2*, 340–349. [CrossRef]

143. Ananya, K.; Sujata, M.; Pande, M.K.; Sanghamitra, N. Essential oils from leaves of micropropagated turmeric. *Curr. Sci.* **2009**, *96*, 1166–1167.

144. Chaaban, A.; Richardi, V.S.; Carrer, A.R.; Brum, J.S.; Cipriano, R.R.; Martins, C.E.N.; Navarro-Silva, M.A.; Deschamps, C.; Molento, M.B. Cuticular damage of *Lucilia cuprina* larvae exposed to *Curcuma longa* leaves essential oil and its major compound α-phellandrene. *Data Br.* **2018**, *21*, 1776–1778. [CrossRef] [PubMed]
145. Chaaban, A.; Gomes, E.N.; Richardi, V.S.; Martins, C.E.N.; Brum, J.S.; Navarro-Silva, M.A.; Deschamps, C.; Molento, M.B. Essential oil from *Curcuma longa* leaves: Can an overlooked by-product from turmeric industry be effective for myiasis control? *Ind. Crops Prod.* **2019**, *132*, 352–364. [CrossRef]
146. Sindhu, S.; Chempakam, B.; Leela, N.K.; Bhai, R.S. Chemoprevention by essential oil of turmeric leaves (*Curcuma longa* L.) on the growth of *Aspergillus flavus* and aflatoxin production. *Food Chem. Toxicol.* **2011**, *49*, 1188–1192. [CrossRef] [PubMed]
147. Parveen, Z.; Nawaz, S.; Siddique, S.; Shahzad, K. Composition and antimicrobial activity of the essential oil from leaves of *Curcuma longa* L. Kasur variety. *Indian J. Pharm. Sci.* **2013**, *75*, 117–122. [CrossRef]
148. Kulpapangkorn, W.; Mai-Leang, S. Effect of plant nutrition on turmeric production. *Procedia Eng.* **2012**, *32*, 166–171. [CrossRef]
149. Dosoky, N.S.; Satyal, P.; Setzer, W.N. Variations in the volatile compositions of *Curcuma* species. *Foods* **2019**, *8*, 53. [CrossRef]
150. Yadav, R.; Lal, R.K.; Chanotiya, C.S.; Shanker, K.; Gupta, P.; Shukla, S. Prediction of genetic variability and character contribution using path analysis in turmeric (*Curcuma longa* L.) germplasm. *Trends Phytochem. Res.* **2019**, *3*, 91–100.
151. Chowdhury, J.U.; Nandi, N.C.; Bhuiyan, M.N.I.; Mobarok, M.H. Essential oil constituents of the rhizomes of two types of *Curcuma longa* of Bangladesh. *Bangladesh J. Sci. Ind. Res.* **2008**, *43*, 259–266. [CrossRef]
152. Pal, K.; Chowdhury, S.; Dutta, S.K.; Chakraborty, S.; Chakraborty, M.; Pandit, G.K.; Dutta, S.; Paul, P.K.; Choudhury, A.; Majumder, B.; et al. Analysis of rhizome colour content, bioactive compound profiling and ex-situ conservation of turmeric genotypes (*Curcuma longa* L.) from sub-Himalayan terai region of India. *Ind. Crops Prod.* **2020**, *150*, 112401. [CrossRef]
153. Kumar Rai, S.; Kumar Rai, K.; Pandey, N.; Kumari, A.; Tripathi, D.; Shashi Pandey, R. Varietal performance of turmeric (*Curcuma longa* L.) with special reference to curcumin and essential oil content under climatic conditions of Indogangetic plains. *Veg. Sci.* **2016**, *43*, 36–43.
154. Shashidhar, M.D.; Hegde, N.K.; Hiremath, J.S.; Kukunoor, J.S.; Srikantprasad, D.; Patil, R.T. Evaluation of turmeric (*Curcuma longa* L.) genotypes for yield, curcumin and essential oil content in northern dry zone of Karnataka. *J. Pharmacogn. Phytochem.* **2018**, *SP3*, 130–134.
155. Sahoo, A.; Kar, B.; Jena, S.; Dash, B.; Ray, A.; Sahoo, S.; Nayak, S. Qualitative and quantitative evaluation of rhizome essential oil of eight different cultivars of *Curcuma longa* L. (Turmeric). *J. Essent. Oil-Bear. Plants* **2019**, *22*, 239–247. [CrossRef]
156. Plotto, A. *Turmeric: Post-Production Management*; Mazaud, F., Röttger, A., Steffel, K., Eds.; Food and Agricultural Organization of the United Nations (FAO): Rome, Italy, 2004; pp. 1–21.
157. Oyemitan, I.A.; Elusiyan, C.A.; Onifade, A.O.; Akanmu, M.A.; Oyedeji, A.O.; McDonald, A.G. Neuropharmacological profile and chemical analysis of fresh rhizome essential oil of *Curcuma longa* (turmeric) cultivated in Southwest Nigeria. *Toxicol. Rep.* **2017**, *4*, 391–398. [CrossRef] [PubMed]
158. Naz, S.; Ilyas, S.; Parveen, Z.; Javed, S. Chemical analysis of essential oils from turmeric (*Curcuma longa*) rhizome through GC-MS. *Asian J. Chem.* **2010**, *22*, 3153–3158.
159. Gonçalves, G.M.S.; Barros, P.P.; da Silva, G.H.; Fedes, G.R. The essential oil of *Curcuma longa* rhizomes as an antimicrobial and its composition by Gas Chromatography/Mass Spectrometry. *Rev. Ciências Médicas* **2019**, *28*, 1–10. [CrossRef]
160. Sandeep, I.; Sanghamitra, N.; Sujata, M. Differential effect of soil and environment onmetabolic expression of tumeric. *Indian J. Exp. Biol.* **2015**, *53*, 406–411.
161. Sandeep, I.S.; Kuanar, A.; Akbar, A.; Kar, B.; Das, S.; Mishra, A.; Sial, P.; Naik, P.K.; Nayak, S.; Mohanty, S. Agroclimatic zone based metabolic profiling of turmeric (*Curcuma longa* L.) for phytochemical yield optimization. *Ind. Crops Prod.* **2016**, *85*, 229–240. [CrossRef]
162. Akbar, A.; Kuanar, A.; Patnaik, J.; Mishra, A.; Nayak, S. Application of artificial neural network modeling for optimization and prediction of essential oil yield in turmeric (*Curcuma longa* L.). *Comput. Electron. Agric.* **2018**, *148*, 160–178. [CrossRef]
163. Garg, S.N.; Bansal, R.P.; Gupta, M.M.; Kumar, S. Variation in the rhizome essential oil and curcumin contents and oil quality in the land races of turmeric *Curcuma longa* of North Indian plains. *Flavour Fragr. J.* **1999**, *14*, 315–318. [CrossRef]
164. Singh, G.; Kapoor, I.P.S.; Singh, P.; de Heluani, C.S.; de Lampasona, M.P.; Catalan, C.A.N. Comparative study of chemical composition and antioxidant activity of fresh and dry rhizomes of turmeric (*Curcuma longa* Linn.). *Food Chem. Toxicol.* **2010**, *48*, 1026–1031. [CrossRef]
165. Gounder, D.K.; Lingamallu, J. Comparison of chemical composition and antioxidant potential of volatile oil from fresh, dried and cured turmeric (*Curcuma longa*) rhizomes. *Ind. Crops Prod.* **2012**, *38*, 124–131. [CrossRef]
166. Monton, C.; Luprasong, C.; Charoenchai, L. Convection combined microwave drying affect quality of volatile oil compositions and quantity of curcuminoids of turmeric raw material. *Rev. Bras. Farm.* **2019**, *29*, 434–440. [CrossRef]
167. Monton, C.; Luprasong, C.; Charoenchai, L. Acceleration of turmeric drying using convection and microwave-assisted drying technique: An optimization approach. *J. Food Process. Preserv.* **2019**, *43*, e14096. [CrossRef]
168. El-Hawaz, R.F.; Grace, M.H.; Janbey, A.; Lila, M.A.; Adelberg, J.W. In vitro mineral nutrition of *Curcuma longa* L. affects production of volatile compounds in rhizomes after transfer to the greenhouse. *BMC Plant Biol.* **2018**, *18*, 122. [CrossRef]
169. De Ferrari, M.P.S.; dos Queiroz, M.S.; de Andrade, M.M.; Alberton, O.; Gonçalves, J.E.; Gazim, Z.C.; Magalhães, H.M. Substrate-associated mycorrhizal fungi promote changes in terpene composition, antioxidant activity, and enzymes in *Curcuma longa* L. acclimatized plants. *Rhizosphere* **2020**, *13*, 100191. [CrossRef]

170. De Ferrari, M.P.S.; da Cruz, R.M.S.; dos Queiroz, M.S.; de Andrade, M.M.; Alberton, O.; Magalhães, H.M. Efficient *ex vitro* rooting, acclimatization, and cultivation of *Curcuma longa* L. from mycorrhizal fungi. *J. Crops Sci. Biotechnol.* **2020**, *23*, 469–482. [CrossRef]
171. Blázquez, M.A. Role of natural essential oils in sustainable agriculture and food preservation. *J. Sci. Res. Rep.* **2014**, *3*, 1843–1860. [CrossRef]
172. Walia, S.; Saha, S.; Tripathi, V.; Sharma, K.K. Phytochemical biopesticides: Some recent developments. *Phytochem. Rev.* **2017**, *16*, 989–1007. [CrossRef]
173. Walia, S.; Saha, S.; Kundu, A. Environment friendly pesticides based on essential oils and their constituents. In *Pesticides and Pests*; Parmar, B.S., Singh, S.B., Walia, S., Eds.; Cambridge Scholars Publishin: Newcastle, UK, 2019; pp. 241–263.
174. Garay, J. Review of essential oils: A viable pest control alternative. *J. Hum. Ecol.* **2020**, *71*, 13–22. [CrossRef]
175. Raveau, R.; Fontaine, J.; Sahraoui, A.L.-H. Essential oils as potential alternative biocontrol products against plant pathogens and weeds: A review. *Foods* **2020**, *9*, 365. [CrossRef]
176. Werrie, P.-Y.; Durenne, B.; Delaplace, P.; Fauconnier, M.-L. Phytotoxicity of essential oils: Opportunities and constraints for the development of biopesticides. A review. *Foods* **2020**, *9*, 1291. [CrossRef] [PubMed]
177. Rawat, S. Food Spoilage: Microorganisms and their prevention. *Pelagia Res. Libr. Asian J. Plant Sci. Res.* **2015**, *5*, 47–56.
178. Salas, M.L.; Mounier, J.; Valence, F.; Coton, M.; Thierry, A.; Coton, E. Antifungal microbial agents for food biopreservation—A review. *Microorganisms* **2017**, *5*, 37. [CrossRef] [PubMed]
179. Amza, J. Seed borne fungi; food spoilage, negative impact and their management: A review. *Food Sci. Qual. Manag.* **2018**, *81*, 70–79.
180. Sawicka, B.; Egbuna, C. Pests of agricultural crops and control measures. In *Natural Remedies for Pest, Disease and Weed Control*; Egbuna, C., Sawicka, B., Eds.; Elsevier Inc.: Cambridge, MA, USA, 2020; pp. 1–16. [CrossRef]
181. WHO. *Aflatoxins*. Available online: https://www.who.int/foodsafety/FSDigest_Aflatoxins_EN.pdf (accessed on 18 September 2020).
182. Abbas, M.; Naz, S.A.; Shafique, M.; Jabeen, N.; Abbas, S. Fungal contamination in dried fruits and nuts: A possible source of mycosis and mycotoxicosis. *Pak. J. Bot.* **2019**, *51*, 1523–1529. [CrossRef]
183. Tournas, V.H. Spoilage of vegetable crops by bacteria and fungi and related health hazards. *Crit. Rev. Microbiol.* **2005**, *31*, 33–44. [CrossRef]
184. Kumar, V.; Iqbal, N. Post-harvest pathogens and disease management of horticultural crop: A brief review. *Plant Arch.* **2020**, *20*, 2054–2058.
185. Schmidt, M.; Zannini, E.; Lynch, K.M.; Arendt, E.K. Novel approaches for chemical and microbiological shelf life extension of cereal crops. *Crit. Rev. Food Sci. Nutr.* **2019**, *59*, 3395–3419. [CrossRef]
186. Brauer, V.S.; Rezende, C.P.; Pessoni, A.M.; De Paula, R.G.; Rangappa, K.S.; Nayaka, S.C.; Gupta, V.K.; Almeida, F. Antifungal agents in agriculture: Friends and foes of public health. *Biomolecules* **2019**, *9*, 521. [CrossRef]
187. Badawy, M.E.I.; Abdelgaleil, S.A.M. Composition and antimicrobial activity of essential oils isolated from Egyptian plants against plant pathogenic bacteria and fungi. *Ind. Crops Prod.* **2014**, *52*, 776–782. [CrossRef]
188. Nikkhah, M.; Hashemi, M.; Najafi, M.B.H.; Farhoosh, R. Synergistic effects of some essential oils against fungal spoilage on pear fruit. *Int. J. Food Microbiol.* **2017**, *257*, 285–294. [CrossRef] [PubMed]
189. Hu, F.; Tu, X.F.; Thakur, K.; Hu, F.; Li, X.L.; Zhang, Y.S.; Zhang, J.G.; Wei, Z.J. Comparison of antifungal activity of essential oils from different plants against three fungi. *Food Chem. Toxicol.* **2019**, *134*, 110821. [CrossRef] [PubMed]
190. Arasu, M.V.; Viayaraghavan, P.; Ilavenil, S.; Al-Dhabi, N.A.; Choi, K.C. Essential oil of four medicinal plants and protective properties in plum fruits against the spoilage bacteria and fungi. *Ind. Crops Prod.* **2019**, *133*, 54–62. [CrossRef]
191. Atif, M.; Ilavenil, S.; Devanesan, S.; AlSalhi, M.S.; Choi, K.C.; Vijayaraghavan, P.; Alfuraydi, A.A.; Alanazi, N.F. Essential oils of two medicinal plants and protective properties of jack fruits against the spoilage bacteria and fungi. *Ind. Crops Prod.* **2020**, *147*, 112239. [CrossRef]
192. Singh, G.; Singh, O.P.; Maurya, S. Chemical and biocidal investigations on essential oils of some Indian *Curcuma* species. *Prog. Cryst. Growth Charact.* **2002**, *45*, 75–81. [CrossRef]
193. Akarchariya, N.; Sirilun, S.; Julsrigival, J.; Chansakaowa, S. Chemical profiling and antimicrobial activity of essential oil from *Curcuma aeruginosa* Roxb., *Curcuma glans* K. Larsen & J. Mood and *Curcuma* cf. *xanthorrhiza* Roxb. collected in Thailand. *Asian Pac. J. Trop. Biomed.* **2017**, *7*, 881–885. [CrossRef]
194. Jena, S.; Ray, A.; Sahoo, A.; Panda, P.C.; Nayak, S. Deeper insight into the volatile profile of essential oil of two *Curcuma* species and their antioxidant and antimicrobial activities. *Ind. Crops Prod.* **2020**, *155*. [CrossRef]
195. Péret-Almeida, L.; da Naghetini, C.C.; de Nunan, E.A.; Junqueira, R.G.; Glória, M.B.A. In vitro antimicrobial activity of the ground rhizome, curcuminoid pigments and essential oil of *Curcuma longa* L. *Cienc. e Agrotecnol.* **2008**, *32*, 875–881. [CrossRef]
196. Kavas, N.; Kavas, G. Use of turmeric (*Curcuma longa* L.) essential oil added to an egg white protein powder-based film in the storage of Çökelek cheese. *J. Food Chem. Nanotechnol.* **2017**, *3*, 105–110. [CrossRef]
197. Jeleń, H.; Wąsowicz, E. Volatile fungal metabolites and their relation to the spoilage of agricultural commodities. *Food Rev. Int.* **1998**, *14*, 391–426. [CrossRef]
198. Hu, Y.; Luo, J.; Kong, W.; Zhang, J.; Logrieco, A.F.; Wang, X.; Yang, M. Uncovering the antifungal components from turmeric (*Curcuma longa* L.) essential oil as *Aspergillus flavus* fumigants by partial least squares. *RSC Adv.* **2015**, *5*, 41967–41976. [CrossRef]

199. Saju, K.A.; Venugopal, M.N.; Mathew, M.J. Antifungal and insect-repellent activities of essential oil of turmeric (*Curcuma longa* L.). *Curr. Sci.* **1998**, *75*, 660–662.
200. Marin, S.; Guynot, M.E.; Sachnis, V.; Arbones, J.; Ramos, A.J. *Aspergillus flavus, Aspergillus niger,* and *Penicillium corylophilum* spoilage prevention of bakery products by means of weak-acid preservatives. *Food Microbiol. Saf.* **2002**, *67*, 2271–2277. [CrossRef]
201. Gougouli, M.; Koutsoumanis, K.P. Risk assessment of fungal spoilage: A case study of *Aspergillus niger* on yogurt. *Food Microbiol.* **2017**, *65*, 264–273. [CrossRef]
202. Mustapha, F.A.; Jai, J.; Raikhan, N.H.N.; Sharif, Z.I.M.; Yusof, N.M. Response surface methodology analysis towards biodegradability and antimicrobial activity of biopolymer film containing turmeric oil against *Aspergillus niger*. *Food Control* **2019**, *99*, 106–113. [CrossRef]
203. Velasco, J.A.C.; Mahecha, P.V.; Andrade-Mahecha, M.M. Essential oil turmeric (*Curcuma longa*) antifungal agent as in edible coatings applied to pumpkin minimally processed. *Rev. Ciências Agrárias* **2017**, *40*, 641–654. [CrossRef]
204. Oranusi, S.; Olarewaju, S. Mycoflora and aflatoxin contamination of some foodstuffs. *Int. J. Biotechnol. Allied Fields* **2013**, *1*, 9–18.
205. Embaby, E.M.; Awni, N.M.; Abdel-galil, M.M.; El-gendy, H.I. Mycoflora and mycotoxin contaminated some juices. *J. Agric. Technol.* **2015**, *11*, 693–712.
206. Saleem, A.R. Mycobiota and molecular detection of *Aspergillus flavus* and *A. parasiticus* aflatoxin contamination of strawberry (*Fragaria ananassa* Duch.) fruits. *Arch. Phytopathol. Plant Prot.* **2017**, *50*, 982–996. [CrossRef]
207. García-Díaz, M.; Patiño, B.; Vázquez, C.; Gil-Serna, J. A novel niosome-encapsulated essential oil formulation to prevent *Aspergillus flavus* growth and aflatoxin contamination of maize grains during storage. *Toxins* **2019**, *11*, 646. [CrossRef]
208. Li, Z.; Tang, X.; Shen, Z.; Yang, K.; Zhao, L.; Li, Y. Comprehensive comparison of multiple quantitative near-infrared spectroscopy models for *Aspergillus flavus* contamination detection in peanut. *J. Sci. Food Agric.* **2019**, *99*, 5671–5679. [CrossRef] [PubMed]
209. Mitchell, N.J.; Bowers, E.; Hurburgh, C.; Wu, F. Potential economic losses to the US corn industry from aflatoxin contamination. *Food Addit. Contam. Part A* **2016**, *33*, 540–550. [CrossRef] [PubMed]
210. Alshannaq, A.F.; Gibbons, J.G.; Lee, M.K.; Han, K.H.; Hong, S.B.; Yu, J.H. Controlling aflatoxin contamination and propagation of *Aspergillus flavus* by a soy-fermenting *Aspergillus oryzae* strain. *Sci. Rep.* **2018**, *8*, 1–14. [CrossRef] [PubMed]
211. Ferreira, F.D.; Mossini, S.A.G.; Ferreira, F.M.D.; Arrotéia, C.C.; Da Costa, C.L.; Nakamura, C.V.; Junior, M.M. The inhibitory effects of *Curcuma longa* L. essential oil and curcumin on *Aspergillus flavus* link growth and morphology. *Sci. World J.* **2013**, *2013*, 34380. [CrossRef]
212. Miedaner, T.; Gwiazdowska, D.; Waśkiewicz, A. *Management of Fusarium Species and Their Mycotoxins in Cereal Food and Feed*; Frontiers Media SA: Lausanne, Switzerland, 2017; p. 8543. [CrossRef]
213. Ibrahim, N.F.; Mohd, M.H.; Nor, N.M.I.M.; Zakaria, L. Mycotoxigenic potential of *Fusarium* species associated with pineapple diseases. *Arch. Phytopathol. Plant Prot.* **2020**, *53*, 217–229. [CrossRef]
214. Munkvold, G.P. *Fusarium* species and their associated mycotoxins. In *Mycotoxigenic Fungi: Methods and Protocols, Methods in Molecular Biology*; Moretti, A., Sueca, A., Eds.; Springer: New York, NY, USA, 2017; Volume 1542, pp. 51–106. [CrossRef]
215. Kumar, K.N.; Venkataramana, M.; Allen, J.A.; Chandranayaka, S.; Murali, H.S.; Batra, H.V. Role of *Curcuma longa* L. essential oil in controlling the growth and zearalenone production of *Fusarium graminearum*. *LWT Food Sci. Technol.* **2016**, *69*, 522–528. [CrossRef]
216. Senouci, H.; Benyelles, N.G.; Dib, M.E.A.; Costa, J.; Muselli, A. Chemical composition and combinatory antifungal activities of *Ammoides verticillata*, *Allium sativum* and *Curcuma longa* essential oils against four fungi responsible for tomato diseases. *Comb. Chem. High Throughput Screen.* **2020**, *23*, 196–204. [CrossRef]
217. Sacchetti, G.; Maietti, S.; Muzzoli, M.; Scaglianti, M.; Manfredini, S.; Radice, M.; Bruni, R. Comparative evaluation of 11 essential oils of different origin as functional antioxidants, antiradicals and antimicrobials in foods. *Food Chem.* **2005**, *91*, 621–632. [CrossRef]
218. Yusof, N.M.; Jai, J.; Hamzah, F. Effect of coating materials on the properties of chitosan-starch-based edible coatings. In *Proceedings of the IOP Conference Series: Materials Science and Engineering*; IOP Publishing: Bristol, UK, 2019; Volume 507, p. 012011. [CrossRef]
219. Bhadoria, P. Allelopathy: A natural way towards weed management. *Am. J. Exp. Agric.* **2010**, *1*, 7–20. [CrossRef]
220. Chappell, M.; Knox, G.; Stamps, R.H. *Alternatives to Synthetic Herbicides for Weed Management in Container Nurseries*; The Institute of Food and Agricultural Sciences (IFAS): Quincy, FL, USA, 2012; pp. 1–6.
221. Busi, R.; Vila-Aiub, M.M.; Beckie, H.J.; Gaines, T.A.; Goggin, D.E.; Kaundun, S.S.; Lacoste, M.; Neve, P.; Nissen, S.J.; Norsworthy, J.K.; et al. Herbicide-resistant weeds: From research and knowledge to future needs. *Evol. Appl.* **2013**, *6*, 1218–1221. [CrossRef]
222. Ramezani, S.; Saharkhiz, M.J.; Ramezani, F.; Fotokian, M.H. Use of essential oils as bioherbicides. *J. Essent. Oil-Bear. Plants* **2008**, *11*, 319–327. [CrossRef]
223. Soltys, D.; Krasuska, U.; Bogatek, R.; Gniazdowska, A. Allelochemicals as bioherbicides—Present and Perspectives. In *Herbicides-Current Research and Case Studies in Use*; Price, A.J., Kelto, J.A., Eds.; IntechOpen: London, UK, 2013; pp. 517–542. [CrossRef]
224. Tworkoski, T. Herbicide effects of essential oils. *Weed Sci.* **2002**, *50*, 425–431. [CrossRef]
225. Hazrati, H.; Saharkhiz, M.J.; Moein, M.; Khoshghalb, H. Phytotoxic effects of several essential oils on two weed species and tomato. *Biocatal. Agric. Biotechnol.* **2018**, *13*, 204–212. [CrossRef]
226. Bessette, S.M. Herbicidal Composition Containing Plant Essential Oils and Mixtures or Blends Thereof. U.S. Patent 2002/0193250 A1, 19 December 2002.
227. Symeonidou, A.; Petrotos, K.; Vasilakoglou, I.; Gkoutsidis, P.; Karkanta, F.; Lazaridou, A. Natural Herbicide Based on Essential Oils and Formulated as Wettable Powder. Patent EP 2684457 A1, 15 January 2014.
228. CABI. *Portulaca oleracea*. Available online: https://www.cabi.org/isc/datasheet/43609 (accessed on 29 September 2020).

229. Jin, R.; Wang, Y.; Liu, R.; Gou, J.; Chan, Z. Physiological and metabolic changes of purslane (*Portulaca oleracea* L.) in response to drought, heat, and combined stresses. *Front. Plant Sci.* **2016**, *6*, 1123. [CrossRef]

230. DiTomaso, J.M.; Kyser, G.B. *Italian and Perennial Ryegrass*; Weed Research and Information Center: Davis, CA, USA, 2013; pp. 1–3.

231. CABI. *Lolium multiflorum* (Italian Ryegrass). Available online: https://www.cabi.org/isc/datasheet/74001 (accessed on 29 September 2020).

232. Bajwa, A.A.; Jabran, K.; Shahid, M.; Ali, H.H.; Chauhan, B.S. Ehsanullah Eco-biology and management of *Echinochloa crus-galli*. *Crops Prot.* **2015**, *75*, 151–162. [CrossRef]

233. Zhang, Z.; Gu, T.; Zhao, B.; Yang, X.; Peng, Q.; Li, Y. Effects of common *Echinochloa* varieties on grain yield and grain quality of rice. *Field Crops Res.* **2017**, *203*, 163–172. [CrossRef]

234. Ibáñez, M.D.; Blázquez, M.A. Ginger and turmeric essential oils for weed control and food crop protection. *Plants* **2019**, *8*, 59. [CrossRef]

235. Prakash, B.; Singh, P.; Kedia, A.; Singh, A.; Dubey, N.K. Efficacy of essential oil combination of *Curcuma longa* L. and *Zingiber officinale* Rosc. As a postharvest fungitoxicant, aflatoxin inhibitor and antioxidant agent. *J. Food Saf.* **2012**, *32*, 279–288. [CrossRef]

236. De Melo, S.C.; De Sá, L.E.C.; de Oliveira, H.L.M.; Trettel, J.R.; da Silva, P.S.; Gonçalves, J.E.; Gazim, Z.C.; Magalhães, H.M. Chemical constitution and allelopathic effects of *Curcuma zedoaria* essential oil on lettuce achenes and tomato seeds. *Aust. J. Crop Sci.* **2017**, *11*, 906–916. [CrossRef]

237. Global Invasive Species Database. *Cortaderia selloana*. Available online: http://www.iucnisd.org/gsid/species.php?sc=373 (accessed on 29 September 2020).

238. Comité Científico; Comité de Flora y Fauna Silvestres. *Solicitud de Dictamen Por Parte de la Junta de Extremadura Sobre la Pertinencia de la Denegación de la Propuesta de una Empresa que Pretende Llevar a Cabo un "Ensayo de Investigación con Nicotiana Glauca- TAPCS1 Como Cultivo Energético Para la Producción*; Ministerio de Agricultura, Alimentación y Medio Ambient: Madrid, Spain, 2016; pp. 1–6.

239. Estrategia de Gestión, Control y Posible Erradicación del Plumero de la Pampa (*Cortaderia selloana*) y otras Especies de *Cortaderia*. Available online: https://www.miteco.gob.es/es/biodiversidad/publicaciones/estrategia_cortaderia_tcm30-478424.pdf (accessed on 29 September 2020).

240. Khattak, S.; Rehman, S.U.; Shah, H.U.; Ahmad, W.; Ahmad, M. Biological effects of indigenous medicinal plants *Curcuma longa* and *Alpinia galanga*. *Fitoterapia* **2005**, *76*, 254–257. [CrossRef]

241. Abbasi, K.; Shah, A.A. Biological evaluation of turemeric (*Curcuma longa*). *Int. J. Curr. Microbiol. Appl. Sci.* **2015**, *4*, 236–249.

242. Akter, J.; Islam, M.Z.; Takara, K.; Hossain, M.A.; Gima, S.; Hou, D.X. Plant growth inhibitors in turmeric (*Curcuma longa*) and their effects on *Bidens pilosa*. *Weed Biol. Manag.* **2018**, *18*, 136–145. [CrossRef]

243. Prakash, B.; Kedia, A.; Mishra, P.K.; Dubey, N.K. Plant essential oils as food preservatives to control moulds, mycotoxin contamination and oxidative deterioration of agri-food commodities—Potentials and challenges. *Food Control* **2015**, *47*, 381–391. [CrossRef]

244. Pandey, A.K.; Kumar, P.; Singh, P.; Tripathi, N.N.; Bajpai, V.K. Essential oils: Sources of antimicrobials and food preservatives. *Front. Microbiol.* **2017**, *7*, 21–61. [CrossRef] [PubMed]

245. Santos-Sánchez, N.F.; Salas-Coronado, R.; Valadez-Blanco, R.; Hernández-Carlos, B.; Guadarrama-Mendoza, P.C. Natural antioxidant extracts as food preservatives. *Acta Sci. Pol. Technol. Aliment.* **2017**, *16*, 361–370. [CrossRef]

246. Hosseini, H.; Tajiani, Z.; Jafari, S.M. Improving the shelf-life of food products by nano/micro-encapsulated ingredients. In *Food Quality and Shelf Life*; Galanakis, C.M., Ed.; Elsevier Inc.: London, UK, 2019; pp. 159–200. [CrossRef]

247. Pezeshk, S.; Rezaei, M.; Hosseini, H. Effects of turmeric, shallot extracts, and their combination on quality characteristics of vacuum-packaged rainbow trout stored at 4 ± 1 °C. *J. Food Sci.* **2011**, *76*, 387–391. [CrossRef]

248. Gul, P.; Bakht, J. Antimicrobial activity of turmeric extract and its potential use in food industry. *J. Food Sci. Technol.* **2015**, *52*, 2272–2279. [CrossRef]

249. Buch, S.; Pinto, S.; Aparnathi, K.D. Evaluation of efficacy of turmeric as a preservative in paneer. *J. Food Sci. Technol.* **2014**, *51*, 3226–3234. [CrossRef]

250. Arulkumar, A.; Ramanchandran, K.; Paramasivam, S.; Palanivel, R.; Miranda, J.M. Effects of turmeric (*Curcuma longa*) on shelf life extension and biogenic amine control of cuttlefish (*Sepia brevimana*) during chilled storage. *CYTA J. Food* **2017**, *15*, 441–447. [CrossRef]

251. Al-Bahtiti, N.H. A study of preservative effects of turmeric (*Curcuma longa*) on mashed potatoes. *Int. J. Appl. Eng. Res.* **2018**, *13*, 4278–4281.

252. Akbar, A.; Ali, I.; Samiullah; Ullah, N.; Khan, S.A.; Rehman, Z.; Rehman, S.U. Drought affects aquaporins gene expression in important pulse legume chickpea (*Cicer arietinum* L.). *Pak. J. Bot.* **2019**, *51*, 1129–1135. [CrossRef]

253. Pop, A.; Muste, S.; Paucean, A.; Chis, S.; Man, S.; Salanta, L.; Marc, R.; Muresan, A.; Martis, G. Herbs and spices in terms of food preservation and shelf life. *Hop Med. Plants* **2019**, *27*, 57–65.

254. Nguyen, M.P. Synergistic effect of turmeric (*Curcuma longa*), galanga (*Alpinia galanga*) powder and lemongrass (*Cymbopogon citratus*) essential oil as natural preservative in chilled storage of white hard clam (*Meretrix lyrata*). *Orient. J. Chem.* **2020**, *36*, 195–200. [CrossRef]

255. Guerra, A.M.S.; Hoyos, C.G.; Velásquez-Cock, J.A.; Acosta, L.V.; Rojo, P.G.; Giraldo, A.M.V.; Gallego, R.Z. The nanotech potential of turmeric (*Curcuma longa* L.) in food technology: A review. *Crit. Rev. Food Sci. Nutr.* **2020**, *60*, 1842–1854. [CrossRef]

256. Priya, R.; Prathapan, A.; Raghu, K.G.; Menon, A.N. Chemical composition and *in vitro* antioxidative potential of essential oil isolated from *Curcuma longa* L. leaves. *Asian Pac. J. Trop. Biomed.* **2012**, *2*, S695–S699. [CrossRef]
257. Ilyasov, I.R.; Beloborodov, V.L.; Selivanova, I.A.; Terekhov, R.P. ABTS/PP decolorization assay of antioxidant capacity reaction pathways. *Int. J. Mol. Sci.* **2020**, *21*, 1131. [CrossRef]
258. Kedare, S.B.; Singh, R.P. Genesis and development of DPPH method of antioxidant assay. *J. Food Sci. Technol.* **2011**, *48*, 412–422. [CrossRef]
259. Cheng, Z.; Li, Y. Reducing power: The measure of antioxidant activities of reductant compounds? *Redox Rep.* **2004**, *9*, 213–217. [CrossRef]
260. Shahwar, D.; Raza, M.A.; Bukhari, S.; Bukhari, G. Ferric reducing antioxidant power of essential oils extracted from *Eucalyptus* and *Curcuma* species. *Asian Pac. J. Trop. Biomed.* **2012**, *2*, S1633–S1636. [CrossRef]
261. Zaki, N.; Fa, M.; Yusof, N.; Jai, J. Turmeric (*Curcuma longa* L.) Oil as Antioxidant Agent in Starch-Based Edible Coating Film for Fresh-Cut Fruits. Available online: https://www.semanticscholar.org/paper/Turmeric-%28-Curcuma-longa-L-%29-Oil-as-Antioxidant-in-NAM-Mustapha/9404fbe5dcc4eef2313abf07d6a9b2d070f331d8?p2df (accessed on 19 November 2020).
262. Ibáñez, M.D.; López-Gresa, M.P.; Lisón, P.; Rodrigo, I.; Bellés, J.M.; González-Mas, M.C.; Blázquez, M.A. Essential oils as natural antimicrobial and antioxidant products in the agrifood industry. *Nereis. Interdiscip. Ibero-Am. J. Methods Model. Simul.* **2020**, *12*, 55–69. [CrossRef]
263. Thakam, A.; Saewan, N. Chemical composition of essential oil and antioxidant activities of *Curcuma petiolata* Roxb. rhizomes. *Adv. Mater. Res.* **2012**, *506*, 393–396. [CrossRef]
264. Dawidowicz, A.L.; Olszowy, M. Influence of some experimental variables and matrix components in the determination of antioxidant properties by β-carotene bleaching assay: Experiments with BHT used as standard antioxidant. *Eur. Food Res. Technol.* **2010**, *231*, 835–840. [CrossRef]

MDPI

St. Alban-Anlage 66

4052 Basel

Switzerland

Tel. +41 61 683 77 34

Fax +41 61 302 89 18

www.mdpi.com

Plants Editorial Office

E-mail: plants@mdpi.com

www.mdpi.com/journal/plants